Engineering rock mechanics

an introduction to the principles

Site Investigation
Borehole Core
Solid Rock
Infilled Fracture
or 'Discontinuity'
The foundation of all concepts is simple unsophisticated experience...
Personal experience is everything...
And logical consistency is not final.
D. T. Suzuki
Former Professor of Philosophy, Otani University, Kyoto, Japan

Engineering rock mechanics

an introduction to the principles

John A. Hudson
Professor of Engineering Rock Mechanics
Imperial College of Science, Technology and Medicine
University of London, UK

and

John P. Harrison
Senior Lecturer in Engineering Rock Mechanics
Imperial College of Science, Technology and Medicine
University of London, UK

Pergamon

An imprint of Elsevier Science

Amsterdam – Lausanne – New York – Oxford – Shannon – Singapore – Tokyo

ELSEVIER B.V.	ELSEVIER Inc.	**ELSEVIER Ltd**	ELSEVIER Ltd
Radarweg 29	525 B Street	**The Boulevard**	84 Theobalds Road
P.O. Box 211, 1000 AE	Suite 1900, San Diego	**Langford Lane, Kidlington,**	London WC1X 8RR
Amsterdam, The Netherlands	CA 92101-4495, USA	**Oxford OX5 1GB, UK**	UK

First edition 1997
Second impression 2000
Third impression 2005

Library of Congress Cataloging in Publication Data
A catalog record is available from the Library of Congress.

British Library Cataloguing in Publication Data
A catalogue record is available from the British Library.

ISBN: 0-08-041912-7 (Hardbound)
ISBN: 0-08-043864-4 (Flexibound)

∞ The paper used in this publication meets the requirements of ANSI/NISO Z39.48-1992 (Permanence of Paper).
Printed in The Netherlands.

For all our past, present and future students and colleagues
at Imperial College

About the authors

Professor J. A. Hudson

John Hudson graduated in 1965 from the Heriot-Watt University and obtained his Ph.D. in 1970 at the University of Minnesota. He has spent his professional career in rock mechanics and rock engineering—as they apply to both civil and mining engineering—in consulting, research and teaching . He has written over 100 scientific papers and been awarded the D.Sc. degree by the Heriot-Watt University for his contributions to the subject.

From 1983 to 1993, Professor Hudson was based at Imperial College where most of the book was written. He is now a Principal of Rock Engineering Consultants, a Visiting Professor at Imperial College, and actively engaged in applying engineering rock mechanics principles to relevant engineering practice worldwide.

Dr J. P. Harrison

John Harrison graduated in civil engineering from Imperial College, University of London and then worked for some years in the civil engineering industry with both contracting and consulting organizations.

This was interspersed by studies leading to a Master's degree, also from Imperial College, in engineering rock mechanics. In 1986 he was appointed Lecturer in Engineering Rock Mechanics at Imperial College, was promoted to Senior Lecturer in 1996 and now directs undergraduate and post-graduate rock mechanics teaching, as well as research there.

His personal research interests are in the characterization and behaviour of discontinuous rock masses and, for his work on novel mathematical methods applied to the analysis of discontinuity geometry, he was awarded the degree of Ph.D. by the University of London in 1993.

Contents

Preface

With the title *Engineering Rock Mechanics*, what is this book about? It is about the discipline, based on mechanics, which is used to design structures built on or in rock masses. These structures, which encompass building foundations, dams, rock slopes, tunnel, caverns, hydroelectric schemes, mines, etc., depend critically on the rock mass properties and the interaction between the rock mass and the engineered structure. As a result, the distinct discipline of engineering rock mechanics has developed. The term 'rock mechanics' refers to the basic science of mechanics applied to rocks; whilst the term 'rock engineering' refers to any engineering activity involving rocks. Thus, the term 'engineering rock mechanics' refers to the use of rock mechanics in rock engineering—within the context of civil, mining and petroleum engineering. Because rock mechanics can also be used to study structural geology, we emphasize through the title that it is the rock mechanics principles in the engineering context that we are presenting.

The book is based on the content of the integrated engineering rock mechanics course given at Imperial College and on the authors' engineering experience. Chapters 1–13 cover rock mechanics, Chapter 14 discusses the principles of rock engineering systems, and Chapters 15–20 cover major applications in rock engineering. The philosophy of the presentation is to provide comprehension of all the subjects discussed. In all aspects, and particularly in the mathematics, we have included some physical explanations of the meaning behind the relations. Also, our philosophy is that although rock mechanics and the associated principles are a science, their application is an art. To paint a good picture, one must know the basic techniques. Knowing these techniques will not necessarily make a good painter, but it will optimize everyone's attempts.

Thus, the book is intended to be an understandable 'across-the-board' source of information for the benefit of anyone involved in rock mechanics and rock engineering: students, teachers, researchers, clients, consulting engineers and contractors. It will be of particular use in the civil, mining and petroleum subject areas: the objectives of the engineering may be different but the principles are the same.

We hope that everyone reading this book not only has a chance to experience the science and art of the subject, but also the romance. Rock engineering occurs deep in the earth, high in the mountains and often in the world's wildest places. We engineer with rocks as we create structures, extract the primary raw materials for mankind and harness the forces of nature. It is the romance and the passion associated with rock engineering that has led us to communicate some of this excitement. 'Personal experience' is everything. So, we hope that you will have the opportunity to experience at first hand some of the principles and applications described in the book.

Lecture notes prepared by the authors for undergraduate and postgraduate students at Imperial College were the basis for the book. Some of the material, especially that of a fundamental nature, is partially based on earlier lecture notes prepared by our predecessors in the rock mechanics section at the college. We acknowledge this general debt with thanks and appreciation. We are also grateful to all our students and recent colleagues at Imperial College who have suggested improvements to the text during the pre-publication 'field-testing' period over the last few years. Finally, we thank Carol and Miles Hudson and Gwen Harrison for painstakingly correcting and compiling the penultimate version. The final text is, of course, our responsibility: if there is anything in the following pages that you do not understand, it is our fault.

J. A. Hudson and J. P. Harrison
Imperial College of Science, Technology and Medicine
University of London

1 Introduction

1.1 The subject of rock mechanics

The subject of rock mechanics started in the 1950s from a rock physics base and gradually became a discipline in its own right during the 1960s. As explained in the Preface, rock mechanics is the subject concerned with the response of rock to an applied disturbance, which is considered here as an engineering, i.e. a man-induced, disturbance. For a natural disturbance, rock mechanics would apply to the deformation of rocks in a structural geology context, i.e. how the folds, faults, and fractures developed as stresses were applied to the rocks during orogenic and other geological processes. However, in this book we will be concerned with rock mechanics applied to engineering for civil, mining, and petroleum purposes.

Thus, rock mechanics may be applied to many engineering applications ranging from dam abutments, to nuclear power station foundations, to the manifold methods of mining ore and aggregate materials, to the stability of petroleum wellbores and including newer applications such as geothermal energy and radioactive waste disposal. Probably, the main factor that distinguishes rock mechanics from other engineering disciplines is the application of mechanics on a large scale to a pre-stressed, naturally occurring material.

In the two photographs in Figs 1.1 and 1.2, we illustrate a typical full-scale rock structure and a closer view of the rock material itself. It is quite clear from these illustrations that the nature of the rock mass and the rock material must be taken into account in both the basic mechanics and the applied engineering. This has been increasingly appreciated since the beginning of the discipline in the early 1960s.

In the civil and mining engineering areas, the subject of rock mechanics flourished in the 1960s. In 1963, a particular landmark was the formation of the International Society for Rock Mechanics which has grown steadily to its current membership of about 7000 from 37 countries. The discipline of rock mechanics is universal in its application and the engineering is especially visible in those countries where the ground surface is predominantly composed of rock, for example, Chile, Finland, Scotland, Spain, and

Figure 1.1 Rock structure illustrating the complex nature of the material.

the former Yugoslavia. In these and other similar 'rocky' countries, rock engineering is a way of life for civil engineering construction: invariably,

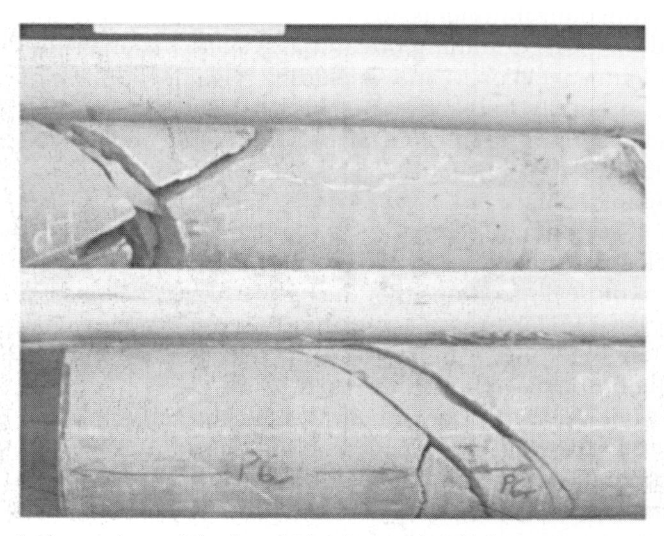

Figure 1.2 A closer view of the breaks in the mechanical continuum, generically termed discontinuities.

Figure 1.3 Example of a bulk blast for production purposes in a quarry.

highways and other features will have been blasted in rock. The two photographs in Figs 1.3 and 1.4 demonstrate this type of engineering.

Naturally, there are many examples of rock engineering underground and these occur in civil engineering projects in rock-dominated countries and during underground mining in all countries. The ubiquitous road and railway tunnels can have quite different appearances depending on the engineering and architectural finish. In the two photographs in Figs 1.5 and 1.6, the contrast is shown between a tunnel that required no lining so the bare rock is visible and a tunnel that required extensive support.

There are often occasions when small or large surface excavations are

Figure 1.4 Example of a rock face made by pre-split blasting to give a stable, relatively smooth finish.

Figure 1.5 Unlined tunnel excavated by a tunnel boring machine (tunnel face to left, tunnel wall to right of photograph).

made in rock. Clearly, it is helpful to be able to evaluate the stability of the excavation—whatever size it may be. This highlights another crucial aspect which has only really been developed since the 1970s, and that is understanding the full role of the rock structure, i.e. not only the intact rock but also the rock fractures and their three-dimensional configuration. In general, the stability of near surface excavations is governed by the rock structure, whereas, deeper excavations can be more affected by the intact rock and pre-existing stresses.

Thus, the rock structure is particularly important in civil engineering and open-pit mines and so it is necessary to be able to characterize and understand the mechanics of a discontinuum. This is well illustrated by the two photographs in Figs 1.7 and 1.8, the first showing how individual rock blocks are formed and the second showing a large open-pit mine with some instabilities.

In fact, there are numerous applications for rock engineering and three are illustrated in Figs 1.9–1.11. Some of the most important are dam abutments and foundations, mining methods (whether as open-pit or as a whole variety of underground techniques) and now non-precedent applications for which there is no previous experience to guide us. These latter projects include geothermal energy, radioactive waste disposal and the general use of underground space for hosting a miscellany of low- and high-technology activities, such as domestic refuse treatment and large

Figure 1.6 Heavily supported tunnel excavated by blasting.

high-energy particle accelerators. For all of these applications, it is essential to understand the rock material and the rock mechanics so that engineering can be conducted in an optimal way.

Figure 1.7 Rock fractures forming rock blocks within the rock structure (with 1 m long white scale).

Figure 1.8 Open-pit mine with slope instabilities.

The three photographs in Figs 1.9–1.11 also illustrate the large scale of some of the existing precedent practice projects: a dam, a mine, and a civil excavation. It is apparent from the pictures that there will be considerable economic benefit in designing and constructing these structures in the optimal way—given that we have the necessary rock mechanics principles and rock engineering experience. It is also evident that one ignores such information at considerable physical and financial peril. A good engineer is one who can do the same job with the same professionalism at a lower price: but this can only be achieved successfully by knowing the rock mechanics principles and applications.

Figure 1.9 Large dam in Portugal.

Figure 1.10 Large open-pit mine in Chile.

All these rock engineering projects, whether we have experience of them or not, can be summarized in the diagram in Fig. 1.12. In this diagram, there is appreciation of the three main aspects of such engineering projects:

- the outer ring represents the whole project complete with its specific objective—and different projects can have widely differing objectives.
- the middle ring represents the inter-relation between the various components of the total problem. For example, there will be relations

Figure 1.11 Hydroelectric cavern in Portugal.

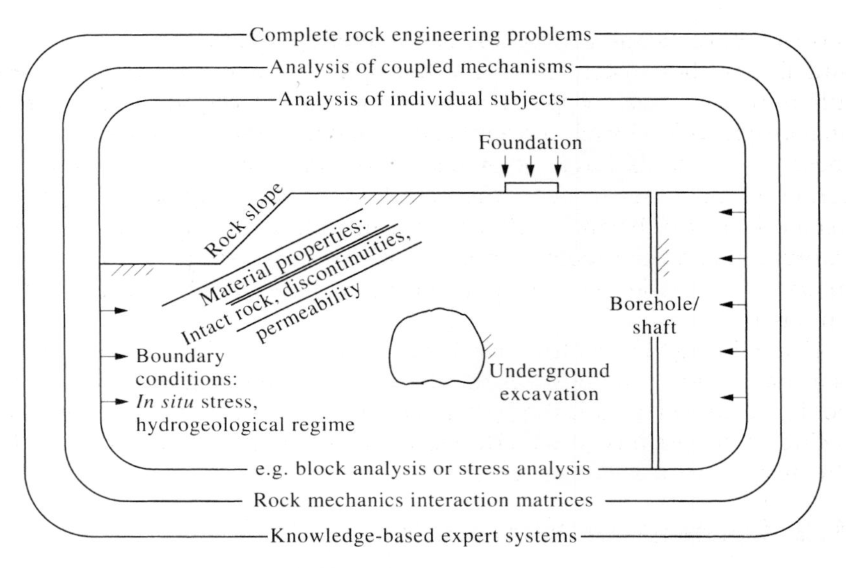

Figure 1.12 Three-tier approach to all rock engineering problems.

between rock stresses and rock structure in the rock mechanics context; and there will be relations between rock support systems and cost implications in the rock engineering context.

- finally, the central ring represents the individual aspects of each project, such as a specific numerical analysis or a specific costing procedure.

In the spirit of this diagram, we discuss the major rock mechanics aspects on an individual basis in Chapters 2–13. The method of studying the interactions between subjects is discussed in Chapter 14. Then, in Chapters 15–20, the main engineering techniques and applications are discussed. In engineering practice, the procedure is to enter the diagram in Fig. 1.12 from the outside having specified the objective, pass through the three rings conducting the necessary analyses at each stage, and then exit synthesizing the most appropriate design.

We have already mentioned that in rock mechanics there was considerable emphasis in the 1960s on intact rock and in the 1970s on discontinuities and rock masses. In the 1980s the emphasis shifted to numerical analysis and we anticipate that, during the remaining part of the 1990s and beyond, there will be combined emphases on material property determination, full-scale *in situ* experiments, enhanced use of the computer, and improved engineering implementation of the principles. Currently, our ability to compute has far outstripped our ability to measure the required input parameters and indeed to know whether the computer modelling is realistic. A good example of this is the theory of elasticity which considers stresses and strains in the rock. The vast majority of elasticity analyses have assumed that the rock is isotropic, i.e. it has the same elastic properties in all directions, which requires two elastic properties. We now recognize that it is more realistic to include further elastic properties, five elastic constants for transverse isotropy (the properties are the same in a plane but different perpendicular to that plane) and nine elastic constants for the orthotropic case (where the properties are different in three perpendicular directions). However, for a fully anisotropic rock, 21 such constants are needed. To the authors' knowledge, these 21 constants have never been used in a numerical analysis and certainly have never been measured in a rock engineering project. Establishing the balance between not including enough rock property information and conducting unnecessarily complex analyses is difficult but made much easier if the engineering rock mechanics principles are understood.

Generalizing the problem described above, one should ask 'What exactly is it that we wish to know for the design of engineering projects?' In this book, we consider both the principles and the applications and we have included discussions which address the question above and will assist in the design process.

1.2 Content of this book

This book is intended for anyone involved in rock mechanics and rock engineering. From the text in the previous section, the reader will have noted that we are making a special attempt to present the principles and then to place them in the engineering context. Thus, the book can be used for both introductory and advanced rock mechanics teaching, and by rock engineers of all persuasions. We feel that the rock mechanics subject does not have to be project-specific and can therefore be generally directed to all types of engineers working on or in rock masses.

The layout follows a logical course from chapters on the basic subjects of rock mechanics such as stress, strain, discontinuities and permeability through the generic rock engineering aspects of excavation and support to specific engineering objectives and projects in the later chapters.

Anyone who has worked in rock engineering will know that all projects have their own idiosyncratic components and are unique. Thus, whether an engineer is involved with a conventional or an unconventional project, it is always vital to understand and apply the principles presented in the first 13 chapters.

This book is about the **principles** of engineering rock mechanics. The book is not intended to be truly comprehensive in the sense of including all information on the rock engineering subject. Readers requiring more information are referred to the five volume compendium *Comprehensive Rock Engineering*, edited by the first author and also published by Elsevier.

Key Papers . . . [illegible]

[illegible]

2 Geological setting

In this chapter, we will set the scene for the consideration of engineering in the natural material, rock. Most of our engineering materials (with the exception of timbers and soil) are manufactured and tested to specification. Subject to correct quality control, we can then be sure that the engineering material in question has a given set of properties which are used in the design process. However, rock is extremely old compared to all other engineering materials: its age is measured in millions of years and it has undergone significant mechanical, thermal and chemical activity.

We will describe in the following sections the ramifications of the rock's history for engineering, starting with rock as an engineering material and following with a discussion of the conditions in natural rock environments. In Section 2.3, we will discuss explicity the influence of geological history on five of the most important aspects of rock mechanics; and, later on, we will explain more directly (in the context of stresses and strains) the concepts of continuity, homogeneity and isotropy of rock material and rock masses.

2.1 Rock as an engineering material

One of the most important, and often frequently neglected, aspects of rock mechanics and rock engineering is that we are utilizing an existing material which is usually highly variable. This is demonstrated in Figs 2.1–2.3. The rock will be used either as a building material so the structure will be made *of* rock, or a structure will be built *on* the rock, or a structure will be built *in* the rock. In the majority of civil engineering cases, rock is removed to form the structure as in, for example, the excavation of rock for a hydroelectric machine hall chamber. In this sense, we are dealing with a reverse type of construction where the rock material is being taken away, rather than added, to form a structure. On the mining side, rock may be excavated in an open pit and we will then be concerned with the stability of the sides of the open pit.

In these examples and all others in rock engineering, the material is natural. As engineers and in the context of mechanics, we must establish

Figure 2.1 Relatively consistent intact rock.

Figure 2.2 'Layered' intact rock.

the properties of the material, the pre-existing stress state in the ground (which will be disturbed by engineering) and consider these in relation to our engineering objective. In civil engineering, the main objective is to create a structure by removing the rock. In mining engineering, it is to obtain the material being removed. A primary information base for these activities is a knowledge of the geological strata, any alteration to the rock material, the presence of large- and small-scale faulting and jointing in the rock, and indeed any geological parameter that is relevant to the engineering. Clearly, the rock type, the rock structure, any alteration to the rock, the *in situ* stress state and hydrogeological regime will be important for all engineering. There are, however, many other aspects of the geological setting which could be of major, if not dominant, significance in the engineering design, construction and subsequent performance. Examples of these are the presence of large natural caverns in karstic regions, the presence of buried valleys encountered during tunnelling, wash-outs in coal seams during mining and the presence of major horizontal fracture zones in granitic masses for radioactive waste disposal.

Figure 2.3 Zone of highly fractured rock.

In the photographs in Figs 2.4 and 2.5, the significance of the likely influence of rock structure on engineering can be imagined. The departure from an 'ideal' material through the existence of this structure can occur on all scales from very large faults to micro-fissures. Similarly, engineering in rock occurs in a variety of sizes and shapes. Examples are the Chuquicamata open-pit copper mine in Chile which is several kilometres long and planned to be 1 km deep, and a petroleum engineering wellbore which is a few tens of centimetres in diameter yet is several kilometres deep. It is the interpretation of this rock structure in conjunction with the size, shape and design requirements of the engineering that make rock engineering a unique discipline.

Thus, rock mechanics applied to engineering is both an art and a science. We will be explaining the principles of engineering rock mechanics in this book, but it should never be forgotten that we cannot specify the rock properties and the rock loading conditions: they already exist because the rock is a natural material and in many cases is significantly stressed naturally before engineering commences. Hence, in the remainder of this chapter, we will develop these concepts by considering the natural rock environments within which the engineering occurs and the specific ways in which the geological setting directly affects both the rock mechanics and the engineering design.

2.2 Natural rock environments

In addition to the direct properties of the rock and rock masses as described above, we have to remember that the natural rock environment can also have a profound effect on the engineering. In general this is basically governed by the location of the engineering, i.e. whether a structure is being built on the surface, whether the structure is being created by

Figure 2.4 Large-scale rock structure.

excavation of the surface rock, or whether the structure is underground. Of course, a particular project may involve two or, indeed, all of these main types, as in many hydroelectric schemes.

It is generally found that the fractures in the rock govern the stability of near surface structures and the natural *in situ* stresses govern the stability of deep structures. For example, the stability of a dam foundation will depend critically on the deformability and permeability of the underlying rocks, which in turn are dictated by the nature and geometrical configuration of the fractures in the rock mass. This is also true for the stability of the side slopes of surface excavations and the roof and sides of near surface underground excavations. However, at medium depths in weak rocks (for example the Channel Tunnel between England and France) and at considerable depths in strong rocks (for example South African gold mines) the natural stress, which is altered by the engineering, can be the dominant problem.

Furthermore, these effects will be influenced by other factors; e.g. whether the rock is wet or dry, cold or hot, stable or squeezing. Typical circumstances where these factors are important are the degradation of chalk and mudstones on either exposure to water movement or desiccation, permafrost engineering, certain Japanese mines in which circulating groundwater can be above boiling point, the difficulty of inducing roof failure during longwall mining operations when the roof is too strong, and loss of tunnel boring machines as they have attempted to cross squeeze ground within major faults. It is the identification of these, and a whole host of other geological factors, which is one of the keys to successful site investigation and correct interpretation of the rock mass environment. Two examples of the effects mentioned are shown in Figs 2.6 and 2.7.

Of course, different projects could be conducted in entirely different rock environments and this would be taken into account utilizing the three-tier

Figure 2.5 Small-scale rock structure.

Figure 2.6 Tunnel in mudstone which has deteriorated over a period of several years after excavation.

approach already shown in Fig. 1.12. Moreover, the explicit site conditions will be taken into account in the project design and analysis. So let us consider now what can be said generally about the influence of geological history on rocks and rock masses, ideas that will apply to all sites and all proposed site investigations, whether for civil or mining engineering.

2.3 The influence of geological factors on rocks and rock masses

Five main subjects are discussed below in terms of the influence of geological factors on rocks and rock masses. In the context of the mechanics problem, we should consider the material and the forces applied to it. We have the intact rock which is itself divided by discontinuities (the latter word being a generic term for all rock fractures) to form the rock structure. We find then that the rock is already subjected to an *in situ* stress. Superimposed on this fundamental mechanics circumstance are the influences of pore fluids/water flow and time. In all of these subjects, the geological history has played its part, altering the rock and the applied forces, and the engineer should be aware that these will have been significantly affected by the geological processes, as explained below.

2.3.1 Intact rock

Intact rock is defined in engineering terms as rock containing no significant fractures. However, on the small scale it is composed of grains with the form of the microstructure being governed by the basic rock forming processes. Subsequent geological events may affect its mechanical

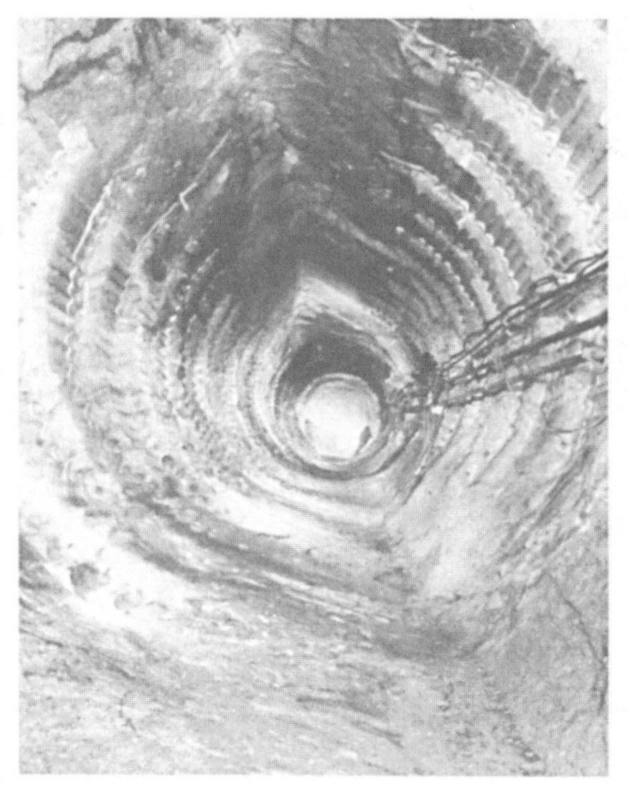

Figure 2.7 Tunnel deformation caused by high rock stresses at the Underground Research Laboratory, Manitoba, Canada.

properties and its susceptibility to water penetration and weathering effects.

The most useful single description of the mechanical behaviour is the complete stress–strain curve in uniaxial compression. This curve will be explained fully in Chapter 6, but is introduced here briefly to illustrate the very significant effect of the rock microstructure and history on the mechanical behaviour. In Fig. 2.8, a general complete stress–strain curve is shown for a sample of rock being compressed in one direction, i.e. in uniaxial compression. The reader should note that: the horizontal axis is strain, which is the relative change in length of the specimen; and the vertical axis is stress, which is the load per unit area.

There are several features of interest, the first of which is the modulus of the rock, represented by the letter E on the diagram. For a high-modulus (i.e. stiff) material, this initial part of the complete stress–strain curve will be steep; for a low-modulus (i.e. soft) material, it will be gentle.

The next feature is the compressive strength which is the maximum stress that can be sustained; this is illustrated by the dotted line in the figure.

The third feature is the steepness of the descending portion of the curve which is a measure of the brittleness, as illustrated in Fig. 2.9. The two main cases shown are the behaviour after the compressive strength is reached

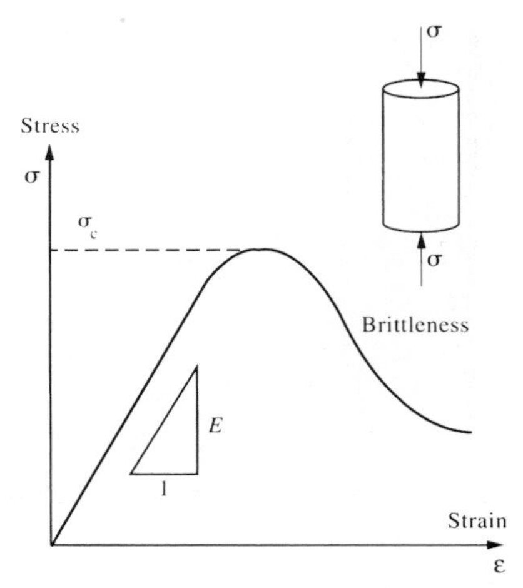

Figure 2.8 Complete stress–strain curve illustrating the stiffness (or modulus, E), the strength, σ_c, and brittleness.

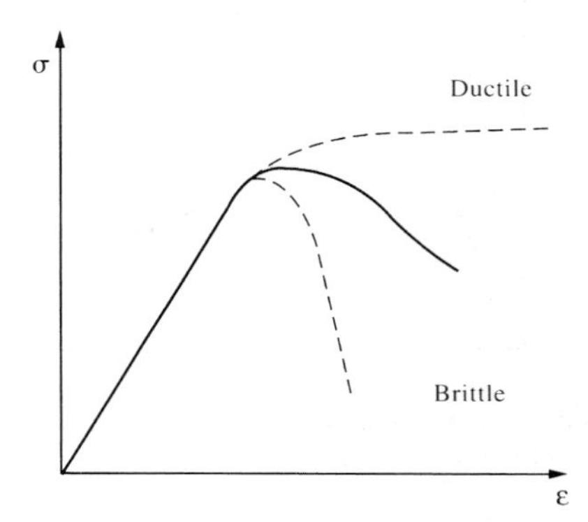

Figure 2.9 Illustration of the difference between a brittle material and a ductile material.

in the form of continuing strain at the same stress level (a ductile material) and a drop in the stress level to zero at the same strain value (a brittle material). The brittleness is indicated by the steepness of the curve between these two limits. In fact, the situation is more complicated than this because it is possible to have strain-hardening materials (a curve above the ductile line) and super-brittle materials (to the left of the brittle line). These cases will be discussed further in Chapter 6.

Possible variation in the three main factors is illustrated for a suite of rocks in Figs 2.10–2.13. The figure legends describe the features qualitatively. As we have mentioned, the form of the complete stress–strain curve is dictated by the nature of the microstructure. For example, a high grain strength, fine grain basalt has a high stiffness, high strength and is very brittle. On the other hand, a limestone rock with a variation in the grain geometry has a medium stiffness, medium strength and a more gentle descending part of the curve caused by the gradual deterioration of the microstructure as it is progressively and increasingly damaged.

There will be variations on this theme for the variety of microstructures that exist and the influence that they have on the shape of the curve—in

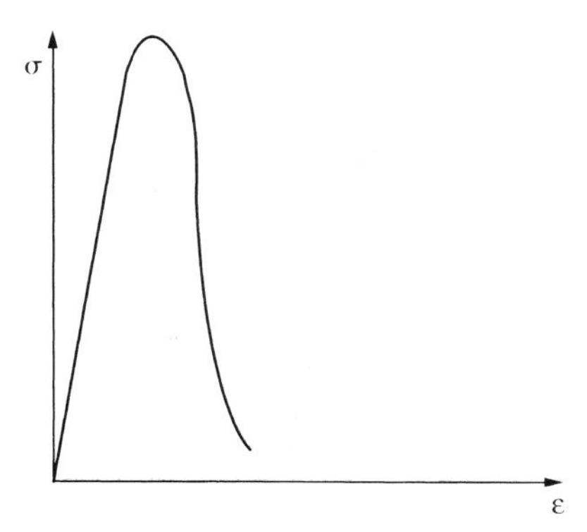

Figure 2.10 Complete stress–strain curve for basalt—high stiffness, high strength, very brittle.

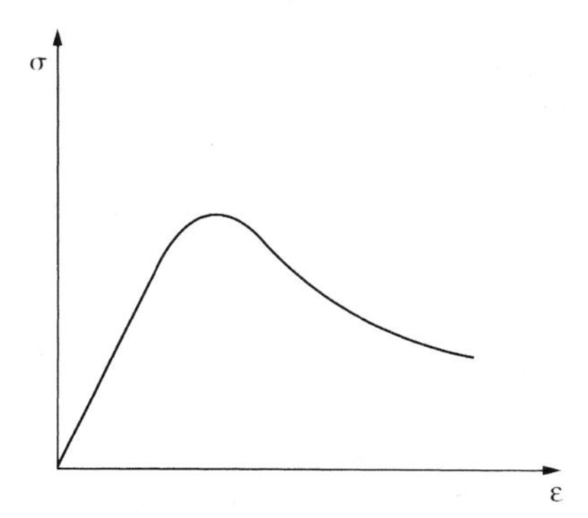

Figure 2.11 Complete stress–strain curve for limestone—medium stiffness, medium strength, medium brittleness.

conjunction with the applied loading conditions and loading rates. The intact rock will also have such characteristics as inhomogeneity and anisotropy: these factors are discussed in Chapter 10.

2.3.2 Discontinuities and rock structure

In the previous section we have indicated one major characteristic of the intact rock, i.e. the stiffness, defined as Young's modulus, E. In the pre-peak portion of the curve in Fig. 2.8, the rock is behaving more or less elastically. When materials are truly elastic they do not absorb energy; they react to the loading instantaneously and they can sustain any stress levels. If rock behaved in this way, and assuming that one were able to excavate it, there would be no problem with either excavation, support or rock failure. However, as we noted in the previous section, the rock does break and it does have post-peak mechanical characteristics. The consequence of this is two-fold:

(a) through natural processes, the *in situ* rock may have already failed and formed faults and joints;
(b) these faults and joints may be the 'weak links' in the rock structure.

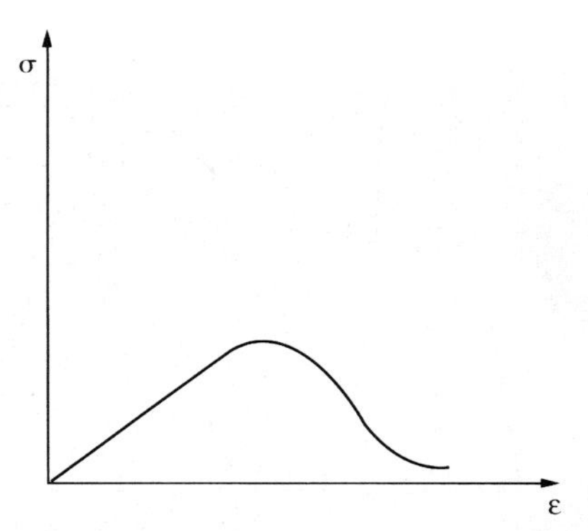

Figure 2.12 Complete stress–strain for chalk—low stiffness, low strength, quite brittle.

During the lithification process and throughout geological history, there have been orogenic periods and other less severe loading processes applied to the rock. The result in terms of the rock fracturing is to produce a geometrical structure (often very complex) of fractures forming rock blocks. An example of such a structure is shown in Fig. 2.14.

Because in the general uses of mechanics and stress analysis it is assumed that a material is continuous, these geological features such as faults, joints, bedding planes and fissures—all of which can be significant mechanical breaks in the continuum—are termed 'discontinuities' for engineering purposes. These discontinuities have many geometrical and mechanical features which often govern the total behaviour of the rock mass. The discontinuities will have certain shapes, certain sizes and be orientated in certain directions. The overall geometrical configuration of the discontinuities in the rock mass is termed **rock structure**. For engineering purposes, it is vital that we understand this geometrical structure, as will be explained further in Chapter 7.

Although the rock engineer is primarily concerned with the mechanical behaviour of the rock, it is very helpful to understand the way in which the discontinuities were formed and hence to have an initial idea of their

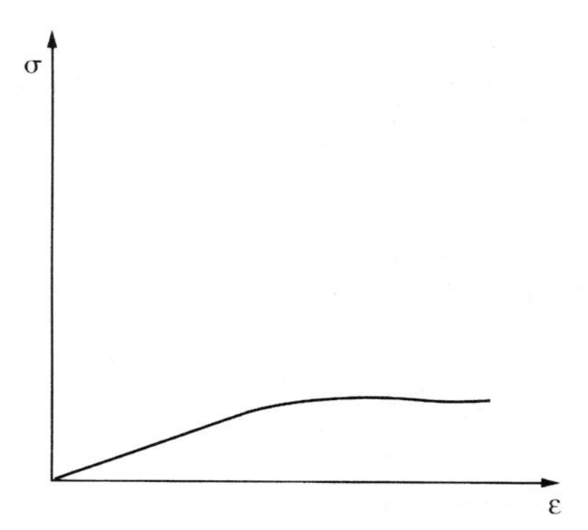

Figure 2.13 Complete stress–strain curve for rock salt—low stiffness, low strength, ductile.

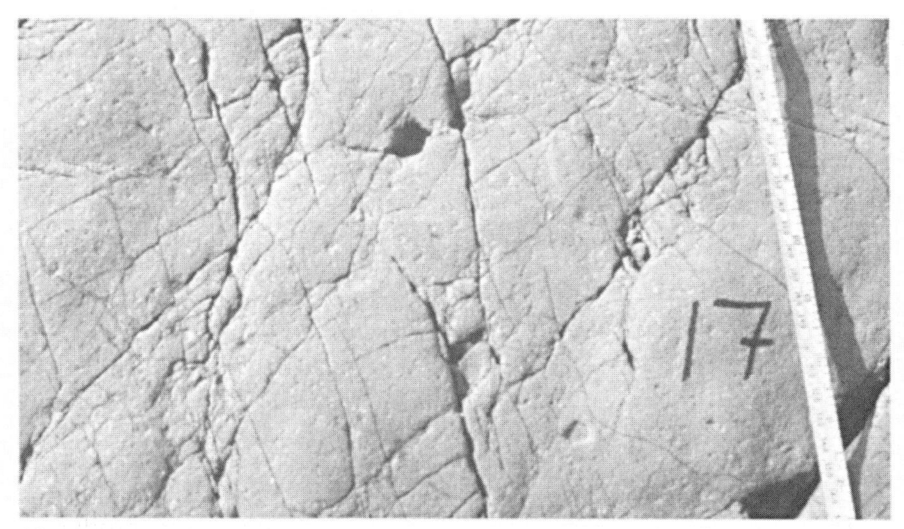

Figure 2.14 Illustration of the complex nature of a rock mass due to successive phases of superimposed fracturing.

likely mechanical characteristics. There are three ways in which a fracture can be formed: one by pulling apart and two by shearing. These are illustrated in Figs 2.15–2.17 showing that this leads to two fundamentally different types of discontinuities: i.e. those which have been simply opened and are termed joints (as in Fig. 2.15); and those on which there has been some lateral movement and are termed shear zones or faults (as in Figs 2.16 and 2.17). Given that such features exist in all rock masses at a variety of scales, it is hardly suprising that they will significantly affect the deformability, strength and failure of rock masses. Moreover, other key characteristics such as the permeability can be governed almost entirely by the rock structure configuration.

It is found in practice that, indeed, the rock discontinuities have implications for all engineering. Failure is very often associated directly with the discontinuities, which are the weak links in our pre-existing,

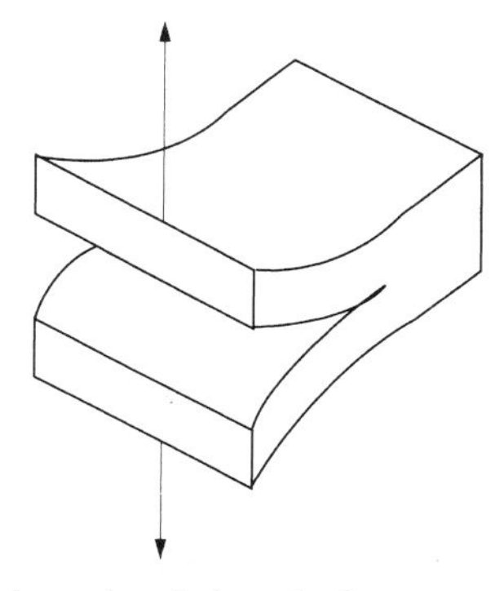

Figure 2.15 Tensile fracturing of rock (mode 1).

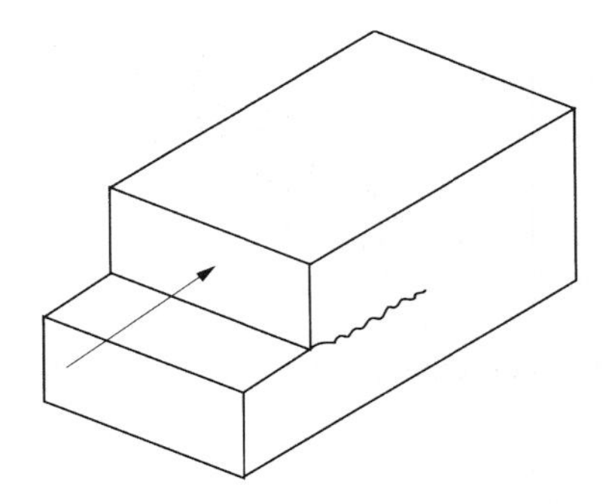

Figure 2.16 Shear fracturing of rock (mode 2).

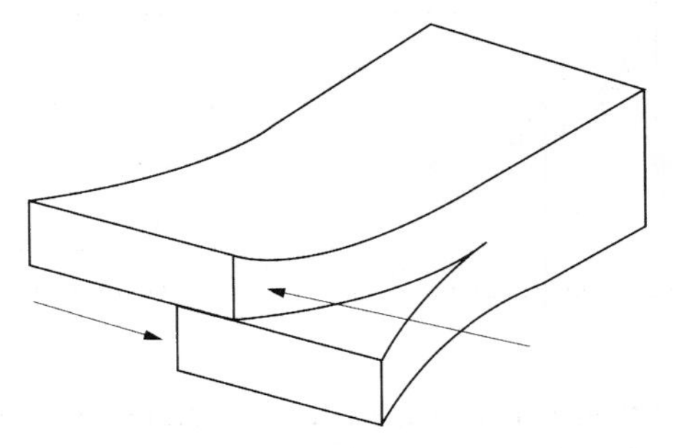

Figure 2.17 Shear fracturing of rock (mode 3).

natural, engineering material. Furthermore, the distinction between joints and faults is important. If the two sides of the fracture have been pushed over one another, as in Figs 2.16 and 2.17, the discontinuities are likely to have a low resistance to any additional shear stresses induced by engineering activities. For these and other reasons, it is most helpful if the engineer has a knowledge of structural geology and rock structure.

Some examples of the way in which the discontinuity genesis leads to differing mechanical properties are illustrated in Figs 2.18–2.20. In Fig. 2.18, an open joint is shown; this is clearly a break in the continuum. As can be seen in the figure, stresses cannot be transmitted across this discontinuity because the two sides are not connected. Moreover, this aperture within the rock mass is an open conduit for water flow with a permeability many orders of magnitude greater than the adjacent intact rock. In Fig. 2.19, a particular type of discontinuity is shown which occurs in limestone and dolomitic rocks and which has a high resistance to shear because of the connecting material across the discontinuity, although this resistance will still be less than the intact rock. Also, such a discontinuity will have a permeability higher than the intact rock. In Fig. 2.20, there is a sketch of the surface of a slickensided fault, i.e. a discontinuity on which there has been slip movment under stress causing the discontinuity surfaces to become altered and, in particular, to have a slippery surface. In some cases, such discontinuities can be pervasive throughout the rock mass with the result that the engineer must expect that, in near surface regions, failure will always occur along the discontinuity surfaces.

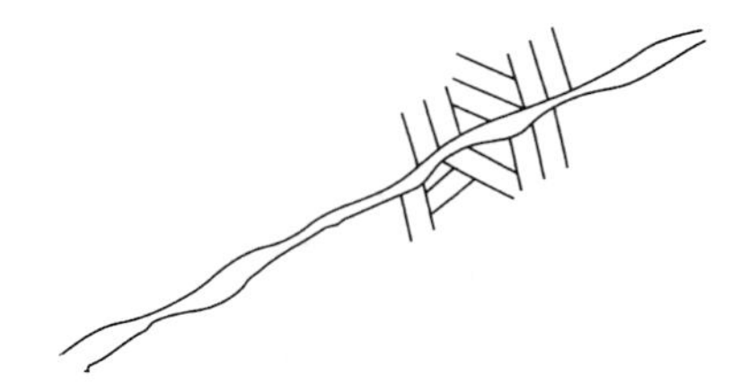

Figure 2.18 Open joint which will allow free flow of water.

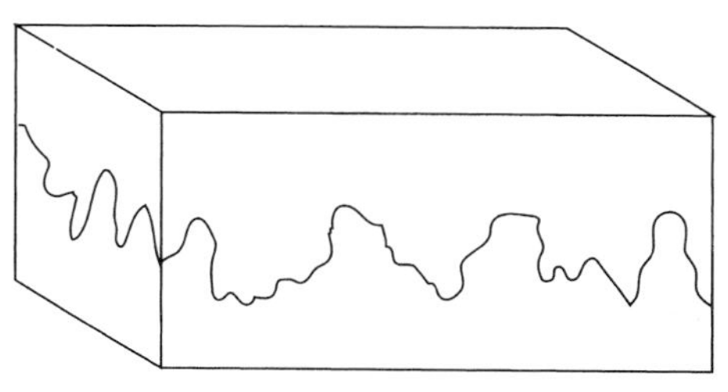

Figure 2.19 Stylolitic discontinuity with high shear resistance.

These are some examples of the way in which the discontinuities can have a dramatic effect on rock behaviour. A comprehensive explanation of the engineering approach to the geometry and mechanical behaviour of discontinuities is presented in Chapter 7. Later on, in Chapter 10, we will be discussing inhomogeneity and anisotropy with reference to the rock structure. It is quite clear from Fig. 2.20 that a slickensided feature in the rock mass will have a very significant effect on the local rock mass properties. In particular, it will cause the rock to have different properties in different directions and hence be a dominant factor causing anisotropy of the rock mass. These topics of inhomogeneity and anisotropy have ramifications throughout the book and for rock engineering in general.

2.3.3 In situ *pre-existing rock stress*

In a mechanics problem, one considers a body with certain mechanical properties and the effect of loading the body with certain forces or stresses. In Sections 2.3.1 and 2.3.2, we have discussed the material properties of the rock in terms of the intact rock and the overall rock structure. We remember the point that rock is a natural material. We now consider the loading conditions and again emphasize that there may already be an *in situ* pre-existing state of stress in the rock. In some cases, such as a dam or nuclear power station foundation, the load is applied as in a conventional mechanics problem (Fig. 2.21). In other cases, such as the excavation of a tunnel or mine stope, no new loads are applied in unsupported excavations: it is the pre-existing stresses that are redistributed by the engineering activity (Fig. 2.22). In all cases, this will result in the stresses being increased in some areas, and decreased in others. Finally, there could

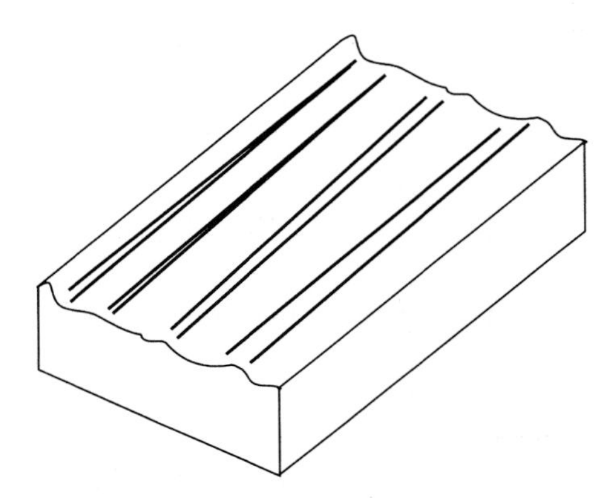

Figure 2.20 Slickensided fault surface with low shear resistance.

be a combination of the two—as in a pressurized water tunnel—where the tunnel is excavated, the rock stress is redistributed and **then** a water pressure applied inside the tunnel (Fig. 2.23). The engineer has to consider the stability of the structure throughout this process.

Thus, it is very important for the engineer to be aware of the types of stress state, both natural and applied, that can be present. In particular, there are two aspects of stress which are perhaps counter-intuitive at first sight:

(a) in the case of a deep underground tunnel, the floor will be affected in the same way as the roof by the stresses around the tunnel;
(b) in the majority of stress states measured throughout the world, one horizontal component of the stress field has greater magnitude than the vertical component.

The result of (a) may be that in addition to rock bolting the roof, the floor may have to be bolted down. The result of (b) is often that our primary

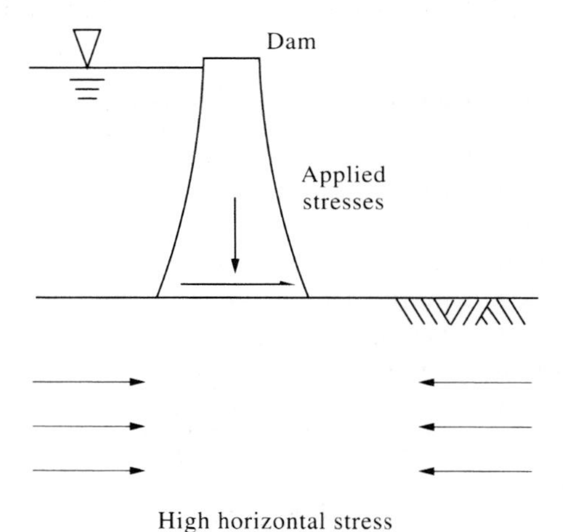

Figure 2.21 Applying loads to a rock mass which may well already contain a pre-existing stress state.

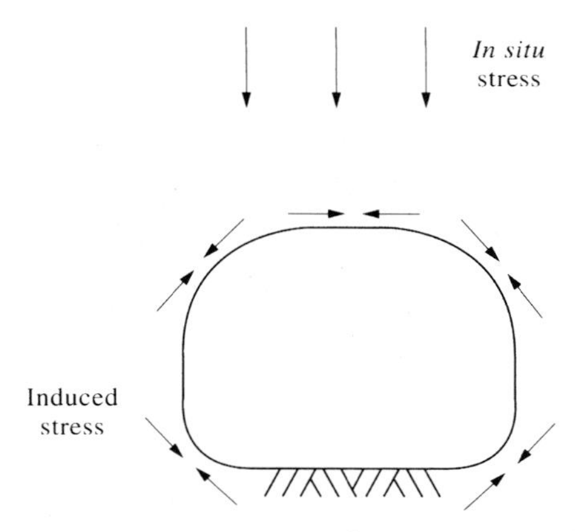

Figure 2.22 Rock engineering construction of an unsupported tunnel in which no loads are applied.

engineering defensive strategy is to support with respect to horizontal stresses rather than vertical stresses. Hence, we need to understand the concept of stress and its generation by natural mechanisms.

Basically, the vertical stress component is caused by the weight of the overlying strata, whereas, the high horizontal stress is mainly due to tectonic forces. In some cases, the horizontal stress can be very large, as is illustrated in Fig. 2.24 showing the subduction zone between the Nazca and Pacific tectonic plates in South America. In Fig. 2.24(a) the location of the Chuquicamata open pit and the El Teniente underground mines in Chile are shown. Both are very significantly affected by the high horizontal stress which acts in a west–east direction: this stress component is perpendicular to the long axis of the essentially elliptical Chuquicamata open pit and has caused problems of stability. Such stress-associated problems have been dramatically manifested underground in the El Teniente mine in the Andes. In 1987, a major rockburst occurred during block caving development at a height of 2700 m **above sea level.** Without a knowledge of the

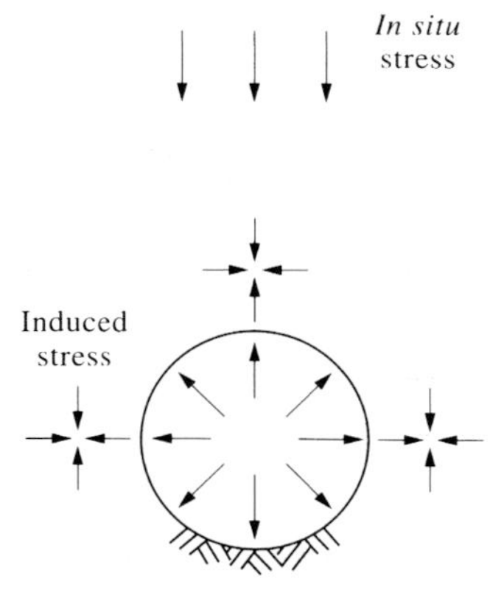

Figure 2.23 Pressurized water tunnel involving load application to a rock mass in which the pre-existing stresses have been redistributed by excavation.

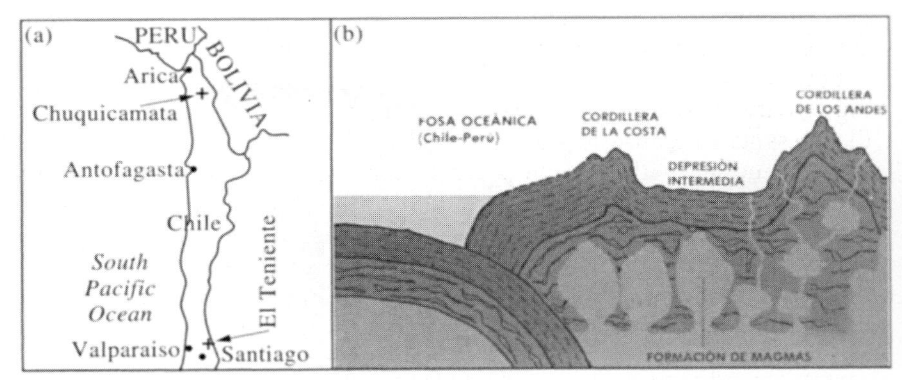

Figure 2.24 (a) Locations of the largest surface mine and largest underground mine in the world. (b) Subduction zone off the coast of Chile.

rock stress environment, it would be impossible to understand the mechanics of this rockburst.

Similar, although not so extreme, circumstances exist throughout the world due to the tectonic activity that is currently underway. Of course, there have been past orogenic events so that the rock has a stress history. Moreover, there are also factors such as surface topography and erosion which affect the *in situ* stress state.

There are ways of estimating the *in situ* stress state from geological indicators and there are ways of measuring the stress directly by engineering techniques. The subject of stress will be explained in detail in Chapter 3. The distribution of *in situ* stress values throughout the world will be discussed in Chapter 4. We cannot over-emphasize the importance of *in situ* stress because of its natural origin, ubiquity and because it is one of the boundary conditions for our mechanics considerations.

2.3.4 Pore fluids and water flow

In soil mechanics, the concept of pore fluid is fundamental to the whole subject. This is because most soils have been formed by the transportation and depostion of discrete particles with significant space around them for water to move through the soil. The water can be under pressure and hence reduce the effect of the applied stresses described in Section 2.3.3. This leads to the concept of effective stresses which have proved to be so important in soil mechanics, both from the theoretical and applied points of view.

However, rock masses have not been reconstituted in the same way as soils, although rock masses are all fractured to a greater or lesser extent. This means that accounting for pore fluids and water flow is much more difficult in rock mechanics than in soil mechanics. Many rocks in their intact state have a very low permeability compared to the duration of the engineering construction, but the main water flow is usually via the secondary permeability, i.e. through the pre-existing fractures. Thus the study of water flow in rock masses will be a function of the discontinuities, their connectivity and the hydrogeological environment.

Both the stress and the water flow are significantly affected by engineering activity. As will be explained later in greater detail, all excavation affects the stress state because all the unsupported excavation surfaces are principal stress planes and all excavations act as sinks because the *in situ* hydraulic pressure is reduced to atmospheric pressure on the excavation boundary. Thus, in the present context we need to understand the nature of permeability and water flow in fractured rock mass systems. Moreover, as will be discussed in Chapter 14, there can be interactions between the stress and the permeability, both naturally and as they are affected by engineering activities.

Depending on the engineering objective, the above considerations may be enough—but there are some applications in which an understanding of fluid flow through intact rock can be critical, e.g. in reservoir engineering for the petroleum industry. Alternatively, a case where the water flow through the fractures is particularly important is in hot dry rock geothermal energy projects where the success of the whole project depends on achieving the required borehole-to-borehole water flow. An extreme example is the case of radioactive waste disposal where the engineer can only state that the design of the waste repository is valid if the radionuclide dosage back to the biosphere can be calculated, and this can only be done if the three-dimensional flow of water through fractured rock masses can be accurately modelled. Because of the long operational life in this latter application, the permeabilities of both the intact rock and the fractured rock mass must be understood as well as other factors such as sorption of radionuclides onto rock fracture surfaces. Many groups have studied the age of the water present in the rock to assist in the approach to this problem, again reinforcing the point that an understanding of the geological setting is fundamental.

There are several other aspects of pore fluids and water flow that may be important in specific cases, such as groundwater chemistry, the formation of caves and rock alteration by fluid movement. The subject of water flow recurs throughout this book.

2.3.5 Influence of time

Another major factor of importance is the influence of time. On the one hand, our engineering material is millions of years old and, on the other, our engineering construction and subsequent activities are generally only designed for a century or less. Thus, we have two types of behaviour: the geological processes in which equilibrium will have been established, with current geological activity superimposed; and the relatively rapid engineering process. Over the millions of years, in some areas, the *in situ* rock stresses will have been continually in a stable state yet, in other areas, the stresses will have been continually altered by tectonic activity. Similarly, the pore pressures even in the most impermeable of rocks will have stabilized, but geological activity could be causing overall hydrogeological changes. In contrast, the reponse of the rock to engineering occurs over a very short time.

Figure 2.25 (a) Joints caused by rapid brittle fracture. (b) Fold caused by slow ductile deformation.

Although geological activity is essentially long term, there exist both brittle and ductile (i.e. rapid and gradual) behaviour which are manifested in quite different geological structures as shown in Fig. 2.25.

In fact, one of our basic tools in mechanics is the theory of elasticity which links stresses and strains by the instantaneous response of the rock. Because there is no time component in elasticity, this theory is unlikely to fully explain geological processes. The theory is, however, likely to be of considerable assistance in engineering when we are interested in the initial redistribution of the stress field upon excavation.

The influence of time is important additionally because of factors such as the decrease in rock strength through time, and the effects of creep and relaxation. Creep is increasing strain at constant stress; relaxation is decreasing stress at constant strain. We might be considering processes which occur very rapidly, in particular, stress waves travelling through the rock. These could be caused by natural processes, as in earthquakes, or by artificial processes such as blasting or mechanical excavation using picks, discs or button cutters. Hence, throughout the time range from milliseconds to millions of years (say, 16 orders of magnitude), the engineer should have some understanding of rate effects. These time aspects will be discussed further in Chapter 13.

decreasing stress [illegible]
which occur very rapidly [illegible]
rock. These could [illegible]
artificial processes such [illegible]
dry or humid [illegible]
to [illegible]
an understanding [illegible]
rock [illegible] in Chapter [illegible]

[illegible]

3.1 Why study [illegible] rock engineering [illegible]

[illegible]

applied, e.g. when [illegible]
case, the [illegible]
rock mass.

[illegible]

surfaces are [illegible]
[illegible]
or the strength of the [illegible]
[illegible]
Stress [illegible]
[illegible] has, for example [illegible]

3 Stress

Stress is a concept which is fundamental to rock mechanics principles and applications. For those encountering stress for the first time, it is not a straightforward concept to grasp—unless explained very clearly. For this reason, and at this stage in the book, we have adopted a key point approach to explaining the concept of stress. This is a direct precursor to Chapter 4 on *in situ* stress and provides a link with strain in Chapter 5. Further explanation of stress is given in Appendix A.

3.1 Why study stress in rock mechanics and rock engineering?

There are three basic reasons for an engineer to understand stress in the context of rock mechanics. These are:

1. There is a pre-existing stress state in the ground and we need to understand it, both directly and as the stress state applies to analysis and design. This has been discussed in Chapter 2 in the context of the geological setting. It is emphasized again here that there can be circumstances when, during the engineering, no new loading is applied, e.g. when driving an unsupported tunnel in rock. In this latter case, the pre-existing stresses are redistributed—which leads us to the next reason.
2. When engineering occurs, the stress state can be changed dramatically. This is because rock, which previously contained stresses, has been removed and the loads have to be taken up elsewhere. In line with this fact, it is also noted in Section 3.9 that all unsupported excavation surfaces are principal stress planes, a concept we will explain. Furthermore, most engineering criteria are related to either the deformability or the strength of the rock or rock mass and the analysis of these subjects involves stresses. For example, almost all failure criteria are expressed as a function of certain stress quantities.
3. Stress is not familiar: it is a **tensor** quantity and tensors are not encountered in everyday life. The second-order tensor which we will be discussing has, for example:

—nine components of which six are independent;
—values which are point properties;
—values which depend on orientation relative to a set of reference axes;
—six of the nine components becoming zero at a particular orientation;
—three principal components; and finally
—complex data reduction requirements because two or more tensors cannot, in general, be averaged by averaging the respective principal stresses.

All this makes stress difficult to comprehend without a very clear grasp of the fundamentals.

3.2 The difference between a scalar, a vector and a tensor

As alluded to above, there is a fundamental difference between a tensor and the more familiar quantities of scalars and vectors. We will explain this first conceptually before the mathematical treatment.

A scalar is a quantity with magnitude only. Examples of scalars are temperature, time, mass and pure colour—they are described completely by one value, e.g. degrees, seconds, kilograms and frequency.

A vector is a quantity with magnitude and direction. Examples of vectors are force, velocity, acceleration and the frequency of fractures encountered along a line in a rock mass—they are described completely by three values, for example, x, y, z components which together specify both direction and magnitude.

A tensor is a quantity with magnitude, direction and 'the plane under consideration'. Examples of tensors are stress, strain, permeability and moment of inertia—they are described completely by six values, as explained in Section 3.7.

It cannot be over-emphasized that a tensor quantity is not the same as a scalar or vector quantity. This applies both in a conceptual sense and in the mathematical sense. The reason why we emphasize this so much is that both mathematical and engineering mistakes are easily made if this crucial difference is not recognized and understood.

3.3 Normal stress components and shear stress components

On a real or imaginary plane through a material, there can be normal forces and shear forces. These are illustrated directly in Fig. 3.1(a). The reader should be absolutely clear about the existence of the shear force because it is this force, in combination with the normal force, that creates the stress tensor. Furthermore, it should be remembered that a solid can sustain such a shear force, whereas a liquid or gas cannot. A liquid or gas contains a pressure, i.e. a force per unit area, which acts equally in all directions and hence is a scalar quantity.

The normal and shear stress components are the normal and shear forces per unit area as shown in Fig. 3.1(b). We have used the notation F_n and F_s for the forces, and σ and τ for the corresponding stresses. However, many

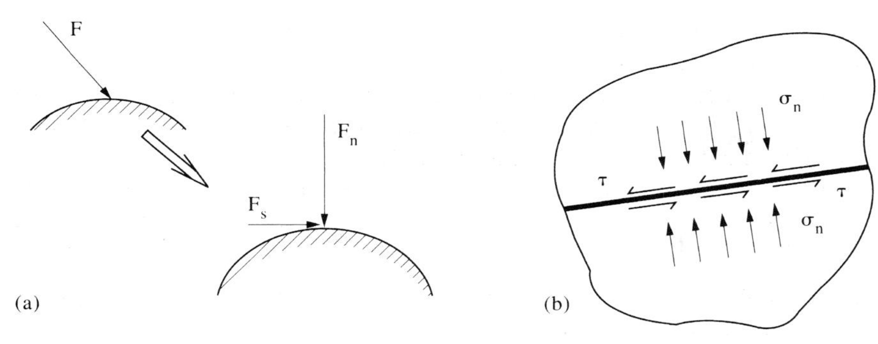

Figure 3.1 (a) Normal forces and shear forces. (b) Normal stresses and shear stresses.

different notations are in use and we encourage the reader not to be disturbed by such differences but to establish which notation is being used and then use it. There is no 'best' notation for all purposes: some types of notation have advantages in specific applications.

We are now in a position to obtain an initial idea of the crucial difference between forces and stresses. As shown in Fig. 3.2(a), when the force component, F_n, is found in a direction θ from F, the value is $F \cos \theta$. **However, and as shown in Fig. 3.2(b), when the component of the normal stress is found in the same direction, the value is $\sigma \cos^2 \theta$.**

The reason for this is that it is only the **force** that is resolved in the first case, whereas, it is both the **force and the area** that are resolved in the second case—as shown in Fig. 3.2(b). This is the key to understanding stress components and the various transformation equations that result. In fact, the strict definition of a second-order tensor is a quantity that obeys certain transformation laws as the planes in question are rotated. This is why our conceptual explanation of the tensor utilized the idea of the magnitude, direction and 'the plane in question'.

3.4 Stress as a point property

We now consider the stress components on a surface at an arbitrary orientation through a body loaded by external forces. In Fig. 3.3(a) a generalized

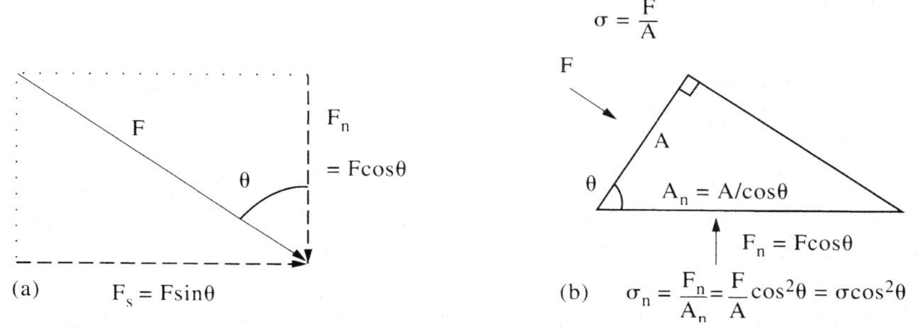

Figure 3.2 (a) Resolution of a normal force. (b) Resolution of a normal stress component.

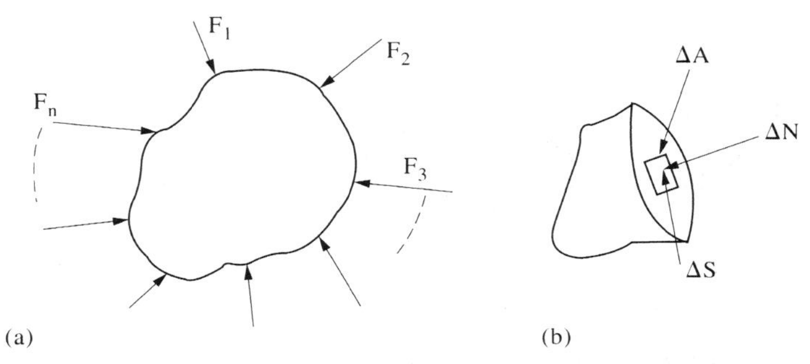

Figure 3.3 (a) Arbitrary loading of any rock shape. (b) The normal force, ΔN, and the shear force, ΔS, acting on a small area, ΔA, anywhere on the surface of an arbitrary cut through the loaded rock.

diagram of a body is shown, in this context a piece of intact rock loaded by the forces F_1, F_2, ..., F_n. This is a generic illustration of any rock loaded in any static way. Consider now, as shown in Fig. 3.3(b), the forces that are required to act in order to maintain equilibrium on a small area of a surface created by cutting through the rock. On any small area ΔA, equilibrium can be maintained by the normal force ΔN and the shear force ΔS. Because these forces will vary according to the orientation of ΔA within the slice, it is most useful to consider the normal stress ($\Delta N/\Delta A$) and the shear stress ($\Delta S/\Delta A$) as the area ΔA becomes very small, eventually approaching zero. In this way, we develop the normal stress σ and the shear stress τ as properties **at a point** within the body.

The normal stress and shear stress can now be formally defined as:

$$\text{normal stress, } \sigma_n = \lim_{\Delta A \to 0} \frac{\Delta N}{\Delta A}$$

$$\text{shear stress, } \tau = \lim_{\Delta A \to 0} \frac{\Delta S}{\Delta A}.$$

There are obvious practical limitations in reducing the size of the small area to zero, but it is important to realize that formally the stress components are defined in this way as mathematical quantities, with the result that stress is a point property.

3.5 The stress components on a small cube within the rock

It is more convenient to consider the normal and shear components with reference to a given set of axes, usually a rectangular Cartesian x–y–z system. In this case, the body can be considered to be cut at three orientations corresponding to the visible faces of the cube shown in Fig. 3.4. To determine all the stress components, we consider the normal and shear stresses on the three planes of this infinitesimal cube.

The normal stresses, as defined in Section 3.4, are directly evident as

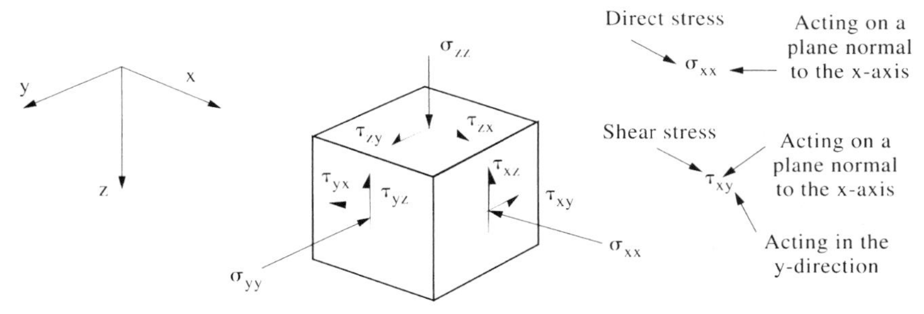

Figure 3.4 The normal and shear stress components on an infinitesimal cube in the rock aligned with the Cartesian axes.

shown in Fig. 3.4; however, the case of the shear stresses is not so direct, because the resulting shear stress on any face will not generally be aligned with these axes. To overcome this problem on any face, the shear force in Fig. 3.3(b) is resolved into two perpendicular components which are aligned with the two axes parallel to the edges of the face. Two components of shear stress are then defined on each of the planes in Fig. 3.4, as the diagrams shown in Fig. 3.5 demonstrate. Thus, we arrive at nine stress components comprised of three normal components and six shear components.

It should be noted that this discussion has been related only to the development and definition of the nine stress components. So far, we have not discussed how these components are affected by rotation of the cube relative to the reference axes: we are only defining them. The next step, therefore, is to list the components in a logical way. We have adopted the convention that the first subscript refers to the plane on which the component acts, and the plane is defined by the reference axis perpendicular to it, e.g. τ_{zy} acts on a plane perpendicular to the z-axis. The second subscript denotes the direction in which the stress component acts, e.g. τ_{zy} acts in the y-direction.

Hence, it is convenient to collate the stress components in a matrix with the rows representing the components on any plane, and the columns representing the components acting in any given direction. This is illustrated as:

$$\begin{bmatrix} \sigma_{xx} & \tau_{xy} & \tau_{xz} \\ \tau_{yx} & \sigma_{yy} & \tau_{yz} \\ \tau_{zx} & \tau_{zy} & \sigma_{zz} \end{bmatrix}.$$

There are many conventions in use for designation of the matrix components.

As an example, the component τ_{xy} in the middle of the top row could be designated as σ_{xy}, σ_{12}, S_{xy}, P_{xy}, or indeed any expression, say Ω_{ab}. The most important aspect of the notation is that the reader should recognize which notation is being used and not be over-concerned about differences of nomenclature.

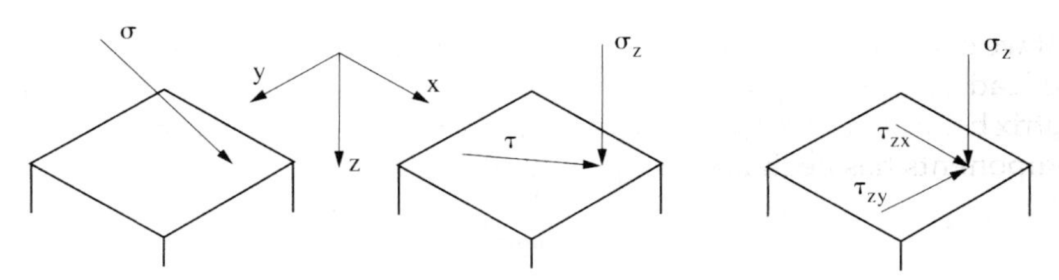

Figure 3.5 Illustration of the development of two shear stresses on each face of an infinitesimal cube.

3.6 The symmetry of the stress matrix

From the text so far, we know that there are nine separate stress components at a point. We also assume that the body is in equilibrium and therefore there should be an equilibrium of forces and moments at all points throughout the body. Thus, after listing the nine components in the matrix above, we should inspect the equilibrium of forces at a point in terms of these stress components.

In Fig. 3.6, we show the four stress components acting on the edges of a small square (which is a cross-section through a cube of edge length Δl) at any given location and in any plane of given orientation in the body. We now define a local Cartesian system of axes, perpendicular and parallel to the edges of the square. Clearly, the forces associated with the normal stress components, σ_{xx} and σ_{yy}, are in equilibrium; however, for there to be a resultant moment of zero, then the two shear stress components have to be equal in magnitude. This is demonstrated by taking moments about the centre of the square:

$$(\Delta l/2) \times (\Delta l)^2 \tau_{xy} - (\Delta l/2) \times (\Delta l)^2 \tau_{yx} = 0.$$

Thus, by considering moment equilibrium around the x, y and z axes, we find that

$$\tau_{xy} = \tau_{yx}, \qquad \tau_{yz} = \tau_{zy}, \qquad \tau_{xz} = \tau_{zx}.$$

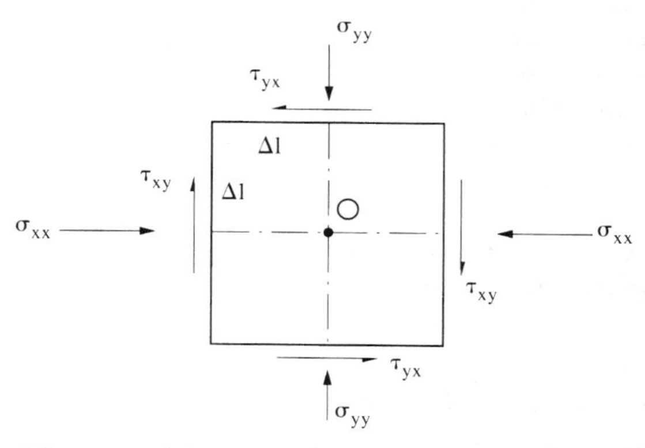

Figure 3.6 Consideration of the rotational equilibrium about the z-axis of a small cubic element at any position in a body.

If we consider the stress matrix again, we find that it is symmetrical about the leading diagonal, i.e. the diagonal from top left to bottom right. The matrix below shows this symmetry after the equality of the respective shear components has been taken into account:

$$\begin{bmatrix} \sigma_{xx} & \tau_{xy} & \tau_{xz} \\ \tau_{xy} & \sigma_{yy} & \tau_{yz} \\ \tau_{xz} & \tau_{yz} & \sigma_{zz} \end{bmatrix}.$$

It should be noted that we have considered only the stress components that exist at a point, their equilibrium, and the method of listing them in the matrix. We have not yet considered how the actual values of these components would change as the reference cube is rotated. We emphasize this because the discussion here is completely general and applies to the stress state at any point anywhere in any rock mass, or for that matter, in any material that can sustain shear stresses.

3.7 The state of stress at a point has six independent components

From our final listing of the stress components in the matrix at the end of Section 3.6, it is clear that the state of stress at a point is defined completely by six independent components. These are the three normal stress components and three shear stress components, i.e. σ_{xx}, σ_{yy}, σ_{zz}, τ_{xy}, τ_{yz} and τ_{xz}.

The fact that the state of stress is completely specified by six independent components is important and has direct ramifications for the stress measurement methods discussed in Chapter 4. Note that a **scalar** quantity can be completely specified by **one** value, and that a **vector** quantity can be completely specified by **three** values. However, the stress state at a point, which is a **tensor** quantity, requires **six** values.

Furthermore, it should be noted that stress is not the same as pressure. The word 'pressure' should be reserved for a specific stress state in which there are no shear components and all the normal components are equal—as exists in a static fluid, which can sustain no shear stress. Pressure is a scalar quantity because it can be completely specified by one value; the stress state, on the other hand, requires six independent components.

The stress state can be specified with reference to a given set of x-, y- and z-axes via the components we have explained, or via the magnitudes and directions of the principal stresses which are explained in Section 3.8. **Whatever method is used to specify the stress state, there must be six independent pieces of information.**

3.8 The principal stresses

The stress components in the stress matrix are the three normal stresses and the three shear stresses. The actual values of these components in a given body subjected to given loading will depend on the orientation of the cube in the body itself. We should consider, therefore, the directions in which the normal stress components take on maximum and minimum values. It is found that in these directions the shear components on all the faces of the cube become zero.

The principal stresses are defined as those normal components of stress that act on planes that have shear stress components with zero magnitude. It is convenient to specify the stress state using these principal stresses because they provide direct information on the maximum and minimum values of the normal stress components—but the orientation of these stresses must also be specified (remembering that six independent values are required to specify a stress state).

The values σ_1, σ_2 and σ_3 in the matrix in Fig. 3.7 are the **principal stresses.** The Arabic subscript notation is used in this book, but it should be noted that other notations can be used, e.g. σ_I, σ_{II} and σ_{III}. In our notation, we make the convention that $\sigma_1 > \sigma_2 > \sigma_3$.

The dramatic significance of this principal stress concept for rock engineering is explained in Section 3.9.

3.9 All unsupported excavation surfaces are principal stress planes

Not only are the principal stresses and their directions of fundamental significance in stress analysis, the concept of a principal stress also has particular significance for rock engineering. This is because *all unsupported excavation surfaces,* whether at the ground surface or underground, have no

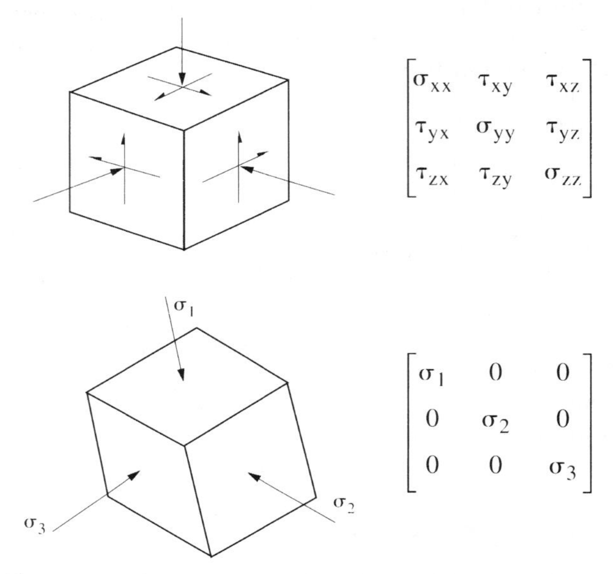

Figure 3.7 The stress components on the reference cube and the principal stress components.

shear stresses acting on them and are therefore principal stress planes. This results from Newton's Third Law ('to every action there is an equal and opposite reaction'). Furthermore, and also from Newton's Third Law, the normal stress component acting on such surfaces is zero. Thus, we know at the outset that the stress state at all unsupported excavation surfaces will be

$$\begin{bmatrix} \sigma_{xx} & \tau_{xy} & \tau_{xz} \\ \tau_{yx} & \sigma_{yy} & \tau_{yz} \\ \tau_{zx} & \tau_{zy} & \sigma_{zz} \end{bmatrix}.$$

or in principal stress notation

$$\begin{bmatrix} 0 & 0 & 0 \\ 0 & \sigma_1 & 0 \\ 0 & 0 & \sigma_2 \end{bmatrix}$$

expressed, respectively, relative to an x-, y-, z-axes system with x perpendicular to the face, and the principal stresses acting as shown in Fig. 3.8.

In Fig. 3.8(a), the pre-existing stress state is shown in terms of the principal stresses. In Fig. 3.8(b) the stress state has been affected by excavation: both the magnitudes and directions of the principal stresses have changed. Neglecting atmospheric pressure, all stress components acting on the air–rock interface must be zero.

It should also be noted that the air–rock interface could be the surface of an open fracture in the rock mass itself. Thus, as we will discuss further in Chapters 4, 7 and 14, the rock mass structure can have a significant effect on the local stress distribution.

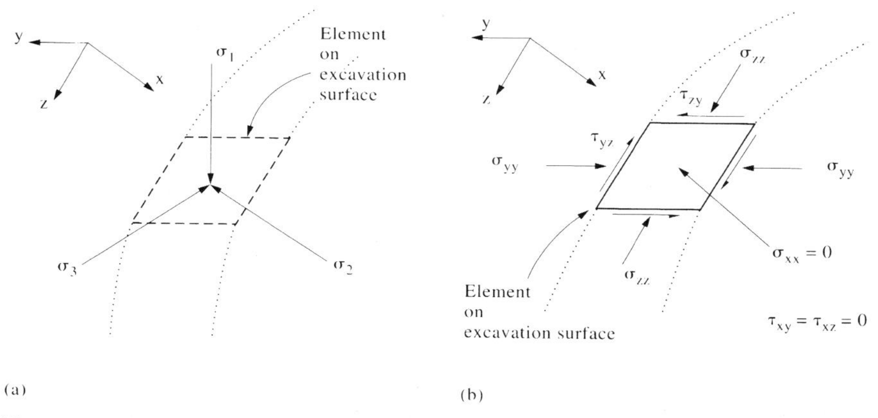

Figure 3.8 (a) Before excavation. (b) After excavation.

3.10 Concluding remarks

We emphasize again that stress is a tensor with six independent components. When a force, F, is resolved through an angle θ, the resulting components are $F \cos \theta$ and $F \sin \theta$. However, when a stress component, σ, contributes to the normal and shear stresses on a plane inclined at an angle θ to the direction in which the stress component acts, the resulting components are $\sigma \cos^2 \theta$ and $\sigma \sin^2 \theta$. It is crucial to note, as we showed in Fig. 3.7, that by suitably orientating the reference cube it is possible to eliminate all shear stresses. Conversely, it is not possible to determine an orientation for the complementary circumstance where all the normal stresses reduce to zero. An elegant method of directly indicating this result, that the normal stresses cannot be reduced to zero, is that the first stress invariant (a property of the second-order tensor),

$$\sigma_{xx} + \sigma_{yy} + \sigma_{zz} = \sigma_1 + \sigma_2 + \sigma_3 = \text{a constant},$$

cannot be made equal to zero whatever the orientation of the cube—because it is a constant. The exception is when the constant is zero, i.e. a state of pure shear, for example, with normal stresses of 3, –1 and –2 MPa, so that the first stress invariant is $3 - 1 - 2 = 0$.

The material that has been presented in this chapter, and that which follows in Chapter 4, is sufficient for a basic understanding of the nature of the state of stress. However, an Appendix on stress analysis has been included. The way in which the stress is taken into account in rock mechanics and rock engineering is described in succeeding chapters.

4 *In situ* stress

In this chapter, we will be describing why a knowledge of *in situ* rock stress is important for rock engineering, how the *in situ* stress data are determined and presented, what we would expect the *in situ* stresses to be, collating stress state data from around the world, and finally commenting on rock stress variability.

4.1 Why determine *in situ* stress?

The basic motivations for *in situ* stress determination are two-fold.

1. To have a basic knowledge of the stress state for engineering, e.g. in what direction and with what magnitude is the major principal stress acting? What stress effects are we defending ourselves and our structures against? In what direction is the rock most likely to break? All other things being equal, in what direction will the groundwater flow? Even for such basic and direct engineering questions, a knowledge of the stress state is essential.
2. To have a specific and 'formal' knowledge of the boundary conditions for stress analyses conducted in the design phase of rock engineering projects.

We have already emphasized that there are many cases in rock engineering where the stresses are not **applied** as such; rather, the stress state is altered by the engineering activities, e.g. in the case of excavating a rock slope or tunnel.

4.2 Presentation of *in situ* stress state data

The stress state at a given point in a rock mass is generally presented in terms of the magnitude and orientation of the principal stresses (remember that the stress state is completely described by six parameters). In Fig. 4.1(a), we recall that the principal stresses have a certain orientation, and in Fig. 4.1(b) that the principal stresses have certain magnitudes. The orientations are often presented as in Fig. 4.1(c) via a stereographic projection.

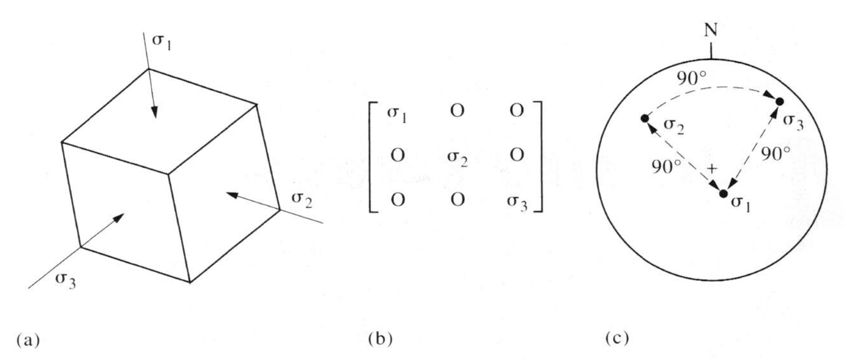

Figure 4.1 (a) Principal stresses acting on a small cube. (b) Principal stresses expressed in matrix form. (c) Principal stress orientations shown on a hemispherical projection.

4.3 Methods of stress determination

Clearly, any system utilized for estimating the *in situ* stress state must involve a minimum of six independent measurements. There are methods of 'direct' stress measurement and there are methods of estimating the stresses via various 'indirect' or 'indicator' methods. In this book, we will concentrate on the four main methods recommended by the International Society for Rock Mechanics (ISRM), while recognizing that there are a multitude of direct and indicator approaches available.

The four direct methods recommended by the ISRM (Kim and Franklin, 1987) are:

(a) the flatjack test;
(b) the hydraulic fracturing test;
(c) the United States Bureau of Mines (USBM) overcoring torpedo; and
(d) the Commonwealth Scientific and Industrial Research Organization (CSIRO) overcoring gauge.
Use of the overcoring method is shown in Fig. 4.2.

Some of the indicator methods are:

(a) borehole breakouts—damage to a borehole indicating principal stress orientations;
(b) fault plane solutions—back analysis of principal stresses causing faults;
(c) acoustic emission—the rock emits low-intensity 'noise' when it is stressed;
(d) anelastic strain relaxation—core exhibits expansion/contraction on removal from the borehole;
(e) differential strain analysis—pressurizing a piece of rock indicates its previous stress state through differential strain effects;
(f) core discing—geometry of stress-induced core fracturing indicates stress components;
(g) observations of discontinuity states, e.g. open discontinuities are not transmitting stress across the gap.

The four direct ISRM recommended methods are described below; for

Figure 4.2 *In situ* stress determination in the Carmenellis granite.

a fuller description of the indirect methods, the reader is referred to Dyke (1988). The key reference for the ISRM methods is *Suggested Methods for Rock Stress Determination*, produced by Kim and Franklin (1987). Here, we now go on to explain these methods in the context of their ability to determine the components of the stress tensor.

In Fig. 4.3, we have shown four stress tensors and indicated the ability of each method to determine the six components of the stress tensor in one application. For the flatjack and with the *x*-axis aligned perpendicular to the flatjack, one normal component—in this case σ_{xx}—can be determined. It immediately follows that, to determine the complete state of stress, six

1. **Flatjack**

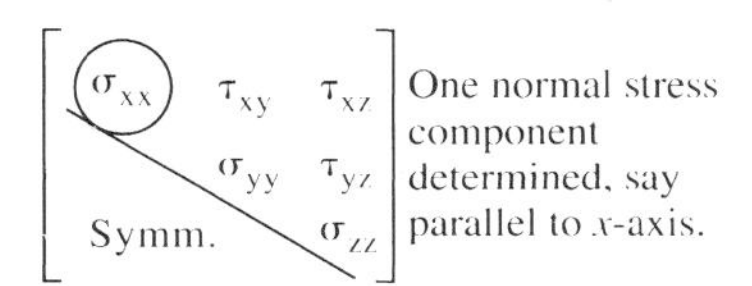

One normal stress component determined, say parallel to x-axis.

2. **Hydraulic fracturing**

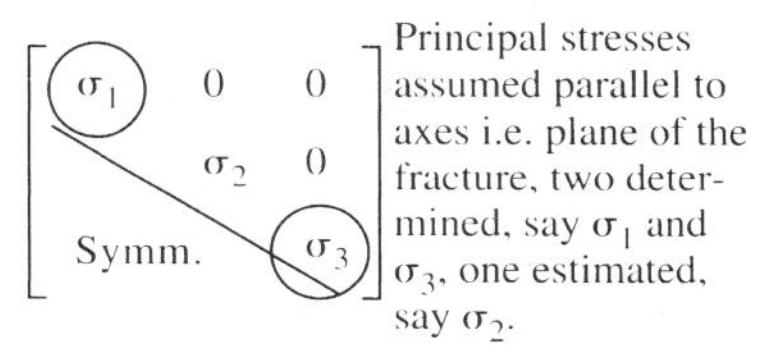

Principal stresses assumed parallel to axes i.e. plane of the fracture, two determined, say σ_1 and σ_3, one estimated, say σ_2.

3. **USBM overcoring torpedo**

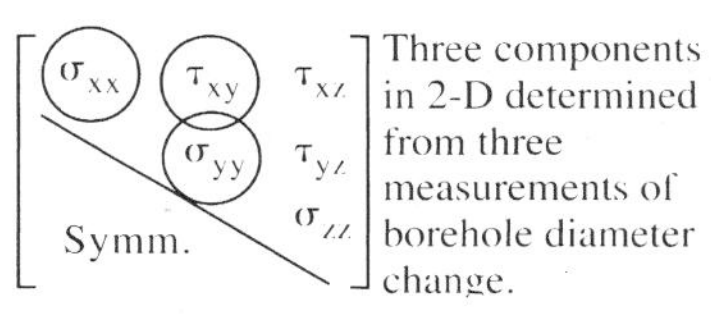

Three components in 2-D determined from three measurements of borehole diameter change.

4. **CSIRO overcoring gauge**

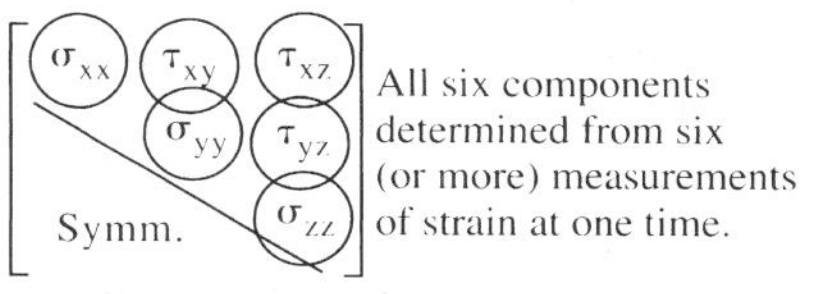

All six components determined from six (or more) measurements of strain at one time.

Figure 4.3 The four ISRM suggested methods for rock stress determination and their ability to determine the components of the stress tensor *with one application of the particular method*.

such flatjack measurements have to be conducted at six different orientations. Note that, in general, the reference axes will not be aligned with the flatjack orientation and separate transformations will have to be used for each flatjack measurement, because it is the normal stress perpendicular to the plane of the flatjack that is being determined, rather than a specific component of the stress tensor. In fact, it is interesting to note that whilst a normal stress can be determined directly, there is no equivalent method of determining a shear stress: the shear components in the tensor must be calculated from the measurements of normal stresses in different directions; they cannot be measured directly. It should also be remembered that this technique determines the stress tensor in an excavation wall and therefore determines the *induced* stress rather than the *field* stress. (A glossary of terms for *in situ* stress can be found in Section 4.10.)

With reference to the top right-hand diagram in Fig. 4.3, the basic hydraulic fracturing method provides only two items of information—the **breakdown pressure** and the **shut-in pressure.** Thus, only two components of the stress tensor can be established by this technique: the shut-in pressure is assumed to give the minor principal stress, σ_3, whilst the major principal stress, σ_1, is given via the breakdown pressure, the value of σ_3 and the magnitude of the tensile strength of the rock.

We have seen that, in the case of the flatjack, the six components can be determined by using the method at six different orientations. In general, this is not possible with hydraulic fracturing, because the tests are conducted deep in a borehole. The major advantage of hydraulic fracturing is that it is the only method of determining part of the stress state more than a few hundred metres from man-access, and, indeed, may be used up to 5 or 6 km depth. However, the major disadvantage is that assumptions have to be made in order to complete the stress tensor. These assumptions are that the principal stresses are parallel and perpendicular to the borehole axis, and that the vertical principal stress can be estimated from the depth of overburden. As a result, in the hydraulic fracturing stress tensor in Fig. 4.3, the two circled components are determined *but* the three zero values for the shear stresses are an assumption, as is the value (of what is taken here to be) σ_2.

In the case of the USBM overcoring torpedo, a two-dimensional state of stress is determined, i.e. the three circled components in the diagram in Fig. 4.3, giving three components of the three-dimensional stress tensor. Thus, two, and preferably three, non-parallel boreholes must be used to determine the complete state of stress. It should be noted that in the cases of the flatjack and hydraulic fracturing, the material properties of the rock have not been used except for the tensile strength which is required in hydraulic fracturing. For the flatjack, only the transformation equations are required; for hydraulic fracturing, only the stress concentration factors for a circular hole are required and these are independent of material properties (assuming ideal elasticity); but, for the USBM overcoring torpedo, in order to convert the measured displacements to stresses, the elastic properties of the rock are required. This introduces a whole new series of assumptions.

Finally, in the case of the CSIRO overcoring gauge, as we have shown in Fig. 4.3, the complete state of stress can be determined from measurements of strain in six or more different directions taken during one application of the method. The material properties of the rock are also required for this method: a device which is equipped with nine or 12 strain gauges can determine the state of stress in a transversely isotropic rock with five elastic parameters.

It is emphasized that the understanding of how the components of the stress tensor are established by these four different methods is crucial to the planning of an optimal strategy for stress measurement. There are other complicating factors which we will be discussing later, but the key is to understand the fundamental basis of the tests as described here. In this context, none of the indicator methods, with the possible exception of differential strain analysis, can estimate the complete stress tensor. It follows that invariably our strategy for stress determination will comprise of integrating all the information to hand.

In the following sub-sections, the four main ISRM methods are outlined and discussed. The diagrams are those used in the ISRM Suggested Methods document.

4.3.1 Flatjack

In Fig. 4.4, the basic principle of the flatjack test is shown. Two pins are drilled and fixed into the excavation boundary. The distance, d, between them is then measured accurately. A slot is cut into the rock between the pins, as shown in the diagram. If the normal stress is compressive, the pins will move together as the slot is cut. A flatjack, which is comprised of two metal sheets placed together, welded around their periphery and provided with a feeder tube, is then grouted into the slot. On pressurizing the flatjack with oil or water, the pins will move apart. It is assumed that, when the pin separation distance reaches the value it had before the slot was cut, the force exerted by the flatjack on the walls of the slot is the same as that exerted by the pre-existing normal stress. There will be some error in this assumption, mainly due to jack edge effects, but these can be taken into account if the jack is suitably calibrated. The test provides a good estimate of the normal stress across the flatjack.

The major disadvantage with the system is that the necessary minimum number of six tests, at different orientations, have to be conducted at six different locations and it is therefore necessary to distribute these around the boundary walls of an excavation. Invariably, these tests will be conducted under circumstances where the actual stress state is different at each measurement location. Hence, to interpret the results properly, it is also necessary to know the likely stress distribution around the test excavation.

4.3.2 Hydraulic fracturing

The hydraulic fracturing method of stress measurement basically provides two pieces of information via the breakdown pressure and the shut-in pressure (cf. the introductory text in Section 4.3 and part 2 of

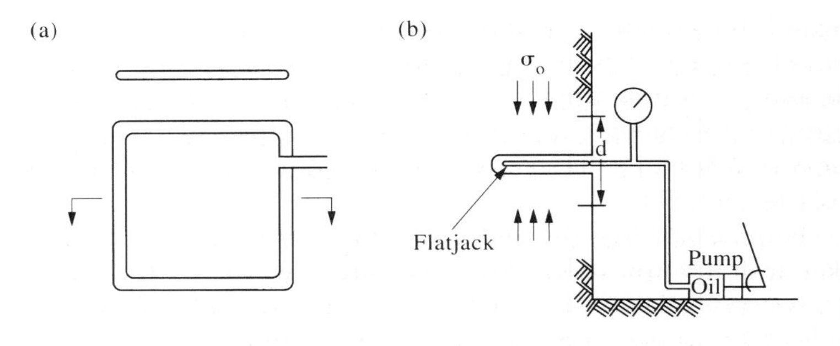

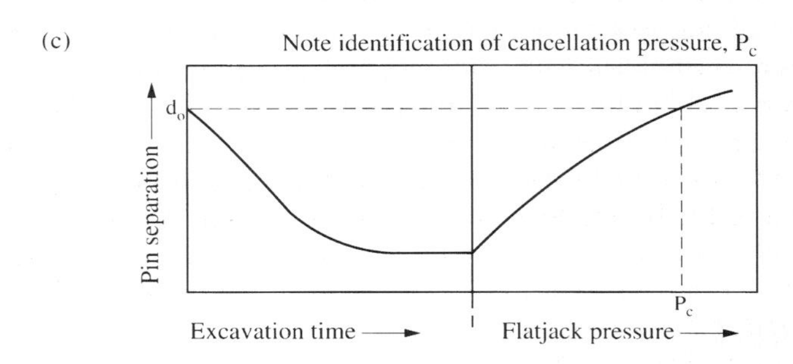

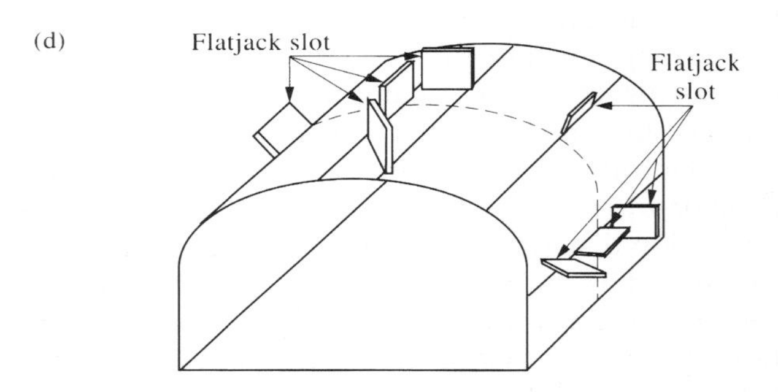

Figure 4.4 The flatjack test (from *Suggested Methods for Rock Stress Determination*, Kim and Franklin, 1987). (a) Flatjack. (b) Test configuration. (c) Pin separation versus slot excavation time and flatjack pressure. (d) The flatjack tests in progress.

Fig. 4.3). A length of borehole is chosen for the stress measurements and an interval, typically 1 m long, is located for the test and isolated using a straddle packer system. The isolated zone is pressurized by water until a fracture occurs in the rock. The two measurements taken are the water pressure when the fracture occurs and the subsequent pressure required to hold the fracture open, known, respectively, as the breakdown and shut-in pressures.

In connection with this method, it is most important to realize the following. First, the packed-off section should be free from fractures so that a **new** fracture is in fact created: a method of establishing this is to use a borehole television camera. Second, it is obviously best if the water pressures are measured at the test section, i.e. downhole rather than at the surface, if possible. Third, it is necessary to use an impression packer or equivalent system to establish the orientation and location of fracture initiation. Finally, it should be remembered that, using the basic technique, it has to be assumed that the borehole is parallel to a principal stress direction.

A schematic representation of the test equipment (consisting of a straddle packer and an impression packer) is shown in Fig. 4.5, together with the interpretative calculations. In Fig. 4.6, an early stage in the hydraulic fracturing stress measurement procedure is shown.

There are several problems inherent in the use of this equipment to measure the stress state. With reference to the four points mentioned earlier, it can often be difficult, if not impossible, to identify a 1 m length of borehole which is fracture free. Furthermore, there can be difficulties in measuring water pressures accurately, and in correctly identifying the breakdown and shut-in pressures. There is the question of whether the crack initiating at the borehole wall in fact propagates in the same direction (e.g. it may curl into the plane normal to the borehole axis). Lastly, it is often a completely unjustified assumption that the borehole is indeed parallel to a principal stress. Against all these points, however, is the fact that the hydraulic fracturing method is the only direct method available for stress measurement at any significant distance from the observer (i.e. distances greater than 100 m), and it has been used to depths of several kilometres.

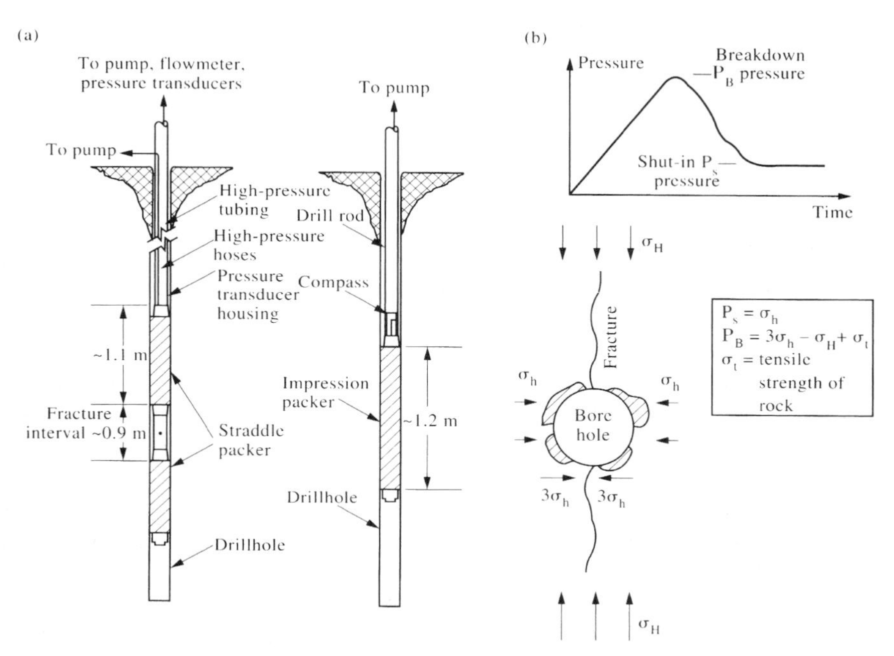

Figure 4.5 (a) The hydraulic fracturing system and (b) associated calculations (from *Suggested Methods for Rock Stress Determination*, Kim and Franklin, 1987).

Figure 4.6 Hydraulic fracturing straddle packer system being lowered into a borehole during stress measurement tests in Cornwall, UK.

In the calculation method shown in Fig. 4.5(b), it is assumed that the stress concentration of a principal stress component around the borehole in the horizontal plane shown has extreme values of –1 and 3. As shown, the shut-in pressure, P_S, is assumed to be equal to the minor horizontal principal stress, σ_h. The major horizontal principal stress, σ_H, is then found from the breakdown pressure. In the formula in Fig. 4.5, the breakdown pressure, P_B, has to overcome the minor horizontal principal stress (concentrated three times by the presence of the borehole) and overcome the *in situ* tensile strength of the borehole rock; it is assisted by the tensile component of the major principal horizontal stress. Note that when a borehole is pressurized with water at a given pressure, a tensile stress component of the same value is induced in the borehole periphery. Moreover, we have also assumed that the crack has propagated in a direction perpendicular to the minor principal stress.

All of these factors contain further tacit assumptions, in particular that the elasticity theory is valid. For this to be true, and the stress concentration factors of –1 and 3 around the circular borehole to be valid, the material of the borehole wall must be continuous, homogeneous, isotropic and linearly elastic. Furthermore, we have also assumed in this basic analysis that the rock is impermeable, so that borehole water has not penetrated the rock and affected the stress distribution.

If the elasticity assumptions are made, we find that the stress concentration factors *do not* depend on the elastic constants of the rock nor the diameter of the borehole. We do, however, need to know the tensile strength of the rock and this is a subject fraught with controversy. Suffice it to say, the best way to measure the tensile strength is under the conditions for which it is required, i.e. by pressurizing a hollow cylinder of rock. This is because the tensile strength (i.e. the stress causing tensile failure) is not a material property. A material property does not depend on the specimen geometry and loading conditions of the test: the tensile strength does.

Against this background of many potential difficulties, a great deal of research effort is being expended on improving confidence in stress measurements made using this technique. There are ways of overcoming all the difficulties mentioned above, see Cuisiat and Haimson (1992).

4.3.3 The USBM borehole deformation gauge

As indicated in part 3 of Fig. 4.3, the USBM technique allows the complete stress state in a plane to be determined from three measurements of the change in different diameters of a borehole when the stresses are released by overcoring the borehole. The instrument is shown in Fig. 4.7. When the torpedo is inserted in a borehole, six 'buttons' press against the borehole wall and their diametral position is measured by strain gauges bonded to the sprung steel cantilevers supporting the buttons. When this borehole is overcored by a larger drill, the stress state in the resulting hollow cylinder is reduced to zero, the diameter of the hole changes, the buttons move, and hence different strains are induced in the strain gauges. From previous calibration exercises, the actual diametral changes are deduced. From these changes, and with the use of elasticity theory, the biaxial stress state in the plane perpendicular to the borehole axis is deduced.

In this test, as in hydraulic fracturing, we are determining far-field stresses which have been concentrated around the measurement borehole. A useful aspect of the USBM technique is that it produces an annular core

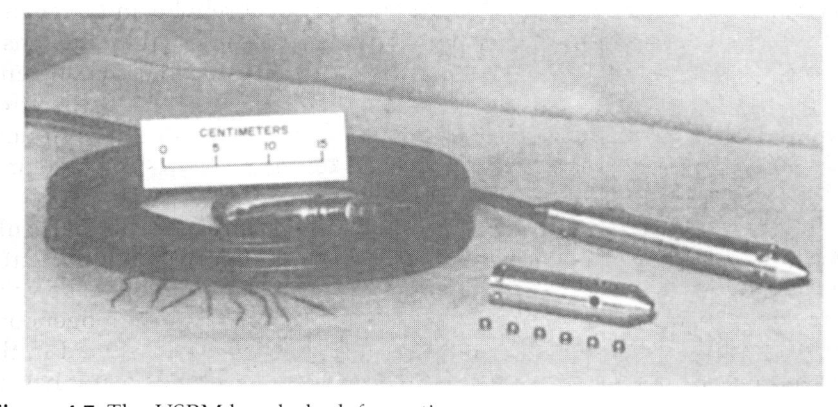

Figure 4.7 The USBM borehole deformation gauge.

which may be used in the laboratory to determine the elastic properties directly at the site where the test was conducted. Given the validity of the assumptions, the USBM gauge and its homologues are efficacious because they are reusable, permit measurements to be made many times within a borehole and are relatively cheap and robust. In Fig. 4.8, the raw data recorded during a USBM overcoring stress measurement test are shown. It can be seen that the effect of removing the pre-existing stress components has been to cause an expansion along all three diameters, with one of the deformations, u_3, in the figure, being more than the other two deformations.

The measurement of a diametral displacement is analogous to the use of a flatjack for measuring the normal stress component. In a similar way to the flatjack, each measurement of displacement effectively permits calculation of one normal strain. Through the use of the stress transformation equations, it is possible to calculate the principal components of the biaxial stress state and their orientations. There is, however, the added complication of the presence of the borehole, which perturbs the stress state from its natural *in situ* state.

4.3.4 The CSIRO overcoring gauge

This device operates on a similar principle to the USBM torpedo except that it is a gauge which is glued into the borehole and can measure normal strains at a variety of orientations and locations around the borehole wall. The gauge is glued into position within the pilot hole, initial readings of strain are taken and the gauge is then overcored. This destresses the resulting hollow cylinder and final strain gauge readings are taken. The

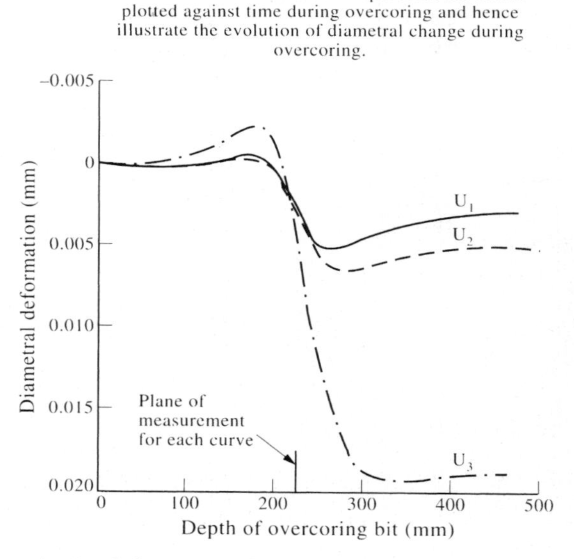

Figure 4.8 Data obtained during a USBM overcoring test.

(a)

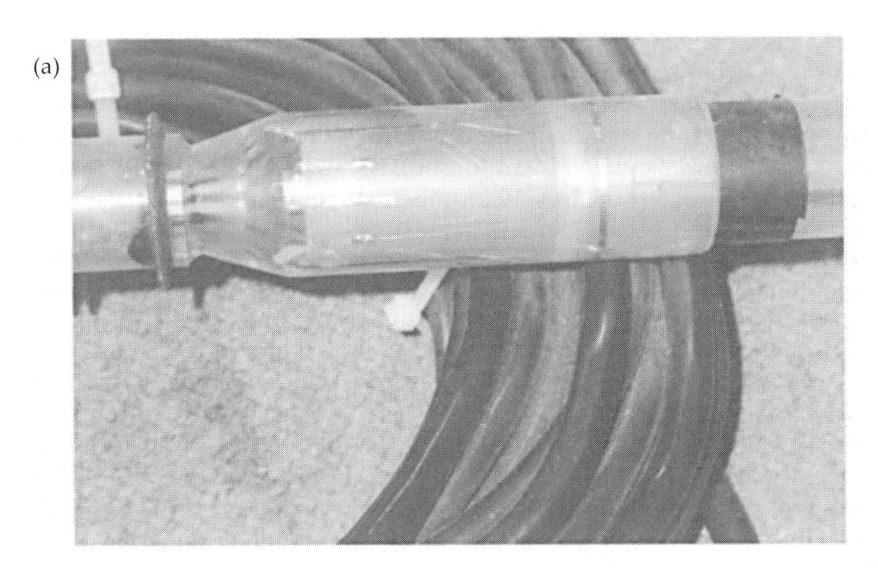

(b)

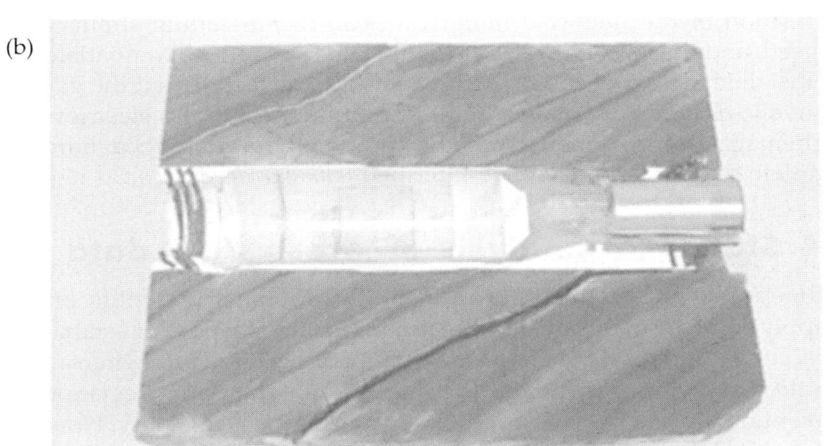

(c)

Figure 4.9 The CSIRO overcoring gauge. (a) The CSIRO gauge. (b) Installation of the gauge. (c) A sectioned hollow cylinder core containing a CSIRO gauge.

gauge has either 9 or 12 separate strain gauges, in rosettes of three, so there is some redundancy in the measurements—thus permitting statistical analysis of the data. Alternatively, if the rock is assumed to be transversely isotropic rather than completely isotropic, then the extra readings allow the stress state to be calculated incorporating the rock anisotropy. For a fuller discussion of anisotropy and the numbers of associated elastic constants, the reader is referred to Chapters 5 and 10.

One major advantage of this and similar gauges is that the resulting hollow cylinder is retrieved from the borehole and can be subjected to laboratory testing under controlled conditions in order to determine both the functionality of the system (e.g. whether any strain gauges have debonded, whether the cylinder is composed of intact rock, etc.) and the necessary elastic constants.

As with all the methods discussed, this technique has its limitations and disadvantages. One major problem is the environment within the borehole: prior to gluing the gauge in place, the surface of the wall can easily become smeared with material deleterious to adhesion; if the drilling fluid is at a different temperature to the rock, then thermal expansion or contraction of the hollow cylinder can lead to misleading strains being induced; and the long-term stability of the glue may not be compatible with the installed life of the gauge. Against this are the factors that the gauge is relatively cheap, it contains built-in redundancy (both electrical and mathematical) and, uniquely of the four methods described here, the complete state of stress can be established *with one installation*.

4.4 Statistical analysis of stress state data

With repeated measurements of a variable, it is customary scientific practice to apply some form of statistical treatment for the purpose of establishing the accuracy and precision of the measurement system. Thus, when a scalar quantity is being measured, the mean and standard deviation are conventionally used as measures of the value and its variability. However, a scalar is defined by only one value, whereas, in the case of the stress tensor, there are six independent values. This has crucial ramifications for averaging a number of stress tensors and for specifying the variability of the stress state.

We have explained that the stress state is normally specified via the magnitudes and orientations of the principal stresses. So, if a number of stress measurements have been made in a particular region, it is very tempting to estimate the average stress field by averaging the principal stresses and their orientations separately, as demonstrated in Fig. 4.10(b). *This is incorrect*: it is wrong to take the average of the major principal stresses in a number of stress tensors—because they may well all have different orientations. The correct procedure is to find all the stress components with reference to *a common reference system*, average these components, and then calculate the principal stresses from the six values of averaged components, as demonstrated in Fig. 4.10(b) and the box in the text. Note also that each of the six independent components of the stress tensor has its own mean and standard deviation: these will generally be different for each of the six stress components. *Thus*, the variability (expressed via the six standard deviations of the components) is in itself a tensor with its own principal values and directions—which may not coincide with the mean principal stress directions. The subject of tensor statistics is, however, beyond the scope of this book.

The point being made is that the correct procedure for determining mean stresses must be utilized. Also, the form of the variability of several measurements made in one location can in itself be diagnostic. The direct procedure for establishing mean principal stresses from two stress tensors, say, the results of two stress determinations, is shown in the box below. The method outlined in the box can, of course, be extrapolated to any number of tensors.

Correct method for averaging two stress tensors

Two principal stress tensors resulting from stress measurement programmes are shown below and identified by the superscripts a and b:

$$\begin{bmatrix} \sigma_1^a & 0 & 0 \\ & \sigma_2^a & 0 \\ \text{Symm.} & & \sigma_3^a \end{bmatrix} \quad \begin{bmatrix} \sigma_1^b & 0 & 0 \\ & \sigma_2^b & 0 \\ \text{Symm.} & & \sigma_3^b \end{bmatrix}.$$

The principal stress components in these tensors will generally have different orientations. Before averaging can proceed, these must be transformed to a common set of reference axes, thus:

$$\begin{bmatrix} \sigma_{xx}^a & \tau_{xy}^a & \tau_{xz}^a \\ & \sigma_{yy}^a & \tau_{yz}^a \\ \text{Symm.} & & \sigma_{zz}^a \end{bmatrix} \quad \begin{bmatrix} \sigma_{xx}^b & \tau_{xy}^b & \tau_{xz}^b \\ & \sigma_{yy}^b & \tau_{yz}^b \\ \text{Symm.} & & \sigma_{zz}^b \end{bmatrix}.$$

When averaged, these tensors give a subsequent tensor ,

$$\begin{bmatrix} (\sigma_{xx}^a + \sigma_{xx}^b)/2 & (\tau_{xy}^a + \tau_{xy}^b)/2 & (\tau_{xz}^a + \tau_{xz}^b)/2 \\ & (\sigma_{yy}^a + \sigma_{yy}^b)/2 & (\tau_{yz}^a + \tau_{yz}^b)/2 \\ \text{Symmetric} & & (\sigma_{zz}^a + \sigma_{zz}^b)/2 \end{bmatrix}$$

from which can be calculated the 'global' average principal stress tensor:

$$\begin{bmatrix} \sigma_1 & 0 & 0 \\ 0 & \sigma_2 & 0 \\ 0 & 0 & \sigma_3 \end{bmatrix}$$

together with the directions of the principal stresses.

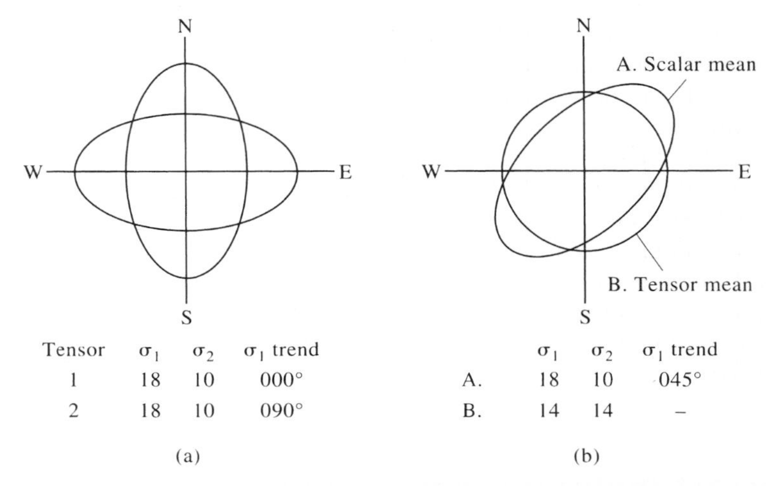

Tensor	σ_1	σ_2	σ_1 trend
1	18	10	000°
2	18	10	090°

	σ_1	σ_2	σ_1 trend
A.	18	10	045°
B.	14	14	–

Figure 4.10 Demonstration of stress tensor averaging. (a) Principal stresses to be averaged. (b) Incorrect (A) and correct (B) methods of averaging.

In Fig. 4.11 we also illustrate the addition of two two-dimensional tensors via the Mohr's circle representation of stress. This figure is particularly interesting because it provides a further intuitive understanding of the tensor concept—as being composed of one scalar component and one vector component. Thus, when adding two tensors one adds the **hydrostatic** components as scalars along the normal stress axis and the **deviatoric** components as vectors in τ–σ space. This representation can also be extrapolated to any number of tensors.

4.5 The representative elemental volume for stress

Later on, and especially in Chapter 9 on permeability, we will be discussing the concept of the representative elemental volume (REV). When tests are conducted on rock there is a spread in the measured values. This spread will occur through natural inhomogeneity of the material but, more importantly in this context, the values will depend on how the pre-existing discontinuities have affected the measured values. The REV is the volume, for any given body, at which the size of the sample tested contains a sufficient number of the inhomogeneities for the 'average' value to be reasonably consistent with repeated testing. This concept is illustrated in Fig. 4.12 where the variability versus volume is generically illustrated.

As shown, with low specimen volumes, the absence or presence of discontinuities is highly variable but, as the specimen volume is increased, the sample of discontinuities becomes more and more statistically representative, until the REV is reached. This concept applies to all rock properties and conditions which are affected by discontinuities, and is especially pertinent (and paradoxical) for stress measurements. The

(a)

$$\sigma_H = (\sigma_1 + \sigma_3)/2$$
$$\sigma_D = (\sigma_1 - \sigma_3)/2$$

(b)

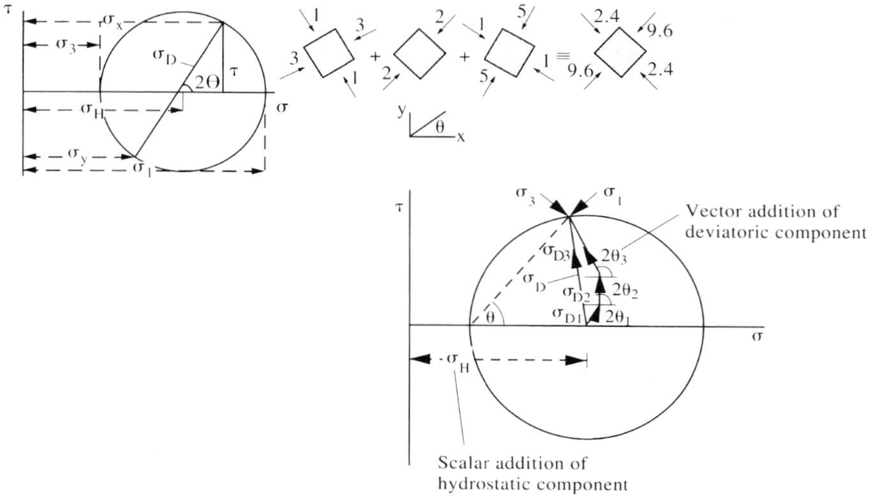

Figure 4.11 Adding two stress tensors using Mohr's circle representation. (a) The hydrostatic and deviatoric components of the stress tensor. (b) Representation of the hydrostatic and deviatoric components in Mohr's circle.

paradox occurs because stress is defined as a property at a point, i.e. the property of a sample with zero volume. The value at zero volume is plotted on the vertical axis in the top diagram of Fig. 4.12. It is immediately apparent that we should expect wide variations in measured *in situ* stress values because of the capricious effect of discontinuities at small volume. It should be recalled in this context that the strain gauges mentioned in Sections 4.3.3 and 4.3.4 measure strains only over distances of about 5 mm. The paradox arises because we are generally attempting to estimate the *in situ* stress which is being applied to a volume greater than the REV, but stress is a property at a point. Certainly, this super-REV stress is the one which we would require for input as a boundary condition to a numerical analysis of an engineering structure. In the design though, it could well be a local maximum in the stress field (the local sub-REV stress) acting on a small volume of rock which is critical for the stability of the structure as a whole.

There are many ramifications of the diagram in Fig. 4.12. The variability of the stress state with sampled volume has strong implications for stress measurement strategies, data reduction and presentation. It immediately suggests the idea of measuring stresses on the super-REV scale through a method such as 'tunnel undercoring' using very long extensometers for strain measurement (Windsor, 1985). Also, the figure suggests that numerical stress analyses of fractured rock should consist of continuum methods for large volumes of rock and discontinuum methods for sub-REV volumes. Moreover, the existence of discontinuities, together with their

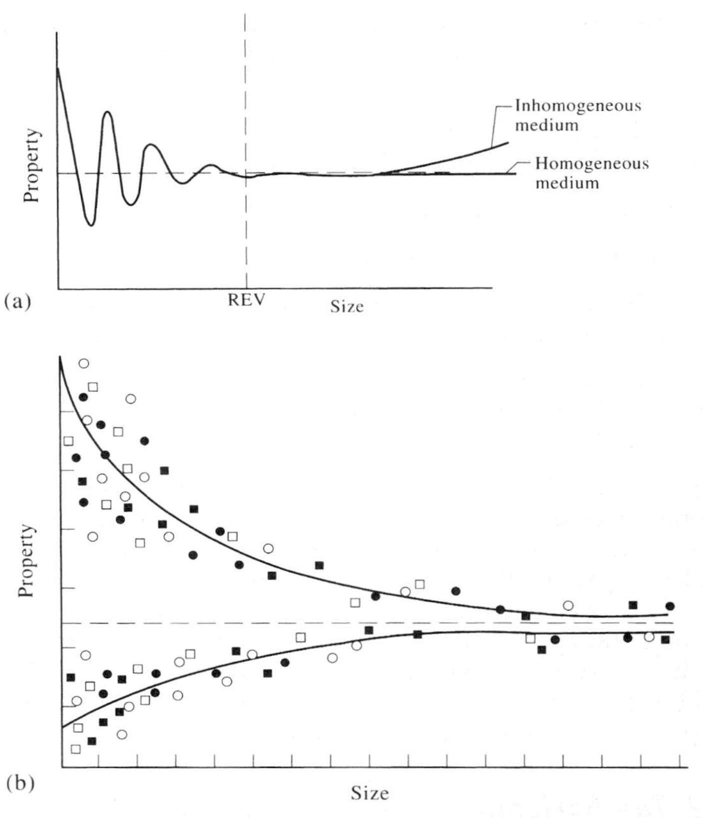

Figure 4.12 Variability in measured values with respect to sample volume, illustrating the REV. (a) General concept. (b) Example data scatter.

past and present effects on the stress state, has led to a plethora of terms describing different types of stress. In order that clarity is preserved here, this chapter concludes with a glossary of terms associated with *in situ* stress.

4.6 Predictions of natural *in situ* stress states based on elasticity theory

We have mentioned that the *in situ* stress field is conveniently expressed via the orientations and magnitudes of the principal stresses. As a first approximation, therefore, let us assume that the three principal stresses of a natural *in situ* stress field are acting vertically (one component) and horizontally (two components). Following this assumption concerning orientations, it becomes possible to predict the magnitudes of these principal stresses through the use of elasticity theory.

4.6.1 *The vertical stress component*

We might expect that the vertical stress component increases in magnitude as the depth below the ground surface increases, due to the weight of the overburden. As rules of thumb, taking the typical density of rock into account:

1 MPa is induced by 40 m of overlying rock, or
1 psi is induced by 1 ft of overlying rock.

More generally, we should use the expression

$$\text{induced vertical stress,} \quad \sigma_v = \gamma z \text{ MPa}$$

where z is the depth, measured in metres, below the ground surface and γ is the unit weight, measured in MN/m^3. Examples of γ are:

$$\begin{aligned} \gamma &= 0.01 \text{ MN/m}^3 && \text{for some coals,} \\ &= 0.023 \text{ MN/m}^3 && \text{for some shales,} \\ &= 0.03 \text{ MN/m}^3 && \text{for gabbro.} \end{aligned}$$

This approach is always used as an estimate of the vertical stress component **unless**, of course, the stress determination programme does include direct measurement of the vertical stress. We have seen, for example, that during the course of data reduction in hydraulic fracturing, the vertical stress component is estimated by this technique. Conversely, using the CSIRO gauge, the complete stress tensor is determined and so it is not necessary to estimate the vertical stress component. We will be discussing later in this chapter whether the measured stress states do correspond to such preconceived notions.

4.6.2 The horizontal stress components

We now consider the magnitudes of the horizontal stress components. Given that the vertical stress has a particular magnitude at a point in a rock mass, we might expect that a horizontal stress would be induced as a result of the vertical compression of the rock. To provide an initial estimate of this stress, based on elasticity theory and assuming isotropic rock, we must introduce the parameters Young's modulus and Poisson's ratio (a more detailed treatment of the elastic constants is given in Chapter 5 and a discussion of the validity of elasticity theory itself is given in Chapter 10).

In Fig. 4.13, an illustration of an element of rock being uniaxially stressed is given—the applied axial stress is σ_a and the resulting axial strain is ε_a. There is also a lateral strain induced, ε_l, because the element expands laterally as it is being axially compressed. From these values, we define the Young's modulus and Poisson's ratio as:

$$\text{Young's modulus, } E = \frac{\text{axial stress}}{\text{axial strain}} = \frac{\sigma_a}{\varepsilon_a}$$

$$\text{Poisson's ratio, } \nu = \frac{\text{lateral strain}}{\text{axial strain}} = \frac{\varepsilon_l}{\varepsilon_a}.$$

Utilizing these parameters, we can derive expressions for the strain along any axis for the small cube at depth in a rock mass illustrated in Fig. 4.13(c).

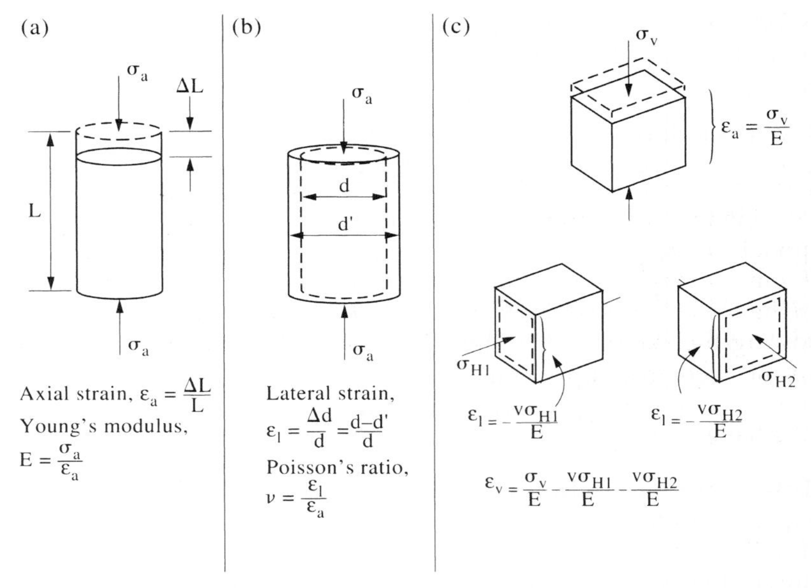

Figure 4.13 Strains on a small element of rock. (a) Axial strain and Young's modulus. (b) Lateral strain and Poisson's ratio. (c) Vertical and horizontal strains.

In this case, the total strain along any given axis may be found from the strain due to the associated axial stress, with the induced strain components due to the two perpendicular stresses being **subtracted**.

For example, the vertical strain, ε_v, is given by the expression

$$\varepsilon_v = \frac{\sigma_v}{E} - \frac{\nu\sigma_{H1}}{E} - \frac{\nu\sigma_{H2}}{E}$$

where σ_{H1} and σ_{H2} are the two principal horizontal stress components.

In the same way, the horizontal strain, ε_{H1}, can be expressed as

$$\varepsilon_{H1} = \frac{\sigma_{H1}}{E} - \frac{\nu\sigma_{H2}}{E} - \frac{\nu\sigma_v}{E}.$$

To provide an initial estimate of the horizontal stress, we make two assumptions:

(a) the two horizontal stresses are equal; and
(b) there is no horizontal strain, i.e. both ε_{H1} and ε_{H2} are zero.

We began this analysis by considering an element within an isotropic rock mass, and so we would expect the two horizontal stresses induced by the vertical stress to be equal. Moreover, the element of rock cannot expand horizontally because it is restrained by similar adjacent elements of rock, each of which is also attempting to expand horizontally. If, therefore, we take ε_{H1} as zero in the second equation above we find

$$0 = \frac{\sigma_{H1}}{E} - \frac{\nu\sigma_{H2}}{E} - \frac{\nu\sigma_v}{E}$$

and so

$$\sigma_H = \frac{\nu}{1-\nu}\sigma_v$$

where $\sigma_H = \sigma_{H1} = \sigma_{H2}$. This relation has been known for some time: according to Turchaninov *et al.* (1979), it was first derived by Academician Dinnik in 1925.

From this analysis, we find that the ratio between the horizontal stress and the vertical stress of $\nu/(1-\nu)$ is only a function of Poisson's ratio. Hence, knowing the extremes of Poisson's ratio for rock-like materials, we can find the theoretical upper and lower bounds for the induced horizontal stress.

We have

$$\nu = 0, \quad \sigma_H = 0$$

$$\nu = 0.25, \quad \sigma_H = 0.33\sigma_v$$

$$\nu = 0.5, \quad \sigma_H = \sigma_v$$

showing that the lower bound is for a value of Poisson's ratio of zero (i.e. the application of a vertical stress does not induce any horizontal strain), when there is no horizontal stress induced. At the other extreme, the upper bound is given for a Poisson's ratio of 0.5 (the value for a fluid) when the induced horizontal stress equals the applied vertical stress. In between, measured values of the Poisson's ratio for intact rock are typically around 0.25, indicating that the induced horizontal stress might be approximately one third of the applied vertical stress.

These calculations have indicated the likely values of the vertical and horizontal natural *in situ* stress components based on the application of elasticity theory to an isotropic rock. It is also implicit in the derivations that gravity has been 'switched on' instantaneously to produce the stresses: this is manifestly unrealistic. Nevertheless, we can now compare these predictions with measured data collated from stress determination programmes worldwide.

4.7 Collated worldwide *in situ* stress data

Because of the need to know the *in situ* stress state for engineering purposes, there have been many measurements made of the *in situ* stress state over the last two or three decades. In some cases, the programmes have been rather cursory and not all components of the stress tensor have been determined; in other cases, the programmes have specifically attempted to estimate all six independent components of the stress tensor. Some of these data were collected by Hoek and Brown (1980) and are presented in the two graphs shown in Figs 4.14 and 4.15.

In Fig. 4.14, the line representing one of the equations intimated in Section 4.6.1, i.e. $\sigma_v = 0.027z$, is also shown (here, the value of 0.027 has been

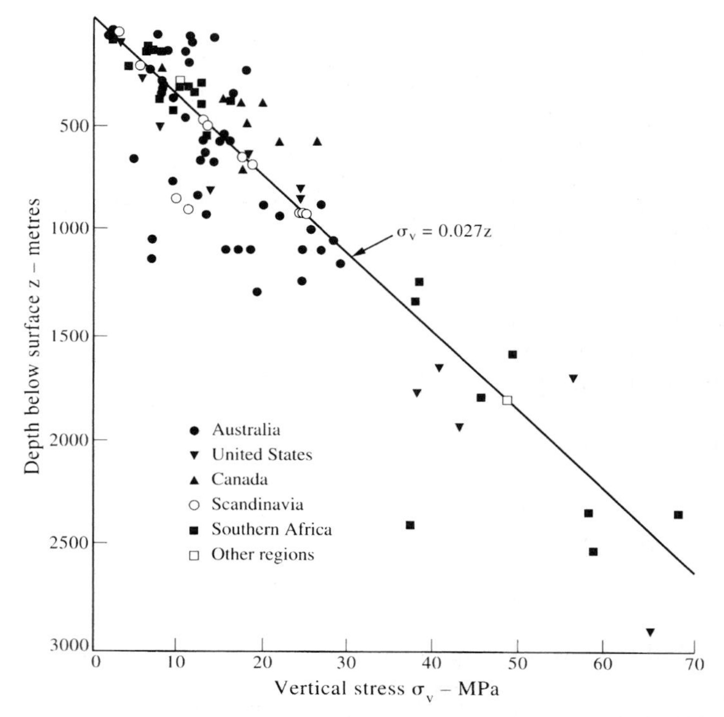

Figure 4.14 Collated worldwide *in situ* stress data: vertical stress component (after Hoek and Brown, 1980).

adopted as a generic unit weight). It can be seen that the estimate of the vertical stress component is basically correct, but only in the sense of a regression, or best fit, line. In some cases, the measured stress component is almost exactly as predicted, but in other cases *and especially at depths less than 1000 m*, the measured stress component can be dramatically different to the predicted component. Note that there are cases near the surface where the measured vertical stress component is about *five times* the predicted component. Also, between depths of 500 and 1500 m, there are cases where the measured stress component is five times less than predicted. We can conclude, therefore, that whilst the equation provides a good predictive estimate of the *average* stress from all the data, it can certainly not be relied upon to provide a correct estimate at any specific location. This implies that, if at all possible, it is best to *measure* rather than *estimate* the vertical stress component.

It should be noted that the horizontal axis in Fig. 4.15 is the mean of the two horizontal stress components, normalized by dividing by the vertical stress component. In this sense, the ratio on the horizontal axis is equivalent to the $\nu/(1-\nu)$ coefficient calculated earlier: in engineering rock mechanics it is generally known as k. A particular point to remember is that by taking the average of the two horizontal stresses, which could well be the major and minor principal stresses, a large element of the more extreme variability may have been suppressed. However, the compilers found this was the

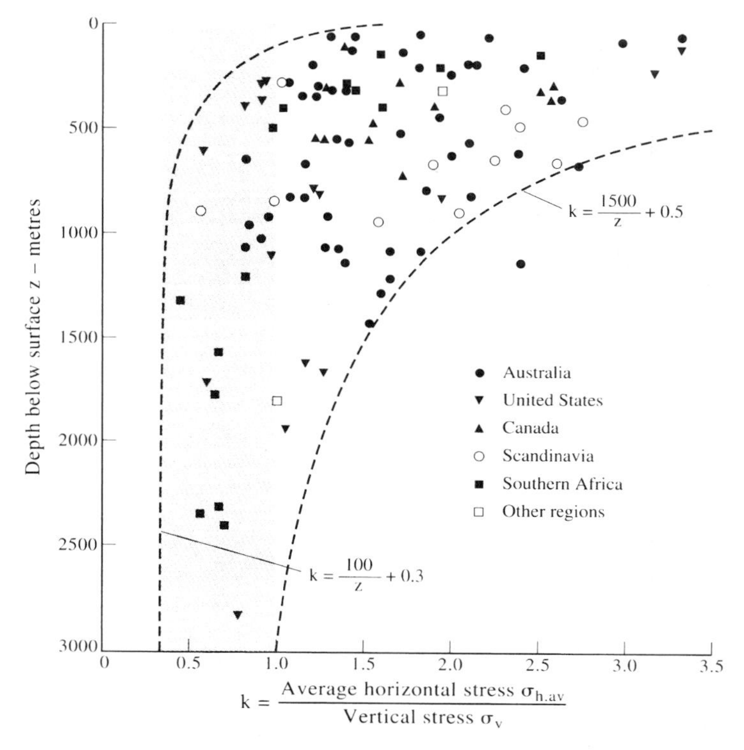

Figure 4.15 Collated worldwide *in situ* stress data: mean horizontal stress component (after Hoek and Brown, 1980).

best approach, because the complete stress tensor was not available in all cases. They suggested two formulae as envelopes for all the data in their compilation, *viz.*

$$\frac{100}{z} + 0.3 < k < \frac{1500}{z} + 0.5.$$

Note that the shaded vertical column in Fig. 4.15 gives the range of k-ratios from 0.33 to 1.00 that was predicted from simple elasticity theory and that, with increasing depth, the k-ratios given by the envelope formulae above tend towards $0.3 < k < 0.5$. Thus, for significant depths, one could argue that the elasticity model provides some indication of the k-value.

It is manifestly clear from the data, however, that it is the rule rather than the exception that the horizontal stress component (defined as the mean of the two horizontal components) is larger than the vertical stress component. For example, at depths likely to be encountered in civil engineering, say 0–500 m, in 92% of the studied cases (100% of the cases outside Canada), the magnitude of the mean horizontal stress exceeds that of the vertical stress component. Also, at typical mining depths (say, anywhere between 0 and 1000 m), the same trend applies. Of course, we would expect the highest ratios to occur very close to, or at, the surface because the vertical stress is zero at the surface.

With reference to Fig. 4.14, we saw that the calculation for the vertical stress component gave a reasonable prediction of the overall trend. This cannot be said for the data presented in Fig. 4.15: the horizontal stress components do not follow the trends predicted by simple elasticity theory, except asymptotically at depths of several kilometres. We should consider the reasons for this large variation in what are universally higher *k*-ratios than predicted. Furthermore, it is likely to be of interest in different regions of the world to plot the orientations of the maximum horizontal principal stress as has been done in Fig. 4.16 for North West Europe. From this map, it can be seen that there is a general trend of north west–south east for the maximum principal stress in the region. This leads us naturally into a discussion of the reasons for high horizontal stresses.

4.8 Reasons for high horizontal stresses

High horizontal stresses are caused by factors which fall into the categories of erosion, tectonics, rock anisotropy, local effects near discontinuities and consequential scale effects.

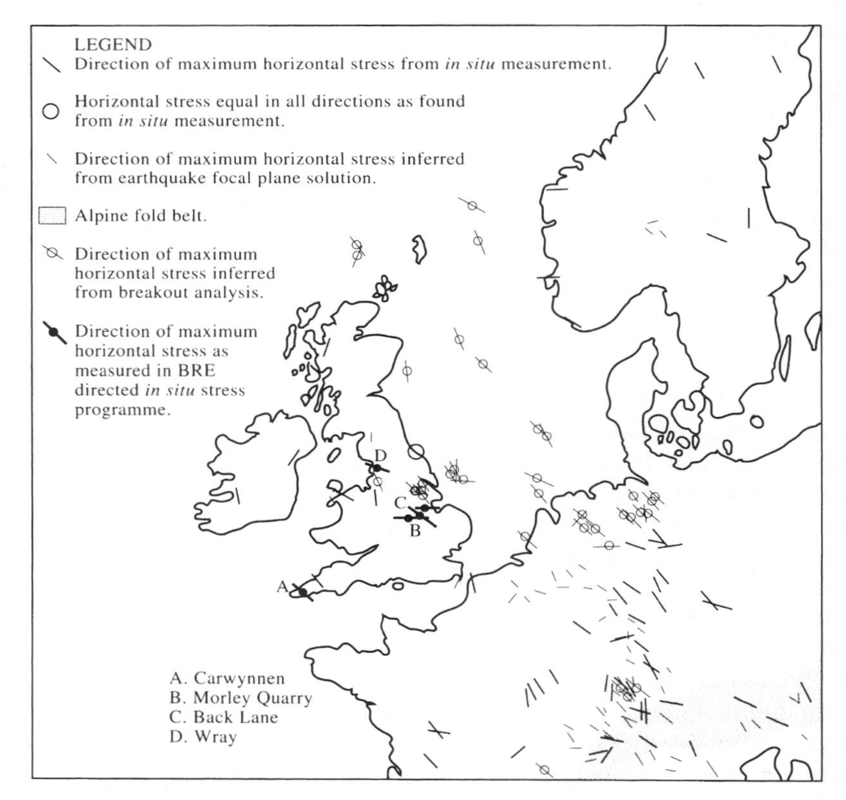

Figure 4.16 Orientation of the maximum horizontal principal stress in North West Europe (after Hudson and Cooling, 1988).

4.8.1 Erosion

The possibility that erosion of the ground surface causes an increase in the k-value is discussed by Goodman (1989). The basic idea is that the formula for $\nu/(1 - \nu)$ is valid for the initial rock mass. On erosion of the ground surface, however, the removal of the overburden and the consequential effect on both the vertical and horizontal stresses will cause an increase in the k-value, including values above unity. Also, if the horizontal stresses become 'locked in', naturally, dividing by a lower vertical stress component will result in a higher k-value. This subject is complex because of the prograde and retrograde modes of geological deformation, the time-dependent effects during this phenomenon, followed by erosion of a possibly uneven ground surface. Suffice it to say that the processes are certain to affect the magnitudes of the horizontal stress components.

4.8.2 Tectonic activity

We discussed under the Geological Setting heading in Chapter 2, that there is significant current stress activity due to tectonic plate movement. Certainly, the data in Fig. 4.16 would indicate that some form of tectonic activity was responsible for a reasonably consistent trend of the maximum horizontal principal stress over such a large region. From the 1906 and 1989 earthquakes along the San Andreas fault in California, USA, we know that

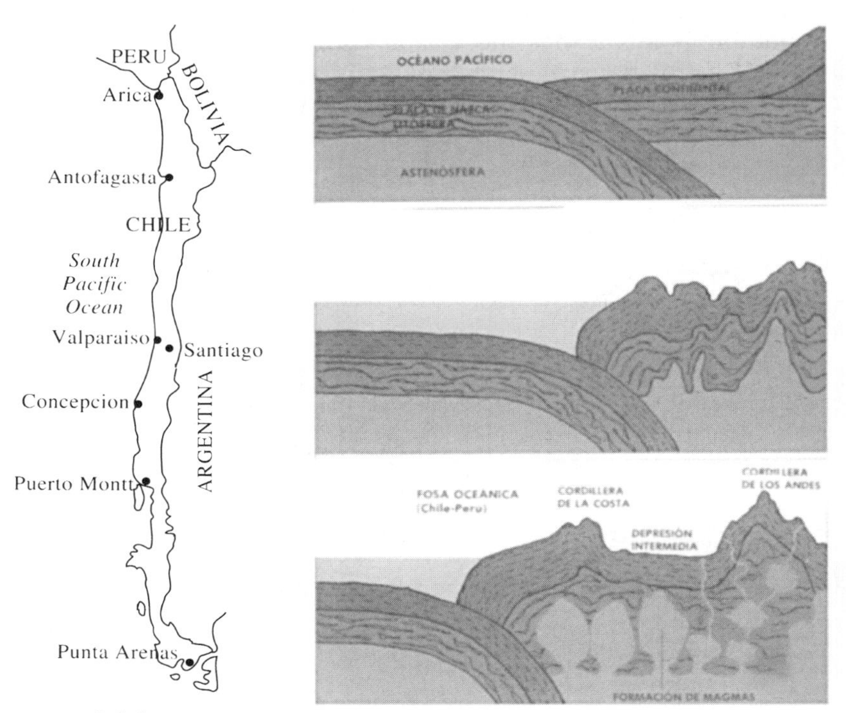

Figure 4.17 Subduction zone off the west coast of South America causing high horizontal *in situ* stresses (from Barros *et al.*, 1983).

high shear stresses can be present and result from tectonic activity. In Fig. 4.17, we illustrate the subduction zone off the coast of Chile, the genesis of the Andes and hence the likely high horizontal normal stresses we would expect from such activity. In fact, both the largest surface and largest underground mines in the World are in Chile—and they both show strong signs of stress-related phenomena in terms of rock slope instability and rockbursts, respectively.

4.8.3 Rock anisotropy

In Section 4.6.2 we derived the expression $\nu/(1-\nu)$ as the ratio between the horizontal and vertical stresses and noted that for ν varying between 0 and 0.5 the corresponding k-ratio varied from 0 to 1. We will be discussing anisotropy in much greater detail in Chapter 5, on strain, and in Chapter 10 on inhomogeneity and anisotropy. It is worth mentioning here, though, that there are three types of isotropy commonly considered in rock mechanics, namely: complete isotropy, transverse isotropy and orthotropy.

Our earlier calculation was for a rock with complete isotropy (having the same properties in all directions). It is possible to calculate the k-value for a material which is transversely isotropic (having different properties in the vertical direction to the horizontal directions), and for the orthotropic case (having different properties in three perpendicular directions). The explanation, both mathematical and intuitive, for these types of isotropy and the associated elastic material constants is given in Chapter 5. The important point is that the k-values for each case are as shown in Fig. 4.18 by the terms in parentheses.

As indicated by the sketches in the figure, transverse isotropy might well represent relatively unfractured sedimentary rocks, whereas, orthotropy

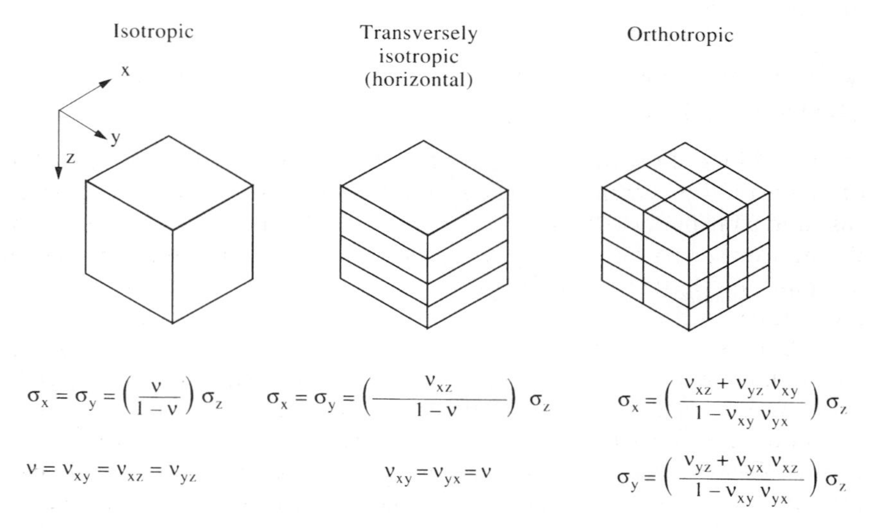

Figure 4.18 Relation between vertical and elastically induced horizontal stresses for the different types of isotropy.

could well be a good representation for rocks containing three mutually perpendicular sets of discontinuities. It can be seen from the expressions for k in Fig. 4.18 that the horizontal stress components can be different for certain values of the elastic constants in the orthotropic case.

It is a consequence of the assumption of transverse isotropy that the two horizontal principal stresses will be equal. However, in the case of orthotropy, the horizontal stresses can take on different values. Thus, it is in this last case that we find the conditions as encountered in the natural rock mass. In fact, there is nothing surprising about having one component of the horizontal stress field being much higher than the other; the apparent inconsistency lies in the oversimplication of the rock mass as a purely isotropic material. This subject will be amplified in Chapter 8 on rock masses.

4.8.4 Discontinuities

The discussion in the earlier sections was about accuracy and precision, i.e. bias and spread in the measurements. We noted that, in the case of the vertical stress component, the prediction based on the overlying rock weight was more or less accurate—in the sense that the prediction was a good fit to the data—but there was a spread in the results. The situation with the horizontal stress component was more complicated because of the unexpectedly high values of the horizontal stress components and the large spread of values. One of the most important factors causing the spread of results in both cases is the fact that the rock is not a continuum. All rocks are fractured on various scales, so the rock mass is a discontinuum and the internal stress distribution reflects this geometry. Thus, we must ask the questions: 'To what extent is the stress state affected in the region of a rock fracture?', 'How is this affected by scale?', and 'How does this affect the results of a stress determination programme?' These are the subject of the discussion presented in the next section.

4.9 Effect of discontinuities on the proximate state of stress

In Fig. 4.19 we show just one example of the influence that a rock fracture can have on the overall stress state, in this case illustrated for a plane strain case and a far-field hydrostatic (i.e. $\sigma_1 = \sigma_2 = \sigma_3$) stress state. It is clear from the figure that both the principal stress orientations and magnitudes are dramatically perturbed by the presence of the fracture. Note also that we have purposely not included any absolute scale in this figure. The elastic modelling used here could represent a fracture of any scale, from a very small flaw in a crystal, through a single rock joint in an otherwise unfractured rock mass, to a fault in a tectonic plate. This has major consequences for stress determination strategies and interpretation of results. Clearly, for a discontinuity of the order of 10 km long, all stress measurements in an adjacent proposed engineering site would be affected by the presence of the discontinuity—but perhaps this is the stress state that should be measured. Conversely, the single rock fracture could

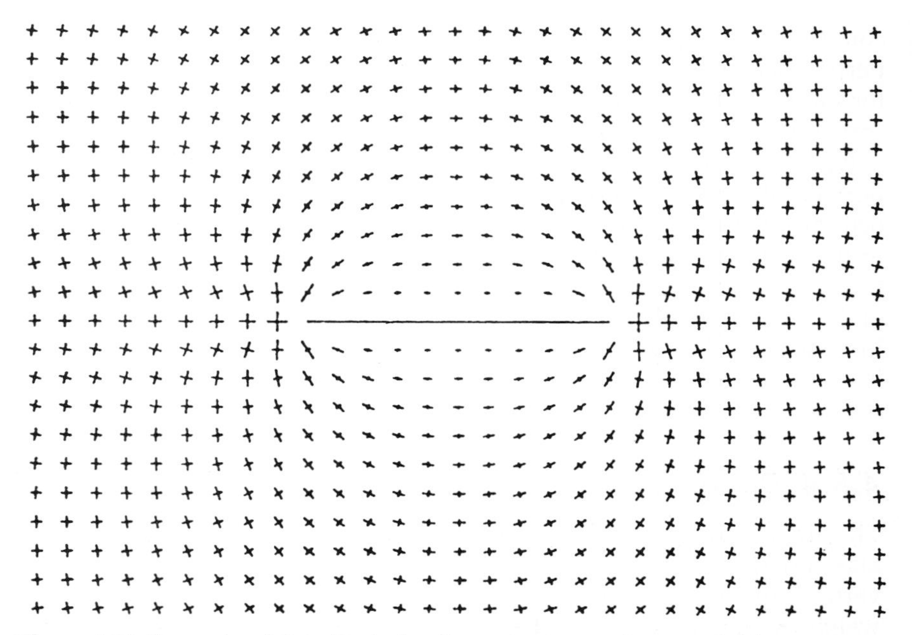

Figure 4.19 Example of the effect of a discontinuity on the near-field stress state, for an applied hydrostatic two-dimensional stress with the discontinuity having a modulus of 10% of the host rock (from Hyett, 1990). The crosses represent the magnitudes and directions of the principal stresses. Note how the stress field close to the discontinuity is quite different from the far-field stress.

be several metres long and only causing a perturbation in the region where the stress determination happens to be made. One can easily imagine the bias and the spread of results that would be obtained by measuring the stresses in boreholes through the rock around the fracture in the diagram. We feel that the large spread of stress state data is mainly due to the combined effects of a hierarchy of such fracture systems, which we know exists in all rocks. It follows that a wide spread in measured values is not necessarily due to bad experimental techniques: on the contrary, the spread itself can indicate a great deal about the *in situ* stress state.

The discontinuity illustrated in Fig. 4.19 has an effective modulus of 10% of the host rock. It is interesting to consider the effect on the stress field when the discontinuity modulus varies from zero to infinity. This is illustrated in Fig. 4.20, in which the principal stresses are altered in the vicinity of the discontinuity. In Case 1, we consider an open discontinuity, similar to the unsupported excavation surface described in Section 3.9. In this case, the major principal stress is diverted parallel to the discontinuity and the minor principal stress takes on a value of zero perpendicular to the discontinuity. In Case 1, the diagram could represent an open discontinuity or an open stope in an underground mine.

In Case 2, in Fig. 4.20, the discontinuity is filled with a material having the same modulus as the surrounding rock. Under these circumstances, and assuming no slip, the discontinuity would be mechanically transparent, with the magnitudes and orientations of the principal

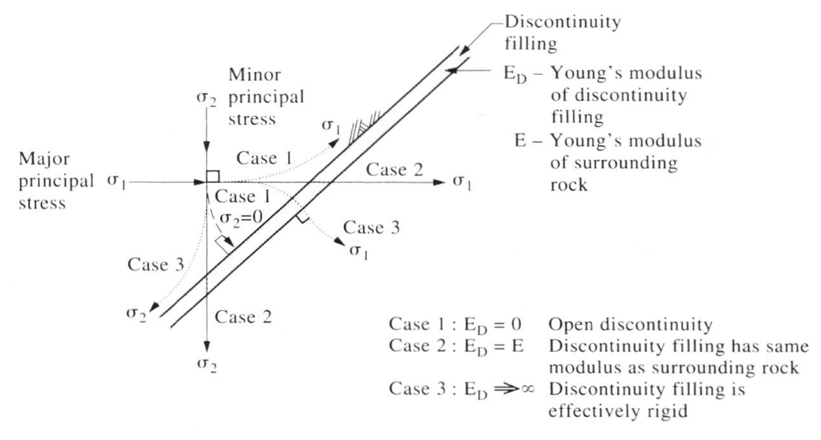

Figure 4.20 Effect of stiffness of discontinuity filling material on the stress state. Two extreme cases are shown—where the discontinuity filling has a modulus of zero (Case 1) and infinity (Case 3)—together with an intermediate case where the discontinuity filling has the same modulus as the surrounding rock (Case 2).

stresses being unaffected. At the other end of the spectrum, in Case 3, the discontinuity is filled by a rigid material. Then, the major principal stress is drawn in perpendicular to the discontinuity and the minor principal stress becomes parallel to the discontinuity.

In general, for most engineering circumstances such as discontinuities or back-filled mine stopes, the stress distribution will be between those shown for Cases 1 and 2. A circumstance between Cases 2 and 3 could arise where a discontinuity in a soft rock is filled with a stiffer material, e.g. quartz veins in soft limestone or a volcanic pipe surrounded by softer rocks.

We have mentioned that the discontinuities being considered could be on a variety of scales. In fact, we would expect effects such as those illustrated in Fig. 4.20 to be superimposed as a result of the existence of

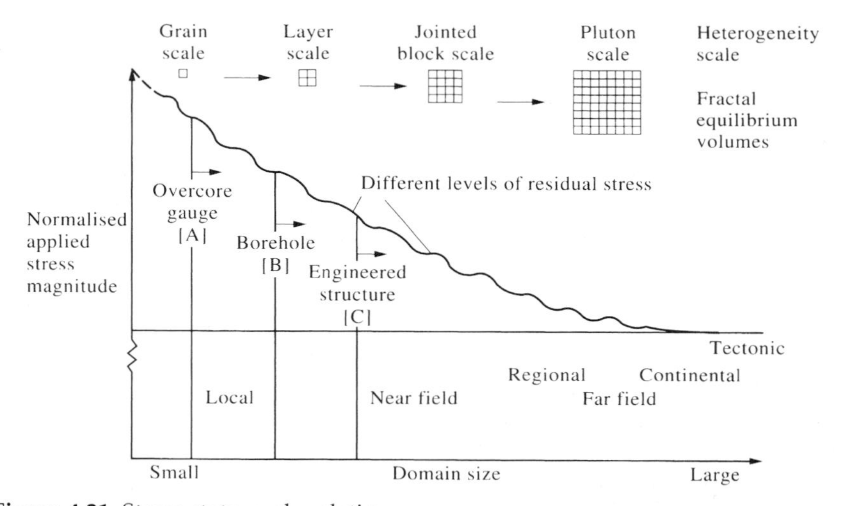

Figure 4.21 Stress state–scale relations.

discontinuities at various scales. Naturally, the zone of influence of a discontinuity will depend on its size. Through such considerations, we again arrive at a concept similar to the representative elemental volume, as shown in Fig. 4.12. It is not possible to provide a coherent diagram covering all eventualities, but we show in Fig. 4.21 the type of relation that could exist between stress state and scale.

In addition to the effects just described, we must expect that considerable deformation of rock masses has occurred during geological history. This results not only in alterations to the existing far-field stress, but also leads to the presence of residual stresses which are superimposed at different scales. We have attempted to illustrate these combined effects in Fig. 4.21, where the horizontal axis is domain size and the vertical axis represents some upper value of a chosen stress component. Indicated on the figure are stresses on the scales of a point, overcore strain gauge, borehole, engineered structure and an intraplate block. The reader should be aware that the curve shown in Fig. 4.21 is an envelope, not an explicit curve. The figure clearly demonstrates that a spread in the results of stress determination programmes must be expected, but this spread will reduce as the size of the sampled volume increases.

4.10 Glossary of terms related to stress states in rock masses

It should be clear by this stage that the determination of stress—and indeed the description of a stress state in words—is fraught with complications. As a result, a plethora of terms is used by many authors. Here, we provide a glossary of terms which are consistent with our explanations and the motivations for determining stress. The descriptive terms are defined for their further use throughout this book and are in part adapted from Hyett *et al.* (1986).

Natural stress. The stress state that exists in the rock prior to any artificial disturbance. The stress state is the result of various events in the geological history of the rock mass. Therefore, the natural stresses present could be the result of the application of many earlier states of stress. Synonyms include: **virgin**, **primitive**, **field** and **active**.

Induced stress. The natural stress state, as perturbed by engineering.

Residual stress. The stress state remaining in the rock mass, even after the originating mechanisms have ceased to operate. The stresses can be considered as within an isolated body that is free from external tractions. A synonym is **remanent stress**.

Tectonic stress. The stress state due to the relative displacement of lithospheric plates.

Gravitational stress. The stress state due to the weight of the superincumbent rock mass. A synonym is **overburden stress**.

Thermal stress. The stress state resulting from temperature change.

Physico-chemical stress. The stress state resulting from chemical and/or physical changes in the rock, e.g. recrystallization, absorption of fluid.

Palaeostress. A previously active *in situ* stress state which is no longer in existence, cf. residual stress which is currently active. Palaeostress can be inferred from geological structures but cannot be determined by stress measurement techniques.

Near-field stress. The natural stress state within the vicinity of, and perturbed by, a heterogeneity (usually caused by engineering activities, e.g. a tunnel as a low-modulus inclusion).

Far-field stress. The stress state that exists in the region beyond the near-field, where no significant perturbation due to the heterogeneity occurs.

Regional stress. The stress state in a relatively large geological domain.

Local stress. The stress state in a small domain—usually with the dimensions of, or smaller than, an engineered structure.

5 Strain

Strain is a change in the relative configuration of points within a solid. One can study finite strain or infinitesimal strain—both are relevant to the deformations that occur in the context of the principles of rock mechanics and their engineering applications. Large-scale strain can be experienced underground as illustrated in Fig. 5.1, where there is severe deformation around a coal mine access tunnel. When such displacements are very small, one can utilize the concept of infinitesimal strain and develop a strain tensor directly analogous to the stress tensor. Thus, we will first discuss finite strain and then infinitesimal strain.

5.1 Finite strain

Strain may be regarded as normalized displacement. If a structure is subjected to a stress state, it will deform. However, the magnitude of the deformation is dependent on the size of the structure as well as the magnitude of the applied stresses. In order to render the deformation as a scale-independent parameter, the concept of strain (which in its simplest form is the ratio of displacement to the undeformed length) is utilized. Such displacements can also occur naturally in rock masses through the application of tectonic stresses resulting from past and present geological processes. Some excellent examples are shown in Ramsey and Huber (1983).

The displacements, whether natural or artificial, can be complex; an example is shown schematically in Fig. 5.2. It should also be noted that strain is a three-dimensional phenomenon that requires reference to all three Cartesian co-ordinate axes. However, it is instructive, in the first instance, to deal with two-dimensional strain: once the fundamental concepts have been introduced, three-dimensional strain follows as a natural progression.

In order to provide a structure for our analysis of two-dimensional strain, we will consider the separate components of strain. There are normal strains and shear strains, as illustrated in Fig. 5.3.

As with normal stress and shear stress components, it is much easier to

Figure 5.1 Large displacements around an originally arch-shaped coal mine access tunnel (from Pan, 1989).

grasp the concept of normal strain than shear strain. This is because the normal displacement and the associated strain occur along one axis. However, in the case of shear strain, the quantity of strain in say, the x-direction, also depends on the position along the y-axis. In other words, normal strain only involves one Cartesian axis, whereas, shear strain involves two (or three), i.e. it involves an interaction between the axes.

One convenient simplification that can be introduced to aid the study of strain is the concept of homogeneous strain which occurs when the state

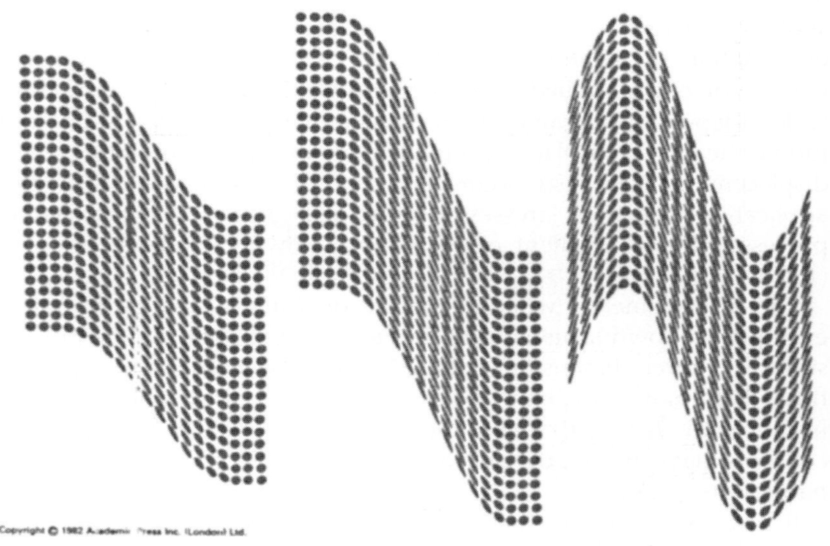

Figure 5.2 Example of the evolution of a complex displacement field (from Ramsey and Huber, 1983).

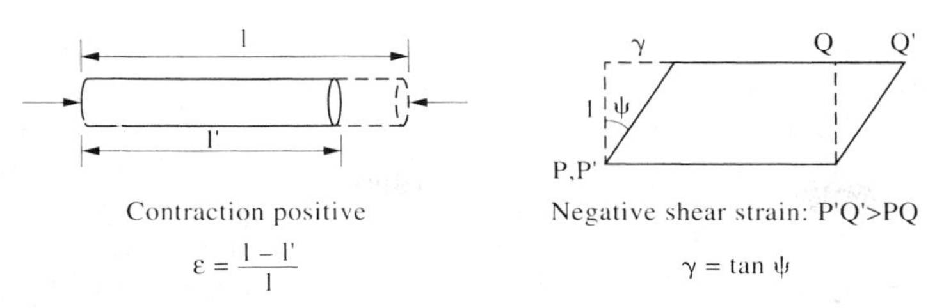

Figure 5.3 Normal strain and shear strain.

of strain *is the same throughout the solid*. Under these circumstances:

(a) straight lines remain straight;
(b) circles are deformed into ellipses; and
(c) ellipses are deformed into other ellipses.

5.2 Examples of homogeneous finite strain

We will now consider four examples of simple homogeneous finite strain. These are all important, both fundamentally and in terms of understanding strain. We will discuss strain components and thence also begin to introduce the notion of strain transformation, in terms of matrices. The four examples are shown in Fig. 5.4.

In each of the examples in Fig. 5.4, we have given the equations relating the new positions (e.g. x') in terms of the original positions (e.g. x) of each point. The coefficients k and γ indicate the magnitudes of the normal and shear strains, respectively. The final case in the figure, pure shear, is a result of extensional and contractional normal strains which will be explored later,

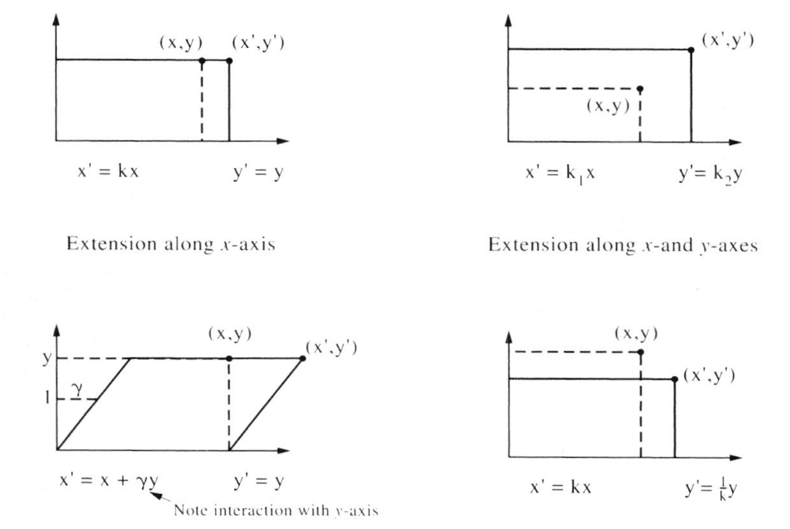

Figure 5.4 Four simple cases of homogeneous finite strain.

especially in terms of the perhaps unexpected relation between the shear modulus and Young's modulus and Poisson's ratio.

We have noted earlier that there may have been successive phases of deformation of the rock mass during geological history. Thus, in decoding such compound deformation into its constituent parts, as illustrated by specific types of simple deformation in Fig. 5.4, we need to know whether strain phases are *commutative*, i.e. if there are two deformational phases A and B, and is the final result of A followed by B the same as B followed by A? Similarly, in engineering, does the sequence of excavation have any influence on the final strain state? Perhaps counterintuitively, the answer is that the final strain state is dependent on the straining sequence in those circumstances where shear strains are involved. An elegant example from Ramsey and Huber (1983) is shown in Fig. 5.5, wherein the non-commutative nature of shear strains is illustrated, both graphically and mathematically. With reference to earlier emphasis on the significance of the off-diagonal terms of the strain matrices, the reader should note that it is these off-diagonal terms which give rise to the non-commutative phenomemon. In Chapter 14, the concept of interactions in the off-diagonal terms is introduced for a matrix with general state variables along the leading diagonal, in the context of rock engineering systems.

It can be helpful to think about these strain operations in general and to be able to identify the components of a general strain transformation matrix for all circumstances. Such affine transformations are used in computer graphics and we mention here the case of distorting any two-dimensional shape. In order to introduce translation, i.e. movement of the entire shape (without rotation) within the plane of the figure, 'homogeneous co-ordinates' are used. These are three co-ordinates, simply the two Cartesian coordinates plus a third which allows translation to be introduced. The transformation of co-ordinates is shown in Fig. 5.6.

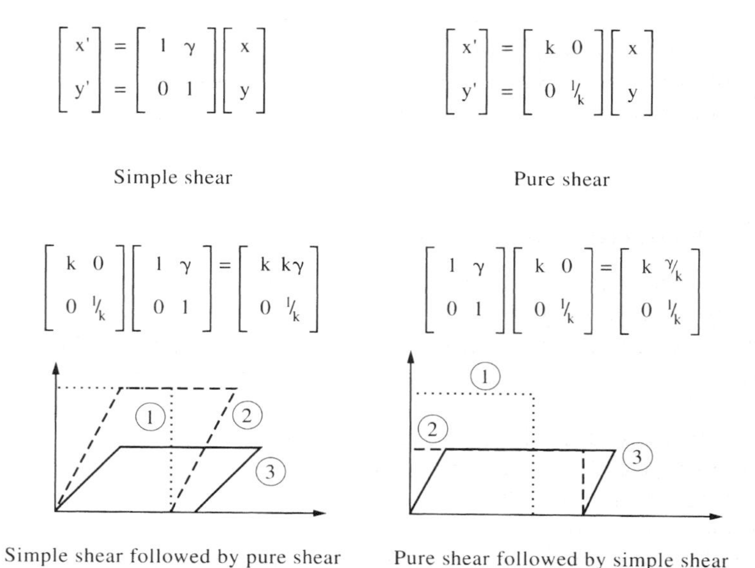

Figure 5.5 Shear strain is not commutative (example from Ramsey and Huber, 1983).

$$\begin{bmatrix} x' \\ y' \\ 1 \end{bmatrix} = \begin{bmatrix} a & c & m \\ b & d & n \\ p & q & s \end{bmatrix} \begin{bmatrix} x \\ y \\ 1 \end{bmatrix} \quad \text{or} \quad \begin{bmatrix} x' & y' & 1 \end{bmatrix} = \begin{bmatrix} x & y & 1 \end{bmatrix} \begin{bmatrix} a & b & p \\ c & d & q \\ m & n & s \end{bmatrix}$$

Figure 5.6 The general transformation of a two-dimensional shape using homogeneous co-ordinates.

Note that in the equation

$$x' = ax + cy + m$$

the coefficient a is related to extensional strain (as shown in Fig. 5.4), the coefficient c is an *interaction* term and related to shear, and m is related to the magnitude of the translation. Through such considerations, we can identify the strain components associated with different parts of the matrix, as shown below.

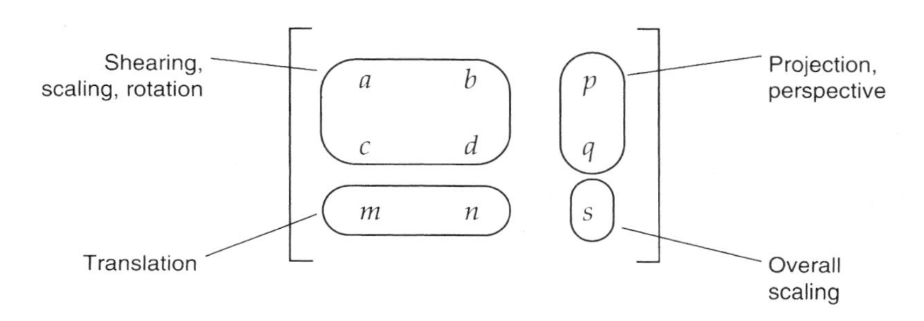

We will see later that the ability to determine which functions are performed by which parts of the transformation matrix is especially helpful when interpreting the compliance matrix. This matrix relates the strains to the stresses for materials with different degrees of anisotropy.

5.3 Infinitesimal strain

Infinitesimal strain is homogeneous strain over a vanishingly small element of a finite strained body. To find the components of the strain matrix, we need to consider the variation in co-ordinates of the ends of an imaginary line inside a body as the body is strained as illustrated in Fig. 5.7. By this means, we can find the normal and shear components in an analogous fashion to the finite case presented above.

In this figure, the point P with co-ordinates (x, y, z) moves when the body is strained, to a point P* with co-ordinates $(x + u_x, y + u_y, z + u_z)$. The components of movement, u_x, u_y and u_z , may vary with location within the body, and so are regarded as functions of x, y and z. Similarly, the point Q (which is a small distance from P), with co-ordinates $(x + \delta x, y + \delta y,$

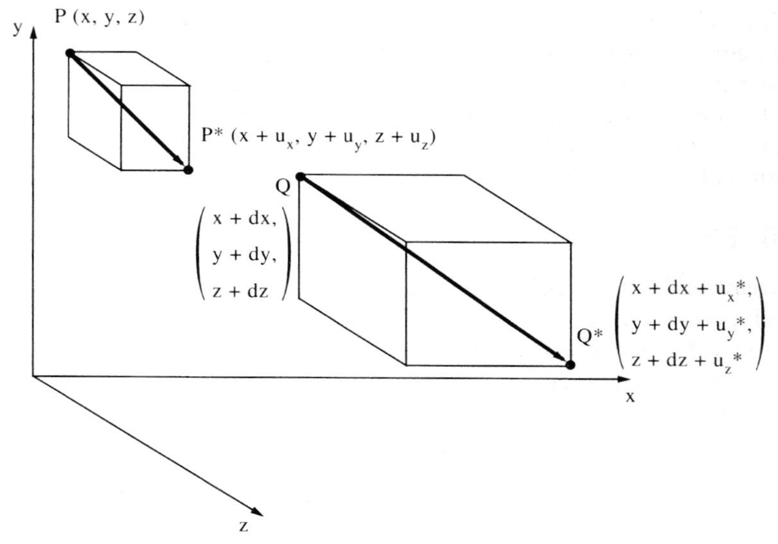

Figure 5.7 Change in co-ordinates as a line PQ is strained to P*Q*.

$z + \delta z$), is strained to Q* which has co-ordinates ($x + \delta x + u_x^*, y + \delta y + u_y^*, z + \delta z + u_z^*$). If we now consider holding P in a constant position and Q being strained to Q*, the normal and shear components of strain can be isolated.

The infinitesimal longitudinal strain is now considered in the x-direction. Because strain is 'normalized displacement' (see Section 5.1), if it is assumed that u_x is a function of x only, as in Fig. 5.8, then

$$\varepsilon_{xx} = du_x/dx \quad \text{and hence} \quad du_x = \varepsilon_{xx} dx.$$

Considering similar deformations in the y- and z-directions, the normal components of the strain matrix can be generated as also shown in Fig. 5.8.

Derivation of the expressions for the shear strains follows a similar course, except that instead of assuming that simple shear occurs parallel to one of the co-ordinate axes, the assumption is made initially that the shear strain (expressed as a change in angle) is equally distributed between both co-ordinate axes, i.e. $du = du_y$ if $dx = dy$. This is graphically illustrated in Fig. 5.9.

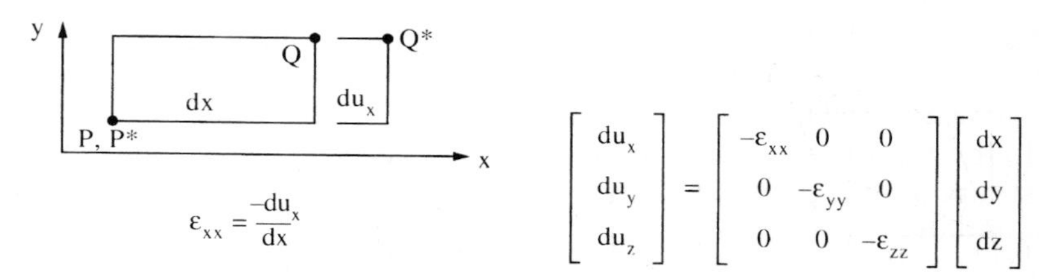

Figure 5.8 Infinitesimal longitudinal strain.

First, we should note that the term γ_{xy}, i.e. 2α, is known as the **engineering shear strain**, whereas the term $\gamma_{xy}/2$, i.e. α, is known as the **tensorial shear strain**. Second, although engineering shear strain is the fundamental parameter by which means shear strain is expressed, it is tensorial shear strain that appears as the off-diagonal components in the strain matrix in Fig. 5.9.

5.4 The strain tensor

Combining the longitudinal and shear strain components which have been developed above, we can now present the complete strain tensor—which is a second-order tensor directly analogous to the stress tensor presented in Section 3.5. The matrix is shown below:

$$\begin{bmatrix} \varepsilon_{xx} & \varepsilon_{xy} & \varepsilon_{xz} \\ \varepsilon_{yx} & \varepsilon_{yy} & \varepsilon_{yz} \\ \varepsilon_{zx} & \varepsilon_{zy} & \varepsilon_{zz} \end{bmatrix}.$$

Note that this matrix is symmetrical and hence has six independent components—with its properties being the same as the stress matrix because they are both second-order tensors. For example, at an orientation of the infinitesimal cube for which there are no shear strains, we have principal values as the three leading diagonal strain components. The matrix of principal strains is shown below:

$$\begin{bmatrix} \varepsilon_1 & 0 & 0 \\ 0 & \varepsilon_2 & 0 \\ 0 & 0 & \varepsilon_3 \end{bmatrix}.$$

The strain component transformation equations are also directly analogous to the stress transformation equations and so the Mohr's circle representation can be utilized directly for relating normal and shear strains on planes at different orientations. Other concepts which we mentioned whilst discussing stress, such as the first stress invariant, also apply because of the mathematical equivalence of the two tensors. Thus, the first strain invariant is

$$\varepsilon_{xx} + \varepsilon_{yy} + \varepsilon_{zz} = \varepsilon_1 + \varepsilon_2 + \varepsilon_3 = \text{a constant.}$$

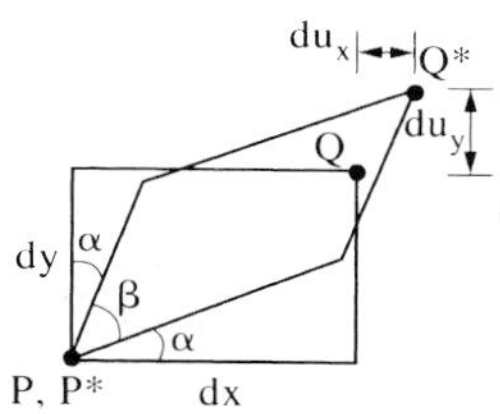

$$\gamma_{xy} = (\tfrac{\pi}{2} - 2\alpha) = \beta \qquad e_{xy} = \gamma_{xy}/2, \text{ etc.}$$

$$\alpha = \frac{\gamma_{xy}}{2}$$

$$\begin{bmatrix} du_x \\ du_y \\ du_z \end{bmatrix} = \begin{bmatrix} 0 & \gamma_{xy}/2 & \gamma_{xz}/2 \\ \gamma_{xy}/2 & 0 & \gamma_{yz}/2 \\ \gamma_{zx}/2 & \gamma_{zy}/2 & 0 \end{bmatrix} \begin{bmatrix} dx \\ dy \\ dz \end{bmatrix} \qquad \begin{bmatrix} du_x \\ du_y \\ du_z \end{bmatrix} = \begin{bmatrix} 0 & e_{xy} & e_{xz} \\ e_{yx} & 0 & e_{yz} \\ e_{zx} & e_{zy} & 0 \end{bmatrix} \begin{bmatrix} dx \\ dy \\ dz \end{bmatrix}$$

Figure 5.9 Infinitesimal shear strain.

The transformation properties of the strain matrix allow us to determine the complete *in situ* or laboratory strain tensor from measurements which are made with strain gauges and which are normal strain measurements only. In the same way that shear stresses cannot be measured directly, neither can shear strains, and hence the complete strain matrix must be established from normal strain measurements.

5.5 The elastic compliance matrix

One may be tempted to ask, given the mathematical equivalence of the strain matrix developed in this chapter and the stress matrix developed in Chapter 3, whether there is any means of linking the two matrices together. Clearly, this would be of great benefit for engineering, because we would then be able to predict either the strains (and associated displacements) from a knowledge of the applied stresses or vice versa. As we will be discussing in Chapter 6, it is often critical to be able to consider whether it is stress, or strain, which is being applied and hence whether it is strain, or stress, which is the result.

A simple way to begin would be to assume that each component of the strain tensor is a linear combination of all the components of the stress tensor, i.e. each stress component contributes to the magnitude of each strain component. For example, in the case of the ε_{xx} component, we can express this relation as

$$\varepsilon_{xx} = S_{11}\sigma_{xx} + S_{12}\sigma_{yy} + S_{13}\sigma_{zz} + S_{14}\tau_{xy} + S_{15}\tau_{yz} + S_{16}\tau_{zx}.$$

Because there are six independent components of the strain matrix, there will be six equations of this type. If we considered that the strain in the x-direction were only due to the stress in the x-direction, the previous equation would reduce to

$$\varepsilon_{xx} = S_{11}\sigma_{xx}$$

or

$$\sigma_{xx} = \varepsilon_{xx}/S_{11} = E\varepsilon_{xx}, \text{ where } E = 1/S_{11}.$$

This form of the relation, where the longitudinal strain is linearly proportional to the longitudinal stress, as is the case for a wire under tension, was first stated by Robert Hooke (the first President of the Royal Society) in 1676. He published the relation as an anagram in *The Times* of London as CEIIINOSSSTTUU and three years later revealed this to mean UT TENSIO SIC UIS, i.e. as the extension so the force. For this reason, the more complete expression where ε_{xx} is related to all six components of stress is known as the generalized Hooke's law.

Hence, the complete set of relations between the strain and stress components is:

$$
\begin{aligned}
\varepsilon_{xx} &= S_{11}\sigma_{xx} + S_{12}\sigma_{yy} + S_{13}\sigma_{zz} + S_{14}\tau_{xy} + S_{15}\tau_{yz} + S_{16}\tau_{zx}\\
\varepsilon_{yy} &= S_{21}\sigma_{xx} + S_{22}\sigma_{yy} + S_{23}\sigma_{zz} + S_{24}\tau_{xy} + S_{25}\tau_{yz} + S_{26}\tau_{zx}\\
\varepsilon_{zz} &= S_{31}\sigma_{xx} + S_{32}\sigma_{yy} + S_{33}\sigma_{zz} + S_{34}\tau_{xy} + S_{35}\tau_{yz} + S_{36}\tau_{zx}\\
\varepsilon_{xy} &= S_{41}\sigma_{xx} + S_{42}\sigma_{yy} + S_{43}\sigma_{zz} + S_{44}\tau_{xy} + S_{45}\tau_{yz} + S_{46}\tau_{zx}\\
\varepsilon_{yz} &= S_{51}\sigma_{xx} + S_{52}\sigma_{yy} + S_{53}\sigma_{zz} + S_{54}\tau_{xy} + S_{55}\tau_{yz} + S_{56}\tau_{zx}\\
\varepsilon_{zx} &= S_{61}\sigma_{xx} + S_{62}\sigma_{yy} + S_{63}\sigma_{zz} + S_{64}\tau_{xy} + S_{65}\tau_{yz} + S_{66}\tau_{zx}.
\end{aligned}
$$

It is not necessary to write these equations in full. An accepted convention is to use matrix notation, so that the expressions above can be alternatively written in the abbreviated form

$$[\varepsilon] = [S][\sigma]$$

$$
\text{where } [\varepsilon] = \begin{bmatrix} \varepsilon_{xx} \\ \varepsilon_{yy} \\ \varepsilon_{zz} \\ \varepsilon_{xy} \\ \varepsilon_{yz} \\ \varepsilon_{zx} \end{bmatrix} \text{ and } [\sigma] = \begin{bmatrix} \sigma_{xx} \\ \sigma_{yy} \\ \sigma_{zz} \\ \tau_{xy} \\ \tau_{yz} \\ \tau_{zx} \end{bmatrix} \text{ and } [S] = \begin{bmatrix} S_{11} & S_{12} & S_{13} & S_{14} & S_{15} & S_{16} \\ S_{21} & S_{22} & S_{23} & S_{24} & S_{25} & S_{26} \\ S_{31} & S_{32} & S_{33} & S_{34} & S_{35} & S_{36} \\ S_{41} & S_{42} & S_{43} & S_{44} & S_{45} & S_{46} \\ S_{51} & S_{52} & S_{53} & S_{54} & S_{55} & S_{56} \\ S_{61} & S_{62} & S_{63} & S_{64} & S_{65} & S_{66} \end{bmatrix}.
$$

The [**S**] matrix shown above is known as the **compliance matrix**. In general, the higher the magnitude of a specific element in this matrix, the greater will be the contribution to the strain, representing an increasingly compliant material. 'Compliance' is a form of 'flexibility', and is the inverse of 'stiffness'.

The compliance matrix is a 6×6 matrix containing 36 elements. However, through considerations of conservation of energy it can be shown that the matrix is symmetrical. Therefore, in the context of our original assumption that each strain component is a linear combination of the six stress components, we find that we need 21 independent elastic constants to completely characterize a material that follows the generalized Hooke's law. In the general case, with all the constants being non-zero and of different values, the material will be completely anisotropic. It is necessary, particularly for practical applications of the stress–strain relations, to consider to what extent we can reduce the number of non-zero elements of the matrix. In other words, how many elements of the compliance matrix are actually necessary to characterize a particular material?

The key to this study is the architecture of the compliance matrix, and especially the off-diagonal terms, which have already been emphasized. For typical engineering materials, there will be non-zero terms along the leading diagonal because longitudinal stresses must lead to longitudinal strains and shear stresses must lead to shear strains. The isotropy of the material is directly specified by the interaction terms, i.e. whether a normal or shear strain may result from a shear or normal stress, respectively. This is illustrated conceptually in Fig. 5.10.

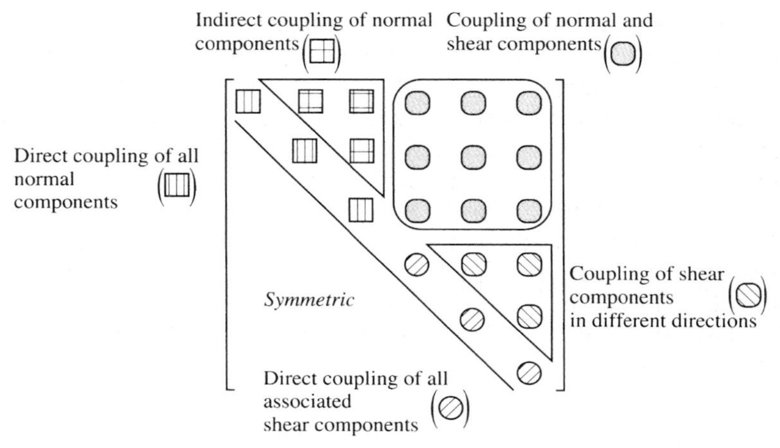

Figure 5.10 The architecture of the elastic compliance matrix.

As a first approximation, and in relation to Fig. 5.10, let us assume that there is no coupling between the normal and shear components and that there is no coupling of shear components in different directions. This means that all of the elements designated by the symbols with dense cross-hatching and left-inclined shading in Fig. 5.10 become zero. We know that the direct relation between a normal strain and a normal stress is given by $1/E$: this is because the definition of Young's modulus, E, is the ratio of normal stress to normal strain. Hence, all the elements designated by the vertical hatching will be reciprocals of Young's moduli. Following the definition for Poisson's ratio given in Section 4.6.2, and recalling that this parameter links orthogonal contractile and extensile strains (which are manifested by a negative sign in equations containing Poisson's ratio), all the elements designated by the wide-cross-hatching will be Poisson's ratios, ν, divided by a Young's modulus. Finally, the elements designated in Fig. 5.10 by the right-inclined shading, being the ratio of shear strain to shear stress, will be the reciprocals of the shear moduli, G.

This results in the reduced elastic compliance matrix shown below:

$$\begin{bmatrix} 1/E_1 & -\nu_{21}/E_2 & -\nu_{31}/E_3 & 0 & 0 & 0 \\ & 1/E_2 & -\nu_{32}/E_3 & 0 & 0 & 0 \\ & & 1/E_3 & 0 & 0 & 0 \\ & & & 1/G_{12} & 0 & 0 \\ & & & & 1/G_{23} & 0 \\ & \text{symmetric} & & & & 1/G_{31} \end{bmatrix}.$$

A material characterized by this compliance matrix has nine independent elastic constants and is known as an **orthotropic material**. The nine material properties are the three Young's moduli, the three shear moduli and the three Poisson's ratios, i.e.

$$E_1 \quad E_2 \quad E_3 \quad G_{12} \quad G_{23} \quad G_{31} \quad \nu_{21} \quad \nu_{32} \quad \nu_{31}.$$

Such a material could arise from the microstructure of the intact rock, or in the case of rock masses when three mutually perpendicular sets of discontinuities with different properties and/or frequencies are present. The double subscripts applied to the Poisson's ratios are required in order to differentiate the effects in the two different axial directions in each case.The reader should note that there are in fact *six* Poisson's ratios: the symmetry of the matrix ensures that there are three relations of the form $\nu_{12}/E_1 = \nu_{21}/E_2$.

We can reduce the elastic compliance matrix even further by considering the case of **transverse isotropy**. This is manifested by a rock mass with a laminated fabric or one set of parallel discontinuities. In the case when the plane of isotropy is parallel to the plane containing Cartesian axes 1 and 2, we can say that

$$E_1 = E_2 = E \quad \text{and} \quad E_3 = E'$$

$$\nu_{12} = \nu_{21} = \nu \quad \text{and} \quad \nu_{13} = \nu_{23} = \nu'$$

$$G_{12} \neq G_{23} \quad \text{and} \quad G_{23} = G_{31} = G'.$$

The associated elastic compliance matrix is then

$$\begin{bmatrix} 1/E & -\nu/E & -\nu'/E' & 0 & 0 & 0 \\ & 1/E & -\nu'/E' & 0 & 0 & 0 \\ & & 1/E' & 0 & 0 & 0 \\ & & & \dfrac{2(1+\nu)}{E} & 0 & 0 \\ & & & & 1/G' & 0 \\ & \text{symmetric} & & & & 1/G' \end{bmatrix}$$

Note that in the above matrix, the term $2(1 + \nu)/E$ has been substituted for $1/G_{12}$ because *in the plane of isotropy* there is a relation between the shear modulus and the Young's modulus and Poisson's ratio. It is vital, however, to realize that this relation, i.e. $G = E/2(1 + \nu)$, only applies for isotropic conditions and so we cannot make a similar substitution for either $1/G_{23}$ or $1/G_{31}$, which are out of the plane of isotropy. Thus, the number of independent elastic constants for a transversely isotropic material is not six but five, i.e.

$$E \quad E' \quad G' \quad \nu \quad \nu'.$$

The final reduction that can be made to the compliance matrix is to assume **complete isotropy**, where

$$E_1 = E_2 = E_3 = E$$

$$\nu_{12} = \nu_{23} = \nu_{31} = \nu$$

$$G_{12} = G_{23} = G_{31} = G.$$

Note that, because we now have complete isotropy, the subscripts can be dispensed with, the shear modulus G is implicit and furthermore the factor $1/E$ which is common to all terms can be brought outside the matrix. Finally, we have

$$1/E \begin{bmatrix} 1 & -\nu & -\nu & 0 & 0 & 0 \\ & 1 & -\nu & 0 & 0 & 0 \\ & & 1 & 0 & 0 & 0 \\ & & & 2(1+\nu) & 0 & 0 \\ & & & & 2(1+\nu) & 0 \\ & \text{symmetric} & & & & 2(1+\nu) \end{bmatrix}.$$

Complete anisotropy was characterized via the elastic compliance matrix through 21 independent constants. By considering the architecture of the full matrix and making all 'cross-coupled' terms zero, we obtained the orthotropic case with nine independent constants. This was further reduced in the case of transverse isotropy to five constants, utilizing the relation between shear modulus and Young's modulus and Poisson's ratio. The ultimate reduction (also using the shear modulus relation) resulted in two elastic constants for the case of a perfectly isotropic material. One is reminded of the quotation given by Jacques Grillo in his book *Form, Function and Design*, that "In anything at all, perfection is finally attained not when there is no longer anything to add, but when there is no longer anything to take away".

5.6 Implications for *in situ* stress

There are many ramifications of the elastic compliance matrix and the possible reductions which we have presented in Section 5.5. One particularly important corollary with reference to Chapter 4 on *in situ* stress relates to the ratio of horizontal to vertical stress, as calculated by the 'switched on gravity' analysis presented in Section 4.6.2. Recall the ratio

$$\sigma_{\mathrm{H}} = \frac{\nu}{(1-\nu)}\sigma_{v}$$

which meant that the horizontal stress could never exceed the vertical stress. Implicit in the derivation of this relation is the fact that the rock was assumed to be isotropic. We can generate similar relations for varying degrees of anisotropy, in particular for transverse isotropy and orthotropy. Using the matrices presented in Section 5.5, and for the case where axis 3 is vertical and plane 12 is horizontal, these are

$$\text{for tranverse isotropy } \sigma_{\mathrm{H}} = \frac{\nu'}{1-\nu}\sigma_{v}$$

and for orthotropy $\sigma_{H_1} = \left(\dfrac{\nu_{13} + \nu_{12}\nu_{23}}{1 - \nu_{12}\nu_{21}} \right) \sigma_v,$

$$\sigma_{H_2} = \left(\frac{\nu_{13}\nu_{21} + \nu_{23}}{1 - \nu_{12}\nu_{21}} \right) \sigma_v.$$

The reader should note that, in order to simplify these relations, use has been made of various complementary Poisson's ratios (e.g. ν_{12} instead of $\nu_{21}E_1/E_2$). These equations are from Amadei *et al.* (1983) and demonstrate conclusively that, for certain combinations of the respective elastic constants, the horizontal components can be significantly different. In fact, an orthotropic model is probably a much better representation of a discontinuous rock mass with three perpendicular discontinuity sets than an isotropic model.

A final point is that, given the 21 independent components of the compliance matrix, the correct engineering approach to the problem of modelling rock masses would be to establish *to what extent the compliance matrix can validly be simplified*. In other words, the logic would be to assume complete anisotropy unless we have reason to assume otherwise. However, because of cost constraints and the practicalities of engineering, of the order of 99% of all analyses that have been conducted have contained the assumption that the rock mass is fully isotropic with only two elastic constants. In the majority of the remaining cases, transverse isotropy has been assumed; and in a few isolated examples, orthotropy (with nine elastic constants) has been assumed. To the authors' knowledge, no one has either measured the 21 constants or conducted an analysis assuming a compliance matrix with non-zero components. There are lessons here concerning the relation between rock mechanics and its application to rock engineering, i.e. the theory and the practice.

6 Intact rock

Having introduced the concepts of stress and strain, we can now consider how the rock reacts to given loads. It is convenient to consider first the intact rock, then the discontinuities and, finally, to consider how they combine to determine the properties of rock masses. Thus, in this chapter, we will discuss the properties of intact rock; in Chapter 7, discontinuities, and in Chapter 8, rock masses.

6.1 The background to intact rock testing

In rock mechanics, since the early 1960s when it began, more attention has been paid to intact rock than to any other feature of the rock mass. This occurred for two main reasons: the subject has relied heavily on the general topic of the mechanics of solid materials (evolving from rock physics); and the main way in which samples are obtained remote from human-access is by diamond drilling to produce cylindrical cores of rock (which are eminently suitable for testing). These two factors caused a concentration of work on intact rock testing because of the 'convenience' of a well-established background and readily available specimens. The circumstances were probably further reinforced by the general practice of engineers to establish the properties and behaviour of the materials with which they work.

In this chapter, we will be concentrating on the deformability, strength and failure of intact rock. The early emphasis on this subject culminated in 1966 with the 'discovery' of the ability to obtain the **complete stress–strain curve**. The curve provided previously unknown information on the behaviour of rocks *after* their peak strength has been reached. The failure region has special significance in rock mechanics and rock engineering because, in some circumstances, we can design an underground structure knowing that the rock will pass into the post-peak region. Such a design is very different from the traditional approach in all other forms of engineering, where the material must be kept in the pre-peak region, i.e. behaving essentially elastically.

In situ, the high stresses that can lead to the material entering the post-peak region either occur directly, as a result of excavation, or indirectly at the corners and edges of rock blocks which have been disturbed by the process of excavation. Thus, the deformability, strength and failure of intact rock *per se* are critically important for understanding the basic mechanisms of excavation, whether by blasting or by machine cutting, and for understanding support requirements—whether to defend against direct stress failure or rock block failure.

6.2 The complete stress–strain curve in uniaxial compression

In Chapter 5, we discussed strain and the complexities of a material which could potentially have 21 independent elastic constants, and could be subjected to any stress state. Here, in considering the behaviour of real rock, we will begin with the simplest form of loading, i.e. uniaxial compression. In the context of the elastic compliance matrix, we will therefore be studying S_{11} for loading uniaxially along the x-axis. The properties of the uniaxial test will be discussed first, before considering triaxial and other multiaxial loading cases.

In its simplest form, the uniaxial compression test is conducted by taking a right cylinder of intact rock, loading it along its axis and recording the displacement produced as the force is increased. In Figs 6.1 and 6.2 we present a typical record of such a test (which also includes the post-peak region obtained using techniques to be discussed in Section 6.3). Note that the force and the displacement have been scaled respectively to stress (by dividing by the original cross-sectional area of the specimen) and to strain (by dividing by the original length). In the curve shown in the figure, the various aspects of the mechanical behaviour of intact rock tested under these conditions can now be identified.

At the very beginning of loading, the curve has an initial portion which is concave upwards (the opposite of typical soil behaviour) for two reasons:

- the lack of perfect specimen preparation, manifested by the ends of the cylinder being non-parallel; and
- the closing of microcracks within the intact rock.

After this initial zone, there is a portion of essentially linear behaviour, more or less analogous to the ideal elastic rock we discussed in Chapter 5.

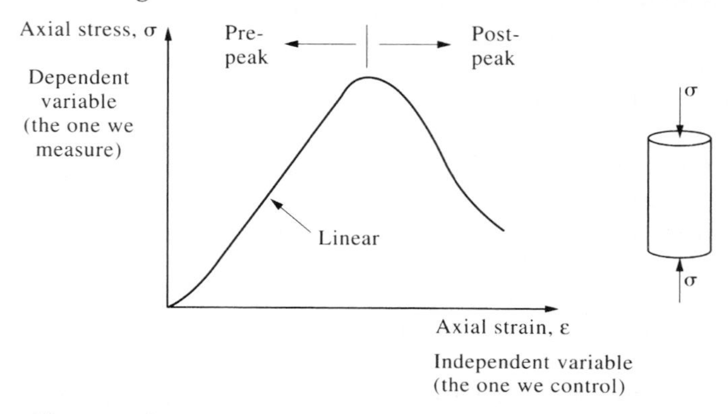

Figure 6.1 The complete stress–strain curve.

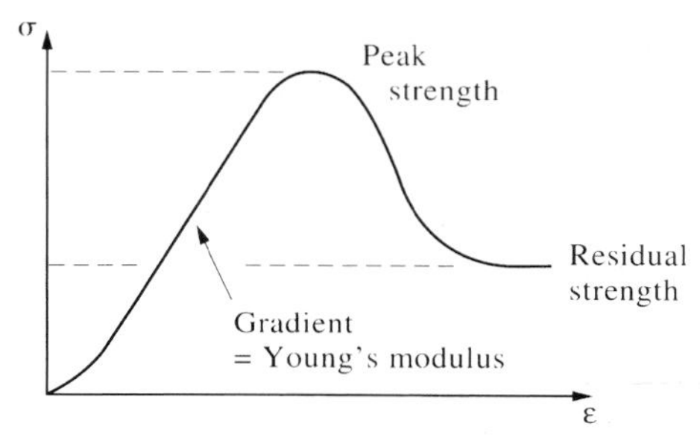

Figure 6.2 The complete stress–strain curve illustrating various mechanical parameters.

Remembering that Young's modulus, E, is defined as the ratio of stress to strain (i.e. $1/S_{11}$), it can be determined in two ways: either by taking the slope of the stress–strain curve at a given point; or by taking the slope of a line connecting two points on this linear portion of the curve (Fig. 6.2). The two slopes are the **tangent modulus** and the **secant modulus**. The tangent modulus is conventionally taken as the gradient of the σ–ε curve at a stress level corresponding to 50% of the peak stress; the secant modulus may be determined anywhere over the entire linear portion. Naturally, both of these are approximations to the real behaviour, but are useful and adequate for simple elastic applications. However, with increasing numerical and computing capabilities, we can represent the complete stress–strain curve more accurately as a piece-wise linear function if required.

The variations in the tangent modulus and the secant modulus throughout the complete stress–strain curve are shown in Fig. 6.3. It should be noted that the portion of the curve after the peak stress has been reached is a **failure locus** and so the negative portion of the tangent modulus curve is not directly meaningful. For this reason, the secant modulus is often more

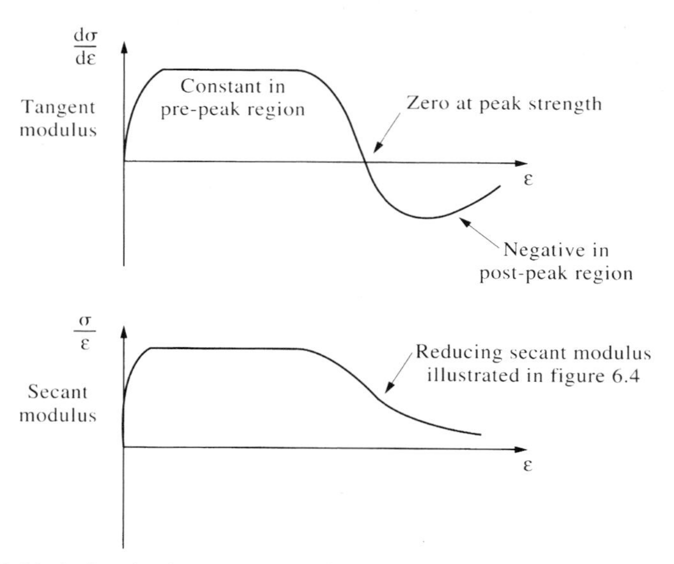

Figure 6.3 Variation in the tangent and secant moduli throughout the complete stress–strain curve.

convenient and can be established by unloading and reloading the specimen from any point on the curve. This is illustrated in Fig. 6.4.

The factors discussed so far have been concerned with the complete stress–strain behaviour and the link with Chapter 5 on the deformability of materials. Another important parameter highlighted in Fig. 6.2 is the maximum stress that the specimen can sustain. Under the loading conditions shown in the diagram, the peak stress is the **uniaxial compressive strength,** σ_c.

It is important to realize that the compressive strength is not an intrinsic material property. Intrinsic material properties do not depend on the specimen geometry or the loading conditions used in the test: the compressive strength does. If a microphone is attached to a specimen during the generation of the complete stress–strain curve, significant acoustic emission is found to occur, starting from a stress level of about 50% of the compressive strength. Through these observations, and by obtaining sections through specimens that have been taken to various points on the complete stress–strain curve, it is observed that microcracking continually increases from this 50% σ_c stress level until the specimen is completely destroyed. The compressive strength is an arbitrary stage in this continual microstructural damage process, representing the maximum sustainable stress. At the peak of the curve, the specimen has had many axial cracks induced within it, but macro-shear (i.e. on the scale of the specimen itself) does not take place until about halfway down the descending portion of the curve.

In other forms of engineering, for example, the strength of concrete in structural engineering, if the applied stress reaches the compressive strength, there can be catastrophic consequences. This is not necessarily the case in rock engineering, which is why we are concentrating the discussion on the characteristics of the complete stress–strain curve, as opposed to the specific value of the compressive strength. However, the compressive strength is probably the most widely used and quoted rock engineering parameter and therefore it is crucial to understand its nature. Also, whether failure beyond the compressive strength is to be avoided at

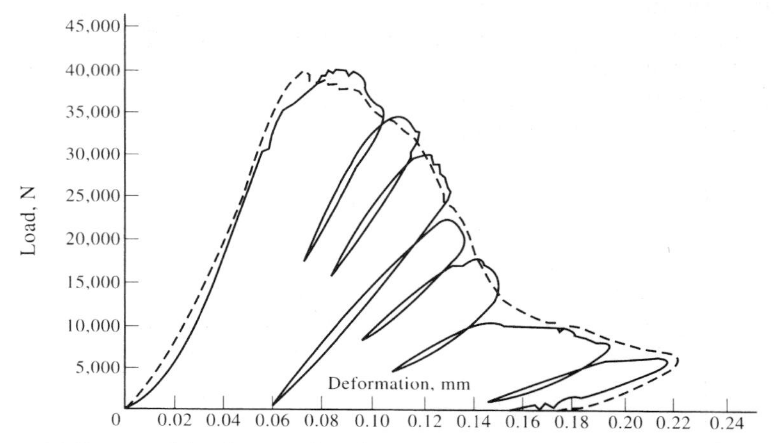

Figure 6.4 Repeated loading and unloading to illustrate the variation in secant modulus throughout the complete stress–strain curve (from Bieniawski, 1968).

all costs, or to be encouraged, is a function of the engineering objective, the form of the complete stress–strain curve for the rock (or rock mass), and the characteristics of the loading conditions. These features are crucial in the design and analysis of underground excavations.

At this stage, we will consider whether the specimen is being loaded at a constant stress rate or a constant strain rate: in other words, is stress the cause and strain the effect, or vice versa? It is customary in science to plot the independent (i.e. controlled) variable along the x-axis and the dependent (or measured) variable along the y-axis. Stress–strain curves are usually plotted with strain on the x-axis, with the implication that the test is strain controlled. Yet, very often, it is the stress rate (say, expressed as MPa/minute) which is specified in standardized testing. It can clearly be seen from the diagrams of the complete stress–strain curve in this section that the inevitable consequence of conducting a test at a constant stress rate will be violent uncontrolled failure at the point of peak strength, when the machine tries to apply more stress than the specimen can sustain. Furthermore, the descending portion of the complete stress–strain curve is difficult to intrepret when stress is considered as the cause of strain, because a reduction in stress apparently causes an increase in strain. Conversely, if strain is considered as the cause of stress, the response of the material in the post-peak region can be interpreted simply as the fact that beyond a certain strain value (corresponding to the maximum stress) the rock continues to suffer further mechanical breakdown with an attendant *loss* of load-bearing capacity. This concept is amplified in Section 6.3.

6.3 Soft, stiff and servo-controlled testing machines

The effects of the two extreme options for loading, i.e. stress control and strain control, are illustrated in Fig. 6.5. Note that in this figure we have chosen the axes such that the independent variable is plotted along the x-axis. The first curve represents the application of an increasing load (for example, a series of weights) to the specimen. When the peak strength is reached, the deadweight causes a continous increase in strain at this peak stress level, i.e. the specimen is uncontrollably crushed. The second curve represents the continual compression of the specimen as the ends are moved together (for example, in a screw-controlled press); the stress associated with this movement can rise or fall without uncontrolled failure.

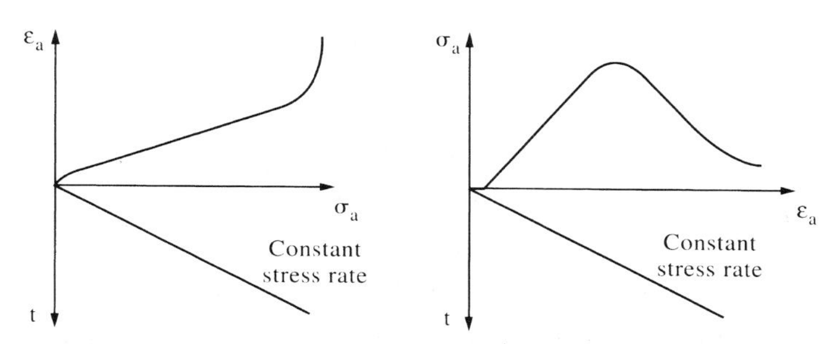

Figure 6.5 Stress- and strain-controlled stress–strain curves.

This situation can be considered as simply shortening the sample and measuring the associated load. The resultant curve, known as the strain-controlled complete stress–strain curve, was first obtained in 1966. A list of the developments in materials science testing leading up to this time is given in Hudson *et al.* (1972).

Because rock often has a higher stiffness than standard testing machines, even under strain control, the complete curve often cannot be obtained without modifying the machine. The testing techniques required for rock mechanics are thus unique, both in their requirements and their methodologies. For example, in soil mechanics testing, the soil usually has a low enough stiffness to allow the complete curve to be obtained as a matter of course. Also, in concrete testing there is not so much importance attached to obtaining the complete curve, because the peak strength is defined as failure. Therefore, we need to consider very carefully the consequences of different methods of testing and different stiffnesses of the applied loading device, whether in the laboratory or in the field.

In Fig. 6.6, we illustrate both schematically and conceptually the testing machine and associated stiffnesses. The specimen has a certain stiffness and the machine has a certain stiffness. Whatever the load in the specimen, an equal and opposite load is applied to the machine. Thus, in the lower diagram in Fig. 6.6, not only can we plot the axial force versus the axial displacement for the specimen, we can also plot the same parameters for the machine. Note that these two curves are drawn on adjacent sides of the axial force axis: this is because compression of the sample is regarded as positive, and the corresponding extension of the machine is negative (another way

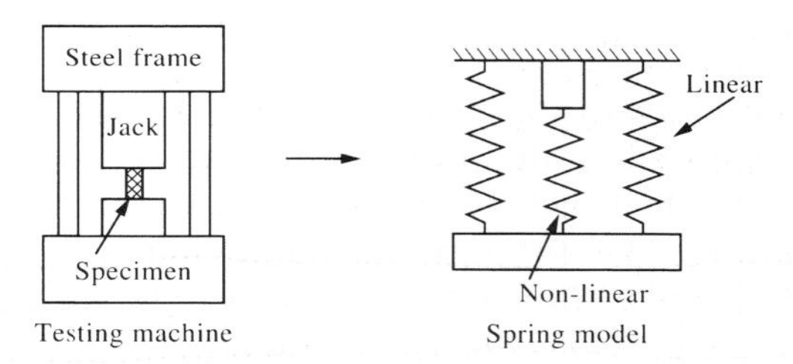
Steel frame
Jack
Specimen
Testing machine
Linear
Non-linear
Spring model
As the jack compresses the specimen, it also stretches the machine

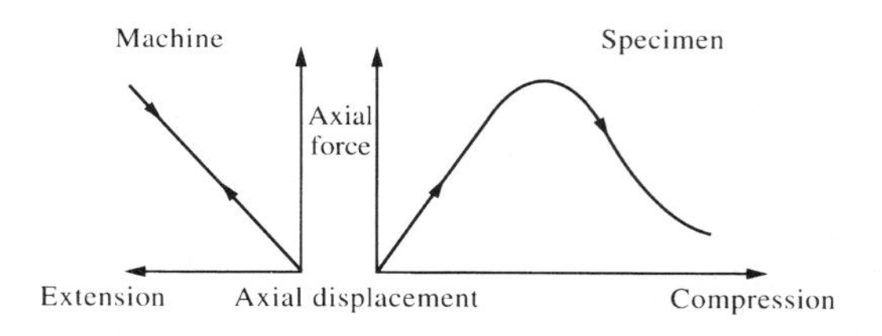
Machine
Specimen
Axial force
Extension
Axial displacement
Compression
While the specimen is compressed to peak strength, the machine stretches. As the specimen is compressed beyond the peak, the machine returns to normal length

Figure 6.6 Schematic and conceptual illustration of specimen and testing machine stiffnesses.

of thinking of the machine stiffness is to consider its axial force–extension curve were we to replace the specimen with a hydraulic jack). As the specimen is loaded, the machine is also loaded, as indicated by the arrows in the diagram. Also, as the specimen's load-bearing capability decreases in the post-peak region, so the machine elastically unloads as the force is being reduced.

Hence, as indicated in Fig. 6.6, the machine can be **soft** or **stiff**, and in a testing machine this stiffness will be a complex function of many of the component parts of the machine: these include the loading platens, the hydraulic system (the fluid, hoses and rams) and the frame. Were we to consider all these as an equivalent cylinder of cross-sectional area A, depth L and modulus E, the stiffness is given as AE/L. Thus, the machine stiffness will increase with increasing area A, decreasing length L and increasing modulus E. This means that the stiffness of the testing machine can be altered via these values.

Figure 6.7 illustrates the same complete stress–strain curve for the rock. Here we have superimposed the assumed linear behaviour of a soft testing machine and a stiff testing machine at the point A, just beyond the peak strength: this is to consider whether the machine can unload purely elastically without any intervention from the operator. In the left-hand diagram in Fig. 6.7, the unloading curve for the machine in the direction AE is very similar to the deadweight mentioned earlier. The machine can unload along this line because at all points the axial force associated with the elastic unloading of the machine is greater than the specimen can sustain, resulting in 'explosive' failure. The failure occurs because, in an increment of axial displacement DC, the machine is capable of performing the amount of work corresponding to the area DCEA, whereas the maximum work the specimen can absorb is given by the area DCBA. This work is utilized in the continuing microstructural disintegration that occurs during the axial displacement increment DC. The work represented by the area AEB is liberated as energy, manifested especially as kinetic energy: particles of the specimen fly in all directions.

We can now compare this with the right-hand diagram in Fig. 6.7, where the testing machine stiffness is represented by the steeper line AE. A similar argument to the previous one can be used to predict the response of the system. In this case, the machine cannot elastically unload of its own volition along AE, because the specimen requires more work to be done than is available. Consequently, the operator will have to increase the strain in order to follow the post-peak portion of the curve.

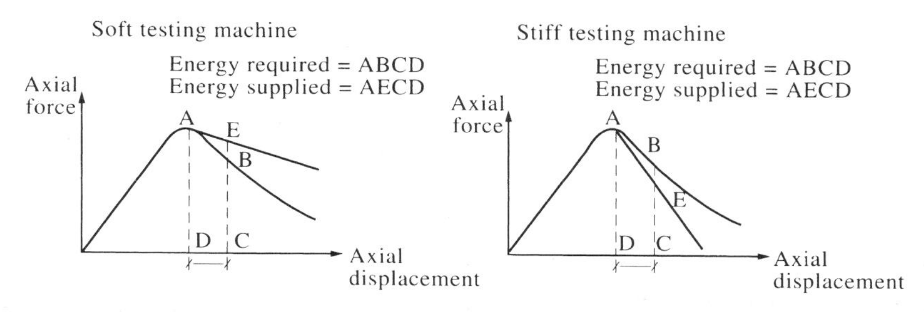

Figure 6.7 Machine stiffness and specimen stiffness in the post-peak region.

We arrive at the conclusion that if the testing machine stiffness is greater than the absolute value of the slope at any point on the descending portion of the stress–strain curve for the rock, the system will be continuously stable and it will be possible to obtain the complete stress–strain curve. Note also that, although we have expressed this argument in terms of a uniaxial compression test, it also applies to tensile tests, shear tests and any other configuration, such as the stability of a three-point loaded beam test.

The most logical method, therefore, of obtaining the complete stress–strain curve (from the *AE/L* equivalent cylinder analogy) is to build a machine which is large, squat and utilizes only high-modulus materials (for example, mercury as the hydraulic fluid). One can go even further and have no moving parts: a solid frame is heated (i.e. expansion takes place), the specimen is inserted and the frame cooled (contraction results). The curves in Fig. 6.8 were obtained by this method.

A further point to note, with respect to the curves in Fig. 6.8, is that they do not all **monotonically** increase in strain. Invoking the argument used previously for soft and stiff testing machines, we could not, therefore, obtain the curve for Charcoal Grey Granite II *even* in an infinitely stiff, i.e. rigid, testing machine. The stiffness of such a machine would be represented by a vertical line in Fig. 6.8. To obtain these curves, it was necessary to further modify the machine using a counter-acting hydraulic jack in the post-peak region. Wawersik and Fairhurst classified complete stress–strain curves into two types: Class I curves monotonically increase in strain; Class II curves do not (Wawersik and Fairhurst, 1970).

Following the pioneering work by Cook, Bieniawski, Fairhurst and Wawersik in the late 1960s, it was realized that such stiff testing machines are inherently cumbersome and functionally inflexible. This led to the introduction of **servo-controlled** testing machines for obtaining the complete stress–strain curve for rock. It is important to note that the means by which a servo-controlled testing machine is able to follow the post-peak curve is *different in principle and implementation* from the stiff testing machines.

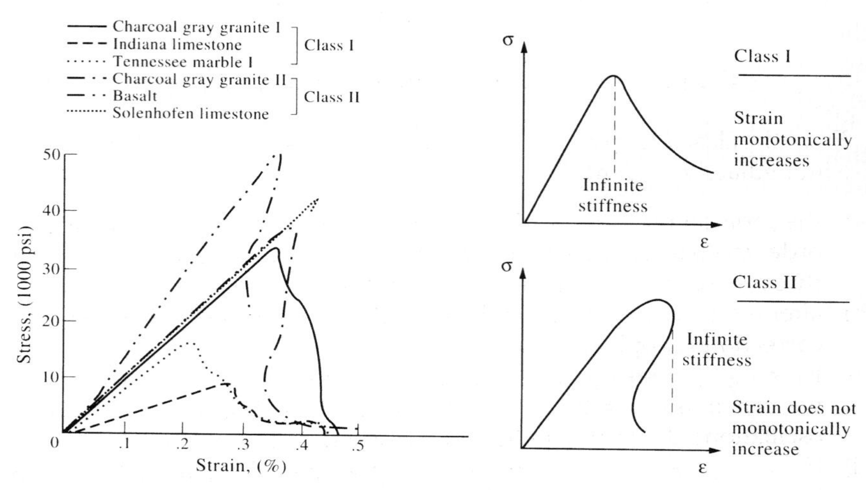

Figure 6.8 Examples of complete stress–strain curves for different rocks (from Wawersik and Fairhurst, 1970).

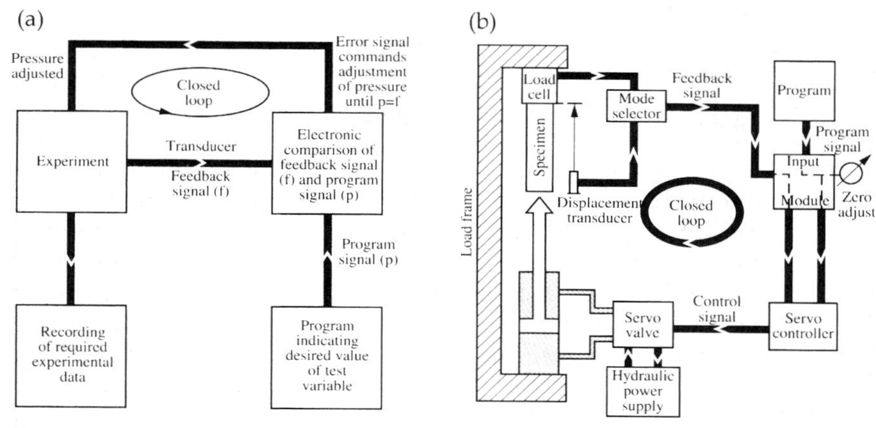

Figure 6.9 (a) Principle of closed-loop control. (b) Schematic of fast–response, closed-loop servo-controlled testing machine (courtesy MTS Systems Corp.).

The basic principle of **closed-loop control** is shown in Fig. 6.9(a) and the specific application for obtaining the complete stress–strain curve is shown in Fig. 6.9(b). The principle of closed-loop control is ubiquitous in mechanical, electrical and biological control systems. In our context, and with reference to Fig. 6.9(a), an experiment is being conducted in the top left-hand box. From this experiment, a choice of **feedback signal** is made, representing the value of a particular parameter at a specific time. In the top right-hand box, the feedback signal, f, is compared with a programmed signal, p, being generated in the lower right-hand box. If there is any difference between the feedback signal and the programmed signal, the hydraulic pressure in the experiment is adjusted to bring the feedback signal to the correct value. Thus there is continuous and automatic closed-loop control by the continual comparison of the signals. The correction signal, c, can be expressed as

$$c = k(p - f)$$

where k is the gain, p is the programmed signal, and f is the feedback signal.

From the closed-loop control equation, several aspects of machine control are immediately evident:

(a) the programmed and feedback signals must be of the same polarity in order to enable the system to be brought to equilibrium (i.e. $p - f = 0$) through application of a control signal of correct sense;
(b) alteration of the gain, k, will alter the magnitude with which the correction is applied;
(c) if $c < c_{min}$ (where c_{min} is a given small difference in the control signal), we can make $c = 0$ in order to avoid 'dithering' (i.e. rapid, small oscillations about an equilibrium position); and conversely
(d) if $c > c_{max}$ (where c_{max} represents, say, a control signal that can only be induced by a system malfunction), the system can be shut down to avoid damage.

Finally, it should be noted that the required experimental data can be monitored independently from the control system, as illustrated by the lower left-hand box in Fig. 6.9(a).

The schematic diagram in Fig. 6.9(b) illustrates this closed-loop control with more direct reference to rock testing. Note the **mode selector** for the feedback signal. If the output from the load cell were to be taken as feedback and the programmed signal were to monotonically increase with time, then we would be programming a stress-controlled test, which would result in explosive failure at the peak of the complete stress–strain curve as the machine attempted to increase the stress beyond the rock's compressive strength. From the arguments already presented, it is the displacement transducer output that would be used as the feedback signal for an axial strain-controlled test.

The tests that can be conducted with the closed-loop control technique are only limited by the imagination. The complete stress–strain curve can be obtained in tension, by using displacement feedback. By utilizing the load cell output and the displacement transducer output, we can program a linear increase in energy to be supplied to the specimen. In fact, any parameter or combination of parameters can be used as feedback.

Note that, in the complete stress–strain curves shown in Fig. 6.8, the Class II curves *do not* monotonically increase in axial strain, and hence *cannot* be obtained utilizing axial displacement (or axial strain) as the feedback signal. To overcome this problem, and as a general principle, one takes as feedback the parameter most sensitive to the failure that will occur in the test in question: in this case the *lateral* displacement—which *does* monotonically increase. The complete Class II stress–strain curve, which *does not* monotonically increase, is then **independently** monitored as it is generated. The lateral displacement is more sensitive to the axial cracking which occurs in a uniaxial compression test. Conversely, the axial displacement is more sensistive to the lateral cracking which occurs in a uniaxial tensile test.

Moreover, as the test configuration can be of any type, we will generally choose the most sensitive indicator of failure as the feedback signal. For example, to consider the mechanics of a hydraulic fracturing test in which a hollow cylinder is internally pressurized to failure, the machine can be programmed to linearly increase the circumference of the internal hole by taking the output from a wire strain gauge bonded circumferentially around the hole as feedback. The hydraulic pressure is then adjusted by the closed-loop control such that the circumference linearly increases and the fracturing is controlled. Figure 6.10 illustrates a suite of rock mechanics tests and the corresponding optimal feedback signals.

With the ability to control failure and generate a failure locus for a variety of testing configurations, the test can be stopped at any time to study stages in failure development. For example, under stress control, if the machine is programmed to 'hold' the stress constant, a creep test is performed: the analogue under strain control with a 'hold' is a relaxation test. Using stress or strain control, the rock can be fatigued with any frequency and stress or strain amplitude. It is even possible to record the three perpendicular components of earthquake motion in the field and apply these through three mutually perpendicular actuators under laboratory conditions. Even

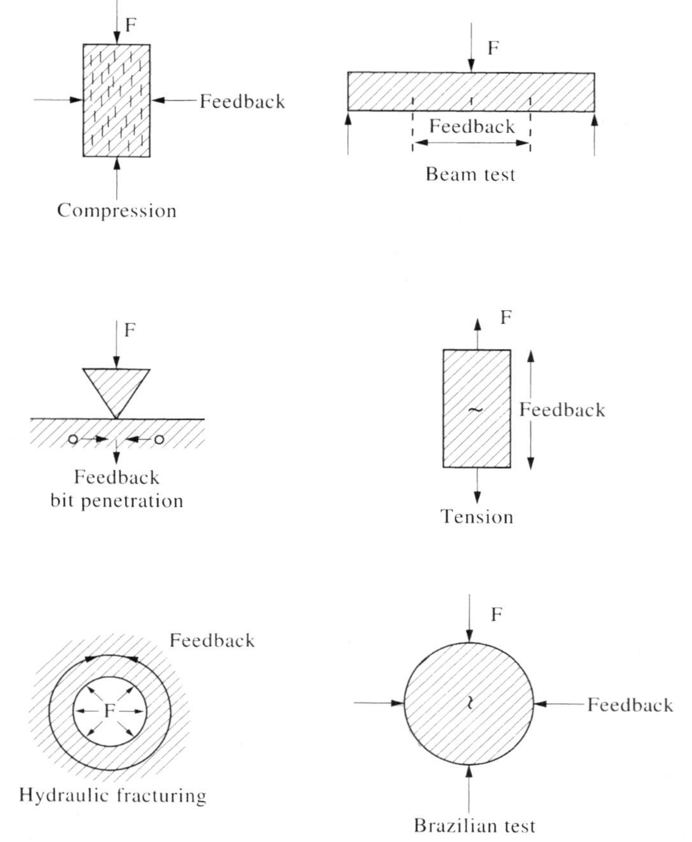

Figure 6.10 Rock testing geometries and optimal feedback for closed-loop control of failure.

today, the full potential of closed-loop control as described here has not been fully exploited in rock mechanics, particularly with respect to discontinuity testing and field testing.

Perhaps the key reason why servo-control is so successful is that the overall response time of the control system is of the order of 5 ms, which is faster than the speed at which any significant cracking can develop. Although terminal crack velocity in rock is about one-third of the acoustic velocity (i.e. very high), it should be remembered that crack propagation must accelerate from zero: the high response time of the control system ensures that the loads applied to a specimen are only sufficient to permit slow crack propagation.

6.4 Specimen geometry, loading conditions and environmental effects

Having described how the complete stress–strain curve can be obtained experimentally, let us now consider the effects of the specimen size and shape, loading conditions, and time and temperature effects.

It is well known that many materials exhibit a **size effect** in terms of strength, with smaller specimens indicating a higher strength than larger specimens. This was probably first recognized by Leonardo da Vinci, who found that longer wires were not as strong as shorter wires of the same diameter. In more recent times, Griffith (1921) showed that thin filaments of glass displayed much higher tensile strengths than thick filaments. Similarly, there are ductility effects as the temperature of a material is increased.

Thus, it is prudent to consider the effects of specimen geometry, loading conditions and environment on the complete stress–strain curve. This is because we need to understand the effects of these variables in order to be able to predict the mechanical behaviour of rock under conditions which may differ from those under which a specimen of the same rock was tested in the laboratory. The discussion below on these effects describes the general trends that have been observed in laboratory tests over the years.

6.4.1 The size effect

In Fig. 6.11, we illustrate how the complete stress–strain curve varies with specimen size, as the ratio of length to diameter is kept constant. The main effects are that both the compressive strength and the brittleness are reduced for larger specimens. The specimen contains microcracks (which are a statistical sample from the rock microcrack population): the larger the specimen, the greater the number of microcracks and hence the greater the likelihood of a more severe flaw. With respect to the tensile testing described earlier, it has been said (Pierce, 1926) that "It is a truism, of which the ramifications are of no little interest, that a chain is only as strong as its weakest link".

The elastic modulus does not vary significantly with specimen size because the relation between overall stress and overall strain is an *average* response for many individual aspects of the microstructure. However, the compressive strength, being the peak stress that the specimen can sustain, is more sensitive to *extremes* in the distribution of microstructural flaws in the sample. A larger sample will have a different flaw distribution and, in general, a more 'extreme' flaw. Also, this statistical effect will influence the form of the post-peak curve.

There have been many attempts to characterize the variation in strength with specimen size using extreme value statistics and, in particular, Weibull's theory, but it should be remembered that this theory is based on fracture initiation being synonymous with fracture propagation, which is not the case in compression. Thus, if extreme value statistics are to be applied to the analysis of compressive strength, then some form of parallel break-down model is required, rather than the weakest-link Weibull approach.

Naturally, a relation needs to be developed between strength and sample size when extrapolating laboratory determined values of strength to site scales.

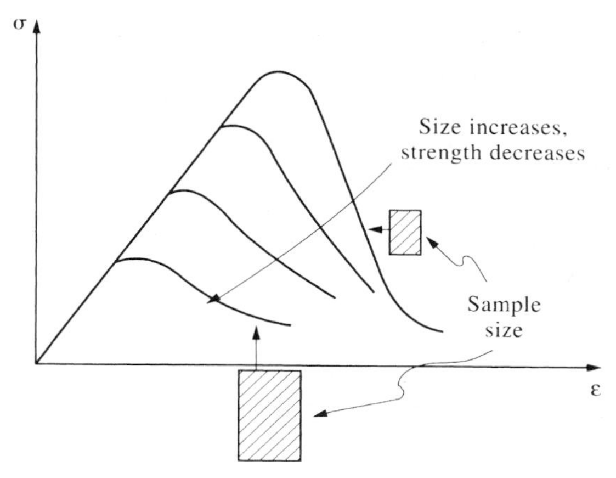

Figure 6.11 The size effect in the uniaxial complete stress–strain curve.

6.4.2 The shape effect

In Section 6.4.1, we discussed the **size effect**, i.e. when the shape of the specimen is preserved but its size changes. Here, we discuss the complementary effect, the **shape effect**, when the size (i.e. volume) of the specimen is preserved but its shape changes. In Fig. 6.12, we illustrate the effect of shape variation in uniaxial compression.

The trends in the curves show that the elastic modulus is basically unaffected by specimen shape, and that both the strength and the ductility increase as the aspect ratio, defined as the ratio of diameter to length, increases. The reason for these trends is different to that in the pure size effect case. When a specimen is loaded in uniaxial compression, end platens made of steel, and preferably of the same diameter as the specimen, are used. Because of an unavoidable mismatch in the elastic properties of the rock and the steel, a complex zone of triaxial compression is set up at the ends of the rock specimen as the steel restrains the expansion of the rock.

This **end effect** has little significance for a slender specimen, but can dominate the stress field in the case of a squat specimen (Fig. 6.12). The same end effect does occur during size effect testing, but the influence is the same for different specimen sizes, because the aspect ratio remains constant.

The effect of a confining pressure during the triaxial test has a dramatic effect on the complete stress–strain curve and it is essentially this confining effect which is causing the shape effect illustrated in Fig. 6.12. The problem is easily overcome in the laboratory by choosing an appropriate aspect ratio, greater than or equal to 2.5, but underground support pillars *in situ* are much more likely to be squat than slender. Thus, the shape effect has the converse effect to the size effect when the results are extrapolated to the field: an *in situ* squat pillar will be stronger than a slender laboratory specimen of the same rock, although there will be different loading conditions in the field which could mitigate the effect.

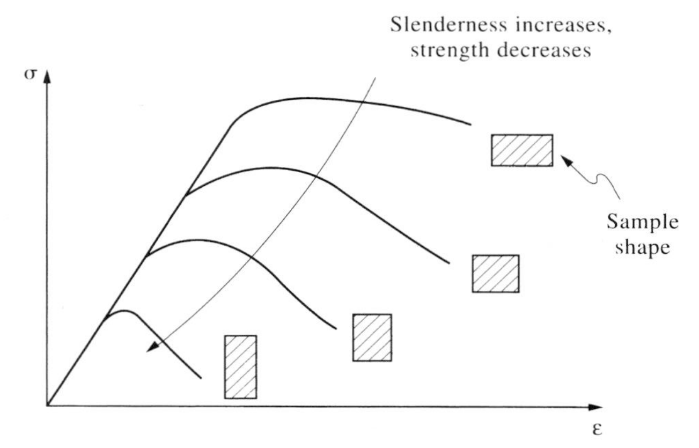

Figure 6.12 The shape effect in uniaxial compression.

To predict the strength of the rock *in situ*, and avoid the influence of the shape effect, we can proceed in one of two ways: improve the laboratory test procedures; or use empirical formulae to account for the shape effect. The main laboratory method is to use platens which reduce the confining effect, either through brush platens (which effectively load the specimen ends over a number of small zones, thereby reducing the volume of rock in triaxial compression) or flatjacks (which prevent shear stresses being transmitted between the platen and the specimen). Other laboratory techniques involve specimen geometries which reduce the effect, for example, axial loading of hollow cylinders. Empirical formulae are the main engineering approach, where a numerical relation is utilized to take into account the shape effect. In fact, these formulae can incorporate the diameter and the length separately and hence it may not be directly clear from the formula how to separate the size and the shape effects, should one wish to do so.

6.4.3 Loading conditions

We have seen, with reference to the shape effect, how the loading conditions can affect the rock behaviour in uniaxial compression. Let us now consider the many possibilities for rock testing and illustrate some of the terms in general use. The sketches in Fig. 6.13 show the loading conditions in the six main testing configurations. A particular point to note is the difference between triaxial and polyaxial compression. Over the years, triaxial compression has come to mean a test conducted using a pressure vessel, with the consequence that $\sigma_2 = \sigma_3$. This is not true triaxial compression in the sense that all three principal stresses can be independently applied: for this latter condition we use the term polyaxial compression. The application of three different principal stresses is quite difficult to achieve in practice, and hence the test is not used routinely in rock mechanics.

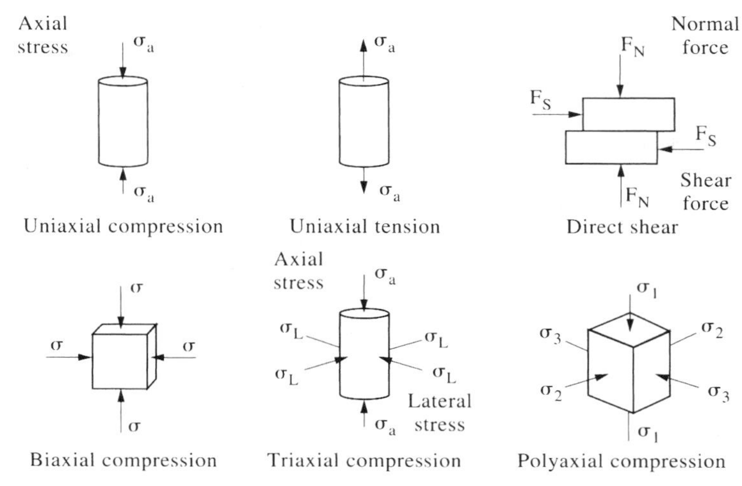

Figure 6.13 Specimen loading conditi ons in general laboratory use.

We have discussed uniaxial compression; let us now consider uniaxial or direct tension, and also the generation of a tensile stress through compressive loading in indirect tensile tests.

The uniaxial tension test, as illustrated in Fig. 6.13, is not as a rule used in engineering practice. There are two reasons for this: first, it is difficult to perform; and, second, the rock does not fail in direct tension *in situ*. Through the servo-controlled testing method, the complete stress–strain curve in tension has been obtained using axial displacement as feedback in the closed loop. This displacement is the most sensitive indicator of failure because a single main crack develops laterally. However, this curve is really only of academic interest because the essentially single crack failure mode leads to ultra-brittle behaviour. Even for establishing the tensile strength of rock itself, a state of pure tension with no applied or induced bending moments is difficult to achieve. Some irregularity in the compression test can be tolerated, but in tension any such irregularity leads to premature failure.

For these reasons, the tensile strength is normally measured by indirect tests in which the tensile stress is generated by compressive loading. (The tensile strength of the rock is very much lower than the compressive strength, so that such indirect tests are possible; for the same reason, it is not possible to have indirect compression tests.)

In Fig. 6.14, two indirect tensile tests are shown, with the point load test being the most widely used test on intact rock. In each case, through the testing configuration, the maximum tensile stress can be calculated from elasticity theory as a function of the compressive force and specimen dimensions. The tensile strength is, therefore, the maximum tensile stress *calculated to be present in the specimen at failure*. Such a calculation is based on ideal material assumptions and does not take account of different critically stressed volumes in each test. As might be expected from our earlier discussion, the tensile strength varies for a given rock type tested in these different ways and hence *is not an intrinsic material property*.

The tensile strength variation is of three main types:

(a) variation with repeated testing;
(b) variation with different volume; and
(c) variation between tests.

The first type of variation occurs because each rock specimen contains a statistical sample of flaws from the total microstructural flaw population. The severity of the worst flaw present in each specimen taken from a set of specimens, all obtained from the same rock block, will be highly variable. Thus, if, say, 50 tensile strength tests are repeated, there could well be a wide variation in the 50 values, from which we can determine the mean and standard deviation.

The second type of variation occurs because the larger the statistical sample, the greater the likelihood of a more severe flaw. Hence, if we were to conduct another 50 tests on larger sized specimens, but utilizing the same type of tensile strength test, we would also obtain a distribution of test results, **but** both the mean and the standard deviation would be lower, as illustrated in Fig. 6.14.

The third type of variation, the inter-test variation, occurs because the critically stressed volume in each test is different. So, if a set of tensile strength results obtained using one test is compared with those obtained from another test, again there is a difference between the histograms of test results, as also illustrated in Fig. 6.14.

The curves in Fig. 6.14 can be related via statistical theories. For example, in Weibull's theory, the probability of failure is integrated over the critically stressed volume, taking the variation in tensile stress into account. This enables the basic probability function for a test to be specified and hence the probability density curves in Fig. 6.14 to be characterized. The variation in the probability density function can be established as a function of test volume and hence the change in density curves with volume

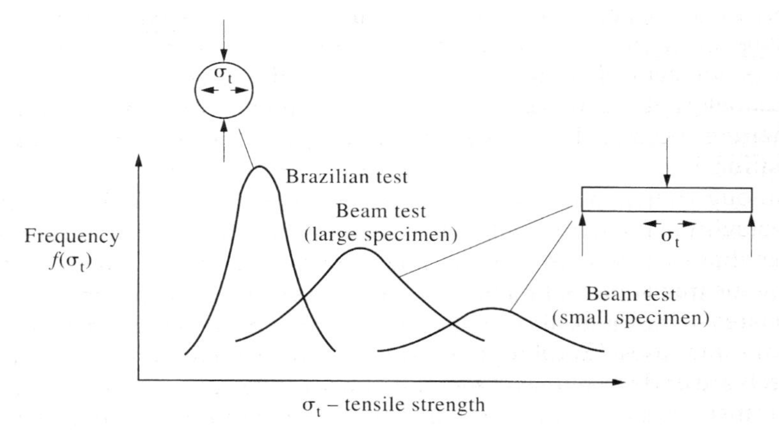

Figure 6.14 Tensile strength variation as a function of specimen volume and type of test.

may be predicted. Indeed, the probability density function for any test condition can be established and hence the inter-test variation can be predicted. One of the most useful formulae to arise from this approach is

$$\sigma_{t1}/\sigma_{t2} = (V_2/V_1)^m$$

where σ_{t1} and σ_{t2} are the *mean tensile strengths* obtained for two sets of samples with different volumes (for any test configuration), V_1 and V_2 are the associated specimen volumes, and m is one of the three material constants used in Weibull's theory. This provides a direct relation between the mean tensile strength and the specimen volume.

At this stage, we should like to caution the reader. Weibull's theory is *solely statistical* and does not include any specific mechanism of fracture or failure. Moreover, the formula above is represented by the ubiquitous straight line in log–log space. There have been several published 'verifications' of the theory, based on straight lines in log–log space, but these results alone do not isolate Weibull's theory. Indeed, any such confirmation for the validity of the formula in compression tests is highly unlikely to be valid because of the distinction between *failure initiation* and *failure propagation* in the compression test.

This cautionary note related to the avoidance of blind acceptance of any particular theory based on power laws (and material constants which can be determined by curve fitting) applies to all rock testing, and particularly to failure criteria (which we will be discussing later in this chapter).

Another factor altering the shape of the complete stress–strain curve in compression is the effect of the confining pressure applied during a test, which can be quite pronounced. The general trend is shown in Fig. 6.15.

The most brittle behaviour is experienced at zero confining pressure: the curve demonstrates less brittle behaviour (or increasing ductility) as the confining pressure is gradually increased. At one stage in this trend, the post-peak curve is essentially a horizontal line, representing continuing strain at a constant stress level; or, in the interpretation of a strain controlled test, the strength is not affected by increasing strain. Below this line, the material **strain softens**: above this line **strain hardening** occurs. The horizontal line is termed the **brittle–ductile transition**.

Although it may be thought that this transition would only be of interest to geologists considering rocks subjected to the high pressures and temperatures that exist at great depths, there can be engineering circumstances where the transition is of importance. This is because the confining pressure associated with the brittle–ductile transition varies with rock type and is low in some cases. Coupling this with the increasing depth at which some projects are undertaken can mean that the transition is important. Note that the transition also represents the boundary between instability with increasing strain (brittle behaviour) and stability with increasing strain (ductile behaviour).

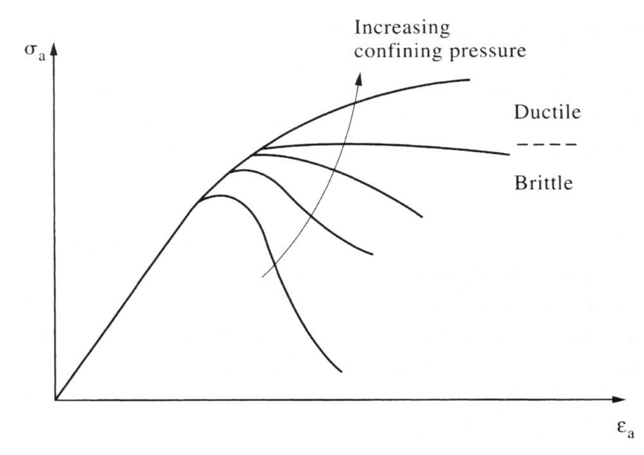

Figure 6.15 The effect of confining pressure in the triaxial test and the brittle–ductile transition.

An idea of the variability in the confining pressure associated with the transition is given in the table below (after Goodman, 1989).

Rock type	Confining pressure (MPa)
Rock salt	0
Chalk	<10
Limestone	20–100
Sandstone	>100
Granite	≫100

It is emphasized that these are representative values, chosen to illustrate the general trend.

One would expect different post-peak behaviour around caverns excavated in a soft rock salt and a hard granite. On construction, the rock at the excavation periphery ceases to be under triaxial compression, because the normal and shear stresses on the rock wall have been reduced to zero. So, the rock will tend to behave in a more brittle fashion. In a rock salt cavern, we could have brittle failure visibly occuring in the walls, with more ductile behaviour occuring out of sight further into the rock. Similarly, in a chalk possessing a 10 MPa brittle–ductile transition, and because the overburden stress is approximately 1 MPa for every 40 m of depth, we might expect ductile behaviour in deep civil engineering in soft rocks, depending of course on a whole host of other factors.

Finally, perhaps the most important aspect of this phenomenon is to understand the reason why the stress–strain curves take the form they do. As we have mentioned earlier, in compression the rock tends to fracture perpendicular to the least principal stress, i.e. parallel to the major principal stress. Consequently, the application of even a small confining pressure has a significant effect on inhibiting the development of these cracks, and indeed, the mechanism of crack formation, which gradually changes to shearing as the confining pressure is increased.

6.4.4 Environmental effects

Other factors which affect rock behaviour, in particular moisture content, time and temperature, can be of importance in engineering and we have grouped them here under the general title of environmental effects.

Moisture content. The moisture content is known to influence the complete stress–strain curve because of its effect, in certain rocks, on the deformability, the compressive strength and post-peak behaviour. For this reason, it is recommended, for example by the ISRM, that the moisture content be measured as an integral part of the compressive strength determination of rocks. It is beyond the scope of this book to provide a comprehensive discussion of all aspects of the influence of moisture content and saturation, but the reader is alerted to the following factors which can be particularly important in rock engineering.

1. Some rocks, and in particular those with high clay mineral contents, may experience **desiccation** when exposed. *In situ*, the rock may possess a stable, but high, moisture content; on exposure after excavation, its properties may change as it dries out and it may become friable and hence crumble with very little applied stress.
2. Similarly, the same types of rock could be saturated on excavation, and be subjected simultaneously to mechanical action as part of the excavation process. This leads to **slaking** and there is an associated **slake durability test** to assess the susceptibility of a rock under these conditions. The rock can then also break down and crumble under a very low applied stress. The reader should be aware that slaking behaviour is not dissolution.
3. Another moisture related effect is the tendency to **swelling** as the moisture content is changed. This can lead to the generation of additional stresses, for example behind tunnel linings. In some cases, the stresses thus generated can be of a similar magnitude to those due to the *in situ* stress field, and can initiate failure.
4. If the pore spaces in the rock are connected and the pore fluid is under pressure, we can subtract this pressure, or a proportion of it, from all the components of normal stress. This leads to the well-known concept of **effective stress**, widely used in soil mechanics and which we will discuss in Chapter 9. If the water pressure is increased sufficiently, the effective stress can be reduced to such an extent that failure occurs. In the case of rocks, the effective stress concept can apply well for materials such as sandstone, but be inappropriate for granites, especially over engineering rather than geological timescales.

These are some of the main effects but there are many others which occur as water (or other pore fluids) move through the rock and cause alterations and effects of different kinds. For example, the chemistry of groundwater can be important, e.g. its acidity. In materials such as chalk and limestone, this results in dissolution of the intact rock with complete removal of the material to produce caves. Freeze–thaw cycles can also degrade intact rock, usually in a similar fashion to slaking.

Time-dependent effects. We have indicated that during the complete stress–strain curve, microcracking occurs from a very early stage in the pre-peak region. For some purposes, it is convenient to assume that much of the pre-peak portion represents elastic behaviour. However, there is no time component in the theory of elasticity; yet, because of the continually increasing microstructural damage even in the 'elastic' region, we would expect some time-dependent behaviour.

There are four main time-dependent effects which are discussed here.

(a) **strain rate**—the total form of the complete stress–strain curve is a function of the applied strain rate;
(b) **creep**—a material continues to strain when the applied stress is held constant;
(c) **relaxation**—there is a decrease in stress within the material when the applied strain is held constant;
(d) **fatigue**—there is an increase in strain due to cyclical changes in stress.

These four effects are shown in Fig. 6.16 and are all manifestations of the time-dependent nature of microcrack development.

The effect of a reduced strain rate is to reduce the overall elastic modulus and the compressive strength. Creep from a point A in Fig. 6.16 is indicated by the line AC. Relaxation is indicated by the line AR. Fatigue is indicated by the stress cycles. The relation between these effects can be seen especially from the form of the complete stress–strain curve at lower and lower strain rates. Depending on whether the control variable is stress or strain, the rock will be continually creeping or relaxing, respectively, during generation of the complete stress–strain curve.

We have noted that stress cannot be used as the control variable to obtain the post-peak region of the curve; nor indeed, as indicated by the line BC in Fig. 6.16, can creep occur in the post-peak region without instantaneous failure. As indicated by the lines AR and BR, relaxation can occur on either side of the curve for a Class I curve. Also indicated in the figure are the lines

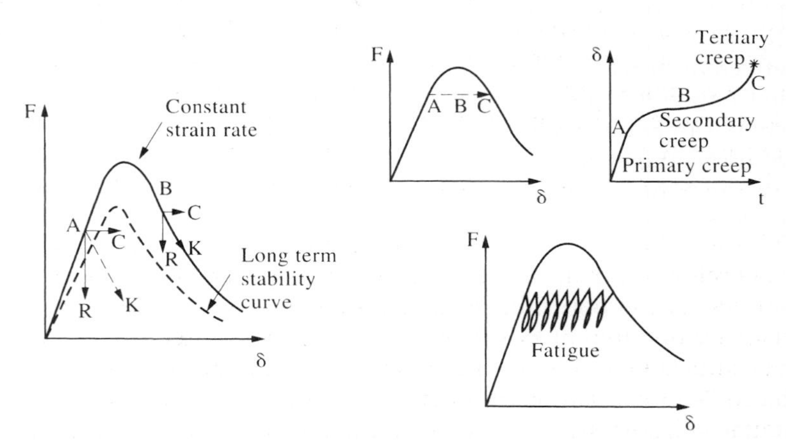

Figure 6.16 Time-dependent effects and the complete stress–strain curve.

AK and BK which represent time-dependent unloading along the stiffness line of the loading configuration, be it a laboratory testing machine or an *in situ* rock structure. The reader should note that the lines AK and BK are the same machine stiffness lines as those shown in Fig. 6.7. Thus, failure along the line BK can also be interpreted as a time-dependent effect, because the specimen cannot sustain the loads associated with BK for any significant length of time.

Furthermore, at high stress levels, creep has often been studied and divided into three types of behaviour: primary, secondary and tertiary creep. These are indicated by the letters A, B and C in the inset diagram in Fig. 6.16. Primary creep is an initial period during which creep occurs at a high rate; secondary creep is a period during which the creep rate is very much diminished; and tertiary creep is a period during which the creep rate accelerates until failure occurs. These periods can be interpreted as the line ABC crossing from the pre-peak portion of the complete stress–strain curve to the post-peak failure locus. In other words, there is an initial period of rapid creep as the displacement moves away from the pre-peak curve; there is a quiescent period next; and finally the creep accelerates as the displacement approaches the post-peak curve.

Finally, fatigue, whether the cycles are in stress or in strain, is a complex process in which the previous types of time-dependent, microstructural cracking described are occuring incrementally at different levels of stress and strain during the cycling process.

In terms of long-term *in situ* structural stability, we would anticipate that for engineering purposes there is a *long-term stability curve* as indicated by the dotted complete stress–strain curve in Fig. 6.16. We know that underground excavations can remain open for thousands of years without any apparent time-dependent collapse. In this case, the stresses and strains associated with the rock around the excavation are on the long-term stability curve, and will have approached it through a combination of creep and relaxation over the years. We would expect different rock types to have different forms of long-term stability curves: the curve for a granite might be similar to the one obtained at relatively high strain rates in the laboaratory, whereas the curve for a rock salt could be very much lower than that obtained in the laboratory. Also, some rocks will suffer mechanical and chemical degradation which will be superimposed on the direct time-dependent effects. Conversely, if stresses applied to a rock structure in the short term are sufficiently high to cause the line AK in Fig. 6.16 to be above the long-term stability curve, then failure will be the inevitable consequence. The consequences for engineering design are manifold.

It is for all these reasons that some degree of standardization is essential in laboratory testing, not only to provide coherency for comparative purposes, but also to be able to extrapolate to field strain rates from a constant worldwide measurement base. This is because the behaviour of rocks differs widely depending on the strain rate to which they are subjected—because of wide variations in the microstructure of rocks. For example, a limestone may exhibit brittle behaviour when subjected to the high strain rates developed by explosives, say 1×10^5 , typical Class I behaviour at a strain rate in the laboratory of 1×10^{-5}, and very much more ductile behaviour when undergoing tectonic movements at strain rates of 1×10^{-16}. Note that this range is through 21 orders of magnitude.

Temperature effects. Only a limited amount of information is available indicating the effect of temperature on the complete stress–strain curve and other mechanical properties of intact rock. The limited test data do however agree with one's intuition, that an increase in temperature reduces the elastic modulus and compressive strength, whilst increasing the ductility in the post-peak region. The complete stress–strain curves shown in Fig. 6.17 illustrate this trend. Also, very high temperatures can result in damage to the microstructure. At the other end of the temperature spectrum, there is increasing interest in the effect of very low temperatures on rock, within the context of liquified natural gas storage.

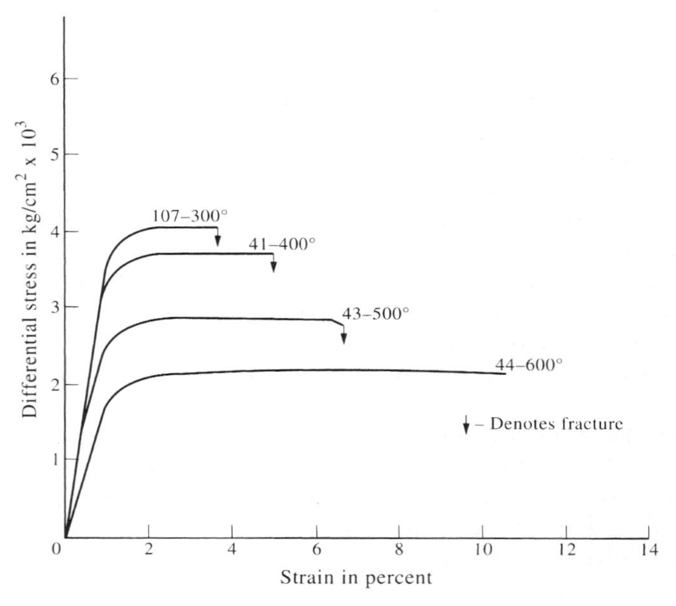

Figure 6.17 The effect of temperature on the complete stress–strain curve.

6.5 Failure criteria

We do not know exactly how a rock fails, either in terms of the precise details of each microcrack initiation and propagation, or in terms of the total structural breakdown as many microcracks propagate and coalesce. In both cases, the process is extremely complex and not subject to convenient characterization through simplified models. Nevertheless, as engineers we should like some measure of the failure properties and the ability to predict when failure will occur. It was mentioned earlier that stress has been traditionally regarded as the 'cause' and strain as the 'effect' in materials testing: as a consequence, early testing and standards utilized a constant stress rate application. It was then natural to express the strength of a material in terms of the stress present in the test specimen at failure. Since uniaxial and triaxial testing of rock are by far the most common laboratory procedures in rock mechanics and rock engineering, the most obvious means of expressing a failure criterion is

$$\text{strength} = f(\sigma_1, \sigma_2, \sigma_3).$$

With the advent of stiff and servo-controlled testing machines and the associated preference for strain rate control, perhaps the strength could be expressed in the form

$$\text{strength} = f(\varepsilon_1, \varepsilon_2, \varepsilon_3).$$

We also discussed the possibility of more eclectic forms of control such as constant rate of energy input, leading to more sophisticated possibilities for strength criteria expressed in the form

$$\text{strength} = f(\sigma_1, \sigma_2, \sigma_3, \varepsilon_1, \varepsilon_2, \varepsilon_3).$$

Despite this possibility, the number and variation of the failure criteria which have been developed, and which are in some degree of everyday use, are rather limited. The Mohr–Coulomb criterion expresses the relation between the shear stress and the normal stress at failure. The plane Griffith criterion expresses the uniaxial tensile strength in terms of the strain energy required to propagate microcracks, and expresses the uniaxial compressive strength in terms of the tensile strength. The Hoek–Brown criterion is an empirical criterion derived from a 'best-fit' to strength data plotted in σ_1–σ_3 space.

We will be presenting outlines of these criteria below; for a full derivation and more complete explanation and discussion, the reader is referred to the text by Jaeger and Cook (1979) for the Mohr–Coulomb and the Griffith criteria, and to Hoek and Brown (1980), Hoek (1990) and Hoek *et al.* (1992) for the Hoek–Brown criterion.

6.5.1 The Mohr–Coulomb criterion

The plane along which failure occurs and the Mohr envelope are shown in Fig. 6.18 for the two-dimensional case, together with some of the key expressions associated with the criterion. From the initial principal stresses, the normal stress and shear stress on a plane at any angle can be found using the transformation equations, as represented by Mohr's circle. Utilizing the concept of **cohesion** (i.e. the shear strength of the rock when no normal stress is applied) and the **angle of internal friction** (equivalent to the angle of inclination of a surface sufficient to cause sliding of a superincumbent block of similar material down the surface), we generate

BASIC EQUATIONS

Rock fails at a critical combination of normal and shear stresses:

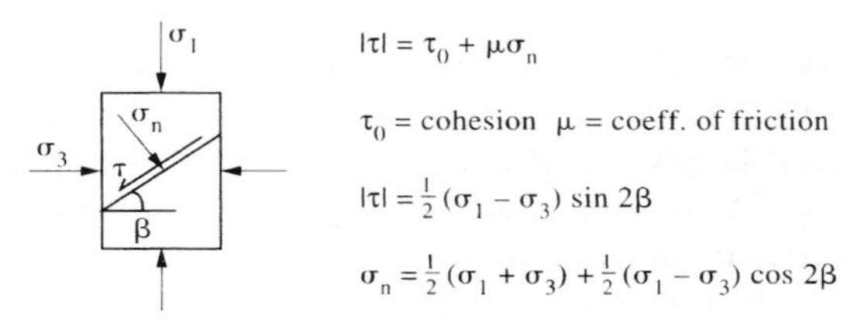

σ_1
σ_n
σ_3
τ
β
$|\tau| = \tau_0 + \mu\sigma_n$
τ_0 = cohesion μ = coeff. of friction
$|\tau| = \frac{1}{2}(\sigma_1 - \sigma_3) \sin 2\beta$
$\sigma_n = \frac{1}{2}(\sigma_1 + \sigma_3) + \frac{1}{2}(\sigma_1 - \sigma_3) \cos 2\beta$

FUNDAMENTAL GEOMETRY

The equation for $|\tau|$ and σ_n are the equations of a circle in (σ, τ) space:

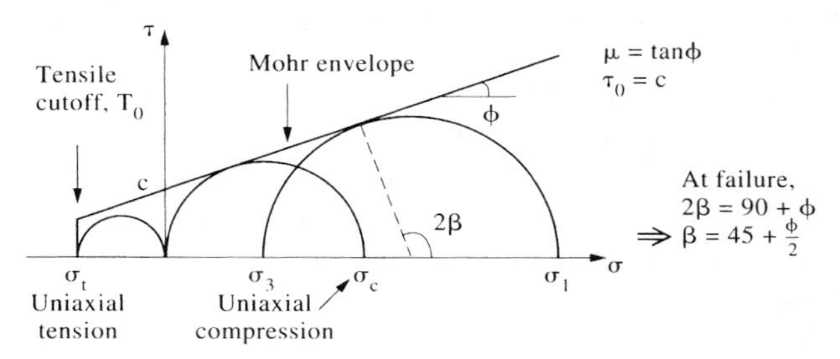

Figure 6.18 The Mohr–Coulomb failure criterion.

the linear Mohr envelope, which defines the limiting size for the Mohr's circles. In other words, σ–τ co-ordinates below the envelope represent stable conditions; σ–τ co-ordinates on the envelope represent limiting equilibrium; and σ–τ co-ordinates above the envelope represent conditions unobtainable under static loading. Because the criterion is developed for compressive stresses, a **tensile cut-off** is usually utilized to give a realistic value for the uniaxial tensile strength.

We anticipate that this criterion is most suitable at high confining pressures when the material does, in fact, fail through development of shear planes. At lower confining pressures, and in the uniaxial case, we have seen that failure occurs by gradual increase in the density of microcracks sub-parallel to the major principal stress, and hence we would not expect this type of frictional criterion to apply directly. However, at the higher confining pressures, the criterion can be useful and it should be noted, with reference to Fig. 6.18, that the failure plane will be orientated at $\beta = 45° + (\phi/2)$.

The influence of a significant water pressure in porous materials (which is deducted from the normal stress components, but not from the shear stress component) is clear as the Mohr's circle is moved to the left by an amount equal to the water pressure, hence introducing the possibility of the Mohr's circle moving from a stable region to be in contact with the Mohr envelope.

Despite the difficulties associated with application of the criterion, it does remain in use as a rapidly calculable method for engineering practice, and is especially significant and valid for discontinuities and discontinuous rock masses.

6.5.2 The plane Griffith criterion

The essence of the Griffith criterion is that for a material to break in tension owing to the presence of an existing microcrack, sufficient energy must be released to provide the necessary new surface energy as the crack propagates. The rate of strain energy release must be equal to or greater than the required surface energy increase. This results in the expression shown in Fig. 6.19 for the uniaxially loaded plate shown. It is possible to extend this criterion from the plane stress case shown to plane strain in both tension and compression, as the figure shows. The basic concept of supplying sufficient surface energy during fracturing also applies during crack propagation, However, the formulae refer only to the **onset** of cracking because the geometry changes during crack propagation. In the case of a tensile test for engineering purposes, fracture initiation and specimen collapse may be considered as synonymous; in the case of compression, however, we have already noted that microstructural cracking occurs throughout the complete stress–strain curve and that the compressive strength is an arbitrary stage in the microstructural breakdown process. Thus, whilst it is interesting to utilize the Griffith criterion for studying microcrack intitiation under compressive loading, it is unlikely that the formula can provide a useful estimate of the engineering compressive strength.

The formula for tensile failure is

$$\sigma_t = (k\alpha E/c)^{0.5}$$

where σ_t is the tensile stress applied to the specimen at failure, k is a parameter that varies with the testing conditions, i.e. $k = 2/\pi$ for plane stress and $k = 2(1 - \nu^2)/\pi$ for plane strain, α is the unit crack surface energy, E isthe Young's modulus, and c is half the initial crack length.

Thus, for a given rock and testing configuration, the tensile strength will vary inversely as the square root of the initial crack length. This provides

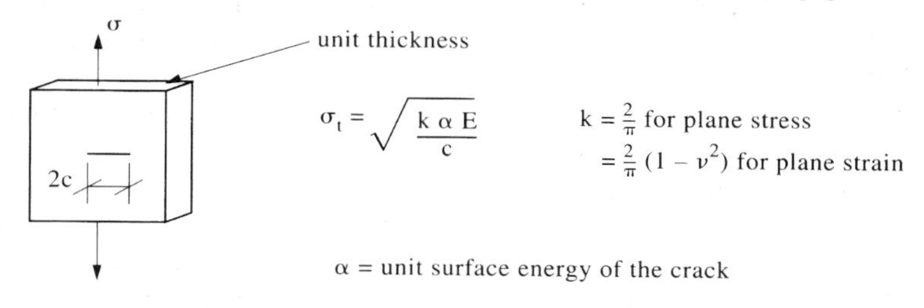

In compression:

$(\sigma_1 - \sigma_3)^2 = 8T_0 (\sigma_1 + \sigma_3)$ when $\sigma_1 + 3\sigma_3 > 0$

$\sigma_3 = -T_0$ when $\sigma_1 + 3\sigma_3 < 0$

Note: compression positive, T_0 positive ($-T_0 = \sigma_t$)

Figure 6.19 The plane Griffith failure criterion.

a direct mechanical explanation of the size effect discussed earlier: the tensile strength decreases with increasing crack length, and larger specimens will tend to contain larger flaws (i.e. larger initial crack lengths).

The Griffith criterion enables a relation to be derived between the uniaxial tensile strength and the triaxial compressive strength as

$$(\sigma_1 - \sigma_3)^2 = 8T_0(\sigma_1 + \sigma_3)$$

which for uniaxial compression with $\sigma_3 = 0$ gives $\sigma_c = 8T_0$ where $T_0 = -\sigma_t$. This relation has been modified by various researchers for a variety of factors, particularly friction across the crack surfaces.

6.5.3 The Hoek–Brown empirical failure criterion

This empirical criterion is derived from a best-fit curve to experimental failure data plotted in σ_1–σ_3 space as shown in Fig. 6.20. Hoek (1990) has noted that "since this is one of the few techniques available for estimating the rock mass strength from geological data, the criterion has been widely used in rock mechanics analysis".

The criterion is expressed as

$$\sigma_1 = \sigma_3 + (m\sigma_c\sigma_3 + s\sigma_c^2)^{0.5}$$

where σ_1 = the major principal stress, σ_3 = the minor principal stress, σ_c = the uniaxial compressive strength of the intact rock, and m and s are constants for a specific rock type.

Although the constants m and s arise from the curve-fitting procedure, there is an element of physical interpretation associated with them which is helpful for the engineer to consider.

The parameter s relates to the degree of fracturing present in the rock sample: it is a representation of the cohesion of the rock. For completely intact rock, it takes the value 1 (which can be demonstrated by substituting $\sigma_3 = 0$ into the criterion: $\sigma_1 = \sigma_c s^{0.5}$ and hence $s = 1$, noting that σ_c is the intercept on the σ_1 axis in Fig. 6.20) and, for rock which is highly fractured, it reduces in value and tends towards zero as the strength is reduced from peak to residual.

The parameter m is related to the degree of 'particle interlocking' present: for intact rock this is high, and reduces as the degree of brokenness increases. There are no clear limits to this parameter; it depends on the rock type and its mechanical quality.

This criterion also provides a relation between the tensile and compressive strengths which can be found by substituting $\sigma_1 = 0$ and $\sigma_t = -\sigma_3$ in the criterion to give

$$\sigma_t = -\sigma_c(m - (m^2 + 4s)^{0.5})/2.$$

Thus, the relation between the two strengths is a function of the rock's mechanical properties: for example, if $s = 1$ and $m = 20$ (a good-quality

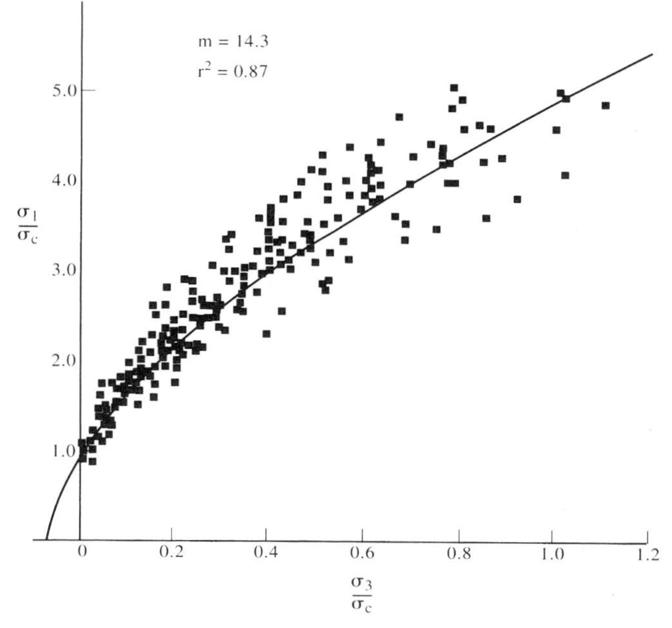

Figure 6.20 The Hoek–Brown empirical failure criterion.

intact granite, say), the compressive strength is found to be about 20 times the tensile strength. Note that these strengths are now the *rock mass* strengths, because not only the intact rock fracturing but also the large-scale rock mass fracturing is incorporated into this criterion via the parameter m. As we will see later in Chapters 8 and 12, relations can be postulated between the parameters m and s and other measures of rock mass quality using classification schemes. Also, inter-criteria relations can be found, in particular linking the Hoek–Brown criterion with the Mohr–Coulomb criterion, i.e. linking m and s with c and ϕ.

The Hoek–Brown criterion has recently been recast (Hoek *et al.*, 1992) to take into account the experience gained over the 10 years since its development.

6.6 Concluding remarks

We have presented three failure criteria that are extensively used in rock mechanics and rock engineering. The reader should be aware of the shortcomings of each of the criteria, and be prepared to make use of whichever is best suited to a particular application. Further criteria are available as listed in Fig. 6.21. None of the criteria take into account the *specific* structure of the rock mass, in particular its fracturing configuration. The occurrence of the natural pre-existing fractures in rock masses is the next subject to discuss because of their profound influence on the deformation, strength and failure of rock masses.

No.	*Authors*	*The criteria*	*Constants involved*	*Development of the criteria*
1	Murrell (1963)	$\tau_{oct}^2 = 8T_0\sigma_{oct}$. or: $J_2 = 4T_0I_1$.	One constant (3D criterion)	Extended 3D Griffith theory.
2	Fairhurst (1964)	if $m(2m-1)\sigma_1 + \sigma_3 \geqslant 0$: $\sigma_1 = K$. if $m(2m-1)\sigma_1 + \sigma_3 < 0$: $\frac{(\sigma_1 - \sigma_3)^2}{(\sigma_1 + \sigma_3)} = -2(m-1)^2 K\left[1 + \frac{2K}{(\sigma_1 + \sigma_3)}\left\{\left(\frac{m-1}{2}\right)^3 - 1\right\}\right]$	Two constants (2D criterion)	Empirical generalisation of 2D Griffith theory for intact rock.
3	Hobbs (1966)	$\sigma_1 = B\sigma_3^b + \sigma_3$, or: $r = K_2\sigma_n^a$.	Two constants (2D criterion)	Empirical test data fitting for intact rocks.
4	Hoek (1968)	$\sigma_1 - \sigma_3 = 2C + A(\sigma_1 + \sigma_3)^B$, or: $\tau_{max} = \tau_{max0} + A\sigma_m^b$.	Three parameters (2D criterion)	Empirical curve fitting for intact rock.
5	Franklin (1971)	$\sigma_1 - \sigma_3 = \sigma_c^{1-B}(\sigma_1 + \sigma_3)^B$.	Two constants (2D criterion)	Empirical curve fitting for 500 rock specimens.
6	Bienlawski (1974)	$\sigma_1 = K'\sigma_3^A + \sigma_c$, or: $\tau = B'\sigma_m^c + 0.1\sigma_c$.	Three constants (2D criterion)	Empirical curve fitting for 700 rock specimens. (5 types)
7	Yoshinaka & Yamabe (1980)	$\sigma_1 - \sigma_2 = \alpha K(q)(\sigma_1 + \sigma_2 + \sigma_3)^\beta$.	Three parameters (3D criterion)	Empirical test data analysis for soft rocks (mudstone, etc).
8	Hoek and Brown (1980)	$\sigma_1 - \sigma_3 = \sqrt{m\sigma_c\sigma_s + s\sigma_c^2}$. or: $\tau = A(\sigma_n + B)^C$.	Three parameters (2D criterion for rocks and rock masses).	Appl. of Griffith theory and empirical curve fitting for rock and rock mass.
9	Kim and Lade (1984)	$\left(\frac{(I_1')^3}{I_s'} - 27\right)\left(\frac{(I_1')}{p_a}\right)^m = \eta_1$.	Three parameters (3D criterion)	Analytical examination on test data (originally for soil and concrete).
10	Johnston (1985)	$\frac{\sigma_1}{\sigma_c} = \left[\frac{M}{B}\frac{\sigma_3}{\sigma_c} + 1\right]^B$.	Three parameters (2D criterion)	Empirical curve fitting for soft rock specimens.
11	Desai and Salami (1987)	$J_2' = \left(-\frac{\alpha}{\alpha_0}(I_1')^n + \gamma(I_1')^2\right)(1 - \beta S_r)^m$.	More than six parameters (3D criterion)	Polynominal expansion in terms of stress invariants to curve fitting.
12	Michelis (1987)	$\ln\left(\frac{q^2}{f_0^2} + a_1 p\frac{q}{f_0} + a_2 p^2\right) = a_4 \ln\left(\frac{2q/pf_0 + a_1 - a_3}{2q/pf_0 + a_1 + a_3}\right) + \ln a_5$.	Four constants (3D criterion)	Analytical and experimental examination on yield surface (true triaxial test).

Figure 6.21 Summary of rock failure criteria up to 1988 (from Pan, 1989).

7 Discontinuities

It is the existence of discontinuities in a rock mass that makes rock mechanics a unique subject. The word 'discontinuity' denotes any separation in the rock continuum having effectively zero tensile strength and is used without any genetic connotation (cf. the words 'joint' or 'fault' which describe discontinuities formed in different ways).

The material comprising the intact rock is natural and has been subjected in most cases to millions of years of mechanical, thermal and chemical action. During these processes, the discontinuities have been introduced into the rock by geological events, at different times and as a result of different stress states. Very often, the process by which a discontinuity has been formed (e.g. a joint which has been pulled open or a fault which has been sheared) may have implications for its geometrical and mechanical properties, and so it is always important to have an understanding of the formation of discontinuities using structural geology principles (see Price and Cosgrove 1990).

In the engineering context here, the discontinuities can be the single most important factor governing the deformability, strength and permeability of the rock mass. Moreover, a particularly large and persistent discontinuity could critically affect the stability of any surface or underground excavation. For these reasons, it is necessary to develop a thorough understanding of the geometrical, mechanical and hydrological properties of discontinuities and the way in which these will affect rock mechanics and hence rock engineering.

In this chapter these subjects will be discussed explicity; in Chapter 8 the discontinuity properties will be incorporated in a study of rock masses; in Chapter 9 discontinuity networks and the associated secondary permeability will be explained; and in Chapter 10 inhomogeneity and anisotropy are highlighted, remembering that these are often caused by variations in the occurrence of discontinuities.

7.1 The occurrence of discontinuities

In Fig. 7.1, we illustrate just two examples of discontinuous rock masses. In fact, all rock masses are fractured, and it is a very rare case where the spacings between discontinuities are appreciably greater than the dimensions of the rock engineering project. In Fig. 7.2, we illustrate the fact

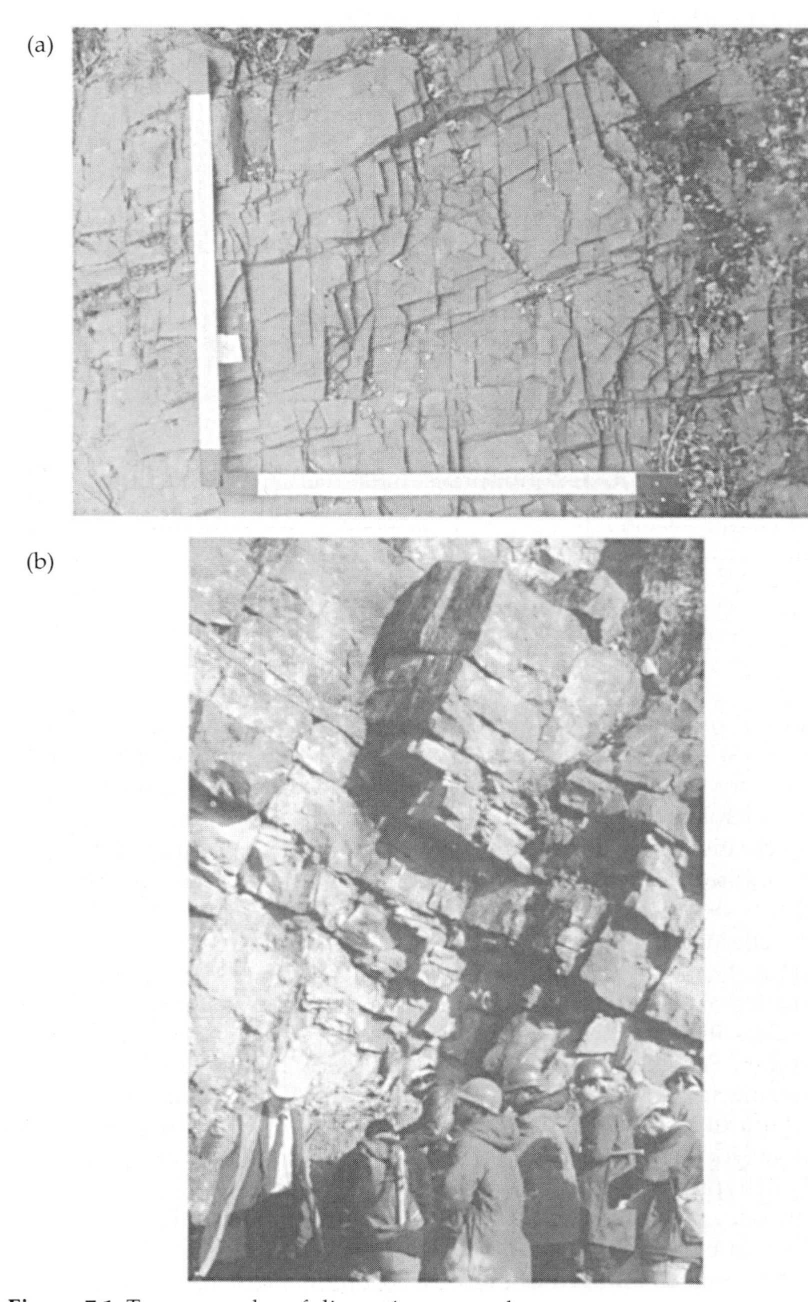

Figure 7.1 Two examples of discontinuous rock masses.

that very often major discontinuities delineate blocks within the rock mass, and within these blocks there is a further suite of discontinuities. Such hierarchial systems of discontinuities may well be much more complex, but this will not affect the ideas presented in this chapter. The engineer should, however, realize the significance of, for example, the likelihood that most of the smaller discontinuities could either terminate within one of the larger blocks or at the boundary of a block. Also, as illustrated by the sketched outlines of the engineered structures shown in Fig. 7.2, in general we might expect that a relation of the form

$$\text{stability} \propto \frac{1}{\text{number of discontinuities}} \propto \frac{1}{\text{engineering dimension}}$$

should exist. This idea is intimated in Fig. 7.2, where schematic outlines of a borehole, tunnel and large excavation have been overlain on a photograph of a rock face.

Another factor which should be considered, and which has influenced development of the study of discontinuities, is to what extent they can be sampled. From Fig. 7.1 it is clear that a great deal of information can be obtained by field measurement on a rock exposure, but even this only provides an essentially two-dimensional sample slice through the three-dimensional rock mass. In practice, we would like to have at least two such exposures at different orientations to feel confident that some estimate of the three-dimensional nature of the rock mass structure was being obtained. Unfortunately, it is often the case that no such exposures are available before the construction of a particular project. In this situation, we must rely on borehole core retrieved during the drilling process, scans using borehole television cameras or indirect methods using geophysical techniques.

By far the most widely used method is to study the borehole core, but this

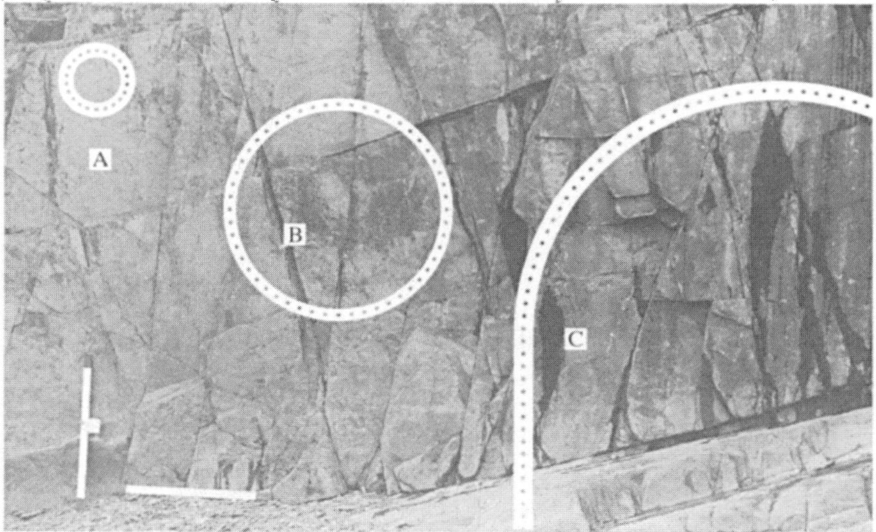

Figure 7.2 Discontinuities in rock and the engineered structure (A, borehole; B, tunnel; C, large excavation).

is an essentially one-dimensional sample through the three-dimensional rock mass, with obvious limitations. For example, a borehole core will provide an excellent means of examining the discontinuity occurrence and hence frequency in the borehole direction, but will provide little information on the lateral extent of the intersected discontinuities. Thus, a key aspect of rock structure interpretation will be the extrapolation to three-dimensional properties from either one- or two-dimensional measurements.

One of the most fundamental aspects of discontinuity occurrence is the mean value and distribution of spacings between discontinuities, and the associated indices of **discontinuity frequency** and **Rock Quality Designation**. These and several other aspects of rock structure geometry will be discussed in the next section.

7.2 Geometrical properties of discontinuities

In Fig. 7.3, we present a schematic representation of two planes within a rock mass. There are no assumptions about whether these planes are real or imaginary surfaces or sections. Also, the borehole or scanline could be real, or postulated solely for the purpose of analysis. This diagram shows the main features of rock mass geometry with, in particular, the following parameters being illustrated:

1. **Spacing and frequency:** spacing is the distance between adjacent discontinuity intersections with the measuring scanline. Frequency (i.e. the number per unit distance) is the reciprocal of spacing (i.e. the mean of these intersection distances).
2. **Orientation, dip direction/dip angle:** the discontinuity is assumed to be planar and so the dip direction (the compass bearing of the steepest line in the plane) and the dip angle (the angle that this steepest line makes to the horizontal plane) uniquely define the orientation of the discontinuity.
3. **Persistence, size and shape:** the extent of the discontinuity in its own

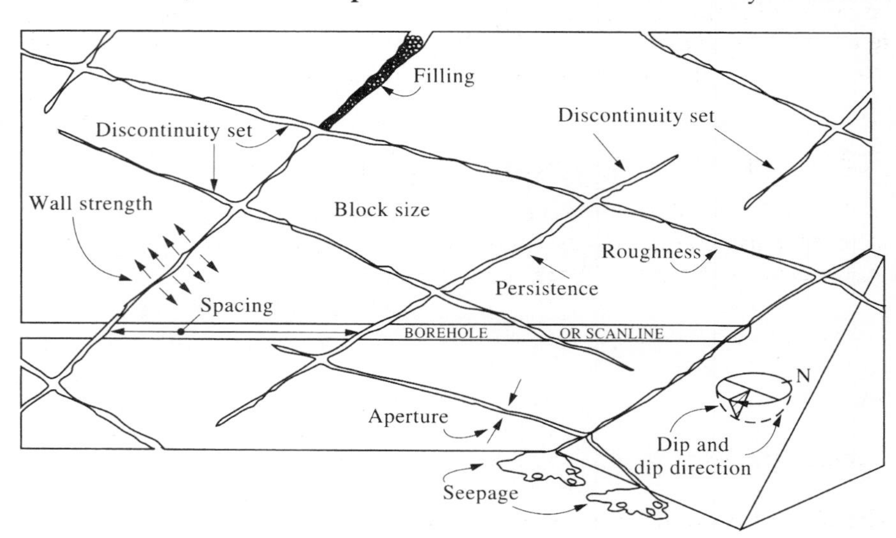

Figure 7.3 Schematic of the primary geometrical properties of discontinuities in rock (from Hudson, 1989).

plane, incorporating factors such as the shape of the bounded plane and the associated characteristic dimensions (e.g. the discontinuities could be assumed to be circular discs for the purpose of analysis and sampling).

4. **Roughness:** although discontinuities are assumed to be planar for the purposes of orientation and persistence analysis, the surface of the discontinuity itself may be rough. Discontinuity roughness may be defined either by reference to standard charts or mathematically.
5. **Aperture:** the perpendicular distance between the adjacent rock surfaces of the discontinuity. This will be a constant value for parallel and planar adjacent surfaces, a linearly varying value for non-parallel but planar adjacent surfaces, and completely variable for rough adjacent surfaces.
6. **Discontinuity sets:** discontinuities do not occur at completely random orientations: they occur for good mechanical reasons with some degree of 'clustering' around preferred orientations associated with the formation mechanisms. Hence, it is sometimes convenient to consider the concept of a discontinuity set (which consists of parallel or sub-parallel discontinuities), and the number of such sets that characterize a particular rock mass geometry.
7. **Block size:** as is illustrated in Fig. 7.3 and depending on the previously described characteristics, rock blocks can be present. In terms of excavation and support, it is helpful to have an estimate both of the mean block size and the block size distribution, which is an *in situ* analogue of the particle size distribution used in soil mechanics.

The degree of work that has been devoted to providing techniques for measurement, data reduction and presentation associated with each of these seven main aspects of the geometrical properties has been highly variable. There is no standardized, or indeed 'correct', method of measuring and characterizing rock structure geometry, because the emphasis and the accuracy with which the separate parameters are specified will depend on the engineering objective. Therefore, initially we will describe the salient features of these parameters, taking each individually, and later invoke this information with respect to specific engineering themes.

7.2.1 Discontinuity spacing and frequency

In Fig. 7.4, we illustrate a sampling line through a rock mass, which intersects a number of discontinuities. The length of the sampling line is L metres, the number of discontinuities it intersects is N, and thus,

$$\text{discontinuity frequency, } \lambda = N/L \text{ m}^{-1}$$

and

$$\text{mean spacing, } \bar{x} = L/N \text{ m.}$$

The discontinuity frequency, being the number of fractures per metre, is the reciprocal of the mean spacing.

We can also consider the distribution of the individual spacings

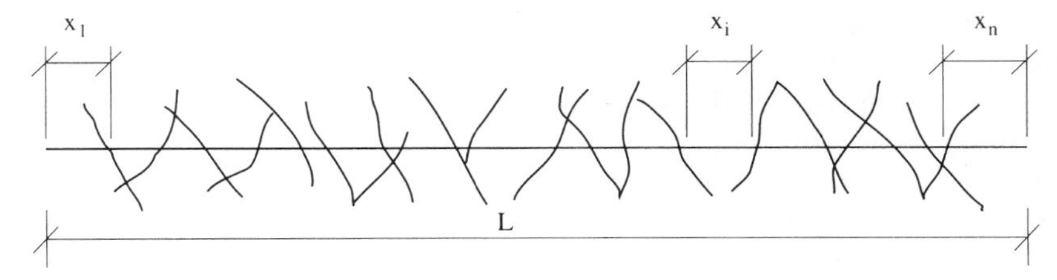

Figure 7.4 Quantifying discontinuity occurrence along a sampling line.

between fractures, denoted by x_i in Fig. 7.4. When a sufficiently large sample of these individual spacing values (preferably more than 200 individual measurements) is plotted in histogram form, a **negative exponential** distribution is often evident, as illustrated in Fig. 7.5. It should be noted that the general trend of this histogram is for there to be many small spacing values and few very large spacing values in the distribution.

We utilize histograms when there is a finite number of values (i.e. the discrete case) and the data can be assigned to chosen class intervals. On approaching the limiting case of an infinite number of spacing values and infinitely small class intervals, the histogram tends to a continuous curve, which can be expressed as the **probability density function**.

$$f(x) = \lambda e^{-\lambda x}.$$

Note that the mean of this distribution is $1/\lambda$, and the standard deviation *is also* $1/\lambda$. It is a one-parameter distribution, with the mean and the standard deviation being equal.

This negative exponential distribution is the spacing distribution associated with the **Poisson process** of random events. However, it must be emphasized that we are not treating the occurence of discontinuities as random events, rather we are using the equation because we expect it to apply and field data support its use. In some statistical cases, such as repeated sampling to determine the mean of a population *with any distribution*, there is convergence of the results to the well-known normal distribution: this is called the **central limit theorem**. The negative exponential distribution is an analogue to the normal distribution, except that it is the distribution to which the spacing values converge when successive spacing distributions *of any type* are superimposed on the sampling line. In other words, the occurrence of the negative exponential distribution is expected as the result of a suite of superimposed geological events, each of which produces fracturing of a given distribution. It should be noted that fracturing is *deterministic* in the sense that it occurs as the result of direct mechanical causes, but that in aggregate a *probabilistic* model is mathematically convenient to use, as will be explained in the next sub-section.

7.2.2 The Rock Quality Designation, RQD

As is evident in Fig. 7.5, a *natural clustering* of discontinuities occurs through this genetic process of superimposed fracture phases, each of which could have a different spacing distribution. An important feature for engineering

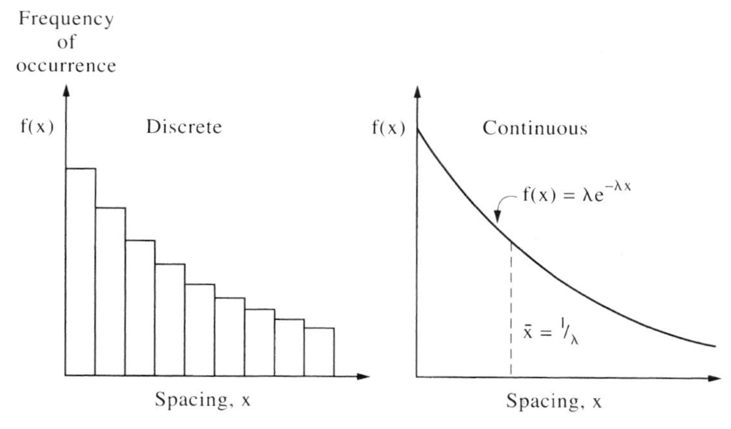

Figure 7.5 The negative exponential distribution of discontinuity spacing values.

is that, although there are more small spacing values than large spacing values, a single large spacing value can be a greater proportion of the scanline length than many small spacing values added together. However, the discontinuity frequency (or mean spacing value) does not give any indication of this phenomenon. For this reason, Deere (1963) developed, for borehole core, the concept of the **Rock Quality Designation**, universally referred to as **RQD**. This is defined as the percentage of the sampling line (or borehole core) consisting of spacing values (or intact lengths of rock) greater than or equal to 4 inches (or 100 mm). Expressed mathematically,

$$\text{RQD} = 100\sum_{i=1}^{n} \frac{x_i}{L}\%$$

where x_i = spacing values greater than 0.1 m, and n is the number of these intersected by a borehole core or scanline of length L.

We are now in a position to relate the discontinuity frequency to the RQD utilizing the negative exponential distribution of spacing values. In Fig. 7.6 the shaded area shows those spacing values above the RQD **threshold value**. We can find the RQD by establishing the percentage of the sampling line that is represented by the spacing values in the shaded area. *This is* **not** *represented just by the shaded area expressed as a percentage of the total area under the curve,* because we must take into account the different contributions made by the different spacing values, as presented below.

The probability of the length of a piece of intact core being between x and $x + \delta x$ is $f(x)\delta x$. Given that the total number of pieces of core is N, then the total number of pieces of core in this interval is $Nf(x)\delta x$ and the length of all of these pieces is $Nxf(x)\delta x$. We can find the total length of all pieces with all values of x by summing:

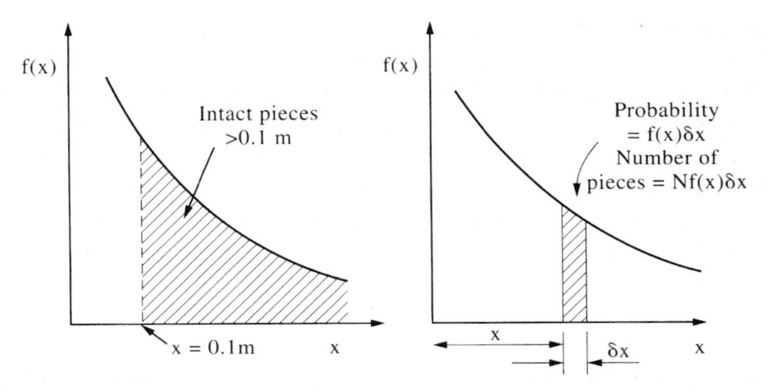

Figure 7.6 The contribution to the core length made by intact pieces of rock greater than 0.1 m in length.

$$\text{total length} = \sum_{i=1}^{n} N x_i f(x)\,\delta x.$$

For RQD, we consider only those pieces of core with a length greater than 0.1 m and so, in the limit for the continuous case, we have

$$\text{total length of pieces} \geq 0.1 \text{ m in length} = \int_{x=0.1}^{x=\infty} N x f(x)\,\mathrm{d}x.$$

Now, substituting this into the previous (discrete) expression for RQD, we find that

$$\mathrm{RQD}^* = 100(1/L) \int_{x=0.1}^{x=\infty} N x f(x)\,\mathrm{d}x$$

where we have used the notation RQD* to represent the theoretical RQD calculated from the distribution of spacing values.

However, we know that $N/L = \lambda$, the discontinuity frequency, so for the general case *with any distribution of spacing values* we have

$$\mathrm{RQD}^* = 100\lambda \int_{x=0.1}^{x=\infty} x f(x)\,\mathrm{d}x.$$

We can evaluate this integral explicitly for the negative exponential distribution. Thus, **for a negative exponential distribution of spacing values,**

$$\mathrm{RQD}^* = 100\lambda^2 \int_{x=0.1}^{x=\infty} x e^{-\lambda x}\,\mathrm{d}x$$

which gives

$$\mathrm{RQD}^* = 100(0.1\lambda + 1)\mathrm{e}^{-0.1\lambda}.$$

Note that by using the negative exponential distribution in this way, the RQD* can be expressed *solely* in terms of the discontinuity frequency. In Fig. 7.7, we have plotted the RQD* against both the mean spacing, $\bar{x}$, and the frequency, λ. From these curves, it can be seen that the RQD* is most sensitive for spacing values in the range 0–0.3 m: above a mean spacing value of 0.3 m, the RQD* is always greater than 95%. It can also be seen, from the plot of RQD* versus λ, that the relation is approximately linear in the range $6 < \lambda < 16$. This leads to the simplified approximate formula

$$\text{RQD}^* = -3.68\lambda + 110.4.$$

To overcome the lack of sensitivity of RQD* for large spacing values, we can use any **threshold value**, t, for calculating RQD*, rather than 0.1 m. This results in the general formula

$$\text{RQD}^* = 100(\lambda t + 1)e^{-\lambda t}.$$

We have plotted this relation for t=0.1 m (the conventional threshold value), t = 0.3 m and t = 1.0 m in Fig. 7.7. For rock masses with large mean discontinuity spacing values, it can be helpful to adopt a two-tier approach to the RQD calculation, utilizing the conventional threshold of 0.1 m and a higher threshold of, say, 1.0 m.

7.2.3 Variation of discontinuity frequency with sampling line direction

From the definition of discontinuity frequency, as illustrated in Fig. 7.4, the measured discontinuity frequency is expected to vary with the direction of the sampling line relative to the orientation of the discontinuities. Because of the complexity of all potential three-dimensional discontinuity patterns, we introduce the method of calculating this variation, firstly with reference to a single set of planar, parallel and persistent discontinuities, and subsequently for any number of discontinuity sets.

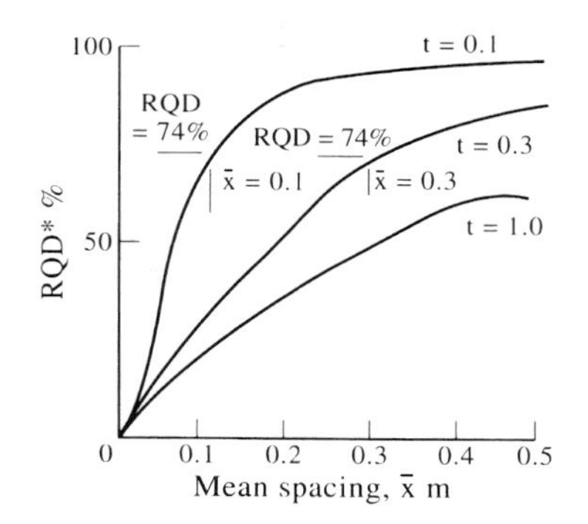

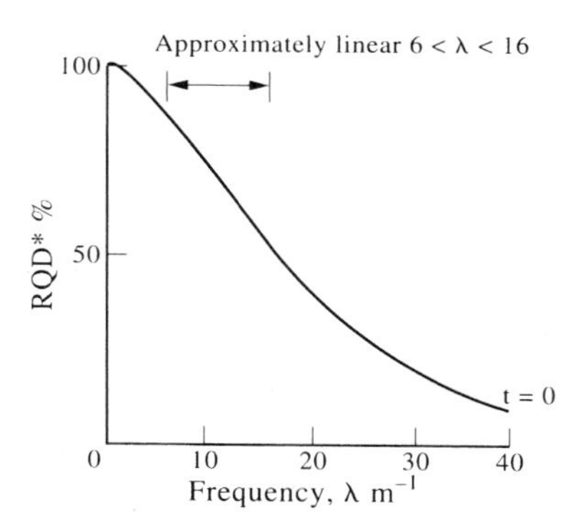

Figure 7.7 The relation between RQD* and both mean spacing and frequency for a negative exponential distribution of spacing values, with different RQD* threshold values.

In Fig. 7.8, a sampling line intersects the traces of one idealized set of discontinuities: this may be a scanline on a rock surface, for example. Note that the discontinuity traces are parallel, persistent and linear, but are not regularly spaced. We assume that the length of the line perpendicular to the discontinuities has length L and intersects N discontinuities. The discontinuity frequency along the set normal, λ, is equal to N/L. Along the scanline, inclined at angle θ to the discontinuity set normal, the discontinuity frequency is calculated by the same method: for the same number of intersected discontinuities, N, the length of the line is $L/\cos\theta$ and the discontinuity frequency along the scanline, λ_s, is given by

$$\lambda_S = \frac{N}{L/\cos\theta} = \frac{N}{L}\cos\theta = \lambda\cos\theta.$$

The discontinuity frequency is always positive and therefore we have

$$\lambda_s = \lambda\,|\cos\theta\,|.$$

Because λ_s is always positive, the discontinuity frequency, having magnitude and direction, may be considered to be two vectors rather than one, as illustrated in the right-hand part of Fig. 7.8. Note that the fundamental set frequency is given by the maximum distance from the origin to the outer dashed circle, i.e. in the 0° and 180° directions.

Apart from the fact that discontinuity frequency must always be positive, it can be resolved like a force as illustrated in Fig. 7.8. As the scanline is rotated from $\theta = 0°$ to $\theta = 90°$, λ_s varies from its maximum value of λ to zero: obviously, $\lambda_s = 0$ occurs when the scanline is parallel to the discontinuities. However, as θ is increased beyond 90°, the discontinuity frequency increases again, to a maximum of λ when $\theta = 180°$. The resultant **cusp** in the locus at $\theta = 90°$ has an important effect as we progress to considering more than one discontinuity set.

The case illustrated in Fig. 7.8 is the most anisotropic case possible for the variation in discontinuity frequency, because the ratio of the greatest to the least frequency is infinite. Note also that the directions of maximum

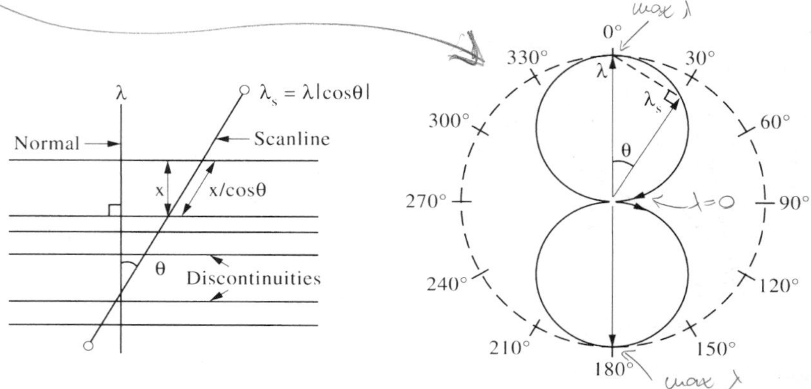

Figure 7.8 Variation in discontinuity frequency for a sampling line passing through a single set of discontinuities—two-dimensional case.

and minimum discontinuity frequency are perpendicular for the case of one set: this is the only circumstance in which the two are perpendicular.

If now two discontinuity sets are considered, as illustrated in Fig. 7.9, the contribution from each set is resolved onto the sampling line as

$$\lambda_s = \lambda_1 \mid \cos \theta_1 \mid + \lambda_2 \mid \cos \theta_2 \mid.$$

It can be seen from the associated polar plot that the rock mass is now far less geometrically anisotropic and that the directions of maximum and minimum discontinuity frequency are no longer orthogonal.

This procedure can be readily extended to any number of discontinuity sets, n, with the result

$$\lambda_S = \sum_{i=1}^{n} \lambda_i \mid \cos \theta_i \mid$$

where λ_i and θ_i are the fundamental set frequency and the angle between the set normal and the sampling direction, respectively, for the ith set. In Fig. 7.10, the progression from one set to an infinite number of sets is illustrated for mutually symmetric set normals and where each set frequency is the reciprocal of the number of sets present (in order to keep the areal density of discontinuities constant).

The progression from anisotropy to isotropy is elegantly demonstrated in the top row of polar plots as the discontinuity frequency locus changes from two circles (for one set), through various lobular shapes, to a single circle (for an infinite number of sets). The diagram shows how a rock mass with, say, four or five similar and equally spaced sets may be regarded as effectively isotropic. The discontinuity frequency variation has been shown for symmetrically orientated sets possessing equal frequencies: there will obviously be more anisotropy in the less-ideal and real cases.

The lower row of stereographic plots in Fig. 7.10 shows the same variation for the same sets, but for the sampling direction being variable in three dimensions. The diagrams are contoured lower-hemisphere projections: a single set results in a single 'peak', and an infinite number

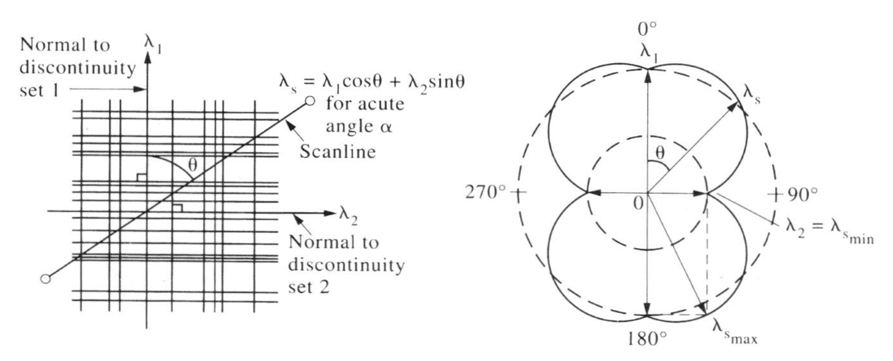

Figure 7.9 Variation in discontinuity frequency for a sampling line passing through two sets of discontinuities—two-dimensional case.

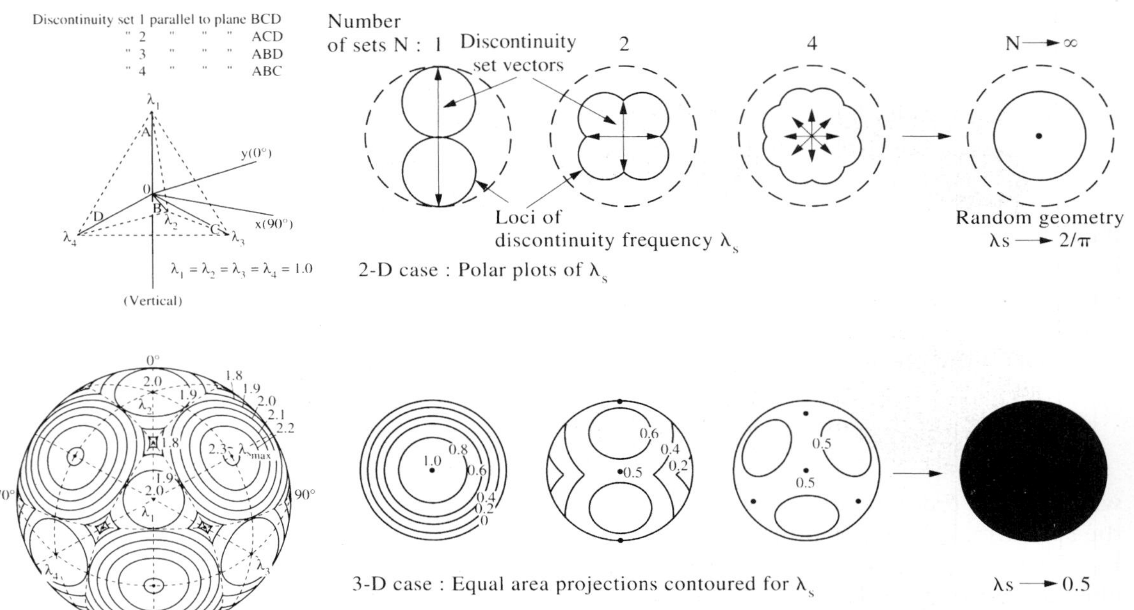

Figure 7.10 Variation in discontinuity frequency for a sampling line passing through multiple sets of symmetrically orientated discontinuities—two- and three-dimensional cases illustrated via polar and lower-hemispherical projection plots.

of sets results in a 'peneplain'. The variation for one, two and three discontinuity sets (which are perpendicular and have equal frequencies) is represented by the three-dimensional polar plots shown as isometric sketches in Fig. 7.11. The format of this diagram is a 3×3 matrix with the individual set loci along the leading diagonal and the binary combinations of sets shown in the off-diagonal positions. The principle of this type of presentation is explained in Chapter 14. The shaded areas in each locus are equivalent to the 'horizontal' two-dimensional loci shown in Figs 7.9 and 7.10.

Because of this variation in discontinuity frequency with direction, the RQD will also vary with direction, bearing in mind the relations given in Section 7.2.2. So, a statement such as 'the rock mass has an RQD of 80%' is inadequate: an RQD value can only apply to measurements made in a specific direction. In fact, the RQD is a vector-like quantity with both magnitude and orientation, directly calculable fom the discontinuity frequency. The variation with direction is particularly relevant to site investigation results containing the ubiquitous parameter RQD with the possibility that the RQD measured in a vertical borehole has been applied to, say, a horizontal tunnel.

7.2.4 Discontinuity orientation, discontinuity sets and block sizes

If we assume that a discontinuity is a planar feature, then its orientation can be uniquely defined by two parameters: dip direction and dip angle.

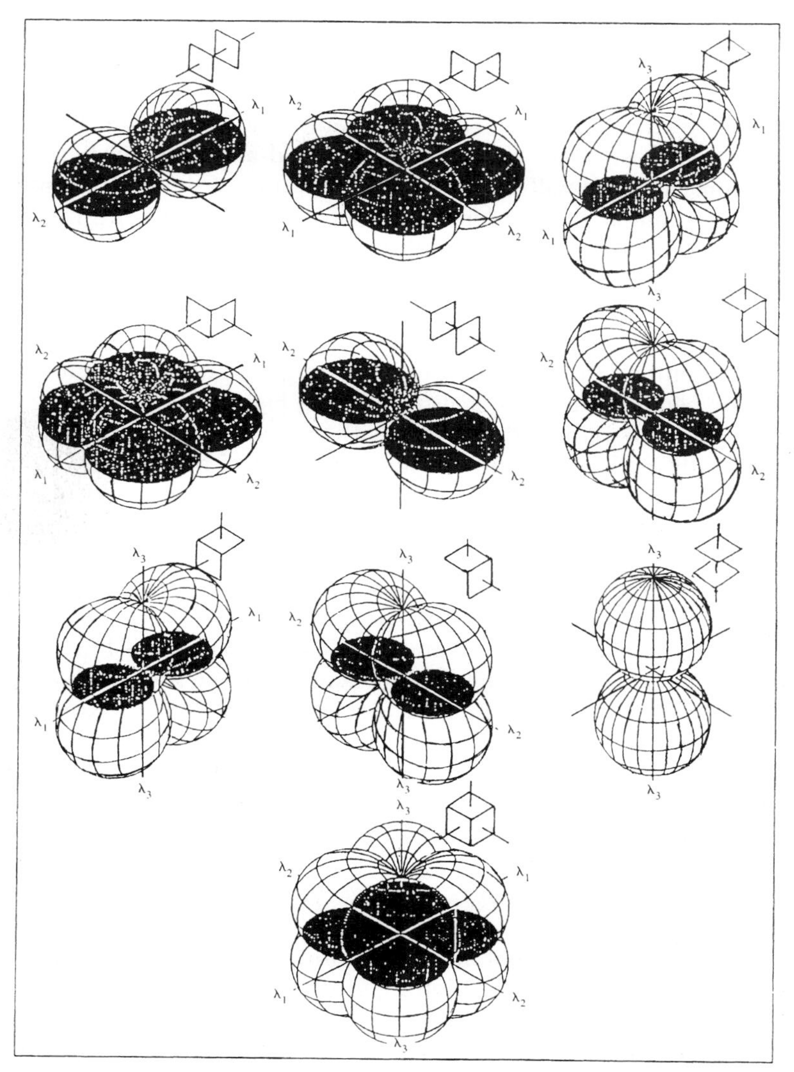

Figure 7.11 Variation in discontinuity frequency for a sampling line passing through one, two and three sets of mutually perpendicular discontinuities—three-dimensional case (from Antonio, 1985).

The dip angle is defined as the steepest line in the plane, i.e. the line down which a ball would roll; the dip direction is the compass bearing of this line, measured clockwise from North. Here, we will use the notation α/β for dip direction/dip angle: the reader should be aware that there are a number of alternative nomenclatures in current usage. In the field, the use of a geological compass will provide direct readings of the dip direction and the dip angle.

In general, a large number of α/β data pairs will be obtained during the course of a field survey. Consequently, it is useful to present these data in a graphical form to aid rapid assimilation and understanding. Thus, we wish to present information about planes in three-dimensional space on a two-dimensional piece of paper. There are many possible techniques for this. In fact, the problem of plotting lines on the Earth's surface on a sheet of paper has been a problem since the early days of navigation.

During the development of rock mechanics and rock engineering, there has been almost total acceptance of **equal-angle lower-hemisphere projection**. Here we will give a basic description of the plotting method, sufficient to allow the reader who is not familiar with the method to follow the discussion; a monograph on the subject has been produced by Priest (1985) and more detail is given here in Appendix B.

In Fig. 7.12, we show the dip direction plotted as the compass bearing and the dip angle plotted *inwards* from the perimeter of the projection. This defines a point on the projection representing the line of maximum dip of the plane being plotted. As is also shown in Fig. 7.12, another line in the plane is the strike line, i.e. the line with zero dip: this is plotted as two diametrically opposed points on the perimeter of the projection. In the same way, all lines in the plane can be plotted using their particular α/β values, resulting in the **great circle** shown in the figure. Thus, a line in the plane is plotted as a point, and the plane itself is plotted as a curve (for equal-angle projection, it is an arc of a circle).

An alternative method of uniquely specifying the plane is to plot the position of a line which is perpendicular to the plane: this line is known as the **normal** and the associated point plotted on the projection is known as the **pole**. The pole of the plane is also plotted in Fig. 7.12. Note that the following two relations exist between the line of maximum dip and the normal:

$$\alpha_{\text{normal}} = \alpha_{\text{dip}} \pm 180^\circ$$

$$\beta_{\text{normal}} = 90^\circ - \beta_{\text{dip}}.$$

Generally, we wish to plot many discontinuity planes, which means that plotting poles is preferred to plotting great circles. Also, once many poles have been plotted on the projection, the basic rock structure can be considered in terms of the clustering of these normals: this is conventionally studied by contouring the projection to locate the densest regions. More advanced techniques involve various clustering algorithms, based either on

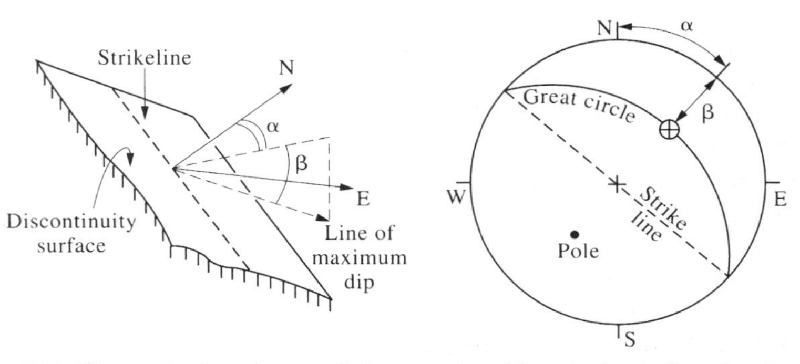

Figure 7.12 Discontinuity plane and the associated hemispherical projection.

statistical or fuzzy-set methods (Harrison, 1993). In Fig. 7.13, we show an example of plotted data, the resulting contoured plot and the main set directions.

It is common to idealize a set of discontinuities as a collection of parallel, persistent and planar features. It is clear from Fig. 7.13 that in practice not only might a set consist of sub-parallel discontinuities, but also there might be difficulty in even distinguishing to which set a particular discontinuity belongs. Moreover, each discontinuity will have other geotechnical parameters of interest apart from its geometry, and it is likely that in the future a more comprehensive analysis of the clustering will evolve.

One subject where the concept of discontinuity sets is important is in the formation of rock blocks and the distribution of their sizes. With the knowledge of the fracture frequency and orientation of the discontinuity sets in a rock mass, it is possible to determine a three-dimensional block volume distribution and associated two-dimensional block area distributions encountered on any plane through the rock mass. An example of the area size distribution is shown in Fig. 7.14, which has been generated by assuming that in one case the discontinuities are randomly orientated and positioned, and in the other case that there are two orthogonal sets, each with negative exponential spacing distributions.

In both cases, it is possible to calculate the probability density function for the tessellated area as illustrated. Note that for these two cases, with the same two-dimensional fracture density, the block area distributions are very similar, indicating that, for this particular case, the orientation of the discontinuities does not significantly affect the sizes of the blocks.

This type of analysis is important in rock engineering design, bearing in mind the earlier discussion on the significance of the scale of the engineering project in relation to the rock mass geometry, whether this be considered along a line (i.e. a borehole or scanline), on a plane (i.e. one of the walls of an excavation) or within the rock volume. (Block volume distributions in the context of excavation are discussed in Chapter 15.)

7.2.5 Persistence, roughness and aperture

To some extent, the parameters of persistence, roughness and aperture reflect the deviation from the assumption of the idealized discontinuities discussed in Section 7.2.4. Note that even the plotting of a single great circle

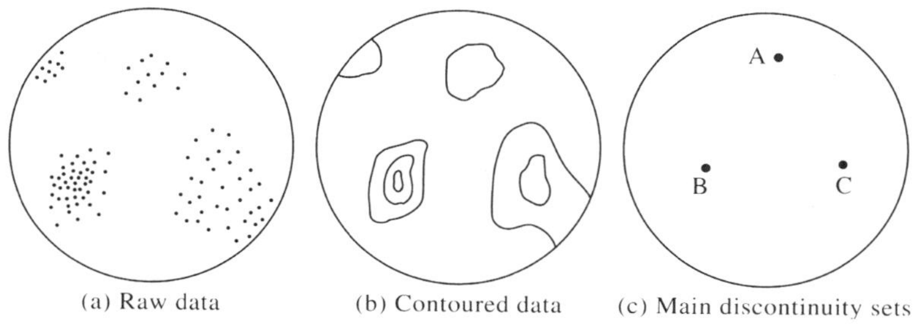

Figure 7.13 Discontinuity orientation data plotted on the lower-hemispherical projection.

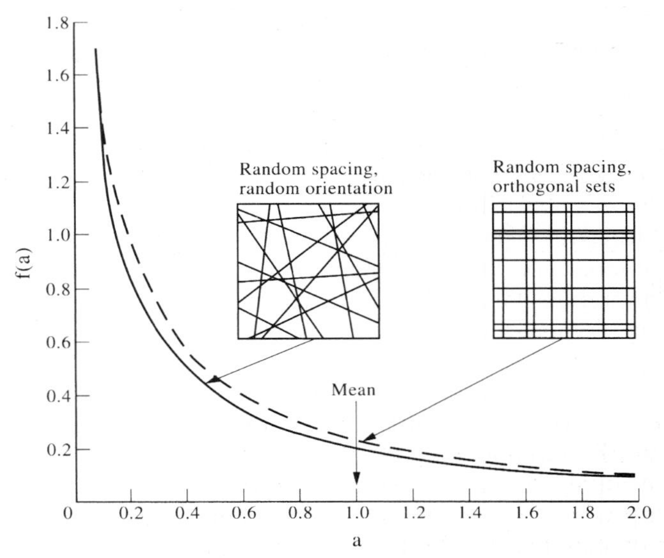

Figure 7.14 Probability density functions of rock block area sizes for rock masses containing random discontinuities and two orthogonal sets.

or pole to represent a discontinuity does assume that it is a perfect plane but that the type of clustering shown in Fig. 7.13 could occur from many measurements made on an undulating surface.

The word **persistence** refers to the lateral extent of a discontinuity plane, either the overall dimensions of the plane, or whether it contains 'rock bridges'. In practice, the persistence is almost always measured by the one-dimensional extent of the trace lengths as exposed on rock faces. This is illustrated in Fig. 7.15. It is clear from this figure that no direct estimation of persistence is possible from borehole core, although geological inference based on the type of discontinuity observed is possible. Also, the distribution of trace lengths obtained from measurements made on an exposure will depend to a great extent on the orientation of the rock face, on the associated orientation of the scanline, and whether the measurements are either *truncated* (values below a certain length are omitted) or *censored* (large values are either unobtainable because of limited rock exposure or because of equipment limitations).

In practice, truncation and censorship always occur, contributing to the **bias** in the measurements. For example, there has been considerable discussion in the literature as to whether the distribution of trace lengths has a similar negative exponential distribution to that of spacings discussed earlier, or is a log-normal distribution. It is likely that some of the differences have arisen from trace length sampling bias, but further work is needed to clarify the situation. At the moment, there is no clear and coherent method of measuring trace length, despite its crucial importance.

The word '**roughness**' is used to denote the deviation of a discontinuity surface from perfect planarity, which can rapidly become a complex mathematical procedure utilizing three-dimensional surface characterization techniques, whether these be by polynomials, Fourier series, noise

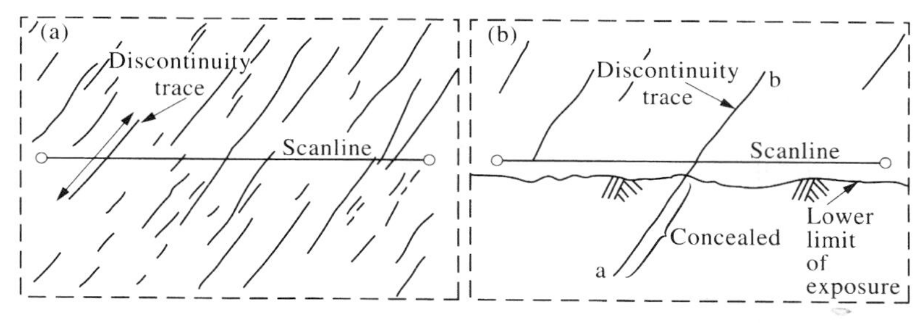

Figure 7.15 Diagrammatic representation of discontinuity traces intersecting a scanline set up on a rock face.

waveforms or fractals. From the practical point of view, only one technique has any degree of universality and that is the **Joint Roughness Coefficient** (JRC) developed by Barton and Choubey (1977). This method involves comparing a profile of a discontinuity surface with standard roughness profiles and hence assigning a numerical value to the roughness. The chart of standard profiles is shown in Fig. 7.16.

Despite the obvious limitations of reducing all roughness information to a single scalar value, the possibly subjective nature of the assessment and its wholly empirical nature, the JRC profiles have proved to be of significant value in rock engineering. The geometrical roughness is naturally related to various mechanical and hydraulic properties of discontinuities. On the purely geometrical side, it is possible to predict the amplitude of asperities from the JRC and profile length. On the mechanical side, shear strength can be predicted via JRC and other properties. Moreover, there are obvious implications for aperture and variation in aperture as a function of discontinuity roughness.

The **aperture** is the distance between adjacent walls of a discontinuity, i.e. it is the openness of the discontinuity. This parameter has mechanical and hydraulic importance, and a distribution of apertures for any given discontinuity and for different discontinuities within the same rock mass is to be expected. A limited estimate of the JRC is possible from borehole core, but in general no indication of aperture is possible from the core itself, except where the fractures are incipient and the core is not separated.

Current research in the hydraulic context indicates that a discontinuity cannot be approximated as two parallel planes because of the phenomenon of channel flow, where the fluid mainly flows through certain channels within the discontinuity created by tracks of larger local apertures.

7.2.6 Statistical analysis and practical examples

The reader will have noted the recurring theme that most of the topics being discussed are not deterministically tractable due to insufficient three-dimensional characterization of the rock mass structure. Consequently, statistical techniques in the data reduction, presentation and analysis are helpful. We may not be able to specify all the discontinuities in the rock mass, but we might be able to provide an excellent engineering approximation via statistical generators which will allow repeated

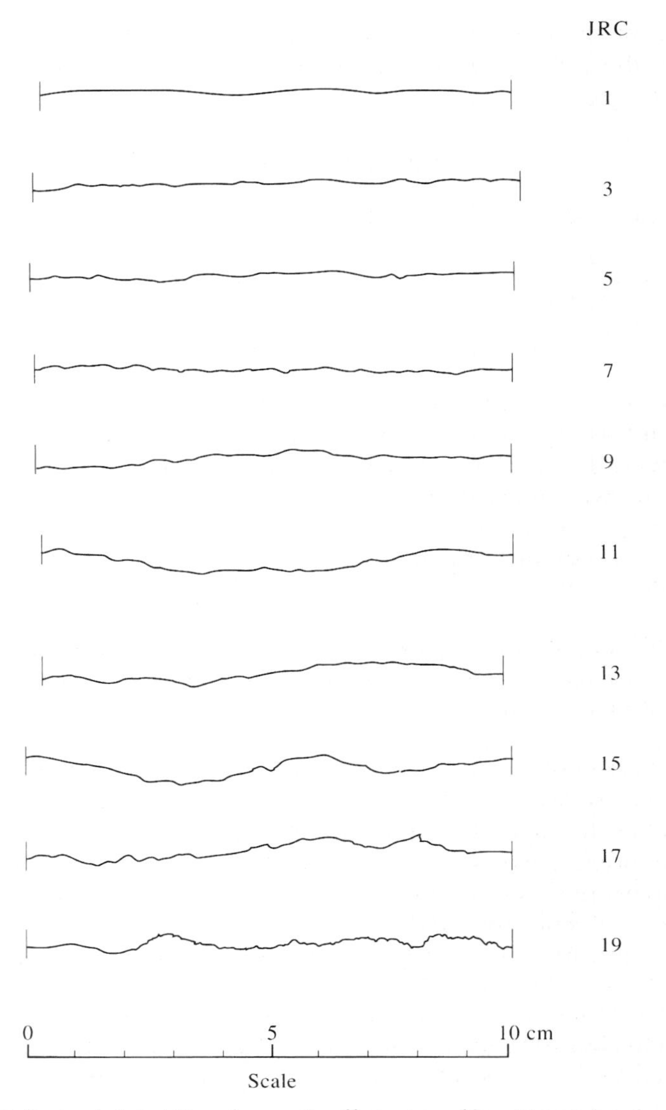

Figure 7.16 Barton's Joint Roughness Coefficient profiles. *Note reduced scale.*

simulations of sufficiently similar synthetic rock masses. Perhaps we can use well-known statistical techniques to answer such questions as, 'How long should a scanline be such that a reliable estimate of the mean discontinuity spacing value is obtained?' We have found, as discussed earlier, that the Poisson process and the associated negative exponential spacing distribution are good theoretical models for discontinuity occurrence, so it is appropriate to begin this explanation of the value of statistical techniques with the Poisson process.

Poisson process. Describing rock mass structure in terms of the Poisson process (and hence implicitly assuming that the position of a discontinuity along a scanline is independent of the position of other discontinuities), we can give the probability that k discontinuities will intersect a scanline interval of length x as

$$P(k, x) = e^{-\lambda x}(\lambda x)^k/k!$$

For example, if the discontinuity frequency, λ, along the scanline is 8.43 m^{-1}, then the probability that *exactly* two discontinuities will be intersected in a 0.3 m scanline interval is given by this equation as

$$P(2,0.3) = e^{-(8.43 \cdot 0.3)}(8.43 \cdot 0.3)^2/2! = 0.255.$$

Hence, in about a quarter of all such possible intervals, two discontinuities will be intersected. Clearly, by repeated use of this formula we can determine the probability of such events as 'two or less', 'less than two' or 'more than two' discontinuities being intersected:

$$P(\leq k,x) = \sum_{l=0}^{k} P(l,x) \quad P(< k,x) = \sum_{l=0}^{k-1} P(l,x) \quad P(> k,x) = 1 - \sum_{l=0}^{k} P(l,x).$$

The usefulness of such calculations in designing, say, the length of rock bolts using criteria such as 'rockbolts should intersect more than three discontinuities' is evident.

Cumulative probability distributions and the central limit theorem. The same type of information is of interest in other areas where we may wish to know, for example, for the design of tunnel boring machine bearings, what is the probability that the compressive strength of the rock will exceed a certain value. This can be established directly from a data set using the **cumulative probability distribution**, as illustrated in Fig. 7.17. In the left-hand part of this figure, the results of a strength testing programme are tabulated, and these results are shown in the form of a cumulative probability distribution in the central part of the figure. Extracting values from the distribution allows statistical questions to be answered: in the example shown here, what is the probability that the strength will be less

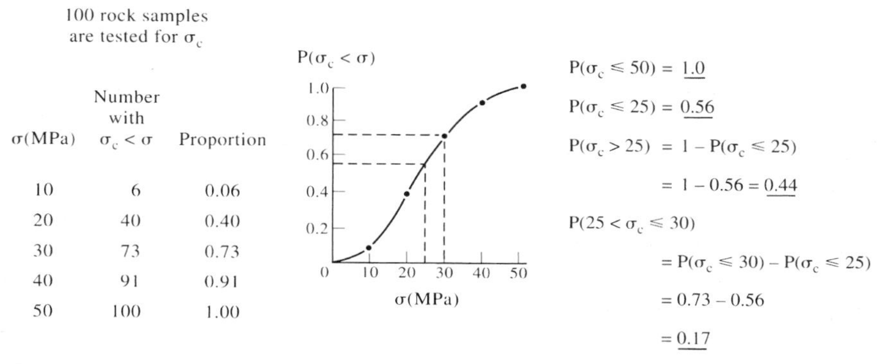

Figure 7.17 Estimating the probability that the compressive strength will be greater or less than certain values.

than or equal to 50 MPa, less than or equal to 25 MPa, greater than 25 MPa, or greater than 25 MPa and less than or equal to 30 MPa?

We can also establish, for example, how long a scanline should be in order to determine the discontinuity frequency, or mean spacing, to within specified tolerances. For this we utilize the **central limit theorem**, which states that the means of random samples of size N taken from a population of any distribution that has a mean $\bar{x}$ and a standard deviation σ will tend to be normally distributed with a mean $\bar{x}$ and a standard deviation of $\sigma/N^{1/2}$. However, in the case of a negative exponential distribution, *the standard deviation is equal to the mean*: they are both $\bar{x}$. Hence, utilizing the concept of **confidence intervals** in the normal distribution, we can find the appropriate scanline length for any desired confidence level.

As an example, let us say that we wish to determine the mean spacing $\bar{x}$ with an 80% confidence that the error will be less than ±20%. The 80% confidence implies that we require our sampled mean to be located within the zone which is 80% of the total area under the standardized normal probability density function. The half bandwidth of this zone for the standardized normal distribution is given by the standardized normal variable, $z = 1.282$ (found from statistical tables), so the half bandwidth for our parameters can then be calculated from $z\sigma/N^{1/2}$. This half bandwidth is equated to the allowable proportional error, ε, which in this example is 0.2, i.e. 20% from the true value of the mean. Then we calculate N from the equality

$$z\sigma/N^{1/2} = \varepsilon\bar{x}.$$

As $\sigma = \bar{x}$ for the negative exponential distribution, we have

$$N = z^2/\varepsilon^2.$$

In the example, $N = 1.282^2/0.2^2 = 41$.

Alternatively, for 90% confidence that we will be within a ±10% error bandwidth, we find that $N = 1.645^2/0.1^2 = 271$. These two examples are highlighted by the dashed lines in Fig. 7.18.

Sampling bias and mean orientation. From Section 7.2.3, we know that the discontinuity frequency, λ_s, along a line subtending an angle θ to the normal of a discontinuity set with frequency λ is $\lambda_s = \lambda \cos\theta$. As θ tends to 90°, so λ_s tends to zero. Consequently, as the scanline is rotated to become nearly parallel to discontinuities, so the number of sampled discontinuities per unit length of scanline reduces. This clearly introduces a **sampling bias** when λ is being estimated, resulting from the relative directions of the discontinuities and the scanline. The bias can be removed by using a weighting factor: this weights the number of intersected discontinuities associated with each set to give an effective number of intersected discontinuities and hence removes the bias. The weighting factor, w, is calculated from the expression $1/\cos\theta$, where θ is the angle between the normal to the discontinuity and the scanline for each set.

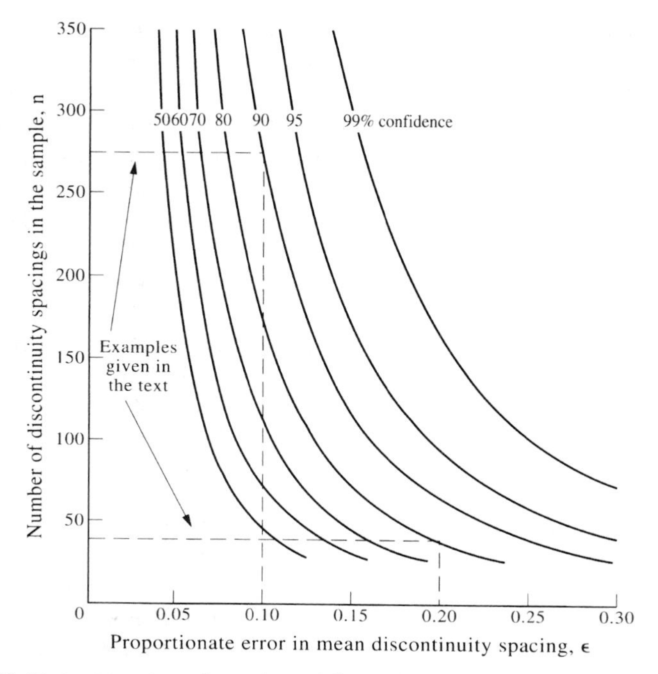

Figure 7.18 Determination of number of discontinuity spacings (and hence scanline length) to estimate the mean spacing value to within a given error at given confidence levels.

In Fig. 7.19, we show the direction cosines associated with a unit vector. These are required for numerical calculation of the mean orientation of a number of discontinuities, using the procedure outlined below. When many discontinuities with different orientations are intersected by a scanline and sampled, the mean dip direction and dip angle may be found by using the procedure outlined in tabular form in Fig. 7.20. This procedure corrects for orientational sampling bias through the introduction of weighted direction cosines. The first two columns are the dip direction and dip angle as measured, α, β, and the following two columns contain α_n, β_n, the trend and plunge of the normal to each discontinuity. The direction cosines of each of the normals are then evaluated in the next three columns using the formulae of Fig. 7.19.

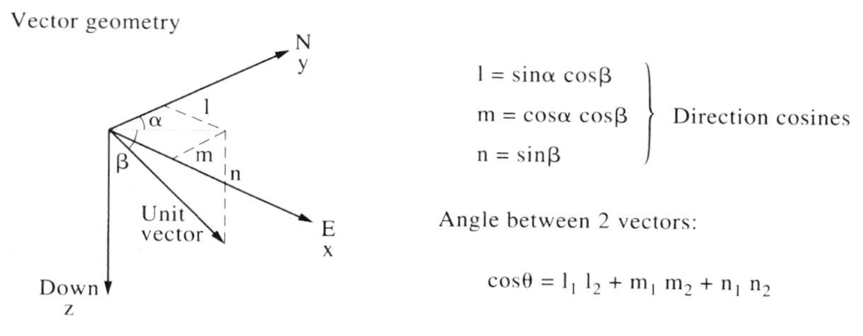

Figure 7.19 Direction cosines of a unit vector.

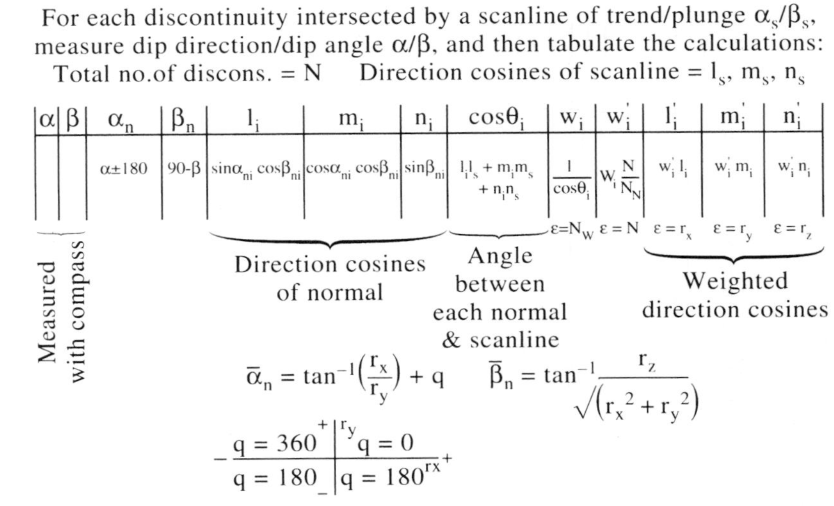

For each discontinuity intersected by a scanline of trend/plunge α_s/β_s, measure dip direction/dip angle α/β, and then tabulate the calculations:
Total no.of discons. = N Direction cosines of scanline = l_s, m_s, n_s

α	β	α_n	β_n	l_i	m_i	n_i	$\cos\theta_i$	w_i	w'_i	l'_i	m'_i	n'_i
		$\alpha \pm 180$	$90-\beta$	$\sin\alpha_{ni}\cos\beta_{ni}$	$\cos\alpha_{ni}\cos\beta_{ni}$	$\sin\beta_{ni}$	$l_i l_s + m_i m_s + n_i n_s$	$\frac{1}{\cos\theta_i}$	$W_i \frac{N}{N_N}$	$w'_i l_i$	$w'_i m_i$	$w'_i n_i$
								$\varepsilon = N_W$	$\varepsilon = N$	$\varepsilon = r_x$	$\varepsilon = r_y$	$\varepsilon = r_z$

Measured with compass (α, β); Direction cosines of normal (l_i, m_i, n_i); Angle between each normal & scanline ($\cos\theta_i$); Weighted direction cosines (l'_i, m'_i, n'_i)

$$\bar{\alpha}_n = \tan^{-1}\left(\frac{r_x}{r_y}\right) + q \qquad \bar{\beta}_n = \tan^{-1}\frac{r_z}{\sqrt{(r_x^2 + r_y^2)}}$$

	$+r_y$	
$-r_x$	$q = 360$	$q = 0$
	$q = 180$	$q = 180$

(upper quadrants: $+r_y$; right: $+r_x$; left: $-r_x$; lower: $-r_y$)

Figure 7.20 Evaluation of the mean orientation of a discontinuity set.

By introducing the direction cosines of the scanline l_s, m_s, n_s we can calculate the corresponding value cos θ_i. The reciprocal of this value is the weighting factor w_i, which is then scaled to w'_i in order that the overall measured discontinuity frequency is maintained. Via this procedure, we arrive at the last three columns, the weighted direction cosines. These values may then be summed, and the mean orientation *of the normal* computed. Obviously, if the aim of the analysis is to determine the mean orientation of a discontinuity set, the procedure should only be attempted on data likely to belong to a specific set of discontinuities.

These last three topics of the Poisson process, central limit theorem, and sampling bias and mean orientation have illustrated the power and *necessity* of applying basic standard statistical techniques to manipulating discontinuity geometry and other rock property data. Indeed, it may be fair to say that no other techniques are available for answering the questions posed at the beginning of Section 7.2.6 and yet these have been solved directly and elegantly as illustrated.

The use of statistical methods should always be an essential part of the rock engineer's analytical capabilities, because we will never have complete knowledge of the geometrical, mechanical and hydraulic properties of rock masses. Readers interested in a wider treatment of discontinuities are referred to Priest (1993).

7.3 Mechanical properties

7.3.1 Stiffness

In Chapter 6, we discussed the stiffness, strength and failure of intact rock via the complete stress–strain curve. We can consider the equivalent properties when a discontinuity is loaded in compression, tension or shear. These are illustrated in Fig. 7.21, with an indication of the type of complete

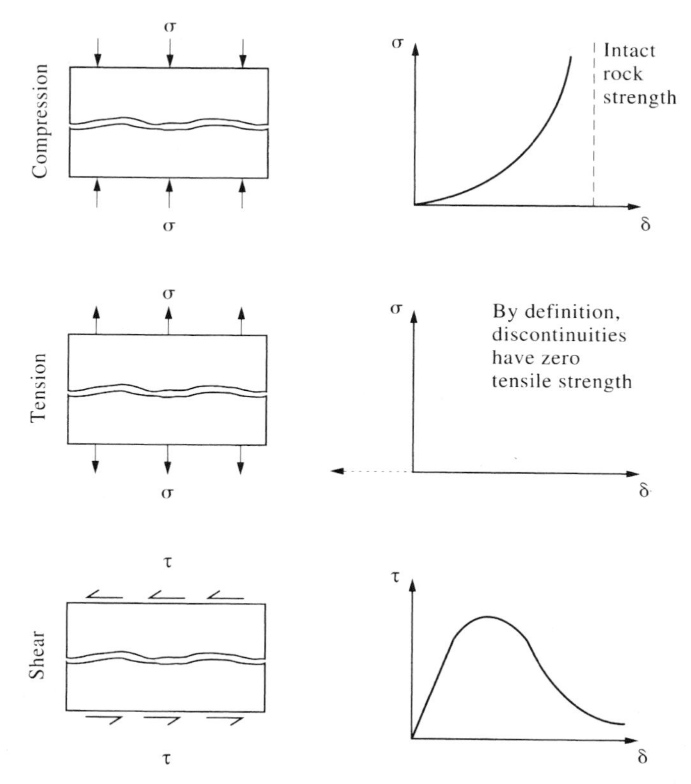

Figure 7.21 A discontinuity loaded in compression, tension and shear.

stress–displacement curve that we would expect. The normal and shear forces applied across a discontinuity can be scaled by the nominal area of the discontinuity to give normal and shear stresses, respectively. However, there is no 'length' for scaling the displacement values to evaluate strain. For this reason, we plot stress–displacement curves, with the result that discontinuity stiffness has units of, say, MPa/m, rather than the MPa units of intact rock stiffness.

The three curves in Fig. 7.21 are of three different types. In compression, the rock surfaces are gradually pushed together, with an obvious limit when the two surfaces are closed. The stiffness associated with this compression process gradually increases with applied stress or displacement, again reaching a limit associated with the strength of the intact rock, as indicated in the figure. In tension, because by definition discontinuities are regarded as having no tensile strength, no tensile stress can be sustained and hence the displacement increases as indicated. Finally, when a discontinuity is subjected to shear stress or shear displacement, the curve is rather like the complete stress–strain curve for compression of intact rock, except of course that all failure is localized along the discontinuity. There is an initial shear stiffness, a shear strength and a post-peak failure locus.

We can consider the stiffnesses for the cases of compression and shear. As we have shown in Fig. 7.21, in neither case is there a linear relation between stress and displacement. However, for analysis it is convenient to assume either one global linear stiffness value or a composite piecewise linear approximation. Goodman has proposed a hyperbolic relation to characterize the normal stress–displacement curve, viz.:

$$v = \frac{\sigma_n}{c + d\sigma_n}$$

where v and σ_n are closure and normal stress, respectively, and c and d are constants. This equation provides a good model for the discontinuity closure curve illustrated in the upper diagram in Fig. 7.21. It is possible to extend this basic concept to consider loading and unloading and many other aspects of the joint behaviour. The comprehensive reference to the mechanical behaviour of a single joint is the *Proceedings of the Rock Joints Symposium* held in Norway (Barton and Stephansson, 1990).

To characterize the shear stress–shear displacement curve in the lower diagram of Fig. 7.21, we can use the expression

$$\tau = \frac{\delta}{a + b\delta}$$

where τ and δ are the shear stress and shear displacement, respectively, and a and b are constants. Again, there are many extensions to this basic formula and the reader is referred to the Rock Joints text referenced above.

It is interesting to note that the expressions for normal displacement and shear displacement are mathematically similar, but have the stress and displacement terms reversed. Whereas the normal displacement must asymptote to a final closure value as the normal stress is increased, the shear displacement can continue indefinitely, usually with a reduced shear stress. Thus, the formula above refers to the deformation behaviour up to the peak shear strength.

Despite the non-linearities of the two curves, as a first approximation we can consider the linear stiffness representations as k_{nn} for the normal case and k_{ss} for the shear case. We can also consider the possibility that a normal stress will cause a shear displacement, using a constant k_{ns}, and that a shear stress will cause a normal displacement, using a constant k_{sn}. These stiffnesses have the dimensions of, for example, MPa/m, because they relate stress to displacement. With these linear approximations for the stiffnesses

$$\sigma_n = k_{nn}\delta_n + k_{ns}\delta_s$$

$$\tau = k_{sn}\delta_n + k_{ss}\delta_s$$

or, in matrix notation,

$$\begin{bmatrix} \sigma_n \\ \tau \end{bmatrix} = \begin{bmatrix} k_{nn} & k_{ns} \\ k_{sn} & k_{ss} \end{bmatrix} \cdot \begin{bmatrix} \delta_n \\ \delta_s \end{bmatrix}$$

or

$$\boldsymbol{\sigma} = \mathbf{k}\boldsymbol{\delta}.$$

This final expression also permits evaluation of the displacements when the stresses are known through use of the inverse of the matrix **k**. It should also be noted that this matrix, containing off-diagonal terms, provides a first approximation to the the coupling of normal and shear stresses and displacements, e.g. a term k_{sn} relating to dilation.

For the displacements of *in situ* discontinuities, the stiffness of the surrounding rock system will also need to be taken into account. There are many other practical aspects, such as the fact that the shear stiffness may be anisotropic in the plane of the discontinuity due to surface features like striations or lineations. We will discuss in Chapter 8 how the total *in situ* rock mass modulus can be predicted from a knowledge of the intact rock stiffness and the discontinuity stiffnesses.

7.3.2 Strength

Deformability has been considered first following the logic of the complete stress–strain or stress–displacement curves for intact rock, and now we consider the strength of discontinuities in shear expressed via the cohesion and angle of friction. It is normally assumed that the shear strength is a function of the angle of friction rather than the cohesion. This is a conservative assumption in the sense that discontinuities possess some, albeit low, cohesion. Basically, we assume that the strength of discontinuities is predicted by the reduced Mohr–Coulomb criterion, $\tau = \sigma \tan \phi$: the basic Mohr–Coulomb criterion was explained in Section 6.5.1. Consideration of any fluid that may be present and the generation of effective stresses will be discussed in Chapter 9. It is by no means clear that effective stresses and effective stress parameters can be used for rocks in this context.

The bi-linear failure criterion illustrated in Fig. 7.22 results from the work of Patton (1966), who introduced the idea that the irregularity of discontinuity surfaces could be approximated by an asperity angle i onto which the basic friction angle ϕ is superimposed. Thus, at low normal stresses, shear loading causes the discontinuity surfaces to *dilate* as shear displacement occurs, giving an effective friction of $(\phi + i)$. As the shear loading continues, the shear surfaces become damaged as asperities are sheared and the two surfaces ride on top of one another, giving a transition zone before the failure locus stabilizes at an angle ϕ. There are many 'complicating' factors in this mechanism, such as the roughness of the surface and the strength of the asperities. This led Barton *et al.* (e.g. 1985) to propose the empirical relation

$$\tau = \sigma_n \tan[\mathrm{JRC}\ \log_{10}(\mathrm{JCS}/\sigma) + \phi_r]$$

where JRC is the Joint Roughness Coefficient illustrated in Fig. 7.16, JCS is the Joint Wall Compressive Strength, and ϕ_r is the residual friction angle. The roughness component, i, is composed of a geometrical component and

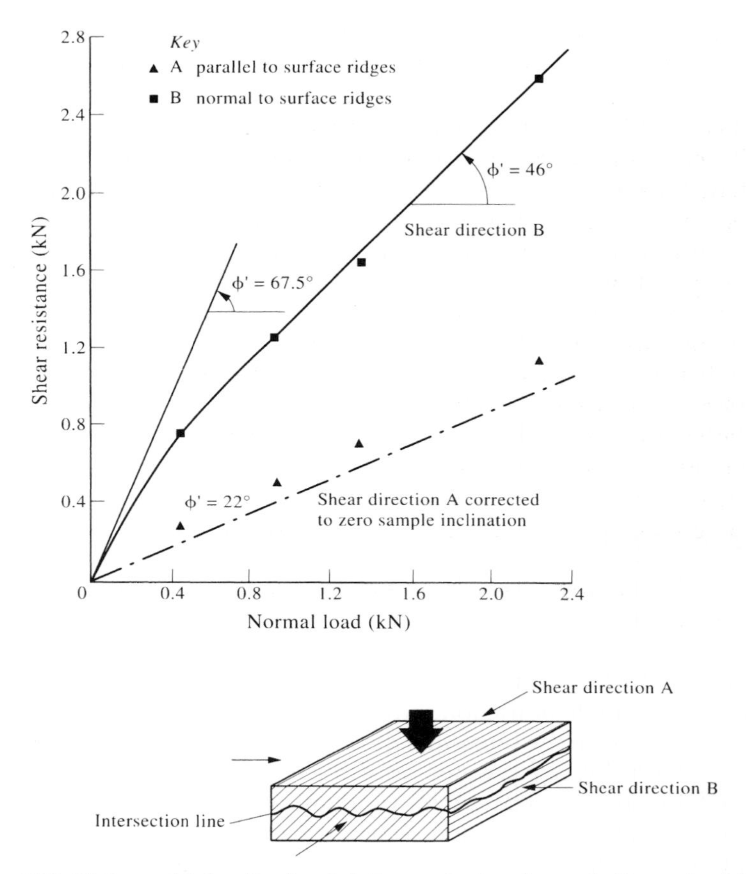

Figure 7.22 Bi-linear Mohr–Coulomb failure criterion for rock discontinuities.

an asperity failure component and ϕ_r is the basic frictional component. We have already discussed the size effect in general, and the shear strength of discontinuities is no exception. Clearly, the roughness is not absolute as the magnitude of the geometrical component will vary according to the sample dimensions.

These ideas have been extended to include mechanical versus hydraulic aperture and a coupling between the mechanical and hydraulic behaviour. Factors such as the degree and type of infilling will also have an effect on the strength of discontinuities. However, these effects have not yet been quantified beyond being able to say that the strength of a discontinuity approaches that of the filling when the filling is thick.

Here we do not consider the post-peak failure of discontinuities, as this is included in the next chapter, particularly with respect to the shear strength of discontinuities.

7.4 Discussion

Developments are possible in both the geometrical and mechanical characterization of discontinuities as a result of research currently underway. We have described just the basic discontinuity geometry and behaviour here, but all the developments are likely to be extensions of conventional techniques (geometrical and mechanical) to allow for the idiosyncrasies of rock fractures. As this work is progressively extended in an attempt to provide more and more realistic representions of prototypical discontinuities, the formulae must become more and more complicated, so that we are progressing into an increasingly labyrinthine cul-de-sac.

The application of some of the newer mathematical and information technology techniques such as fractals, chaos theory and object-oriented programming for characterizing rock features may prove to be more elegant, succinct and practical means of achieving the rock mechanics and rock engineering objectives. It is essential in the long run that some form of simplified approach is developed, because all rock engineering problems will involve some degree of coupling of the rock discontinuities, the stress field, the water flow and the effects of construction (the interactions will be discussed in Chapter 14). In some cases, it may be sufficient to have a qualitative idea of the coupling; in other cases, such as radioactive waste disposal, it is essential to have a quantitative prediction of the coupled behaviour. This prediction will not be achieved by a complex extension of current methods.

8 Rock masses

In this chapter, we will concentrate on extending the ideas discussed in the previous chapter on discontinuities to provide a predictive model for the deformability and strength of rock masses. In Chapter 12, we will be discussing rock mass classification (which is a method of combining selected geometrical and mechanical parameters) to semi-quantitatively provide an overall characterization, mainly for assessing excavation support requirements.

8.1 Deformability

Consider first, as an initial step in the overall development of a deformability model, the deformation of a set of parallel discontinuities under the action of a normal stress, assuming linear elastic discontinuity stiffnesses. This circumstance is illustrated in Fig. 8.1. To calculate the *overall* modulus of deformation, the applied stress is divided by the *total* deformation. We will assume that the thickness of the discontinuities is negligible in comparison to the overall length under consideration, L. Additionally, we will assume that the deformation is made up of two components: one due to

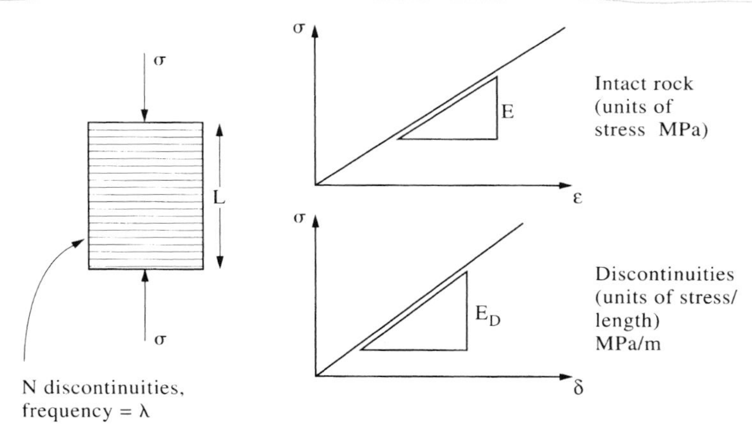

Figure 8.1 The modulus of deformation of a rock mass containing a discontinuity set.

deformation of the intact rock; the other due to the deformability of the discontinuities.

The contribution made by the intact rock to the deformation, δ_I, is $\sigma L/E$ (i.e. strain multiplied by length). The contribution made by a single discontinuity to the deformation, δ_D, is σ/E_D (remembering that E_D relates stress to displacement directly). Assuming a discontinuity frequency of λ, there will be λL discontinuities in the rock mass and the total contribution made by these to the deformation will be δ_D^t, which is equal to $\sigma\lambda L/E_D$. Hence, the total displacement, δ_T, is

$$\delta_T = \frac{\sigma L}{E} + \frac{\lambda \sigma L}{E_D}$$

with the overall strain being given by

$$\varepsilon = \frac{\delta_T}{L} = \frac{\sigma}{E} + \frac{\lambda\sigma}{E_D}.$$

Finally, the overall modulus, E_{MASS}, is given by

$$E_{MASS} = \sigma/\varepsilon = 1/[(1/E) + (\lambda/E_D)].$$

A suite of curves illustrating this relation is given in Fig. 8.2 for varying discontinuity frequencies and stiffnesses. It is simple to extend this formula for multiple intact rock strata with differing properties, discontinuity frequencies and discontinuity stiffnesses, and hence model stratified rock with discontinuities parallel to the bedding planes.

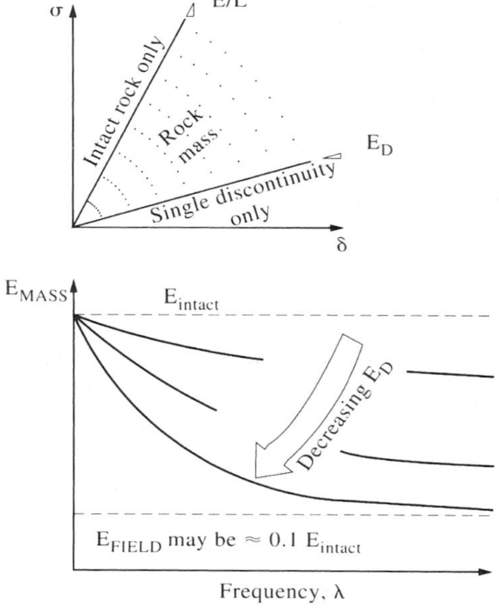

Figure 8.2 Variation of *in situ* rock deformability as a function of the discontinuities (idealized case for a single set of discontinuities).

The case illustrated via the mathematics above and shown in Fig. 8.2 only involves loading parallel to the discontinuity normals. Clearly, even in these idealized circumstances, we need to extend the ideas to loading at any angle and the possibility of any number of non-parallel sets. An argument similar to that given above can be invoked in the derivation of shear loading parallel to the discontinuities, as succinctly described by Goodman (1989), to give

$$G_{MASS} = \tau/\gamma = 1/[(1/G) + (\lambda/G_D)].$$

The mathematics associated with further extensions to account for discontinuity geometry rapidly becomes complex. A complete solution has been provided by Wei (1988), which can incorporate the four stiffnesses of a discontinuity (normal, shear and the two cross terms), any number of sets and can approximate the effect of impersistent discontinuities.

In the stress transformations presented in Chapter 3, the resolution of the stress components involves only powers of *two* in the trigonometrical terms, because the *force* is being resolved and the *area* is also being resolved. However, for the calculation of the deformability modulus, powers of *four* are necessary because of the additional resolution of the *discontinuity frequency* (explained in Chapter 7) and the *displacements*. An example equation from Wei's theory, the roots of which provide the directions of the extreme values of the modulus, is

$$(A \tan^4 \alpha - B \tan^3 \alpha - C \tan^2 \alpha - D \tan \alpha - F) \cos^4 \alpha = 0$$

where A, B, C, D, E and F are constants formed by various combinations of the discontinuity stiffnesses and α is the angle between the applied stress and one of the global Cartesian axes. The reader is referred to Wei's work for a complete explanation.

The utility of this type of analysis is illustrated by the polar diagrams in Fig. 8.3 representing the moduli variations for two discontinuity sets in two dimensions. (It is emphasized that this figure is one example of a general theory.) When k is high, as in the left-hand diagram, the lowest moduli are in a direction at 45° to the discontinuity sets, and the highest moduli are perpendicular to the sets. Conversely, when k is low, as in the right-hand

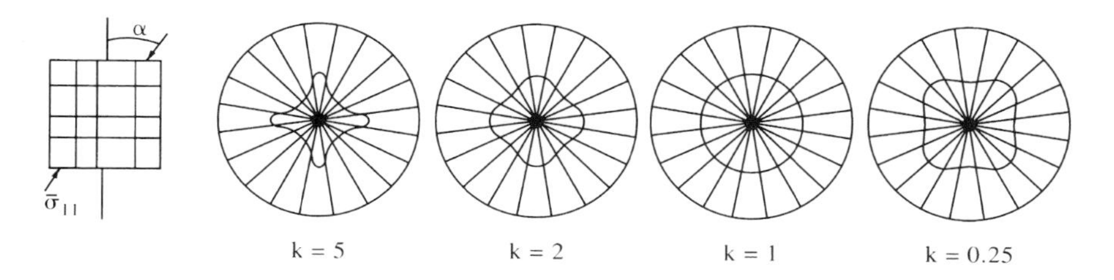

k is the ratio of the shear stiffnesses to normal stiffnesses

Figure 8.3 Variation in rock mass modulus for two orthogonal discontinuity sets with equal frequencies and equal stiffnesses (from Wei, 1988).

diagram, the minimum moduli are in a direction perpendicular to the sets, and the maximum moduli are at a direction of 45° to the sets. Like the discontinuity frequency, the directions of maximum and minimum moduli are not perpendicular.

A most interesting case occurs when $k = 1$, i.e. the normal and shear stiffnesses are equal, and the modulus is isotropic. The significance of even this very simple case of rock mass deformability for *in situ* testing and numerical modelling is apparent.

8.2 Strength

In the same way as we have considered the deformability of a rock mass, expressions can be developed indicating how strength is affected by the presence of discontinuities, starting with a single discontinuity and then extending to any number of discontinuities. The initial approach is via the 'single plane of weakness' theory, attributable to Jaeger, whereby the strength of a sample of intact rock containing a single discontinuity can be established. Basically, the stress applied to the sample is resolved into the normal and shear stresses on the plane of weakness and the Mohr–Coulomb failure criterion (discussed in Chapter 6) applied to consider the possibility of slip.

The strength of the sample depends on the orientation of the discontinuity. If the discontinuity is, for example, parallel or perpendicular to the applied loading, it will have no effect on the sample *strength*. At some angles, however, the discontinuity will significantly reduce the strength of the sample. This is illustrated in Fig. 8.4 which shows that the lowest strength occurs when the discontinuity normal is inclined at an angle of $45° + (\phi°/2)$ to the major applied principal stress. The formula for the reduction in strength is found by establishing the normal and shear stress on the plane passing through the specimen and substituting these into the Mohr–Coulomb failure criterion.

Given the geometry of the applied loading conditions in Fig. 8.4,

$$|\tau| = \tfrac{1}{2}(\sigma_1 - \sigma_3)\sin 2\beta_w$$

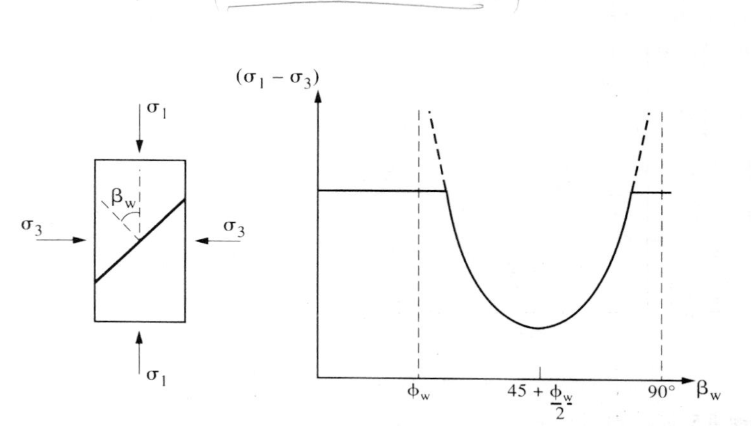

Figure 8.4 Effect of a discontinuity on the strength of a rock sample.

$$\sigma_n = \tfrac{1}{2}(\sigma_1 + \sigma_3) + \tfrac{1}{2}(\sigma_1 - \sigma_3)\cos 2\beta_w.$$

Substituting these into the Mohr–Coulomb criterion, $|\tau| = c_W + \sigma_n \tan \phi_w$, and rearranging gives

$$(\sigma_1 - \sigma_3) = \frac{2(c_W + \sigma_3 \tan\phi_W)}{(1 - \cot\beta_W \tan\phi_W)\sin 2\beta_W}$$

where c_W and ϕ_W are the cohesion and the angle of friction for the discontinuity (i.e. plane of weakness), and β_W is illustrated in Fig. 8.4. The plot of the equation in Fig. 8.4 shows the minimum strength and also the angles at which the sample strength becomes less than that of the intact rock.

An alternative presentation is via the Mohr's circle representation, as shown in Fig. 8.5. The Mohr–Coulomb failure loci for both the intact rock and the discontinuity are shown. We also show three Mohr's circles, A, B and C, representing the lowest strength, an intermediate case and the highest strength.

- Circle A represents the case when the failure locus for the discontinuity is just reached, i.e. for a discontinuity at the angle $2\beta_W{}^\circ = 90^\circ + \phi_w{}^\circ$.
- Circle B is a case when failure can occur along the discontinuity for a range of angles, as indicated in the figure.
- Circle C represents the case where the circle touches the intact rock failure locus, i.e. where failure will occur in the intact rock if it has not already done so along the discontinuity.

The importance of these different failure mechanisms will be made clear in later chapters, when we consider the stresses around excavations in rock containing discontinuities. According to the circumstances, failure can either occur along the discontinuities or through the intact rock, depending on the relative orientations of the principal stresses and the discontinuities.

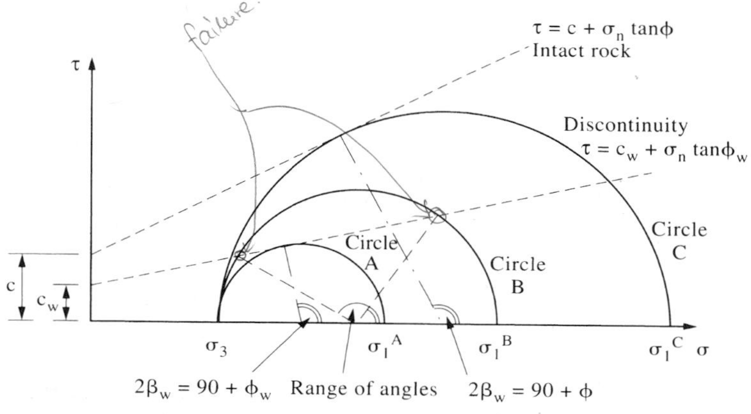

Figure 8.5 Mohr's circle representation of the possible modes of failure for rock containing a single plane of weakness.

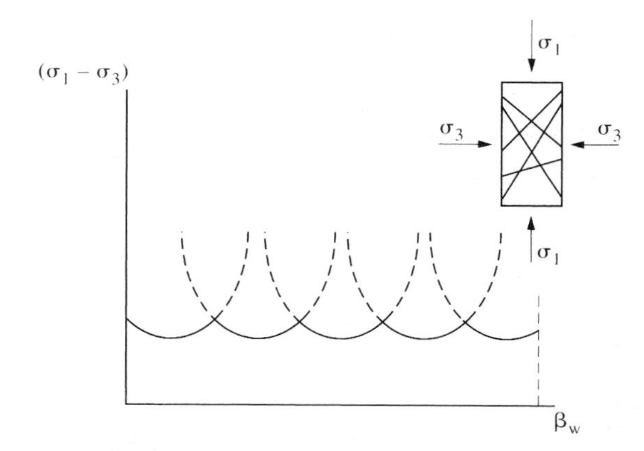

Figure 8.6 Strength of a rock mass containing multiple discontinuity sets.

We can consider, on the basis of this single plane of weakness theory, what would happen if there were two or more discontinuities at different orientations present in the rock sample. Each discontinuity would weaken the sample as indicated in Fig. 8.4, but the angular position of the strength minima would not coincide. As a result the rock is weakened in several different directions simultaneously as shown in Fig. 8.6. The material tends to become isotropic in strength, like a granular soil. When plotting the superimposed curves, care should be taken in interpreting the magnitude of β_w correctly for each of the discontinuities concerned.

The main advantage of the single plane of weakness theory is its simplicity and helpfulness in the interpretation of rock mass failure. We have presented here the two-dimensional case (applicable to plane stress) and one can imagine an extension to the general three-dimensional loading case in which none of the principal stresses is perpendicular to the discontinuity normal. In reality, the situation is rather complicated because the stresses will not be transmitted directly through the discontinuity. However, despite these shortcomings the authors feel that the advantages gained by understanding such idealized models do help in interpreting the far more complex behaviour of real rock masses.

We have already discussed the Hoek–Brown failure criterion in Chapter 6 in terms of its application to intact rock. The criterion is especially powerful in its application to rock masses due to the constants m and s being able to take on values which permit prediction of the strengths of a wide range of rock masses. Recent publications, i.e. Hoek and Brown (1988), Hoek (1990) and Hoek *et al.* (1992), provide an update of the failure criterion as it applies to rock masses, together with the relation between the Hoek–Brown and Mohr–Coulomb criteria.

From the first of these publications, we have included Table 8.1 which indicates the relation between rock mass quality and the m and s material constants. The table also provides a guide to the relation between these constants and two commonly used rock mass characterization values, i.e. the CSIR and NGI classification ratings (the latter being described in Chapter 12)

Table 8.1 Approximate relations between rock mass quality and the material constants in the Hoek–Brown failure criterion (from Hoek and Brown (1988)). Undisturbed values are in italics

Approximate relationship between rock mass quality and material constants

Disturbed rock mass m and s values — *undisturbed rock mass m and s values*

EMPIRICAL FAILURE CRITERION

$\sigma_1 = \sigma_3 + \sqrt{m\sigma_c\sigma_3 + s\sigma_c^2}$

σ_1 = major principal stress
σ_3 = minor principal stress
σ_c = uniaxial compressive strength of intact rock, and
m and s are empirical constants.

		CARBONATE ROCKS WITH WELL DEVELOPED CRYSTAL CLEAVAGE *dolomite, limestone and marble*	LITHIFIED ARGILLACEOUS ROCKS *mudstone, siltstone, shale and slate (normal to cleavage)*	ARENACEOUS ROCKS WITH STRONG CRYSTALS AND POORLY DEVELOPED CRYSTAL CLEAVAGE *sandstone and quartzite*	FINE GRAINED POLYMINERALLIC IGNEOUS CRYSTALLINE ROCKS *andesite, dolerite, diabase and rhyolite*	COARSE GRAINED POLYMINERALLIC IGNEOUS & METAMORPHIC CRYSTALLINE ROCKS – *amphibolite, gabbro gneiss, granite, norite, quartz-diorite*
INTACT ROCK SAMPLES						
Laboratory size specimens free	m	7.00	10.00	15.00	17.00	25.00
from discontinuities	s	1.00	1.00	1.00	1.00	1.00
CSIR rating: RMR = 100	*m*	*7.00*	*10.00*	*15.00*	*17.00*	*25.00*
NGI rating: Q = 500	*s*	*1.00*	*1.00*	*1.00*	*1.00*	*1.00*
VERY GOOD QUALITY ROCK MASS						
Tightly interlocking undisturbed rock	m	2.40	3.43	5.14	5.82	8.56
with unweathered joints at 1 to 3m.	s	0.082	0.082	0.082	0.082	0.082
CSIR rating: RMR = 85	*m*	*4.10*	*5.85*	*8.78*	*9.95*	*14.63*
NGI rating: Q = 100	*s*	*0.189*	*0.189*	*0.189*	*0.189*	*0.189*
GOOD QUALITY ROCK MASS						
Fresh to slightly weathered rock, slightly	m	0.575	0.821	1.231	1.395	2.052
disturbed with joints at 1 to 3m.	s	0.00293	0.00293	0.00293	0.00293	0.00293
CSIR rating: RMR = 65	*m*	*2.006*	*2.865*	*4.298*	*4.871*	*7.163*
NGI rating: Q = 10	*s*	*0.0205*	*0.0205*	*0.0205*	*0.0205*	*0.0205*
FAIR QUALITY ROCK MASS						
Several sets of moderately weathered	m	0.128	0.183	0.275	0.311	0.458
joints spaced at 0.3 to 1m.	s	0.00009	0.00009	0.00009	0.00009	0.00009
CSIR rating: RMR = 44	*m*	*0.947*	*1.353*	*2.030*	*2.301*	*3.383*
NGI rating: Q = 1	*s*	*0.00198*	*0.00198*	*0.00198*	*0.00198*	*0.00198*
POOR QUALITY ROCK MASS						
Numerous weathered joints at 30-500mm,	m	0.029	0.041	0.061	0.069	0.102
some gouge. Clean compacted waste rock	s	0.000003	0.000003	0.000003	0.000003	0.000003
CSIR rating: RMR = 23	*m*	*0.447*	*0.639*	*0.959*	*1.087*	*1.598*
NGI rating: Q = 0.1	*s*	*0.00019*	*0.00019*	*0.00019*	*0.00019*	*0.00019*
VERY POOR QUALITY ROCK MASS						
Numerous heavily weathered joints spaced	m	0.007	0.010	0.015	0.017	0.025
<50mm with gouge. Waste rock with fines.	s	0.0000001	0.0000001	0.0000001	0.0000001	0.0000001
CSIR rating: RMR = 3	*m*	*0.219*	*0.313*	*0.469*	*0.532*	*0.782*
NGI rating: Q = 0.01	*s*	*0.00002*	*0.00002*	*0.00002*	*0.00002*	*0.00002*

8.3 Post-peak strength behaviour

In Chapter 6, on intact rock, and in Chapter 7, on discontinuities, we have demonstrated that it is possible to describe the complete mechanical behaviour from initial deformation, via the peak strength, to the failure process. In the case of intact rock, the post-peak strength behaviour can be characterized using the shape of the complete stress–strain curve. Similarly, in the case of discontinuities, we can discuss the residual frictional value that is reached after the discontinuity has been fully sheared. However, it is much more difficult to provide any simple characterization of the total failure behaviour of a rock mass: this is because of the presence of the discontinuities and the manifold ways in which the rock mass structure can break down.

For example, a block might fall out of the rock mass during excavation and lead to a ravelling-type failure. Alternatively, discontinuities might be sheared in one direction in the plane but not in another direction. Individual blocks may fail due to high stresses. The collapse of the rock mass may occur through many such events, or through failure on one significant through-going discontinuity such as a fault.

It is impossible to summarize all these possibilities in any simple quantitative fashion. Also, the failure of the rock mass and its interaction with the excavation process has links with the objective of rock engineering. The way in which one characterizes the failure will depend on the engineering objective. In civil engineering, the objective may be, for example, that the displacement in the rock mass nowhere exceeds a certain amount. In mining engineering, we may *wish* the rock to be in a continual state of imminent failure as, for example, in the block caving method of mining, where a large block of ore is undercut and allowed to collapse in a controlled way via ore passes.

The only generic way of quantitatively characterizing rock mass failure is through the Hoek–Brown criterion and/or the rock mass classification systems. We have already described the former; the latter will be described in Chapter 12. The discussion of rock mass failure, with specific reference to applications, will be continued in Chapters 17–20.

9 Permeability

The subject of permeability is concerned with fluid flow through a material, or rocks and rock masses in our current context, and is one of the most difficult topics facing the practising rock engineer. There are formal definitions of permeability and the associated characteristics for continuous materials but, as is emphasized throughout this book, a rock mass contains discontinuities, and discontinuities are preferential flow paths. In this chapter, therefore, we will initially present the definitions of permeability and hydraulic conductivity and discuss the tensorial nature of permeability. We will then consider fluid flow in discontinuities and the associated ideas of primary and secondary permeability. The subject of the permeability of the rock mass can be studied in terms of the effective permeability of discontinuity networks: there is then the corollary of a natural scale effect and the representative elemental volume. Finally, we will discuss the difficulty of dealing with effective stresses in rock mechanics and the permeability-related applications of grouting and blasting.

We should note in passing that the phenomenon of water flowing through rock has been observed since antiquity. In Creech's (1683) English translation of Lucretius' six books of Epicurean philosophy on the nature of things, in Book I are the lines,

> Tho free from Pores, and Solid Things appear,
> Yet many Reasons prove them to be Rare:
> For drops distill, and subtle moisture creeps
> Thro hardest Rocks, and every Marble weeps... .

9.1 Fundamental definitions

Permeability, k_{ij}, is a mathematical quantity directly analagous to stress and strain which we have already described. It is a second-order tensor, meaning that it is a quantity with magnitude, direction and a reference plane. Permeability has six independent components and follows mathematical transformation equations when the orientation of the reference plane is changed. Permeability is formally defined as

$$q_i = -\frac{k_{ij}}{\mu}\frac{\partial P}{\partial x_j}$$

where q_i is the specific discharge,
$\partial P/\partial x_j$ is the pressure gradient causing flow,
μ is the fluid viscosity and
k_{ij} are the components of the permeability tensor.

These components are schematically illustrated in Fig. 9.1 to show the analogy of permeability with stress and strain. Within the context of this book, it is inappropriate to pursue the full mathematical development of this subject further, because permeability is almost always regarded as a scalar value in engineering practice: the interested reader is referred to Raudkivi and Callander's (1976) book on groundwater flow for an excellent treatment of the subject. Consequently, we will concentrate on the semi-empirical approach that is utilized in engineering. As we will be describing later, there are major problems in considering a fractured rock mass as an effectively continuous permeable medium.

Because in practice permeability has rarely been regarded in its full tensorial state, and because we will be considering one-dimensional flow through discontinuities, it is convenient here to consider the reduced forms of the above equations. Assuming permeability to be a scalar, we have

$$q = -\frac{k}{\mu}\frac{\partial P}{\partial x}.$$

The permeability, k, is independent of the fluid under consideration having the dimensions L^2.

Very often in rock engineering the percolating fluid is water and so we can alter the form of the above equation to

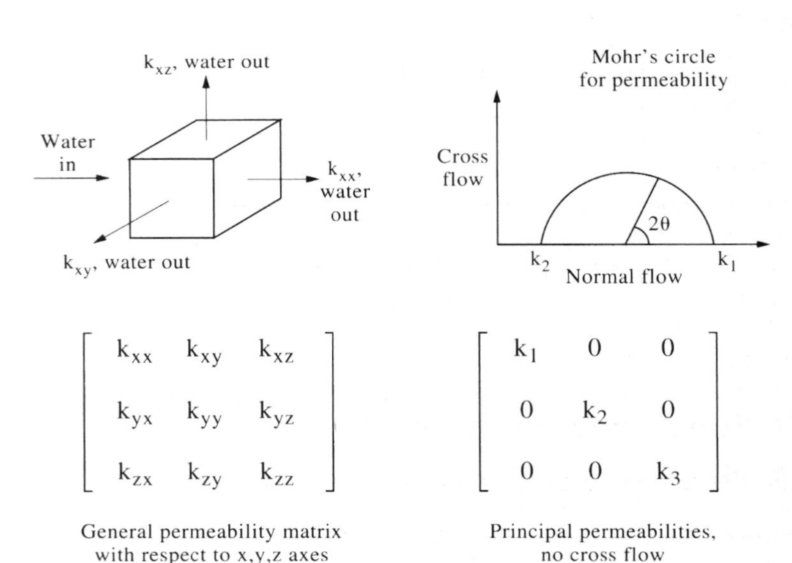

Figure 9.1 Illustration of permeability as a tensor quantity.

$$q = -\frac{K \partial P}{\gamma_f \partial x}.$$

The term, K, is the **coefficient of permeability** (or **hydraulic conductivity**) with dimensions L/T, and the term γ_f is the **fluid specific weight**. The relation between k and K is therefore

$$\text{permeability, } k = \mu K/\gamma_f \quad (L^2) \text{ or}$$
$$\text{hydraulic conductivity, } K = \gamma_f k/\mu \quad (L/T).$$

In its most basic form, and for the case of laminar flow, **Darcy's law** links the water flow rate to the pressure gradient, i.e.

$$Q = KAi$$

where Q is the flow rate (dimensions of L^3T^{-1}),
A is the cross-sectional area of the flow, and
i is the hydraulic gradient, $\Delta h/\Delta l$.

9.2 Primary and secondary permeability

Because of the presence of discontinuities in a rock mass, we have the concepts of **primary permeability** and **secondary permeability**. Primary permeability refers to the rock matrix permeability, whereas, the secondary permeability refers to the rock mass permeability. In some circumstances, e.g. petroleum engineering, we will be specifically interested in the primary permeability, but in most rock engineering it is the secondary permeability which dominates the design and construction procedures. It has already been mentioned that there are interrelations between most of the rock properties: the flow of fluid through a fractured rock mass is no exception, as it will depend on:

(a) the aperture of the fractures, which in turn will depend on
(b) the normal stress acting across the fractures, which in turn will depend on
(c) the depth below the ground surface.

In the extreme case, at great depth, all the fractures may be effectively closed, so that the primary and secondary permeabilities are similar.

Figure 9.2 illustrates the variabilities of primary and secondary hydraulic conductivity for different rock types. A key aspect of the primary hydraulic conductivity diagram is the extreme range—through at least 8 orders of magnitude. Similarly, for the secondary hydraulic conductivity, there is an even greater range—of 11 orders of magnitude—with limestones, dolomites and basalts covering the entire range.

9.3 Flow through discontinuities

The development of the theory for considering fluid flow through a discontinuity is described by Hoek and Bray (1977) and is based on the flow between a parallel pair of smooth plates. Darcy's law can be rewritten as

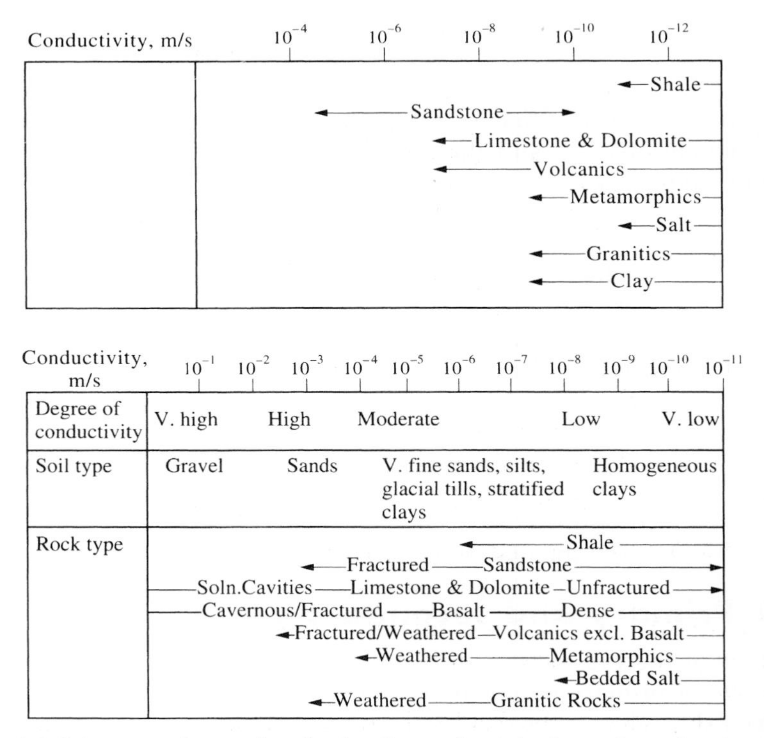

Figure 9.2 Primary and secondary hydraulic conductivity for rocks and rock masses (after Isherwood, 1979).

$$Q = cH_L$$

where c is the conductance, given by $ge^3/12vL$ and
H_L is the head loss between the two end sections.

In the expression for c, e is the aperture between the pair of plates, v is the kinematic viscosity of the fluid (which for water may be taken as $1.0 \times 10^{-6} \text{m}^2/\text{s}$) and L is the length of the plates in the direction of flow. Figure 9.3 illustrates this equation.

Note that in the formula for the conductance given above, the flow rate is proportional to the cube of the discontinuity aperture. The flow rate is thus extremely sensitive to small changes in aperture: a doubling of aperture results in an eight-fold increase of flow rate. As a consequence, one very open discontinuity through a rock mass can totally dominate the water flow conditions.

A natural extension of this formula for conductance is to consider a set of parallel discontinuities. The hydraulic conductivity parallel to the set is given by

$$K = \frac{\lambda g e^3}{12v} \quad (L/T)$$

where λ is the discontinuity frequency.

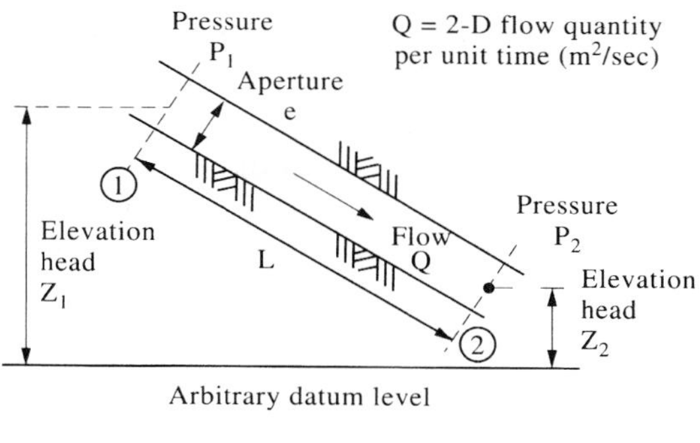

Flow is due to head:

Pressure head = P/γ

Elevation head = z

Kinematic head = $v^2/2g$ *Ignore*

At ①: $H_1 = \frac{P_1}{\gamma} + z_1$ At ②: $H_2 = \frac{P_2}{\gamma} + z_2$

Figure 9.3 Flow between two parallel surfaces.

The variation in K with both the aperture and the discontinuity frequency for flow parallel to a single set of discontinuities is shown in Fig. 9.4.

With reference to Fig. 9.4, Hoek and Bray pointed out that the permeability of such a system is sensitive to the discontinuity apertures, and hence also to the presence of any *in situ* stress. They extend the idea to two orthogonal sets of discontinuities, and discuss pumping tests in a borehole traversing such an array.

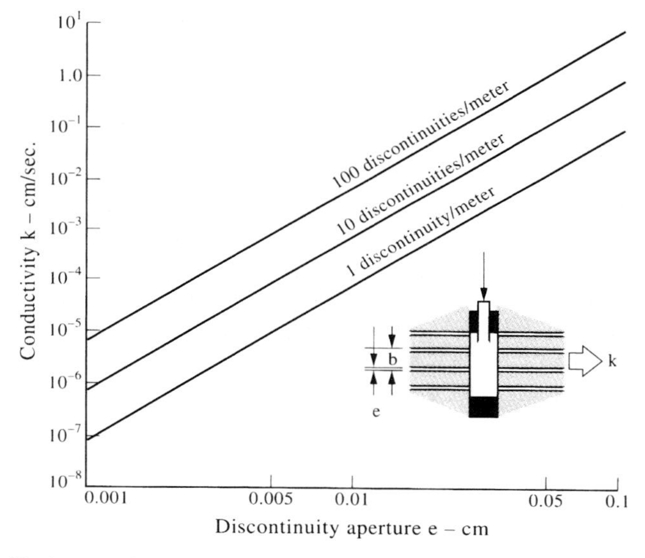

Figure 9.4 Variation of discontinuity set permeability as a function of the aperture and discontinuity frequency (after Hoek and Bray, 1977).

9.4 Flow through discontinuity networks

Often, in discontinuity arrays, one discontinuity will terminate against another and it is therefore of interest, not only to be able to compute the permeability of a set of parallel discontinuities, but also to analyse conditions where two discontinuities meet, and indeed, to study the complex discontinuity networks that are contained within rock masses.

To start, we can consider the flow at a node in a simple network, as illustrated in Fig. 9.5. This figure indicates the notation for node, channel and flow numbering, such that application of the continuity equation (i.e. 'what goes in must come out') gives

$$Q_{14} + Q_{24} + Q_{34} = 0.$$

The equation given earlier for flow through a single discontinuity, i.e. $Q = cH_L$, can be generalized as $Q_{ij} = c_{ij}(H_i - H_j) = c_{ij}H_i - c_{ij}H_j$ so that the hydraulic head at the jth node can be expressed as

$$H_j = \frac{\Sigma c_{ij} H_i}{\Sigma c_{ij}}.$$

Assuming the flow in the network is laminar, Bernoulli's equation

$$\text{total head} = \frac{P}{\gamma} + z + \frac{v^2}{2g}$$

may be applied. Generally, the velocity of the flow will be sufficiently low to permit the velocity head term, $v^2/2g$, to be ignored, giving

$$\text{total head} = \frac{P}{\gamma} + z.$$

Thus, for a more complex discontinuity array, and applying this type of analysis, we can establish the hydraulic heads at nodal points by solving

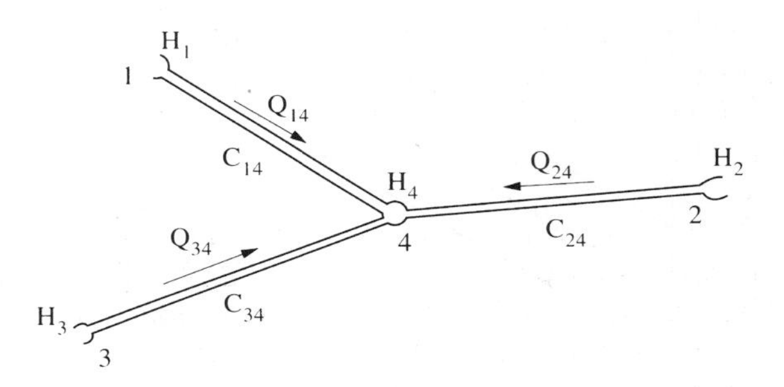

Figure 9.5 Flow at a network node.

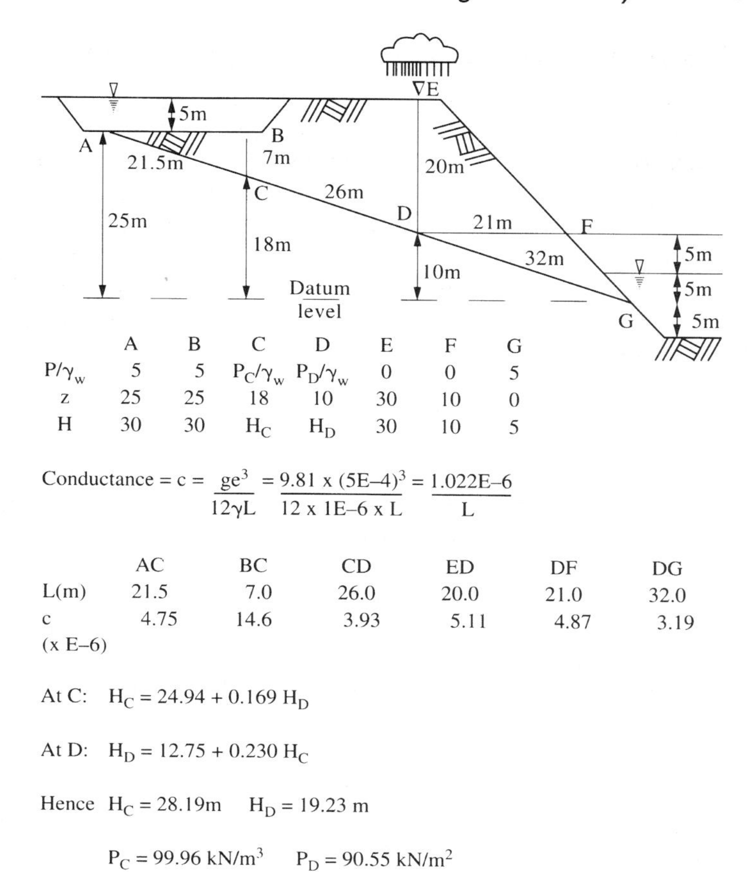

Figure 9.6 Example of network flow calculations.

the resulting set of simultaneous equations, and finally computing the flow through each of the individual channels. An example of this calculation is given in Fig. 9.6.

Figure 9.7 illustrates the results of an analysis of a simulated discontinuity array. The numbers on the diagram indicate the total head at each node. From these it can be seen that the boundary conditions are such that overall flow is from left to right across the network: however, local idiosyncratic flows can be in an opposing direction, as the figure demonstrates.

Obviously, for more complex networks, the use of a suitable computer-based numerical solution is necessary. It should be noted that the analysis presented here is for an essentially two-dimensional network: the analysis cannot be simply extended into three dimensions because two discontinuity planes will meet along an intersection line, along which the hydraulic head may be changing. However, commercial computer programs are available for studying fluid flow through three-dimensional fracture networks (noting that the word 'fracture' is used instead of 'discontinuity' in hydrogeological literature).

9.5 Scale effect

There are no length dimensions in Fig. 9.7: the discontinuity array could represent small fractures over lengths of a few centimetres or master joints over lengths of many tens of metres. Imagine that a borehole had been drilled into this array to estimate the flow rate through the rock. In the case of short fractures, it may well be that the borehole would be approximately the same size as the diagram and hence the result be fairly reliable. In the case of joints, the borehole could well intersect no discontinuities, or perhaps one or two, *at a number of discrete locations*. Moreover, the hydraulic heads and flow directions at these points might in no way reflect the overall pattern of flow. This is an important practical consideration, and is generally termed the **scale effect**.

The scale effect for fluid flow has been studied via computer simulation by Long (1983). In Fig. 9.8, we present one of her most illuminating diagrams illustrating the connectivity within a fracture network and the associated scale effect. The column of diagrams on the left-hand side of the figure shows different sized samples of the same simulated discontinuity network. The column of diagrams on the right-hand side shows the connected network within the samples to the left, i.e. those discontinuities through which water can flow throughout the network. The diagrams dramatically illustrate the effect of scale. In the top right-hand diagram, water can only flow from top to bottom through the sample. In the fourth diagram down, water can only flow laterally. Progressing through the suite of diagrams, one can see the permeability stabilizing as the number of discontinuities in the sample increases. So, estimation of the permeability from small samples can give almost any result but, as the sampled

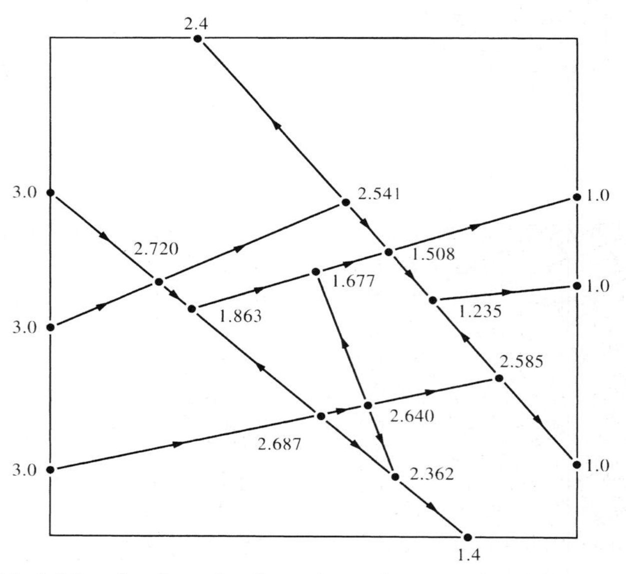

Figure 9.7 Nodal head values for flow through a simulated discontinuity array (from Samaniego and Priest, 1985).

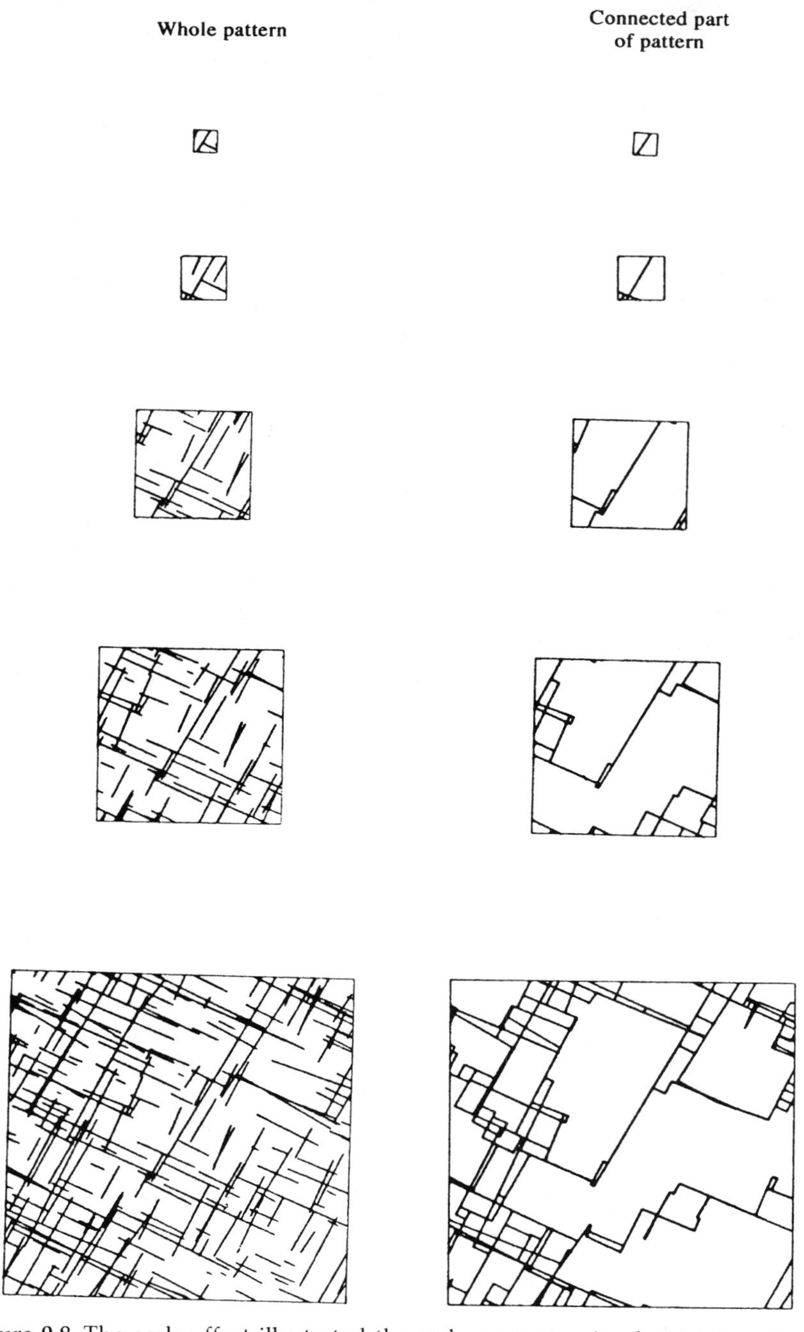

Figure 9.8 The scale effect illustrated through computer simulation (from Long, 1983).

volume increases, so the measured values become more representative. This leads automatically to the concept of the **representative elemental volume**, or **REV**.

In Fig. 9.9, we present the generalized figure foreshadowed in the discussion referring to Fig. 4.12 presented in Section 4.5. This illustrates that wide variations in permeability are expected when the measured sample volume is small. Also, we have seen directly in Fig. 9.8 how the permeability stabilizes with an increase in volume, despite the fact that earlier we mentioned that permeability is a tensor, and hence the property at a point, i.e. at 0 on the volume (horizontal) axis in Fig. 9.9. Clearly, like *in situ* rock stress, this presents a problem because the tensorial permeability concept can only apply to primary permeability: by definition, the secondary permeability must involve a non-zero volume.

All this is summarized concisely in Fig. 9.9 and the REV introduced, *inter alia*, by Bear (1972), is the position on the volume axis when the between-test variability of the permeability measurements is acceptably low. It has been tacitly assumed in this discussion that the discontinuity occurrence is statistically homogeneous in the region being measured: discontinuity inhomogeneity can also be a problem if the discontinuity geometry is changing within the REV volume scale. This is indicated by showing that it is possible for the mean permeability to be changing above the REV volume when the rock mass is inhomogeneous. A question that naturally arises is whether the mean of many measurements performed on sub-REV samples is in fact equal to the REV value of permeability. The answer will depend on the type of discontinuity inhomogeneity.

We have intimated here some of the difficulties that will arise in any attempt to use the permeability concepts in a practical arena. However, the key to good rock engineering is to understand the principles of rock mechanics and then to make engineering decisions on the optimal course of action. For example, we can predict that, when a tunnel is constructed in rock, there will be local increases in the water flow rate into the

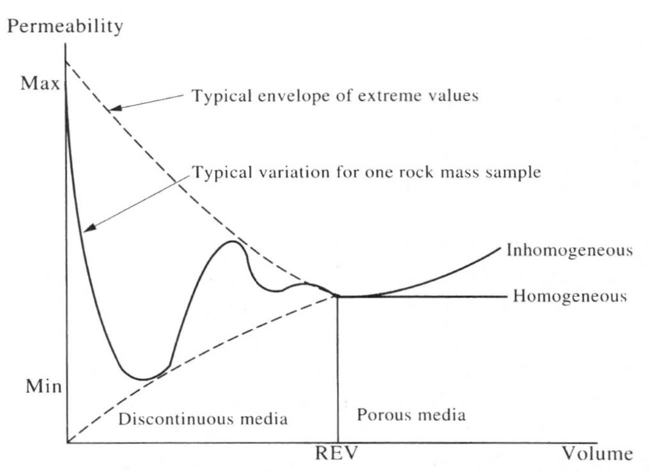

Figure 9.9 The representative elemental volume (REV) for permeability.

excavation as different discontinuities are traversed. We would predict that for almost all rock tunnels there will be lengths where there will be little inflow, and lengths where there could be high inflow. In general, we will not be able to predict the specific local water inflow: consideration of the diagrams in Fig. 9.8 makes this clear. Additionally, it will generally be unknown how such a network is connected to the regional hydrogeological regime. Thus, engineers know that during tunnel construction they should have defensive strategies against major local high water inflows, but that the precise location of these flows cannot be predicted.

Given the discussion in this chapter, would the reader include permeability measurements in a site investigation for a particular project? If so, should the necessary tests be conducted using boreholes? If so, how are the results to be interpreted? Questions of scale effect and permeability anisotropy will not be answered using a borehole strategy and, as a consequence, many engineers have used full-scale prototype excavations to determine permeability, e.g. the water inflow into a section of tunnel. Much, of course, depends on the engineering objective.

Of all of the subjects presented in this book, perhaps permeability and its corollaries are the prime examples of the fact that rock engineering is an art. We think we understand the scientific principles, we understand the difficulties of dealing with a natural rock mass, we may have large resources, but there is no simple procedure for establishing 'the' permeability of a rock mass.

9.6 A note on effective stresses

In soil mechanics, wide use is made of the concept of effective stress, as developed by Terzaghi (1963). We recall the explanation in Chapter 3 that stress is a tensor, comprised of three normal and three shear components. If fluid is present in the material matrix, the pressure, u, exerted by the fluid will effectively reduce the normal components of stress in the stress tensor, because the fluid has a hydrostatic pressure acting in all directions. This hydrostatic pressure has no effect on the shear components of the stress tensor. Thus, when the fluid is present, we can modify the stress tensor to an effective stress tensor as follows:

$$\begin{bmatrix} \sigma_{xx} & \tau_{xy} & \tau_{xz} \\ \tau_{yx} & \sigma_{yy} & \tau_{yz} \\ \tau_{zx} & \tau_{zy} & \sigma_{zz} \end{bmatrix} - \begin{bmatrix} u & 0 & 0 \\ 0 & u & 0 \\ 0 & 0 & u \end{bmatrix} = \begin{bmatrix} \sigma_{xx} - u & \tau_{xy} & \tau_{xz} \\ \tau_{yx} & \sigma_{yy} - u & \tau_{yz} \\ \tau_{zx} & \tau_{zy} & \sigma_{zz} - u \end{bmatrix}.$$

Here we have considered the simple case where the full hydrostatic pressure has been subtracted; the reader should note that many suggestions have been made for modifying the full value by coefficients to account for the material microstructure and degree of saturation.

In Fig. 9.10(a), the water pressure is acting within the material microstructure, i.e. in the context of primary permeability, resulting in the effective stress tensor given above. In Fig. 9.10(b) we show, via the Mohr circle diagram, the effect on strength of introducing water. Before water is introduced, the stress condition is as in Case 1 in the diagram. When a

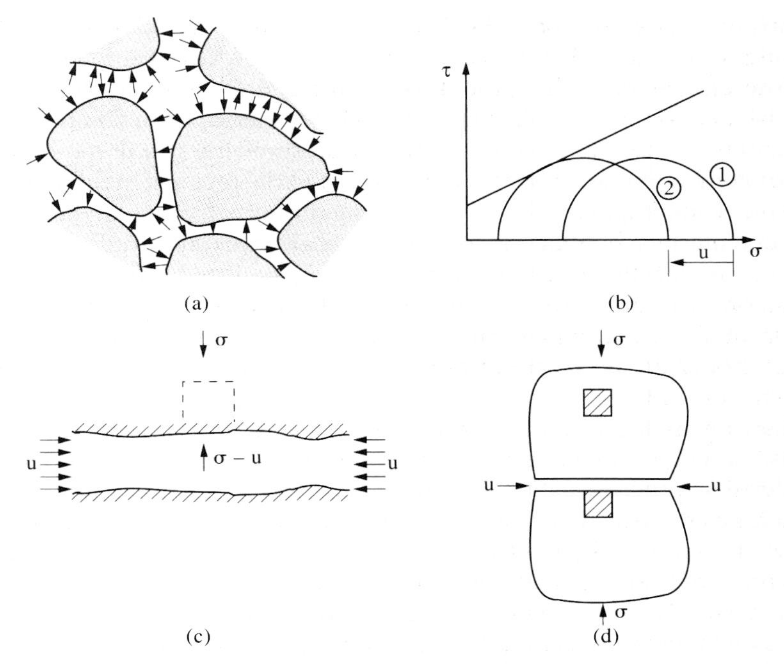

Figure 9.10 Effective stresses for intact rock and discontinuities.

water pressure of u is introduced, all normal components of the stress tensor are effectively reduced by u. In the Mohr's circle presentation, this results in the circle being moved to the left by an amount u, which could result in the circle then reaching the failure locus, which is Case 2.

In Fig. 9.10(c), we show the more complex problem of dealing with a water pressure u in a rock discontinuity, i.e. within the context of secondary permeability. There are two problems as compared to the primary permeability: first, the water pressure does not, depending on the location of the element in question, act on all normal components of the stress tensor; second, the water pressure is a local phenomenon, i.e. it is only acting in the discontinuities (within the timescales of engineering changes). Thus, the presence of the water could well have a profound effect on the mechanical behaviour of the discontinuities, but a much lesser effect on the behaviour of the intact rock. In fact, we have two effective stress concepts: one for the intact rock and one for the discontinuities. It is difficult to integrate these into a *global* effective stress law, as is illustrated in Fig. 9.10 (d), showing elements in proximity to a discontinuity: the stress tensors are different for each of these elements.

9.7 Some practical aspects: grouting and blasting

One of the main engineering solutions for reducing the permeability of a fractured rock mass is to inject a grout, which may be a suspension (e.g. cement grout), an emulsion (e.g. bitumens) or a solution (e.g. a silicate), which subsequently blocks the flow paths through the rock mass. Luckily, during such a grouting process, the grout will follow the path of least resistance, generally along the discontinuities with greatest aperture and persistence—which are the very ones which conduct the most water. We will wish to optimize such a grouting process, bearing in mind the types of discontinuity array that may be present—as, for example, that illustrated in Fig. 9.8.

Generally, such optimization will involve tailoring the location and orientation of the injection boreholes, together with the grout type, injection rates, pressures and volumes, to the discontinuity geometry. This is essentially an empirical process, but obviously considerations such as those described above are of great help in establishing the fundamental design principles.

Blasting technology has also tended to be almost entirely empirically based: the type of blasting round, the quantity of explosive and the detonation procedures have been established by trial and error. A rock which has no discontinuities has to be fractured by the blasting; in a rock which is very heavily fractured, it may only be necessary to disaggregate the rock mass without actually inducing any failure of intact rock. This leads us to utilize the fact that, as the blasting causes firstly a 'stress wave' to travel through the rock followed by a buildup of gas pressure within the borehole, we ought to tailor the type of explosive to the discontinuity geometry. By varying the explosive, the proportionate energy associated with the stress wave and gas pressure can be varied according to whether we are trying to break the rock with the stress wave or disaggregate it with the gas pressure.

There is an elegant extrapolation of this concept whereby, via the engineering, we can effectively *create a large artificial discontinuity* exactly where it is required. This technique is known as pre-split blasting and will be discussed in Chapter 15. The aim of the method is to create the final excavation surface *before* bulk blasting the remaining rock: this artificial discontinuity then prevents stress wave damage of the intact rock and disaggregation of the rock mass behind the final excavation surface. New and innovative engineering techniques can be developed if the principles of rock mechanics are known and understood. Here, within the pre-existing discontinuity pattern, an artificial discontinuity has been introduced having a greater, and yet more beneficial, effect than any others. One can imagine the extension of this concept to other subjects, such as controlled drainage and controlled rockbursts.

10 Anisotropy and inhomogeneity

We have already emphasized the natural history of the rock material which is being used for engineering purposes. A consequence of the millions of years of mechanical, chemical and thermal processes to which the rock mass has been subjected is that it may well be **anisotropic** and **inhomogeneous.** In this chapter, we will first define these two terms *in the context of rock engineering* and then explain two useful acronyms, **CHILE** and **DIANE**, representing respectively the assumptions required for modelling, as opposed to the actual rock properties. We then discuss the concepts of anisotropy and inhomogeneity, and conclude with a section on the ramifications of the ideas for rock engineering analysis and design.

10.1 Definitions

An anisotropic rock has different properties in different directions. These properties may be of any type: for example, deformability modulus, strength, brittleness, permeability and discontinuity frequency. In some cases, the ability to deal with anisotropy is built into the fundamental definition of the property itself, e.g. the compliance matrix for intact rock can contain up to 21 elastic constants which represent all possible types of elastic anisotropy. In other cases, for example, the compressive strength, there is no such in-built capability and the engineers are left to decide on the optimal characterization of anisotropy commensurate with their requirements. If we measure the compressive strength parallel and perpendicular to the laminations in a metamorphosed rock, is this sufficient to characterize the anisotropic variation?

An inhomogeneous rock has different properties at different locations. Again, this refers to any of the properties which we may be measuring. However, the ability to characterize inhomogeneity is not built-in to *any* of the fundamental definitions of the properties and we have to have recourse to statistical techniques. Later on, we will distinguish between 'point properties' and 'volume properties', the former being essentially the properties of intact rock, and the latter being essentially the properties governed by the structure of the rock.

Thus, the type of anisotropy and/or inhomogeneity variation could be gradual within the intact rock or sudden as a discontinuity is crossed. In fact, there can be variation on all scales: within grains or crystals, within the microstructure, within laboratory-sized samples of intact rock, within engineering structure-sized volumes of rock, and so on. These scales will be incorporated in our discussions in the succeeding three sub-sections.

However, an immediate distinction can be made now between the assumptions that are traditionally required for modelling and the real properties of the rock. This can be remembered by two acronyms: CHILE and DIANE.

A *Continuous, **H**omogeneous, **I**sotropic and **L**inearly-**E**lastic* (**CHILE**) material is one that is most commonly assumed for the purposes of modelling. Traditional stress analysis techniques are formulated in terms of these four attributes, simply for necessity and/or convenience for obtaining closed-form solutions. In the past, limited computational techniques precluded any more sophisticated analysis. Nowadays, however, especially in consulting and research organizations, there are computer codes available which will routinely deal with violation of any of these traditional assumptions. This leads directly to the second acronym.

A ***D**iscontinuous, **I**nhomogeneous, **A**nisotropic, **N**on-**E**lastic* (**DIANE**) rock is the material with which the engineer has to deal.* We should therefore consider the significance of the difference between the CHILE material being modelled and the DIANE rock being engineered, and the likely error arising from the direct application of a model based on a CHILE material. Alternatively, the specific attributes of the DIANE rock can be modelled. Superb examples of the latter procedure are the development of block theory and the use of distinct element techniques in numerical analysis.

There is a connection between each of the characteristics of discontinuousness, inhomogeneity, anisotropy and non-elasticity. For example, in a cyclothem of repeating sandstone, mudstone and limestone strata, containing bedding plane separations and exhibiting time dependency, there will be inter-relations between the attributes—because of the physical characteristics and mechanisms. The rock is discontinuous because of the bedding plane separations and any other fracturing that may be present. It is inhomogeneous because of the existence of the different rock types. It is anisotropic because of its sedimentary nature. It is not elastic because there is hysteresis and time dependency, perhaps related to the presence of the bedding planes. *Moreover*, the rock is anisotropic *because* it is inhomogeneous, e.g. the deformation modulus for a suite of strata will be different parallel and perpendicular to the bedding planes. *Also*, the rock is anisotropic *because* it is discontinuous. The matrix showing example connections between these four main attributes is shown in Fig. 10.1.

There is little chance of any modelling based on CHILE assumptions being realistic. There are a few circumstances where all the assumptions would apply simultaneously, to the extent where the modelling would produce results of value in rock engineering analysis and design.

*The acronym DIANE was coined by Branko Vukadinovic of Energoprojekt, Belgrade, Yugoslavia.

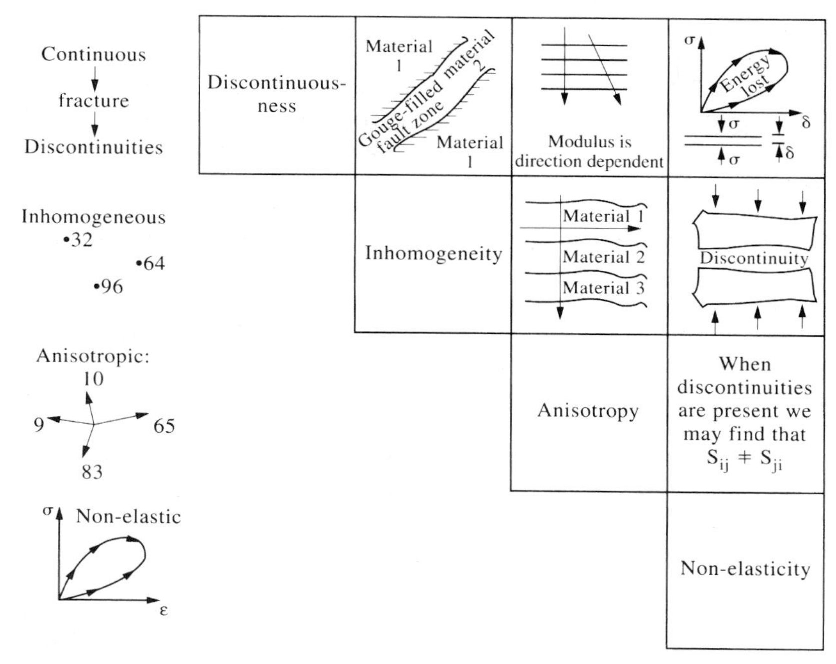

Figure 10.1 Connections between the attributes discontinuousness, inhomogeneity, anisotropy and non-elasticity.

Examples are the South African goldmines and well-bores at great depth: in these circumstances the high stresses can effectively close all discontinuities and the rock is more or less homogeneous and isotropic within the scale of the engineering being considered.

It follows, therefore, that engineers *must always question the results of all modelling in rock mechanics and rock engineering* to consider for themselves to what extent they consider the DIANE rock is well represented by the CHILE material in the model. In some circumstances such representation may be valid, in others, it may be wrong, misleading and dangerous to use. The following discussion is designed to provide the reader with some background knowledge in order to address this problem with confidence.

10.2 Anisotropy

The word 'anisotropy' is derived from the two Greek words *anisos* (meaning unequal) and *tropos* (meaning turning or direction). Directionality is one of the key aspects of rock engineering. If, for example, we conduct a site investigation using a vertical borehole, will the fracture data be useful for designing a horizontal tunnel? The answer to this question is that if the rock is fractured isotropically, the results will be most useful. However, if the rock is anisotropic, the results could well be misleading if used without care. We have already demonstrated in Chapter 7 that the discontinuity frequency can vary significantly with direction, and so the value derived from measurements made in the vertical borehole might be different to that in a specific horizontal direction. We also noted that properties such as the rock mass deformability and permeability will be functions of the discontinuity frequency, and hence will be anisotropic in nature. In the case of discontinuity frequency, we showed explicitly the variation with direction. For deformability, the architecture of the elastic compliance matrix takes into account the linking between stresses and strains and hence also explicitly quantifies some anisotropy. It was also explained in Chapter 9 that permeability is a second-order tensor with three principal permeabilities, again explicitly characterizing some anisotropy via the tensor. Figures illustrating these concepts have been included in Chapters 7–9.

However, not all rock properties have anisotropy incorporated into their characterization. For example, as was asked at the beginning of the chapter, how do we characterize the anisotropy of compressive strength? Compressive strength is *usually* assumed to be a scalar value, which is by definition directionless: measurements of compressive strength should be qualified with information on the direction of loading relative to the rock structure.

Figure 10.2 demonstrates the anisotropy of compressive strength recorded for a series of tests performed on a slate. In this case, the anisotropy can be characterized through application of the single plane of weakness theory (discussed in Chapter 8), which does have directionality built into its formulation.

One should be very careful with the measurement of any *assumed* scalar property in rock mechanics and rock engineering, because there is no inbuilt directionality in the characterization of such a property. The three most frequently measured parameters in rock mechanics and rock engineering are discontinuity frequency, RQD and point load strength. These are almost always (but incorrectly) assumed to be scalar properties (*and hence imply isotropy*), whereas, they are actually higher-order parameters (*implying anisotropy*).

Where it is economically viable, rock masses should always be assumed to be anisotropic unless it can be demonstrated that isotropy is a sufficiently accurate representation for the particular rock mass and engineering objective.

10.3 Inhomogeneity

The word 'inhomogeneity' is derived from the two Greek words *homos* (meaning the same, with the Latin prefix *in-* forming the negative) and *genos* (meaning kind). Anisotropy means having different properties in different directions at a certain location, *with the location unspecified*. Now we consider inhomogeneity, which means having different properties at different locations *given a certain measurement direction*. If the measurement direction is not specified, then a compound of the two aspects could occur.

We saw that anisotropy is intrinsic to the very definition of many geotechnical parameters. This is not the case for inhomogeneity, and so we must have recourse to statistical and geostatistical techniques. Understanding the inhomogeneity of rock can be important. Indeed, in many cases, we may be interested in the extreme values rather than the mean

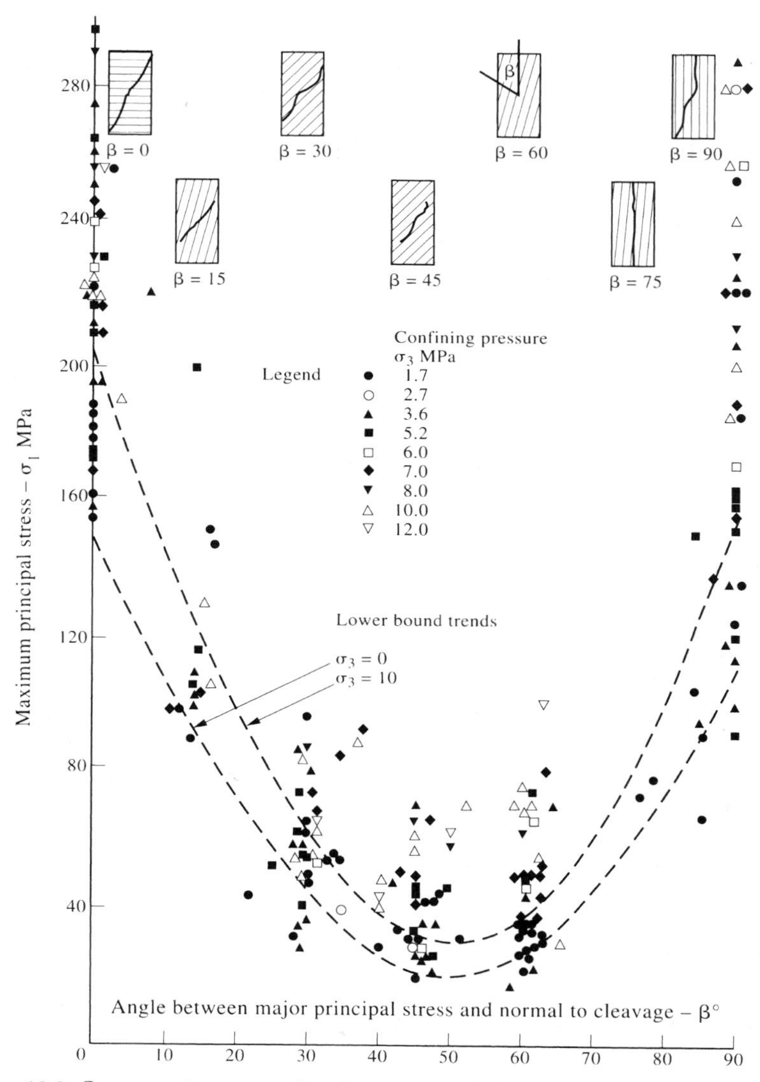

Figure 10.2 Compressive strength anisotropy in dark grey slate (after Brown *et al.*, 1977).

values of a rock property. For example, in choosing the type of cutters for a tunnel boring machine, not only the mean compressive strength would be required but also the range within which the top, say, 10% of strengths occur.

In Fig. 10.3, we illustrate both the standard statistical procedure for characterizing a parameter through the full probability density curve (*which does not explicitly take the distances between sample locations into account*) and the **semi-variogram** of geostatistics (*which does take these distances into account*).

Where we encounter rock properties varying with location within a rock mass, there are three main approaches to the characterization procedure:

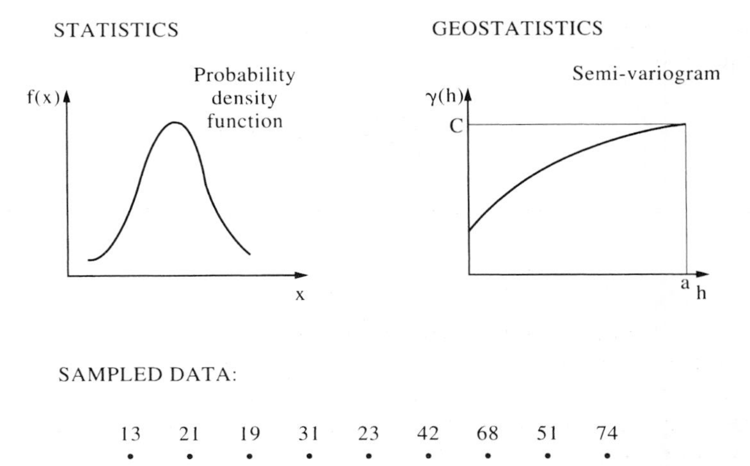

Figure 10.3 Methods of quantifying inhomogeneity through statistics and geostatistics.

(a) lump all the data into one histogram;
(b) separate the site into a number of discrete 'structural regions' and create a histogram for each; and
(c) use the techniques of geostatistics, specifically semi-variograms and kriging.

Note that the use of these three different techniques represents three ways in which the locations of the sampling points are taken into account. When the data are lumped together, the information on location is suppressed, except that all samples came from an assumed single universe. In the second case, the location information is similarly suppressed, except that different probability density functions can now be distinguished in the different sampled areas, and hence statements can be made about whether there is any variation between the regions of the sampled universe. In the third case, the distances between the specific sampling points are explicitly taken into account in the creation of a semi-variogram, as shown in the top right of Fig. 10.3 and explained below.

In this context of inhomogeneity, we should distinguish between accuracy and precision—which are our two main parameters for assessing the level of inhomogeneity using the lumped histogram approach. **Accuracy** is the ability to obtain the correct answer 'on the average': the sampled mean is, on the average, the true mean, i.e. there is no **bias** in the measurements. **Precision**, commonly measured by the standard deviation, is the degree of spread of the measurements, *whether or not they are accurate*. Considering the probability density function at the top left of Fig. 10.3, the material will be inhomogeneous if the spread of results is greater than that which would result from sampling error alone. In fact, this is our sole measure when the data are lumped together. However, if different probability density functions are constructed for the different structural regions, we can utilize the differences between their means as a measure of the inter-'structural region' inhomogeneity and the spread of individual histograms as a measure of the intra-'structural region' inhomogeneity.

The semi-variogram illustrated in the top right of Fig. 10.3 is derived from the equation

$$\gamma(h) = \frac{1}{2n}\sum_{i=1}^{n}[p(x) - p(x+h)]^2$$

where $\gamma(h)$ is the semi-variogram statistic for samples a distance h apart,
n is the number of sample pairs,
$p(x)$ is the rock property at location x, and
$p(x + h)$ is the rock property at location $x + h$.

Each statistic, $\gamma(h)$, refers to the overall variation for samples taken at a distance h from each other. In Fig. 10.3, this statistic is plotted against h to indicate the rock property variation as a function of distance between the observations. Naturally, when h tends to zero, we might expect $\gamma(h)$ to tend to zero (although this is not always the case because of measurement inaccuracies and sudden differences in the rock property, the so-called 'nugget effect'). More interestingly, as h increases, so $\gamma(h)$ will increase until it reaches a constant value—indicating no correlation between the data points making up each pair. This occurs at a distance $h = a$, which is the **range of influence** of a sample, and is at the value of $\gamma(h) = C$, termed the **sill** of the semi-variogram.

Although the techniques of geostatistics and Geographical Information Systems have not been fully exploited in rock mechanics, it is clear that the approach does take the location of the sample into account and does provide a method for quantifying inhomogeneity. The concept of the range of influence is important in establishing the distance to which one can extrapolate borehole information. Also, one can examine anisotropy by constructing semi-variograms in different directions.

Figure 10.4 shows simulated discontinuity patterns for both statistically homogeneous and statistically inhomogeneous cases. These patterns illustrate the need to account for inhomogeneity in order to develop a correct understanding of rock mass variability or rock mass structure at any site.

10.4 Ramifications for analysis

The overall validity of models has been discussed in terms of the CHILE and DIANE assumptions and, at this stage, it will be helpful to consider anisotropy and inhomogeneity in the modelling procedures. The models are either solutions for continuous materials or solutions for discontinuous materials, and in a few cases a combination of the two.

In the first case, with 'classical' solutions we have little room for manoeuvre. For example, as illustrated in Fig. 10.5, from Daemen's work, a set of 'laminations' (i.e. one set of parallel, planar and persistent discontinuities) has been included. Daemen has assumed, by applying the Mohr–Coulomb failure criterion for potential slip along the discontinuities, that the presence of the discontinuities *does not* affect the fundamental stress distribution around the tunnel, *but does* affect the strength of the material. This approach provides a useful indication of the likely areas subject to failure under these circumstances, and hence also provides guidance on support

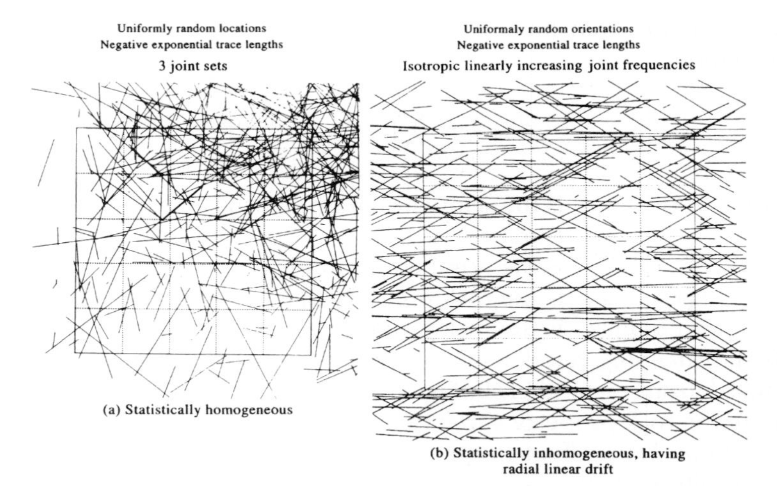

Figure 10.4 Computer-generated rock jointing patterns (from LaPointe and Hudson, 1985). (a) Statistically homogeneous. (b) Statistically inhomogeneous, having radial linear drift.

requirements, but the possibilities for extending the classical solutions in this way are limited. This not only applies to the anisotropy of the material properties, but also to anisotropy of the problem geometry.

With reference to the plane strain solution for the stresses around underground excavations, there are 'classical' solutions for circular and elliptical openings and, through the use of complex variables, various pseudo-rectangular shapes. However, extensions to, for example, the interaction between two parallel openings are not possible—this is the realm of engineering approximations. Thus, it is unlikely that our four main 'problem attributes' can be incorporated as extensions of classical solutions.

Over the last two decades, there has been development in computer-based numerical solutions which are specifically designed to deal with more complex geometry and material properties. These techniques include finite difference, finite element, boundary element and distinct element formulations, providing the capability of incorporating discontinuousness, anisotropy, inhomogeneity and more complex constitutive behaviour. With this capability, the types of rock properties that can occur need to be studied further.

An initial step in dealing with the four attributes is to consider the distinction between rock properties at a point and rock properties over a volume. In other words, there are some properties, such as density, which can be considered as essentially point properties and do not depend on the discontinuites. There are other properties, such as secondary permeability, which are *dictated* by the presence of discontinuities and cannot be considered as point properties: these are associated with a certain volume of rock. In Table 10.1, we present examples of both point and volume properties.

The distinction between the two types of property is not cut and dried. For example, the state of stress in a rock mass is, of course, influenced by the discontinuites; but considering the definition of stress at a point (which

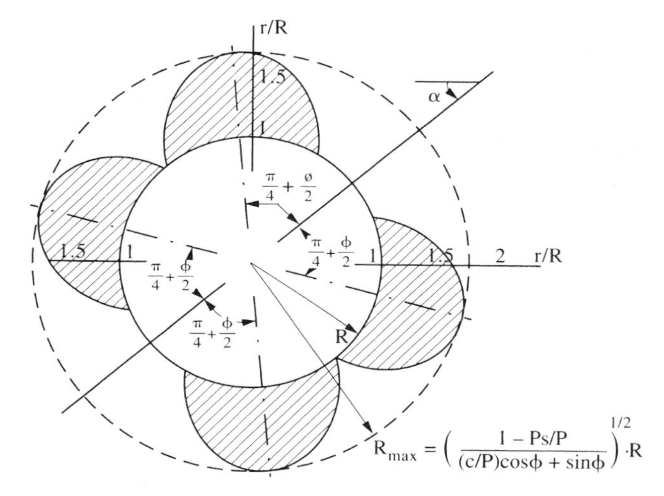

Figure 10.5 Use of the classical Kirsch solution for plane strain stresses around a circular opening for studying the effect of anisotropy of rock strength (from Daemen, 1983).

was given in Chapter 3), it is included as a point property. The key point about these two types of property is that, within the numerical analysis techniques which have been developed, variations in point properties can be accommodated relatively easily (although not necessarily comprehensively). For example, using the finite element technique we could incorporate variations in density, i.e. a form of inhomogeneity.

Of the four attributes, all the numerical techniques can, to a greater or lesser extent, accommodate wide variations in problem geometry and the presence of discontinuites. This is not the case for the inhomogeneity, anisotropy and constitutive behaviour relating to *volume properties*, because the individual elements in these numerical formulations should not be assigned a single value relating to a volume property **which may be varying on a scale commensurate with the elements themselves**.

Table 10.1 Examples of rock properties classified according to whether they are point properties or volume properties

Point Property *(not dependent on discontinuites)*	**Volume Property** *(dependent on discontinuities)*
Density	Modulus of deformation
Primary porosity	Secondary porosity
Permeability of intact rock	Permeability of the rock mass
Point load strength	Discontinuity frequency
Cuttability	RQD
State of stress	Rock mass classification indices

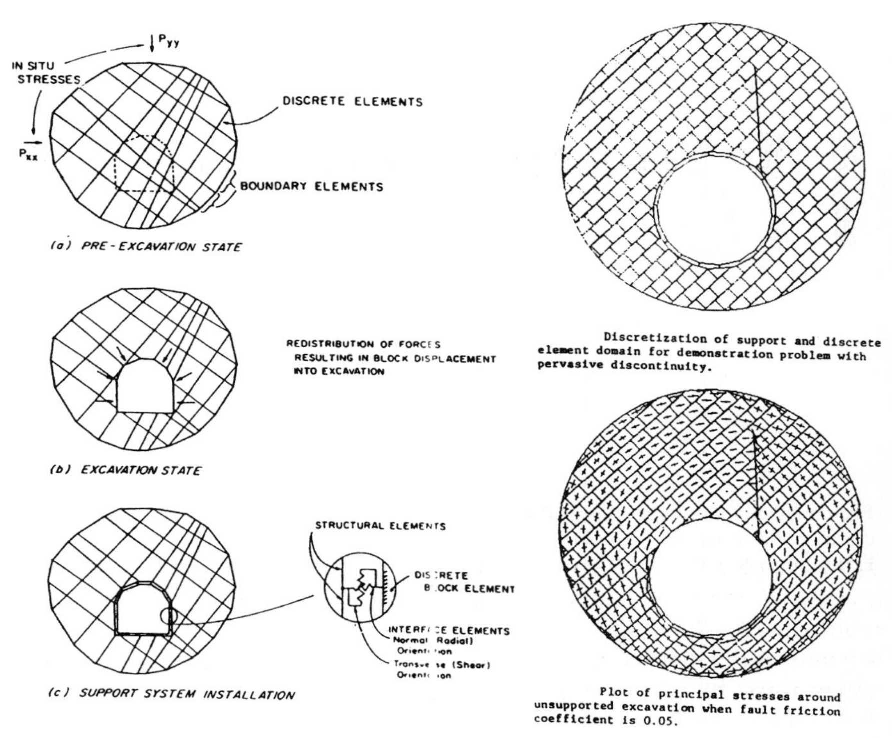

Figure 10.6 Schematic of a hybrid computational scheme for excavation and support design in jointed rock media (from Lorig and Brady, 1984).

Developments aimed at overcoming these difficulties are the use of hybrid numerical formulations which recognize the advantages of the continuum and discontinuum component methods. An example is shown in Fig. 10.6, from the work of Lorig and Brady (1984). Here, a boundary element solution has been utilized for analysing the **far-field**, and a distinct element model used for the **near-field**. There are advantages in assuming the material to be a continuum in the far-field while simultaneously modelling the discontinuities explicitly in the region of interest around the excavation. Via such a hybrid technique, we are able to model the discontinuities, tailoring the modelling to the engineering and design objectives.

We leave further discussion of inhomogeneity and anisotropy until later chapters and concentrate now on the implications of all the subjects covered so far on rock mechanics testing techniques.

11 Testing techniques

In this chapter, the first two sections concern the practical aspects of obtaining access to the rock mass for testing, and the general philosophy of the testing requirements—bearing in mind the engineering objective. In the three subsequent sections, there are overviews of tests on intact rock, discontinuities and rock masses. We conclude with a discussion on standardized tests.

11.1 Access to the rock

In Fig. 11.1, the main ways in which there is physical access to a rock mass are indicated. These are a rock exposure (whether at the surface or underground), lengths of borehole core, and the borehole wall itself after the hole has been drilled. Because the amount of exposed rock is limited and possibly locally altered, testing has tended to be concentrated on the cylindrical lengths of borehole core that are obtained by drilling during a site investigation and by measurements made within the borehole. When studying the illustrations in Fig. 11.1, recall our discussions about intact rock, discontinuities, rock masses, stresses, and the inhomogeneity and anisotropy factors in the previous chapters. What, exactly, is it that we wish to know about the rock mass in order to design and construct a rock engineered structure?

From a rock exposure, it is relatively easy to measure any property of the intact rock. Similarly, the rock mass structure is evident and a good estimate of most discontinuity properties can be obtained. Also, flat jack stress determination tests and larger-scale modulus and permeability tests can be conducted.

Considering borehole core, it is again evident that any mechanical property of the intact rock can be measured. The measurements of discontinuity properties are, however, immediately severely constrained. An excellent estimate of the discontinuity frequency in the direction of the borehole can be obtained, but virtually no information on the persistence of the discontinuities is revealed. One cannot measure the *in situ* stress from the borehole core (techniques such as anelastic strain recovery, differential strain analysis and the Kaiser effect are still questionable). Also, there is no possibility of measuring rock *mass* properties from the rock core.

Figure 11.1 Access to the rock mass via a rock exposure (top), borehole core (middle) and the borehole wall (bottom—which is an image derived from video tape).

What about tests that can be conducted on a borehole wall? It is evident that, apart from directly viewing the rock, the testing will usually consist of indirect methods of assessing the rock mass properties—due to limitations imposed by the borehole size. An example is shown in the lower part of Fig. 11.1, where the dark sinusoidal wave indicates the trace of an inclined discontinuity intersecting the borehole wall, being viewed via the rotating mirror of a borehole camera.

There are always constraints on resources, and so it is necessary when optimizing the rock characterization procedures to consider the requirements and to choose the rock access method and testing techniques in accordance with the engineering objective. **Because there are many different rock engineering objectives, there can be no standardized site investigation**. Individual tests can—and indeed should—be standardized, but the total programme and number of tests cannot be specified independently of the objective. For example, the information requirements for designing a block caving mining operation and a radioactive waste repository are different.

11.2 Tailoring testing to engineering requirements

There are the three main methods of accessing the rock, and there are many tests that can be conducted. The objective is to tailor the testing to the engineering objective. In Fig. 11.2, there is a histogram of the rock parameters that have been studied in association with the design of pressure tunnels for hydroelectric schemes worldwide. The parameters have been plotted in order of frequency of occurrence along the horizontal axis and indicate the perceived first 10 (for example) most important parameters involved in the design of pressure tunnels (this histogram has been constructed from a study of the literature, but could equally well have been obtained from existing design practice or numerical modelling requirements). It follows that we should design our testing programme accordingly in line with the design requirements. In this example, *in situ* stress is regarded as being of prime importance and should, therefore, definitely be determined. Conversely, it is assumed that in most cases the *in situ* stress is not an important parameter for surface blasting and slope stability, and therefore would not be determined to support these objectives.

Standardized procedures are advantageous for measuring rock properties and site conditions, but we must decide whether we are going to make particular measurements on the basis of the overall objective. If a slope is being designed, there is a great deal of experience for guidance; if a new structure is being designed, such as the tunnels housing a superconducting energy storage magnet, the most relevant parameters have to be established. In the latter case, since the magnet expands and contracts on charging and discharging, it could well be the fatigue properties of the excavation-peripheral discontinuities that are of paramount importance, and there are no standardized tests for these.

Finally, while discussing the overall strategy of approach, one has to consider whether emphasis is to be placed on index tests, fundamental tests, or a combination of the two. An index test is one that can be

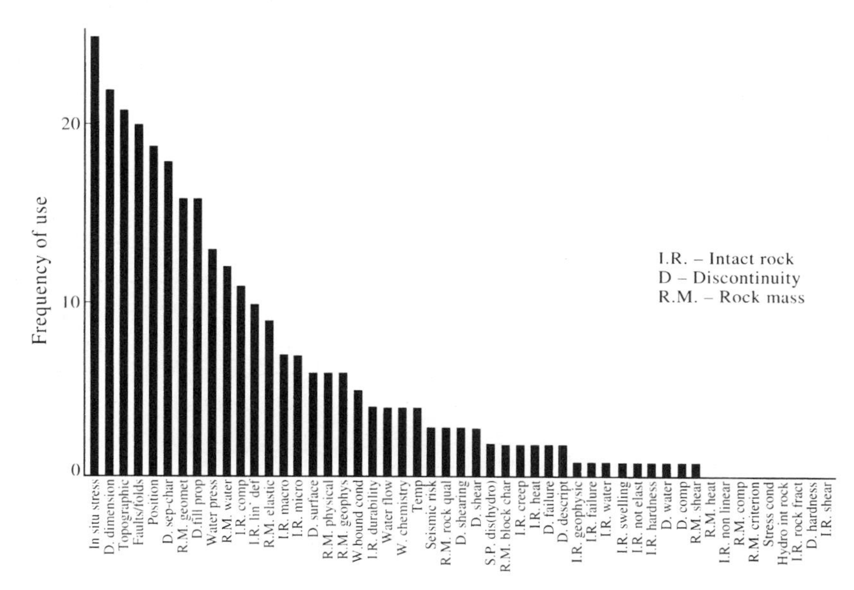

Figure 11.2 Importance of rock mechanics parameters for hydroelectric scheme pressure tunnel design (as established from the literature and compiled by Arnold, 1993).

performed quickly, but may not determine an intrinsic property—the point load test and Schmidt hammer rebound tests are examples. A fundamental test, on the other hand, measures a conventionally accepted intrinsic property, such as the compressive strength. One can estimate the compressive strength via the point load test in the field, or one can conduct direct compressive strength tests in the laboratory. The latter are more expensive and time consuming but do measure the property directly. Alternatively, one could conduct many tests using the point load apparatus, with fewer direct tests in the laboratory, or rely mainly on the point load test calibrated occasionally against direct tests in the laboratory.

The example in the previous paragraph is illustrative of all rock mechanics testing. Consider the measurement of the *in situ* rock deformation modulus. Should one use a dilatometer in a borehole, or a tunnel jacking test, or estimate the modulus using an analytical model and laboratory-determined values for the component parameters? In Fig. 11.3, we illustrate a 6 MN tunnel boring machine tunnel jacking test in chalk. The modulus determined by this means was between 7 and 10% of the laboratory, determined value, and of the same order as that determined from a finite element analysis back-calculation of ground settlement, back-analysis being yet another method available to the engineer.

In the majority of cases, the rock mechanics information is obtained from tests on borehole core, so it is essential that the drilling report and borehole core logs are correctly completed and available. In this book we will not deal with borehole core logging: instead we refer the reader to the Geological Society of London recommended procedures, the relevant British Standard, and, of course, all other relevant National Standards. With the advent of significantly increased microcomputing power, there is a move to provide more automation and immediate data acquisition and presentation in the field. In Fig. 11.4, one idea for automated discontinuity location recording is presented; this was developed by Nordqvist (1984).

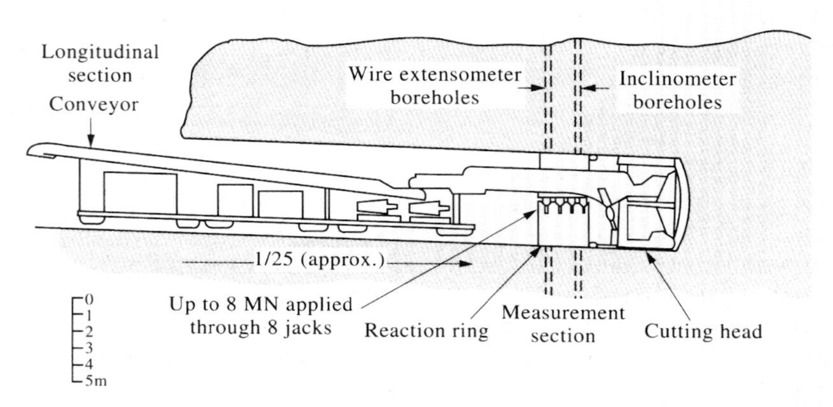

Figure 11.3 A 6 MN tunnel loading test for estimating *in situ* rock modulus, from Hudson *et al.* (1977).

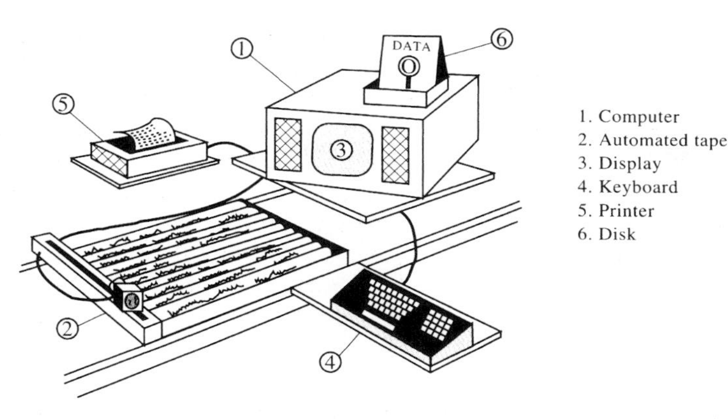

Figure 11.4 Automated discontinuity location measuring equipment (after Nordqvist, 1984).

11.3 Tests on intact rock

Tests on intact rock are associated with describing the character of the rock material, measuring the mechanical properties, and measuring other properties such as permeability. As rock mechanics has developed, following geological guidelines and soil mechanics experience, the description of rock has tended to be of a 'field-book entry' type of approach during borehole logging. One wishes to know the lithology, colour, etc, of the rock, but in classifying for rock engineering the mechanical properties are also required.

In Fig. 11.5, there are photographs of the two most common field tests conducted in association with exposure examination and borehole logging in the field: these are the **Schmidt rebound hammer** and the **point load test**. Both of these are *index tests* in that they provide a measure of the 'quality' of the rock; the fundamental properties are estimated from tables using the measured index values.

The Schmidt rebound hammer is a portable device, by which a spring-driven cylindrical hammer rebounds off the rock surface; the rebound distance is considered to be a measure of the rock quality. The hammer can be used directly on a rock surface, or on a rock core: in the latter case, a special support cradle is required. When the *in situ* block size is large, the Schmidt hammer does measure the properties of the intact rock; when the rock is fragmented, the use of the Schmidt hammer on the exposed rock surface will be a measure of the rock mass quality rather than the intact rock *per se*. In addition, by the very nature of the test, the condition of the tested rock surface will have a significant effect on the result, because of geometrical irregularities or because the surface itself has deteriorated and is not representative of the fresh, intact rock. For this reason, it is recommended that the Schmidt hammer is used repeatedly within the immediate vicinity of a measurement location. If it is suspected that variation in the results is occurring because of geometrical irregularity, the lower readings should be discarded. If the surface is weathered, then all the values will be significant. When measurements are made on

Figure 11.5 The Schmidt hammer and point load test equipment.

discontinuity surfaces, the condition of the surfaces is of particular importance. In Fig. 11.6, we show a series of empirically determined curves relating the Schmidt hammer readings for various hammer types and orientations to the unconfined compressive strength. Despite its apparent simplicity, the Schmidt hammer has proved to be one of the most useful indicators of rock strength.

The point load test is used on small pieces of rock which are either borehole core or irregular lumps. The test is derived from the so-called Brazilian* test in which a disc is compressed diametrally between two loading platens and the tensile stress at failure, σ_t, calculated from the breakdown load as

$$\sigma_t = (2/\pi)\frac{P}{Dt}$$

where P is the load on the disc at failure,
D is the disc diameter, and
t is the disc thickness.

When the effect of the geometry of the disc is studied, it is found that the outside perimeter shape has little effect on the above formula: for example, a square shape loaded across opposite corners gives a similar result for the tensile strength. For this reason, and because the testing may be conducted on irregular lumps (as shown in Fig. 11.5), a **point load index**, I_S, was developed by Franklin (1985):

$$I_S = \frac{P}{D^2}\text{(units of stress).}$$

This formula is directly related to the Brazilian test formula given earlier, except that the term D^2 has replaced the term Dt because the specimen could be an irregular lump. The reader should note that a coherent set of units must be used in evaluating I_S: in order to produce a result with units of MPa, units of N and mm are required for P and D, respectively.

The size effect was discussed earlier and the point load test is no exception to this phenomenon. The standard test is conducted on a core of 50 mm diameter, with correction methods being available to convert the measured index to the I_{S50} index if a different sized core or lump is used. There are also methods of characterizing the results for tests conducted parallel and perpendicular to the structure in anisotropic rock. In Fig. 11.6 the calibration curve and a set of results for converting the point load index number to unconfined compressive strength (UCS) are given.

The point load test is useful because hundreds of tests can be easily completed in a day with minimal sample preparation. Also, the prediction of compressive strength has proved to be remarkably accurate over a wide range of rocks. Bearing in mind that the sample can be irregular, that elasticity theory is unlikely to be the correct predictive model, that failure will probably occur under the loading platen, and that the compressive

*Readers may be interested to know the apocryphal tale surrounding the Brazilian test—that it was developed from the observation of a church being moved in Brazil on concrete rollers: when the rollers split, the idea of the test was born.

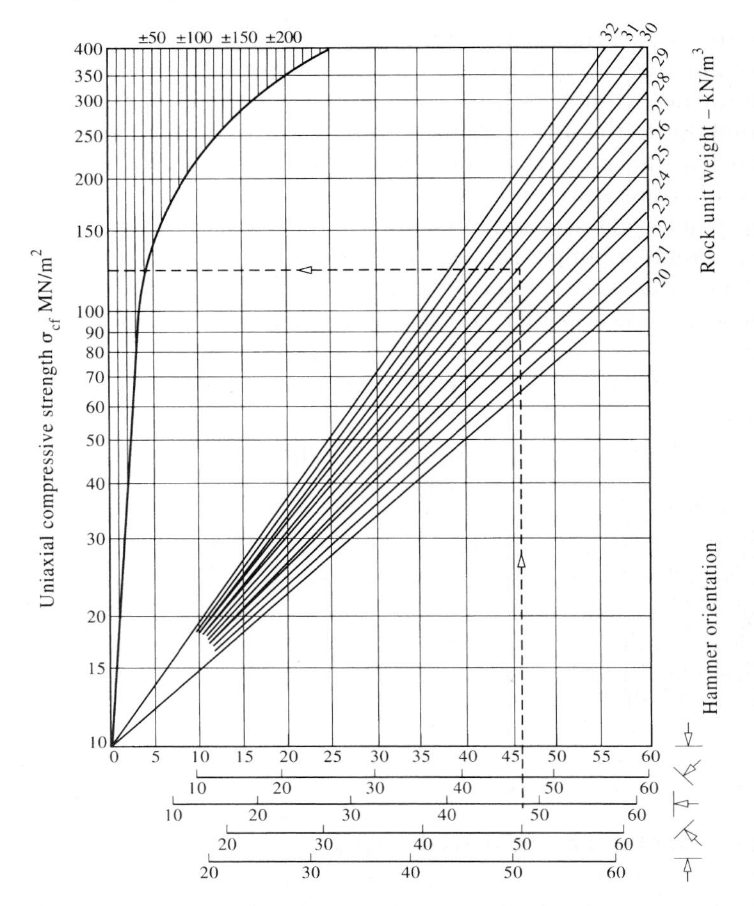

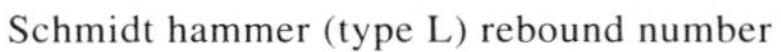

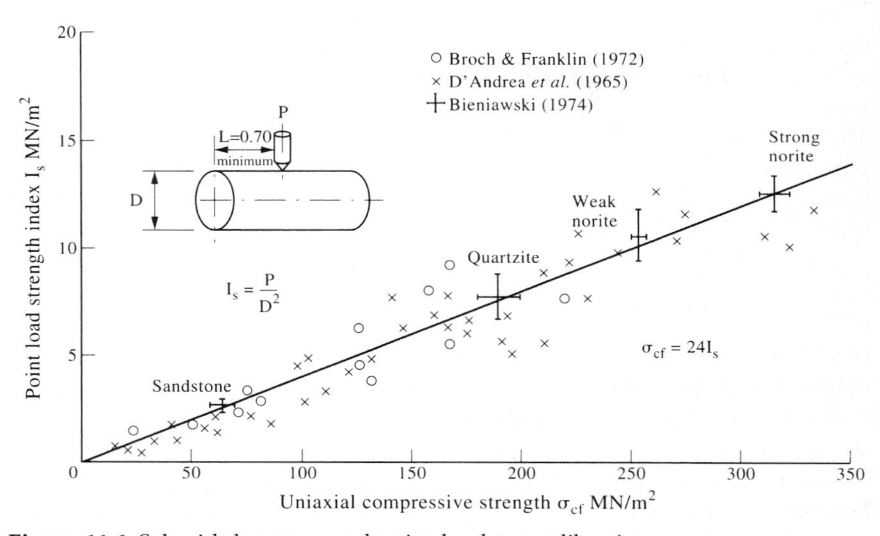

Figure 11.6 Schmidt hammer and point load test calibration curves.

strength is predicted from a calibration curve, then we should be grateful as engineers that the test does indeed prove to be useful. Over the years, the point load test has become the most widely used test for measuring the strength of intact rock.

Other examples of index tests are the measurement of acoustic velocity and slake durability. The acoustic velocity can be measured easily using portable equipment, the *P*- and *S*-wave velocities either being used as an index in their own right (indicating anisotropy and/or inhomogeneity), or Young's modulus and Poisson's ratio can be estimated using relations developed from the theory of elasticity. Since the stress wave velocities depend on the elastic moduli of the rocks and there is an empirical correlation between rock moduli and rock strength, the UCS can also be estimated from acoustic testing.

The degradability of rock is important when engineering in 'soft' rocks. A typical material that degrades is shale: a tunnel excavated in shale may initially be stable, only to collapse a few days later. It is useful, therefore, to be able to assess the degradability of the rock, for which purpose the slake durability test was developed by Franklin (1979). A piece of rock is placed in a copper cage which is rotated under water under specified testing conditions. The loss of sample weight is a measure of the susceptibility of the rock to the combined action of slaking and mechanical erosion.

Fuller descriptions of these and other tests on intact rock are given in the standardized testing procedures published on behalf of the ISRM and National Standards bodies, such as the American Society for Testing and Materials (see Table 11.1).

11.4 Tests on discontinuities

The discussion in Section 11.3 is concerned with the rock material, i.e. the solid blocks in the rock exposure shown in the top photograph and the solid core pieces shown in the middle photograph of Fig. 11.1. Now we discuss testing the breaks in the continuum, the discontinuities.

11.4.1 Geometrical attributes

Referring to the middle photograph of Fig. 11.1, certain geometrical properties of the discontinuities can be measured easily from a length of core. The properties can be determined more accurately using a scanline set up on a rock surface, as in the top photograph of Fig. 11.1. A purpose-designed logging sheet for borehole core, which can also be used for scanline work, is presented in Fig. 11.7. The specific contents of this logging sheet which should be noted are:

(a) the fact that there is a photograph of the core;
(b) there is an associated sketch of the discontinuties present within it;
(c) the discontinuities are numbered;
(d) the distance to each discontinuity is measured;
(e) the angle of each discontinuity to the core axis is measured;
(f) and there are comments on the discontinuity type (genesis) and surface condition and/or coating.

Table 11.1 ISRM and ASTM published testing methods (as at 1992)

ISRM TEST METHODS	ASTM TEST METHODS
Petrographic Description	Laboratory Determination of Pulse Velocities and Ultrasonic Elastic Constants
Hardness and Abrasivity	Creep in Uniaxial Compression
Monitoring Rock Movements using Borehole Extensometers	Creep in Triaxial Compression
Determining Sound Velocity	Direct Tensile Strength
Quantitative Description of Discontinuities	Modulus of Deformation using Flexible Plate Loading
Tensile Strength	Modulus of Deformation using Rigid Plate Loading
Uniaxial Compressive Strength and Deformability	Rockbolt Anchor Pull Test
Water Content, Porosity, Density, Absorption	Rockbolt Long-Term Load Retention Test
In Situ Deformability of Rock	*In Situ* Deformability and Strength in Uniaxial Compression
Pressure Monitoring using Hydraulic Cells	Dimensional and Shape Tolerances of Rock Core Specimens
Geophysical Logging of Boreholes	*In Situ* Creep
Strength in Triaxial Compression	*In Situ* Shear Strength of Discontinuities
Surface Monitoring of Movements across Discontinuities	Modulus of Deformation using a Radial Jacking Test
Rock Anchorage Testing	Permeability Measured by Flowing Air
Point Load Strength	Thermal Expansion using a Dilatometer
Deformability using a Large Flat Jack	Elastic Moduli of Intact Rock in Uniaxial Compression
Deformability using a Flexible Dilatometer	*In Situ* Stress by USBM Borehole Deformation Gauge
Rock Stress Determination	Rock Mass Monitoring Using Inclinometers
Fracture Toughness	Specific Heat
Seismic Testing within and between Boreholes	Splitting Tensile Strength of Intact Rock Core
Laboratory Testing of Argillaceous Swelling Rocks	Transmissivity and Storativity of Low Permeability Rocks using the Constant Head Injection Test
Large Scale Sampling and Triaxial Testing of Jointed Rock	Transmissivity and Storativity of Low Permeability Rocks using the Pressure Pulse Technique
	Triaxial Compressive Strength
	Undrained Triaxial Compressive Strength
	Unconfined Compressive Strength
	Thermal Diffusivity

PHOTOGRAPH OF CORE

Sketch of discontinuities

Nominal 3m length

Corebox Logsheet: Tan-y-Bryn
Borehole 42: Weeping Granite Cliff

Corebox logged by: | Noted by JAH:

Corebox No:	Depth: (from collar, m)	To: (m)
Core dia. (mm)		
Date:	Depth: (AOD, m)	To: (m)

List of Discontinuities

Discontinuity No. and Type	Collar Depth to discontinuity centre	AOD depth to discontinuity centre	Angle to core, ie assumed dip	Comments on discontinuity type (incipient, discrete, compound) and surface coating (clean, green, red, white, slickensided, etc)

Overview Statistics

Length of core in the corebox (m):	Number of natural discontinuites in the corebox:	Fracture frequency/m: Rock Quality Designation (RQD):

Characteristics of the intact granite

Colour/ Grain size	Strength (MPa)	Any other comments:

Note: This rock engineering logsheet has been developed specifically for the Tan-y-Bryn Weeping Granite Cliff hard rock boreholes using the REMIT system (Rock Engineering Mechanisms Information Technology). With this system, the objective is to match the informationbeing obtained with the engineering objective. These corebox logsheets should not, therefore, be used for any other purpose, unless modified appropriately.

Figure 11.7 Example of a borehole core logging sheet.

From the record, and the overview statistics, one can determine the discontinuity frequency (and hence mean spacing), the discontinuity spacing histogram and the RQD.

This style of log sheet is designed for direct input to a computer and hence to be able to take advantage of modern developments in databasing and information technology. The log sheet is also designed to assist a structural geologist to interpret the information, because we advocate the use of a 'geological approach' whereby intelligence can be incorporated into the sampling process, resulting in a staged approach and hence optimal use of resources. Databasing the information contained on such log sheets is still in its infancy, as is the automated recording of values illustrated in Fig. 11.4. However, both these subjects will develop in the future.

Once the information is contained within a database, a multitude of analyses can be conducted in various interrogative modes. For example, from the log sheets compiled from granite core, one can assess the occurrence of, say,

$$\text{(slickensided discontinuities)} \cap (\approx 30^\circ \text{ dip}) \cap \text{(green coating)}$$

on the assumption that thrust faulting may be present. This is just one example of how the speed associated with interrogation of a computer database can assist—by showing the relevant relations contained within it, as determined by the engineer and geologist. Moreover, algorithms can be developed to produce the semi-variograms and associated criteria for partitioning the rock mass into different structural regions.

We explained earlier that it is generally necessary to have 50 or so discontinuities for a reasonable estimate of discontinuity frequency, and of the order of 200 to provide a reasonably coherent histogram. If the information is contained within a database, algorithms can be used to assess directly the significance of the variety of discontinuity statistics that may be output. These ideas also apply to measurements on exposures, where other parameters such as trace length can be measured.

We are not discussing in any further detail the use of borehole television cameras, as illustrated in the lower photograph of Fig. 11.1, except to say that we anticipate greater use of these as the technology of the video images improves.

11.4.2 Mechanical attributes

There is a variety of testing procedures for the mechanical attributes of discontinuities, ranging from the **tilt test**, through the **field shear box** and standard triaxial procedures to sophisticated tests on servo-controlled equipment.

The tilt test, illustrated in Fig. 11.8, is used to estimate the angle of friction between the discontinuity surfaces, or indeed any two rock surfaces. The test can be either carried out simply by hand, or with increased sophistication using an elevating cradle. The test is simple, with the angle of inclination when slip occurs directly indicating ϕ_j, but the process by which

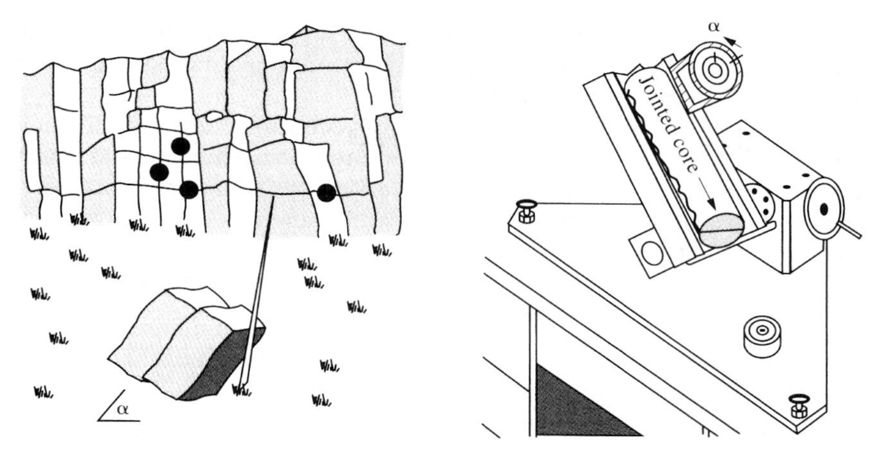

Figure 11.8 The tilt test for measuring the angle of friction between discontinuity surfaces (after Barton *et al.*, 1985).

one irregular surface slides over another is complex. For good practical reasons, in the past it has been assumed that there is only one friction angle, i.e. there is no anisotropy in the roughness of the discontinuity. In fact, because of the mode of formation of most discontinuities, there will be anisotropy in the friction angle. The tilt test should be conducted in several different directions with respect to any visible directional structure. Also, if there is any stepping on the discontinuity surface, the friction angle will vary with the direction of relative slip across the discontinuity. The direction of faulting causing natural slickensiding on discontinuity surfaces can be detected by the roughness of the surface in different directions.

The most widely used apparatus for rapid determination of discontinuity strength parameters is the field shear box, as illustrated in Fig. 11.9. A sample of rock containing a discontinuity is cast in plaster and set in the apparatus such that the discontinuity can be sheared between the two halves of the box. This can be conducted at varying levels of normal stress

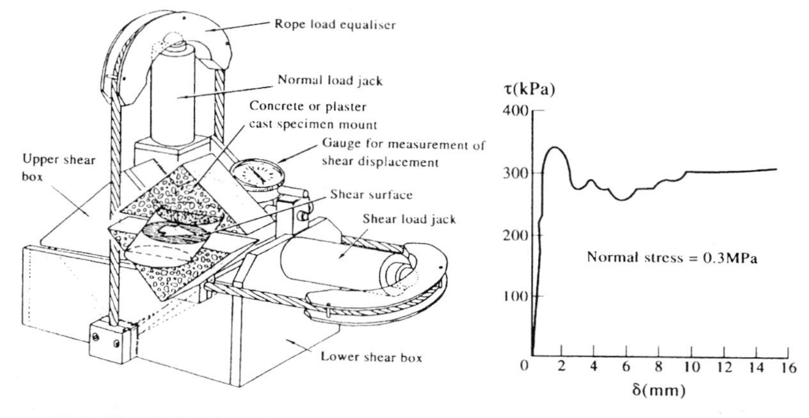

Figure 11.9 The field shear box and typical results.

and rates of shearing, allowing a wide range of parameters to be obtained (c, ϕ, curvature of the failure locus, residual strength, variation with shearing rate, and so on).

In order to measure the more complex discontinuity stiffness behaviour described in Section 7.3.1 and to obtain more precise information on the strength and failure behaviour, it is necessary to use a laboratory-based shear or triaxial testing machine. Such tests are difficult to conduct, because:

(a) of the possibility of sampling disturbance causing premature failure of the discontinuity;
(b) a rock specimen containing a discontinuity does not manifest uniform behaviour in the triaxial cell, causing relative rotation of the specimen halves or puncturing of the sleeve, as illustrated in Fig. 11.10;
(c) the discontinuity properties are likely to be anisotropic and so it is time consuming and difficult to establish, for example, the 3×3 stiffness matrix relating the normal and two shear stresses with the normal and two shear displacements.

Notwithstanding these difficulties, it is important in many rock engineering projects to understand the mechanical behaviour of discontinuities in detail, and to study the combined thermo-hydro-mechanical properties. In the Hot Dry Rock geothermal energy project (where cold water is pumped down one borehole, passes through a fractured zone within the hot rock, and returns via another borehole), there is little experience to guide the work so it must be driven by numerical analysis. Determination of the discontinuity properties for these complex conditions is a vital ingredient in supporting the analysis process. The TerraTek testing machine illustrated in Fig. 11.11 has been used at Imperial College to study such behaviour.

11.5 Tests on rock masses

The determination of rock mass properties can be approached in two ways:

(a) via the properties of the intact rock and the properties of the discontinuities, which together make up the rock mass properties; or
(b) via the properties of the rock mass as measured or estimated directly.

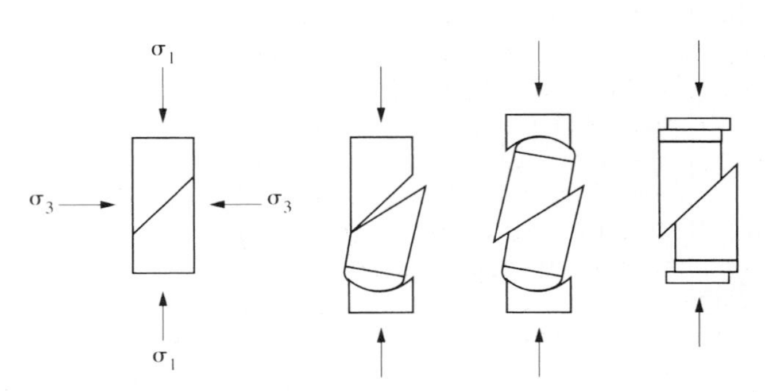

Figure 11.10 Triaxial testing of discontinuities (after Brady and Brown, 1985).

Figure 11.11 TerraTek servo-controlled triaxial testing machine at Imperial College as used for thermo-hydro-mechanical testing of discontinuities.

Sections 11.3 and 11.4 have separately included the items in (a) above, therefore direct testing of rock masses is now discussed. Always remember, though, that the measured properties and overall characterization will vary according to the project requirements.

Testing rock masses is a large subject: we provide here an overview of deformability, strength and permeability testing: for more advanced treatment the reader is referred to the book by Dunnicliff (1988).

A logical extension of the use of the site investigation borehole is to install some form of loading within the borehole, so that a force–displacement curve is obtained and the associated elastic parameters of the rock estimated. In soil mechanics, pressure meters and dilatometers are used extensively and attempts are continually being made to develop similar instrumentation for rocks with higher moduli. Rock masses are usually anisotropic and so an essential feature of such a device must be the ability to both apply load and measure displacement in different radial directions. One of the best of these types of device is the Goodman Borehole Jack, shown in Fig. 11.12 together with example results. The ability to estimate the modulus, varying both the position and direction of application of the load, is a major advantage.

The development of this, and other similar, devices has not been without difficulties. A salutory paper published by Heuze and Amadei (1985) lists the interpretative problems encountered by several investigators and documents the evolution of the Goodman jack. For example, imagine estimating the overall rock mass modulus from a series of measurements made on a borehole wall, often in close proximity to discontinuities. There will be a range of moduli values as the jack alternately measures within intact rock blocks and at locations where discontinuities intersect the borehole wall.

A similar circumstance occurs with a plate loading test conducted either on a surface rock exposure or underground. In this test, a large steel plate is set on a cement grout pad and loaded, usually by the application of dead weights or by means of a hydraulic ram reacting against an opposing tunnel wall or a system of rock anchors, as illustrated in Fig. 11.13. A force–displacement curve can be generated from the hydraulic pressure

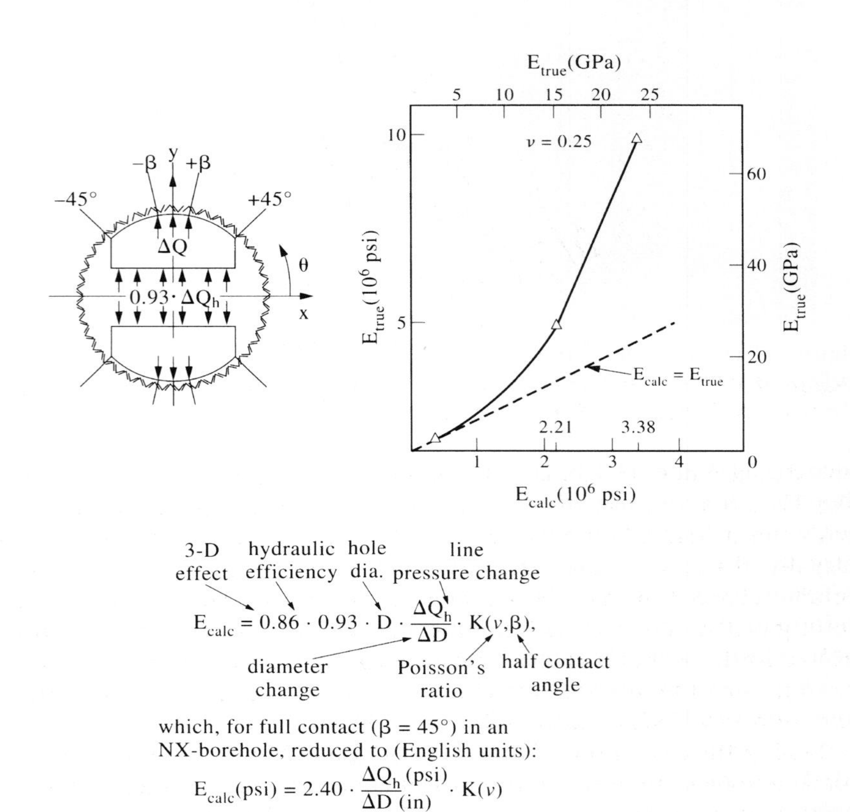

$$E_{calc} = 0.86 \cdot 0.93 \cdot D \cdot \frac{\Delta Q_h}{\Delta D} \cdot K(\nu,\beta),$$

$$E_{calc}(\text{psi}) = 2.40 \cdot \frac{\Delta Q_h\,(\text{psi})}{\Delta D\,(\text{in})} \cdot K(\nu)$$

Figure 11.12 The Goodman Borehole Jack and example results.

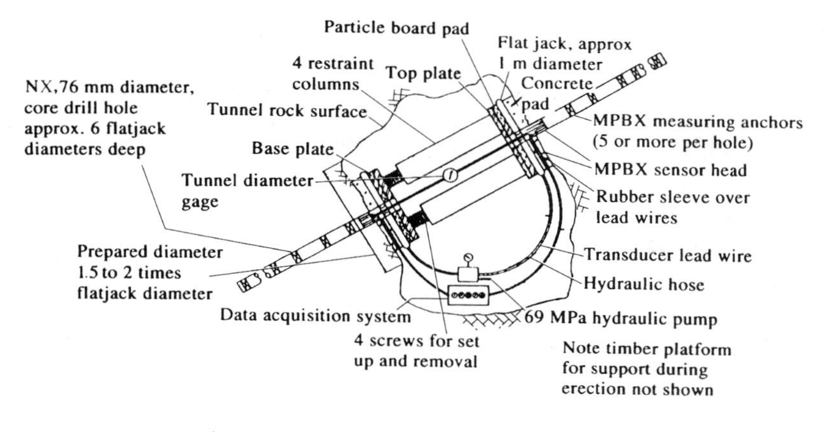

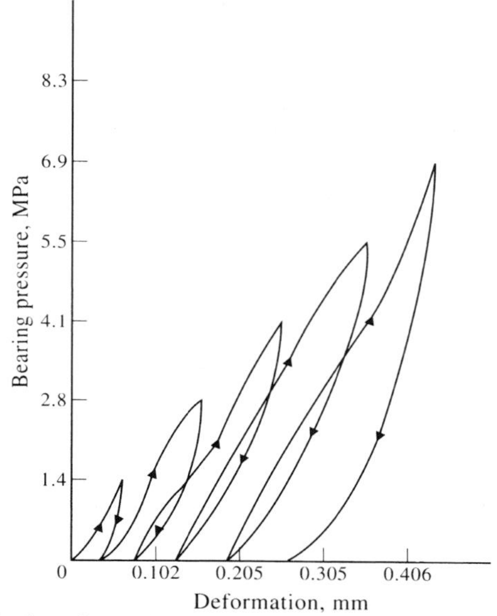

Figure 11.13 The plate loading test for estimating rock mass deformability (after Brown, *ISRM Suggested Methods*, 1981).

and displacement transducers located beneath the plate. It can be seen in Fig. 11.13 that there is hysteresis present in the loading–unloading cycles, with the attendant problem of establishing the actual modulus value. Usually, the hysteresis is directly associated with the presence of discontinuities. Large tests have been attempted, such as the surface chalk moduli testing programme conducted using a water tank loading the rock surface at Munford, UK (Burland and Lord, 1969) or the underground chalk moduli testing using the reaction ring of a tunnel boring machine at Chinnor, UK (Hudson *et al.*, 1977), illustrated in Fig. 11.3.

Testing the strength of a rock mass is also difficult because of the high loads involved. One seminal paper, illustrating the existence of the size effect in coal pillars up to 6 feet wide by testing them to destruction, was published by Bieniawski (1968). In this type of test, rock pillars in an under-

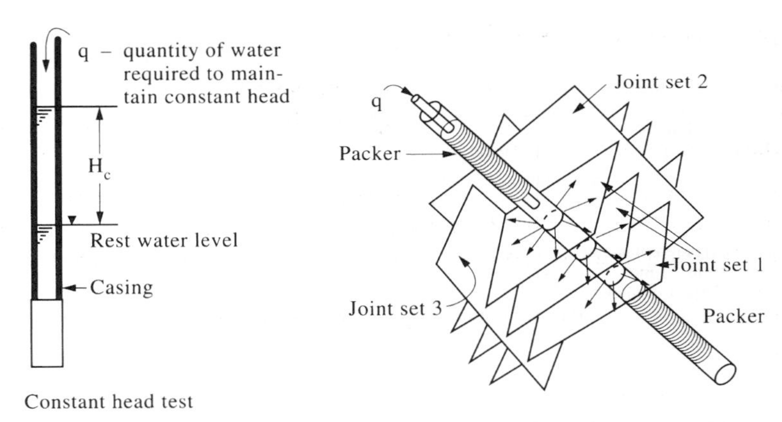

Figure 11.14 Borehole permeability testing (after Hoek and Bray, 1977).

ground facility are formed to the desired size and loaded using hydraulic flat jacks or by further excavation, using the excavation periphery for reaction.

Traditional methods of estimating permeability are summarized in Hoek and Bray (1977). These include the borehole falling-head test, the borehole recharge test and the borehole packer permeability test, as illustrated in Fig. 11.14. Although such estimations of the permeability suffer from all the deficiencies described in Chapter 9, the tests can be useful as indices. If, however, the permeability is required at the REV size, then one method is to isolate a large underground excavation, circulate air through it, and measure the change in air moisture content.

We discussed stress measurement in Chapter 4, with devices for determining *in situ* stress being discussed in Section 4.3. As an exciting analogue to the large-scale deformation and permeability tests, large *in situ* stress determination tests are now being attempted by shaft and tunnel 'undercoring', in which an excavation is driven through an instrumented zone of rock and the stress back-calculated from the measured responses. This is one way to deal with the REV problem, but there are limitations to the number of such tests that can be conducted in estimating the stress over the region of a large structure such as a hydroelectric scheme or radioactive waste repository.

Because of the difficulties associated with rock mass testing which we have highlighted here, the subject is one of the most important research areas in rock mechanics and rock engineering. There are several surface and underground facilities around the world where tests are being conducted to solve the basic problem of rock mass characterization. The Underground Research Laboratory (URL) at Pinawa, Canada, operated by the Atomic Energy of Canada Ltd, has had an on-going research programme addressing these problems for the last decade. More stress measurements have been conducted at URL (at least 800 tests) than at any other single site in the world, and extensive work is being conducted on permeability estimation. We await the conclusions of these and other such programmes before recommendations can be made on 'the way ahead'.

11.6 Standardized tests

Although the strategy of rock characterization is a function of the engineering objectives, the tactical approach to individual tests can be standardized. The advantages in doing this are that:

(a) the standardization guidance is helpful to anyone conducting the test;
(b) the results obtained by different organizations on rocks at different sites can be compared in the knowledge that 'like is being compared with like'; and
(c) there is a source of recommended procedures for use in contracts, if required.

Also, there is an increasing move towards paying contractors according to the quality of the rock, and it is only through the use of some form of standardized procedures that one can hope to determine the quality objectively.

These are the practical advantages. There is no intention in the minds of those producing these standards that they should in any way inhibit the further development of rock mechanics and rock engineering. In fact, most research projects and many engineering projects will take the testing procedures beyond these standards. The International Society for Rock Mechanics Commission on Testing Methods has been producing Suggested Methods for rock testing and characterization since 1978, and these are widely used. There are also national bodies which produce standards for their own countries. In particular, the American Society for Testing and Materials (ASTM), via Committee D18.12, has produced an extensive series of methods for rock testing. There are many other countries which have their own wide range of standards. To illustrate the ISRM and ASTM test methods that are available for testing rock, we have compiled Table 11.1 (in which the publications are listed chronologically).

11.6 Standardized tests

12 Rock mass classification

In earlier chapters there was discussion about the influence of the rock mass structure on the rock mass properties necessary for both the theory and practice of rock engineering. In Chapter 20, we will refer to block theory and to the fact that there is now a complete topological solution to the rock block geometry. This validity of the theory depends critically on the persistence of the discontinuities. We also mentioned that, given the discontinuity geometry and all the associated stiffnesses, the deformability of a rock mass can be calculated. But the ability to make this calculation depends on the availability of data on the discontinuity geometry and stiffnesses.

It is evident that even with the most generous resources available for site investigation, there remain problems in applying the theories in practical engineering circumstances. As a consequence, several engineers have developed rock mass classification schemes—which are essentially a compromise between the use of a complete theory and ignoring the rock properties entirely. All the classification schemes consider a few of the key rock mass parameters, and assign numerical values to the classes within which these parameters lie for a given rock type. As we will see, the schemes provide a short-cut to the rock mass properties that are more difficult to assess (e.g. the prediction of rock mass deformability) and provide direct guidance for engineering design (e.g. in predicting the amount of support required for a tunnel). One of the pioneers of rock mass classification, Professor Z. T. Bienawski, has recently written a book *Engineering Rock Mass Classifications* (1989) on the subject.

Here we will provide a brief review of the two main classification schemes which have been widely used. We will also be discussing the classifications within the overall philosophy of this book, noting their strengths and deficiencies.

12.1 Rock Mass Rating (RMR) system

The advantage of the Rock Mass Rating (RMR) system, in common with the Q-system described in the next section, is that only a few basic parameters relating to the geometry and mechanical condition of the rock mass are used. In the case of the RMR system, these are:

(a) the uniaxial compressive strength of the intact rock;
(b) RQD;
(c) discontinuity spacing;
(d) condition of discontinuity surfaces;
(e) groundwater conditions; and
(f) orientation of discontinuities relative to the engineered structure.

The way in which these parameters are used to provide an overall rating is shown in Table 12.1.

Table 12.1 The rock mass rating system (after Bieniawski, 1989)

Rock Mass Rating System

A. CLASSIFICATION PARAMETERS AND THEIR RATINGS

	Parameter		Ranges of values						
1	Strength of intact rock material	Point-load strength index (MPa)	>10	4 - 10	2 - 4	1 - 2	For this low range, uniaxial compressive test is preferred		
		Uniaxial compressive strength (MPa)	>250	100 - 250	50 - 100	25 - 50	5 - 25	1 - 5	<1
		Rating	15	12	7	4	2	1	0
2	Drill core quality RQD (%)		90 - 100	75 - 90	50 - 75	25 - 50	<25		
		Rating	20	17	13	8	3		
3	Spacing of discontinuities		>2m	0.6 - 2m	200 - 600mm	60 - 200mm	<60mm		
		Rating	20	15	10	8	5		
4	Condition of discontinuities		Very rough surfaces Not continuous No separation Unweathered wall rock	Slightly rough surfaces Separation <1mm Slightly weathered wall rock	Slightly rough surfaces Separation <1mm Highly weathered wall rock	Slickensided surfaces or Gouge <5mm thick or Separation 1 - 5mm Continuous	Soft gouge >5mm thick or Separation >5mm Continuous		
		Rating	30	25	20	10	0		
5	Groundwater	Inflow per 10m tunnel length (l/min)	None	<10	10 - 25	25 - 125	>125		
		ratio (joint water pressure)/(major principal stress)	0	<0.1	0.1 - 0.2	0.2 - 0.5	>0.5		
		General conditions	Completely dry	Damp	Wet	Dripping	Flowing		
		Rating	15	10	7	4	0		

B. GUIDELINES FOR CLASSIFICATION OF DISCONTINUITY CONDITIONS

Parameter	Ratings				
Discontinuity length (persistence)	<1m	1 - 3m	3 - 10m	10 - 20m	>20m
	6	4	2	1	0
Separation (aperture)	None	<0.1mm	0.1 - 1.0mm	1 - 5mm	>5mm
	6	5	4	1	0
Roughness	Very rough	Rough	Slightly rough	Smooth	Slickensided
	6	5	3	1	0
Infilling (gouge)	Hard filling			Soft filling	
	None	<5mm	>5mm	<5mm	>5mm
	6	4	2	2	0
Weathering	Unweathered	Slightly weathered	Moderately weathered	Highly weathered	Decomposed
	6	5	3	1	0

Table 12.1 (cont)

C. EFFECT OF DISCONTINUITY ORIENTATIONS IN TUNNELLING

Strike perpendicular to tunnel axis			
Drive with dip		Drive against dip	
Dip 45 - 90	Dip 20 - 45	Dip 45 - 90	Dip 20 - 45
Very favourable	Favourable	Fair	Unfavourable

Strike parallel to tunnel axis		Irrespective of strike
Dip 20 - 45	Dip 45 - 90	Dip 0 - 20
Fair	Very unfavourable	Fair

D. RATING ADJUSTMENT FOR DISCONTINUITY ORIENTATIONS

Orientations of Discontinuities		Very Favourable	Favourable	Fair	Unfavourable	Very Unfavourable
Ratings	Tunnels & mines	0	-2	-5	-10	-12
	Foundations	0	-2	-7	-15	-25
	Slopes	0	-5	-25	-50	-60

E. ROCK MASS CLASSES DETERMINED FROM TOTAL RATINGS

Rating	100 - 81	80 - 61	60 - 41	40 - 21	<20
Class no.	I	II	III	IV	V
Description	Very good rock	Good rock	Fair rock	Poor rock	Very poor rock

F. MEANING OF ROCK MASS CLASSES

Class no.	I	II	III	IV	V
Average stand-up time	20yr for 15m span	1yr for 10m span	1wk for 5m span	10h for 2.5m span	30min for 1m span
Cohesion of rock mass (kPa)	>400	300 - 400	200 - 300	100 - 200	<100
Friction angle of rock mass (deg)	>45	35 - 45	25 - 35	15 - 25	<15

RMR = $\sum$(classification parameters) + discontinuity orientation adjustment

In Section A of Table 12.1, with the first five of the classification parameters and their ratings, the parameters are grouped in five classes, each one covering a range of values appropriate to that parameter. When assessing a given rock mass, one establishes into which of these groups the parameter lies, and then sums the resulting numerical ratings for the five parameters.

In Section B of Table 12.1, there are ratings for discontinuity characteristics. The orientation of the discontinuities becomes progressively more important from tunnels and mines, through foundations, to slopes, Sections C and D.

In Sections E and F of the table, the rock mass classes are given with a description from 'very good rock' through to 'very poor rock', with estimates for tunnel stand-up time and the Mohr–Coulomb strength parameters of cohesion and friction angle for the rock mass.

Despite the simplicity of approach when dealing with complex rock masses, considerable engineering benefit has accrued through the application of this classification scheme *and the resultant thinking that it has provoked.* Bieniawski (1989) mentions 351 case histories covering 15 years.

12.2 Q-system

In a similar way to the RMR system, the Q-rating is developed by assigning values to six parameters. These are:

(a) RQD;
(b) number of discontinuity sets;
(c) roughness of the 'most unfavourable' discontinuity;
(d) degree of alteration or filling along the weakest discontinuity;
(e) water inflow; and
(f) stress condition.

The *Q*-value is expressed as

$$Q = \frac{\text{RQD}}{J_n} \cdot \frac{J_r}{J_a} \cdot \frac{J_w}{\text{SRF}}$$

where

RQD = rock quality designation,
J_n = joint set number (related to the number of discontinuity sets),
J_r = joint roughness number (related to the roughness of the discontinuity surfaces),
J_a = joint alteration number (related to the degree of alteration or weathering of the discontinuity surfaces),
J_w = joint water reduction number (relates to pressures and inflow rates of water within the discontinuities), and
SRF = stress reduction factor (related to the presence of shear zones, stress concentrations and squeezing and swelling rocks).

The motivation in presenting the *Q*-value in this form is to provide some method of interpretation for the three constituent quotients.

The first, RQD/J_n, is related to the rock mass geometry: *Q* increases with increasing RQD and decreasing number of discontinuity sets. RQD increases with decreasing number of discontinuity sets, so the numerator and denominator of the quotient mutually reinforce one another. Basically, the higher the value of this quotient, the better the 'geometrical quality' of the rock mass. We noted, in Chapter 7, that RQD determined using the conventional threshold of 0.1 m is insensitive to discontinuity frequencies less than about 3 m^{-1}, so this quotient may also be insensitive. Moreover, there is also the problem (which is, in fact, common to both the RMR system and the *Q*-system) that RQD generally exhibits anisotropy, yet anisotropy is not considered.

The second quotient, J_r/J_a, relates to the 'inter-block shear strength' with high values of this quotient representing better 'mechanical quality' of the rock mass: the quotient increases with increasing discontinuity roughness and decreasing discontinuity surface alteration. The different discontinuity sets in the rock mass may have different roughnesses and degrees of alteration, so the *Q*-system uses the worst case.

The third quotient, J_w/SRF, is an 'environmental factor' incorporating water pressures and flows, the presence of shear zones, squeezing and swelling rocks and the *in situ* stress state. The quotient increases with decreasing water pressure or flow rate, and also with favourable rock mass strength to *in situ* stress ratios.

The *Q*-system is more complex to use than the RMR system. We are therefore including, in Table 12.2, the full range of classes for the six parameters involved in the system.

Table 12.2 Q-system parameters

Q-system of Rock Mass Classification

$$Q = \frac{RQD}{J_n} \times \frac{J_r}{J_a} \times \frac{J_w}{SRF}$$

Rock Quality Designation, RQD		RQD	
	a: Very poor	0-25	*Where RQD is reported or measured as 10 (including 0), a nominal value of 10 is used to evaluate Q*
	b: Poor	25-50	
	c: Fair	50-75	
	d: Good	75-90	*RQD intervals of 5, i.e., 100, 95, 90, etc., are sufficiently accurate*
	e: Excellent	90-100	

Joint Set Number, J_n		J_n	
	a: Massive, none or few joints	0.5-1.0	*For intersections, use ($3.0 \times J_n$). For portals, use ($2.0 \times J_n$)*
	b: One joint set	2	
	c: One joint set plus random	3	
	d: Two joint sets	4	
	e: Two joint sets plus random	6	
	f: Three joint sets	9	
	g: Three joint sets plus random	12	
	h: Four or more joint sets, random, heavily jointed 'sugar cube', etc.	15	
	j: Crushed rock, earthlike	20	

Joint Roughness Number, J_r		J_r	
	(i) Rock wall contact and		*Add 1.0 if the mean spacing of the relevant joint set is greater than 3m.*
	(ii) Rock wall contact before 10cm shear		*J_r = 0.5 can be used for planar slickensided joints having lineation, provided the lineations are favourably orientated.*
	a: Discontinuous joint	4	*Descriptions b to g refer to small-scale features and intermediate-scale features, in that order.*
	b: Rough or irregular, undulating	3	
	c: Smooth, undulating	2.0	
	d: Slickensided, undulating	1.5	
	e: Rough or irregular, planar	1.5	
	f: Smooth, planar	1.0	
	g: Slickensided, planar	0.5	
	(iii) No rock wall contact when sheared		
	h: Zone containing clay minerals thick enough to prevent rock wall contact	1.0	
	j: Sandy, gravelly, or crushed zone thick enough to prevent rock wall contact	1.0	

Joint Alteration Number, J_a		J_a	ϕ_r (approx.) *Values of ϕ_r are intended as an approximate guide to the mineralogical properties of the alteration products, if present*
	(i) Rock wall contact		
	a: Tightly healed, hard, nonsoftening, impermeable filling, i.e., quartz or epidote	0.75	
	b: Unaltered joint walls, surface staining only	1.0	25°-35°
	c: Slightly altered joint walls. Nonsoftening mineral coatings, sandy particles, clay-free disintegrated rock, etc.	2.0	25°-30°
	d: Silty or sandy clay coatings, small clay fraction (nonsoftening)	3.0	20°-25°
	e: Softening or low-friction clay mineral coatings, i.e., kaolinite, mica. Also chlorite, talc, gypsum, and graphite, etc., and small quantities of swelling clays (discontinuous coatings, 1-2mm or less in thickness)	4.0	8°-16°
	(ii) Rock wall contact before 10cm shear		
	f: Sandy particles, clay-free disintegrated rock, etc.	4.0	25°-30°
	g: Strongly over-consolidated, nonsoftening clay mineral fillings (continuous, <5mm in thickness)	6.0	16°-24°
	h: Medium or low over-consolidation, softening, clay mineral fillings (continuous, <5mm in thickness)	8.0	12°-16°
	j: Swelling clay fillings, i.e., montmorillonite (continuous, <5mm in thickness). Value of J_a depends on percentage of swelling clay-sized particles, and access to water, etc.	8.0-12.0	6°-12°

When making estimates of the Rock Mass Quality (Q), the following guidelines should be followed, in addition to the notes in the tables:

1. When borehole core is unavailable, for the case of clay free rock masses *RQD* can be estimated from $RQD = 115 - 3.3J_v$ (approx.) where J_v = total number of joints per m³ (RQD=100 for J_v<4.5). J_v is evaluated as the sum of the number of joints per metre for each joint set.
2. The parameter J_n, representing the number of joint sets, will often be affected by foliation, schistosity, slaty cleavage or bedding, etc. If strongly developed, these features should be counted as a complete joint set: if they are poorly developed or rarely visible, then it will be more appropriate to count them as 'random joints' when evaluating J_n.
3. The parameters J_r and J_a (representing shear strength) should normally be relevant to the weakest significant

Table 12.2 (cont)

Q-system of Rock Mass Classification $Q = \frac{RQD}{J_n} \times \frac{J_r}{J_a} \times \frac{J_w}{SRF}$

Parameter	Description	Value			Remarks
Joint Alteration Number, J_a	*(iii) No rock wall contact when sheared*				
	k: Zones or bands of disintegrated or rock and clay (see g:, h:, j: for description of clay condition)	6.0, 8.0 or 8.0-12.0			6°-24°
	l: Zones or bands of silty clay, small clay fraction (nonsoftening)	5.0			
	m: Thick, continuous zones or bands of clay (see g:, h:, j: for description of clay condition)	10.0, 13.0 or 13.0-20.0			
Stress Reduction Factor, SRF		SRF			*Reduce these SRF values by 25-50% if the relevant shear zones only influence but do not intersect the excavation*
	(i) Weakness zones intersecting excavation, which may cause loosening of rock mass when tunnel is excavated				
	a: Multiple occurrences of weakness zones containing clay or chemically disintegrated rock, very loose surrounding rock (any depth)	10.0			
	b: Single weakness zones containing clay or chemically disintegrated rock (excavation depth <50 m)	5.0			
	c: Single weakness zones containing clay or chemically disintegrated rock (excavation depth >50 m)	2.5			
	d: Multiple shear zones in competent rock (clay-free), loose surrounding rock (any depth)	7.5			
	e: Single shear zones in competent rock (clay-free) (depth of excavation <50 m)	5.0			
	f: Single shear zones in competent rock (clay- free) (depth of excavation >50 m)	2.5			
	g: Loose open joints, heavily jointed or 'sugar cube', etc. (any depth)	5.0	σ_c/σ_1	σ_t/σ_1	
	(ii) Competent rock, rock stress problems				*Few case records available where depth of crown below surface is less than span width. Suggest SRF increase from 2.5 to 5 for such cases*
	h: Low stress, near surface	2.5	>200	>13	
	j: Medium stress	1.0	200-10	13-0.66	
	k: High-stress, very tight structure (usually favourable to stability, may be unfavourable for wall stability)	0.5-2.0	10-5	0.66-0.33	
	l: Mild rock burst (massive rock)	5-10	5-2.5	0.33-0.16	
	m: Heavy rock burst (massive rock)	10-20	<2.5	<0.16	
	(iii) Squeezing rock; plastic flow of incompetent rock under the influence of high rock pressures				*For strongly anisotropic stress field (if measured): when $5 \leq \sigma_1/\sigma_3 \leq 10$, reduce σ_c and σ_t to $0.8\sigma_c$ and $0.8\sigma_t$; when $\sigma_1/\sigma_3 > 10$, reduce σ_c and σ_t to $0.6\sigma_c$ and $0.6\sigma_t$ (where σ_c = unconfined compressive strength, σ_t = tensile strength (point load), σ_1 and σ_3 = major and minor principal stresses)*
	n: Mild squeezing rock pressure	5-10			
	p: Heavy squeezing rock pressure	10-20			
	(iv) Swelling rock; chemical swelling activity depending on presence of water				
	q: Mild swelling rock pressure	5-10			
	r: Heavy swelling rock pressure	10-15			
Joint Water Reduction Factor, J_w		J_w	Approx. water pressure (kg/cm²)		
	a: Dry excavations or minor inflow, e.g. 5 l/min locally	1.0	<1		
	b: Medium inflow or pressure, occasional outwash of joint fillings	0.66	1.0-2.5		
	c: Large inflow or high pressure in competent rock with unfilled joints	0.5	2.5-10.0		*Factors c to f are crude estimates. Increase J_w if drainage measures are installed*
	d: Large inflow or high pressure, considerable outwash of joint fillings	0.33	2.5-10.0		
	e: Exceptionally high inflow or water pressure at blasting, decaying with time	0.2-0.1	>10.0		*Special problems caused by ice formation are not considered*
	f: Exceptionally high inflow or water pressure continuing without noticeable decay	0.1-0.05	>10.0		

joint set or clay filled discontinuity in a given zone, but the value of J_r/J_a should relate to the surface most likely to allow failure to initiate. Thus, if the joint set or discontinuity with the minimum value of J_r/J_a is favourably orientated for stability, then a second, less favourably orientated joint set or discontinuity may sometimes be more significant, and its higher value of J_r/J_a should be used when evaluating Q.

4. When a rock mass contains clay, the factor *SRF* appropriate to 'loosening loads' should be evaluated. In such cases the strength of the intact rock is of little interest. However, when jointing is minimal and clay is completely absent, the strength of the intact rock may become the weakest link, and the stability will then depend on the ratio rock-stress/rock-strength. A strongly anisotropic stress field is unfavourable for stability and is roughly accounted for as in the note in the table for *SRF* evaluation.
5. The compressive and tensile strengths (σ_c and σ_t) of the intact rock should be evaluated in the saturated condition if this is appropriate to present or future *in situ* conditions. A conservative estimate of strength should be made for those rocks that deteriorate when exposed to moist or saturated conditions.

12.3 Applications of rock mass classification systems

Using either of the classification systems described in the previous two sections, the engineering quality of a rock mass can be assessed. The RMR system gives a number between 0 and 100, and the Q-system gives a number between 0.001 and 1000. By these approaches, we are able to produce a description of the rock mass based on classes defined by the

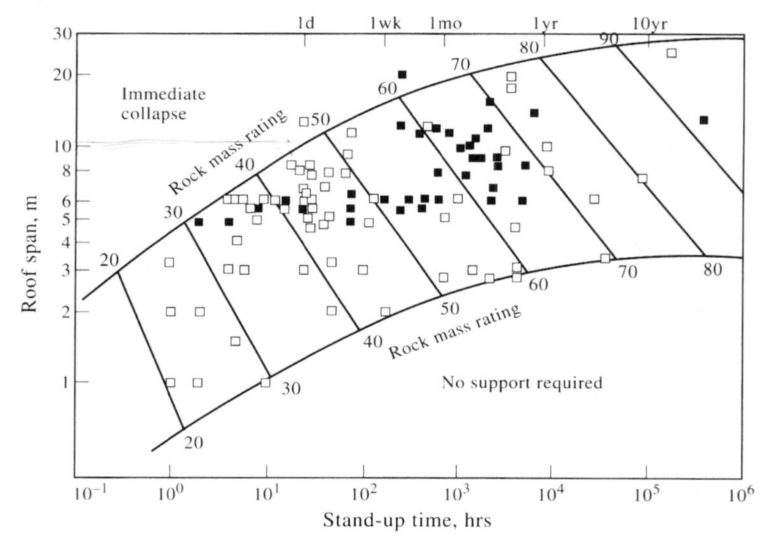

Figure 12.1 Excavation stand-up time for the RMR system.

numbers in the classifications. For example, an RMR value of 62 is a 'good rock': similarly, a Q-value of 20 indicates a 'good rock'. The RMR value provides five such quality classes and the Q-system provides nine.

Both the classifications described were developed for estimating the support necessary for tunnels excavated for civil engineering schemes. The engineer should be careful when using classification schemes for other projects. It is one thing to utilize the rock mass parameters in a taxonomic system for classifying and describing the rock; it is quite another to extrapolate the information to the general design of excavations and their support. Bieniawski (1989) has noted "it is important that the RMR system is used for the purpose for which it was developed and not as the answer to all design problems".

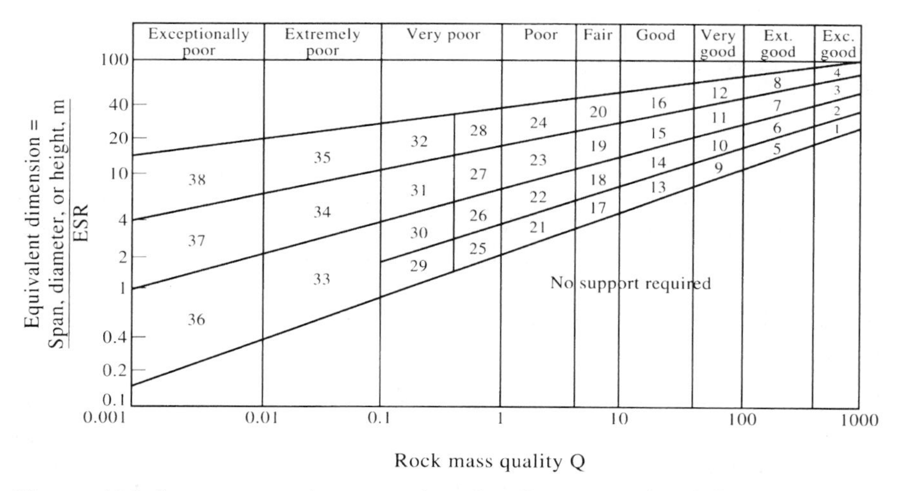

Figure 12.2 Support requirements for the Q-system (for fuller details see Bieniawski, 1989).

Using the rock mass parameters in each case to provide a quantitative assessment of the rock mass, and utilizing experience gained from previously excavated stable and unstable tunnels, design charts have been constructed, as shown in Figs 12.1 and 12.2, for estimating 'stand-up time' or support requirements. For a description of the complete technique for establishing the support requirements, the reader is referred to Bieniawski (1989), which expands on the fundamentals given here of the two systems.

Attempts have been made to extend the classification system to slopes (Romana, 1985). Naturally, the six parameters utilized in the RMR system are relevant to slope stability, but the classification value needs to be adjusted for different engineering circumstances. The way in which Professor Romana numerically adjusted the RMR value was by considering the following factors:

(a) F_1 associated with parallelism between the slope and the discontinuity strike direction;
(b) F_2 related to the discontinuity dip for plane failure;
(c) F_3 concerning the slope angle compared to the discontinuity dip angle; and
(d) F_4 relating to the method of excavation.

The classification value is then found from the formula

$$\mathrm{RMR_{SLOPE}} = \mathrm{RMR_{BASIC}} - (F_1 \times F_2 \times F_3) + F_4.$$

Table 12.3 indicates the numerical values of the four factors required to adjust $\mathrm{RMR_{BASIC}}$ to $\mathrm{RMR_{SLOPE}}$, together with the SMR classes, the types of failure anticipated, and any remedial measures necessary to improve stability.

Table 12.3 The SMR rating system (from Romana, 1985 and Bieniawski, 1989)

Case		Very Favorable	Favorable	Fair	Unfavorable	Very Unfavorable
P	$\lvert\alpha_j - \alpha_s\rvert$	>30°	30–20°	20–10°	10–5°	<5°
T	$\lvert\alpha_j - \alpha_s - 180°\rvert$					
P/T	F_1	0.15	0.40	0.70	0.85	1.00
P	$\lvert\beta_j\rvert$	<20°	20–30°	30–35°	35–45°	>45°
P	F_2	0.15	0.40	0.70	0.85	1.00
T	F_2	1	1	1	1	1
P	$\beta_j - \beta_s$	>10°	10–0°	0°	0°–(−10°)	<−10°
T	$\beta_j + \beta_s$	<110°	110–120°	>120°		
P/T	F_3	0	−6	−25	−50	−60

P = plane failure. T = toppling failure. α_s = slope dip direction. β_s = slope dip. α_j = joint dip direction. β_j = joint dip.

Method	Natural Slope	Presplitting	Smooth Blasting	Regular Blasting	Deficient Blasting
F_4	+15	+10	+8	0	−8

$SMR = RMR - (F_1 \times F_2 \times F_3) + F_4$

Tentative Description of SMR Classes

Class No.	V	IV	III	II	I
SMR	0–20	21–40	41–60	61–80	81–100
Description	Very poor	Poor	Fair	Good	Very good
Stability	Very unstable	Unstable	Partially stable	Stable	Fully Stable
Failures	Large planar or soil-like	Planar or large wedges	Some joints or many wedges	Some blocks	None
Support	Reexcavation	Extensive corrective	Systematic	Occasional	None

In the same vein, extensions to the RMR and Q-classification systems have been made to estimate rippability, dredgeability, excavatability, cuttability and cavability (Bieniawski, 1989).

12.4 Links between the classification systems and rock properties

The rock mass classification systems have some common parameters, so we might expect a link to exist between the RMR and Q-systems. The most well-known correlation is given by the equation

$$\text{RMR} = 9\log_e Q + 44.$$

Naturally, this has to be an approximation, because *in situ* stress and rock strength are not common to the two systems.

Since the rock mass properties, e.g. deformability and strength, are also functions of the intact rock properties and the discontinuity properties, it follows that we may be able use the classification scheme values to estimate the modulus and strength of rock masses. Empirical relations are available for both of the systems described.

Bieniawski (1989) has suggested that

$$E_{\text{MASS}} = 2 \times \text{RMR} - 100 \text{ GPa (for RMR} > 50).$$

Another relation has been proposed by Serafim and Pereira (1983), which covers the entire RMR range, and is

$$E_{\text{MASS}} = 10^{(\text{RMR} - 10)/40} \text{ GPa}.$$

For the Q-system, Barton *et al.*, (1985) proposed that

$$10 \log_{10} Q < E_{\text{MASS}} < 40 \log_{10} Q, \text{ with } E_{\text{MEAN}} = 25 \log_{10} Q.$$

Other expressions have been developed (Hoek and Brown, 1988). For the Hoek–Brown strength parameters m and s (see Section 6.5.3) as:

$$\text{undisturbed rock}: \; m = m_i \exp\left(\frac{\text{RMR} - 100}{28}\right) \quad s = \exp\left(\frac{(\text{RMR} - 100)}{9}\right)$$

$$\text{disturbed rock}: \; m = m_i \exp\left(\frac{(\text{RMR} - 100)}{14}\right) \quad s = \exp\left(\frac{\text{RMR} - 100}{6}\right)$$

where RMR is the unadjusted rating from the system.

12.5 Discussion

It is important to place the value of rock mass classification schemes and the estimations described above within the context of practical rock engineering. It is easy to point to the value of the classifications when, often inexperienced, personnel have to make assessments of rock mass quality and support requirements, especially when faced with no other clear alternative. Similarly, it is easy to say that none of the techniques has any solid scientific foundation and can quite clearly be dangerously misleading if the potential failure mechanism is not identified within the classification system. Stress is not included in the RMR system; the intact strength of rock is not included in the Q-system. Either of these parameters could be a fundamental cause of failure in certain circumstances. Even more severely, a shear or fault zone in the rock could exist which dominates the potential failure mechanism of, say, a cavern or slope.

Because the perceived main governing parameters for rock engineering have been included in the RMR and Q-systems, their use must provide some overall guidance. However, the use of these systems as the sole design tool cannot be supported on scientific grounds. For example, the fact that the measured values of discontinuity frequency and RQD depend on the direction of measurement has been clearly explained in Chapter 7, yet this is not accounted for in either of the systems described. Similarly, because the rock mass modulus depends on the discontinuity stiffnesses to a large extent, the modulus is also anisotropic, yet the predictions of E only provide a single (i.e. isotropic) value.

We feel, therefore, that despite its past contributions, the rock mass classification approach will be supplemented by other methods in due course, as the correct mechanisms are identified and modelled directly. Moreover, it is an unnecessary restriction to use the same classification parameters without reference to either the project or the site. For example, in a hydroelectric scheme pressure tunnel, the *in situ* stress and proximity of the tunnel to the ground surface are two of the most important parameters. The RMR system cannot help under these circumstances. The Q-system cannot be used for predicting E below a dam if the stratified nature of the rock mass means that there is significant anisotropy of stiffness.

12.6 Extensions to rock mass classification techniques

Given our comments in the previous section, we believe that there are two main ways in which the rock mass classification approach can be improved. The first is a straightforward extension of the current systems, but incorporating fuzzy mathematics to account for variations in the individual component parameters. The second is to choose those parameters that are most relevant to a particular engineering objective and hence the classification systems for different projects would involve different constituent parameters—using the RES (rock engineering systems) approach briefly described in Chapter 14 (Hudson, 1992).

12.6.1 Use of fuzzy mathematics

Engineers may encounter problems in using the current rock mass classification systems because the inherent variability of rock masses is difficult to take into account—for example, if mean discontinuity spacing varies from 0.3 to 2.0 m, what value should be used in the system? By assigning a **fuzzy number** to such parameters, and then using the techniques of fuzzy arithmetic to combine the numbers, it is possible to generate a fuzzy number representing the classification value. Such a number then embodies the 'most certain' classification value, together with information regarding its maximum and minimum values, and the manner in which it varies between the two. Hence, fuzzy mathematics permits the 'uncertainty' surrounding the assessment of parameters to be included. Also, the application of this technique in rock mass classification is straightforward and direct, because fuzzy numbers may be assigned easily to the parameters in a rock mass classification scheme.

Some examples of fuzzy numbers and fuzzy arithmetic are shown in Fig. 12.3. It is important to realize that the distributions of *uncertainty* illustrated in Fig. 12.3 *are not probability density distributions*. The parameters A and B illustrated in Fig. 12.3 are *uncertain* numbers: we know that A varies between 3 and 10, with a most likely value of 7—but we are uncertain as to which precise value it will take; similarly, B will vary between 12 and 20 with a most likely value of 14. The fuzzy sum and fuzzy multiplicand of two fuzzy numbers are shown in Fig. 12.3.

In Fig. 12.4, as an example we have applied fuzziness to the six parameters of the Q-system, presenting illustrations of the nature of fuzzy assessment of parameters. The assessment of RQD and J_r are straightforwardly analogous to A and B in Fig. 12.3. However, in the case of parameter J_w, its most likely value coincides with its maximum value—with the result that the skewed number shown is generated. Similarly, for J_n its most likely value also coincides with its minimum value. The two parameters J_a and SRF have been assessed as having only one value: these are *crisp*, i.e. conventional, numbers.

Applying fuzzy arithmetic to the basic formula of the Q-system, given in Section 12.2, results in the fuzzy classification value shown in Fig. 12.4. Taking all of the most likely values of the individual parameters, and combining them, gives a value of 5.8, which is the most likely value of the classification value. Similarly, the maximum and minimum values of the

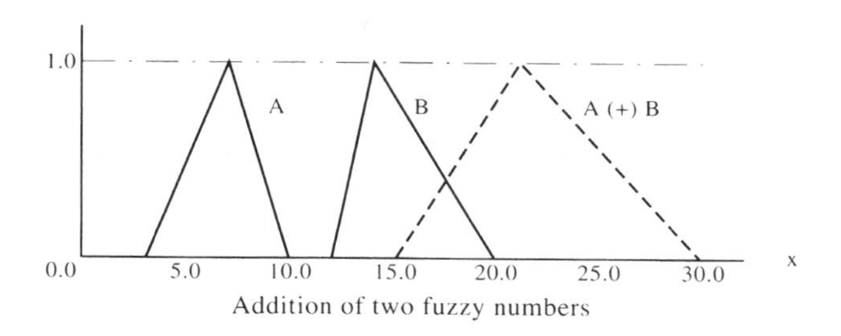

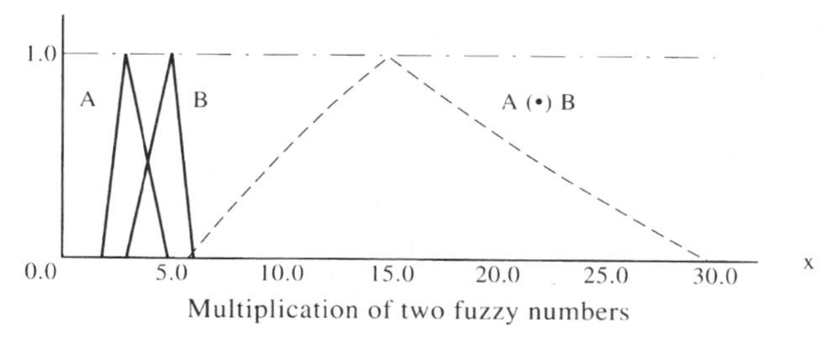

Figure 12.3 Fuzzy numbers and their additive and multiplicative forms.

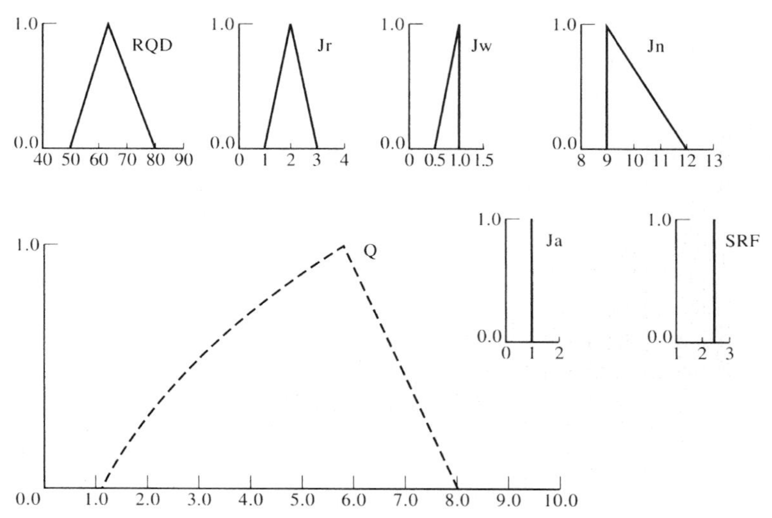

Figure 12.4 Application of fuzzy methods to the assessment of Q.

classification of 8.0 and 1.1, respectively, are found from the corresponding values of the individual parameters. The distribution of the remaining values of the number are found by combining the values of the individual parameters at membership values of 0.1, 0.2 and so on. It is interesting to see that the result is a number in which the distribution of values is non-linear: the 'flanks' of the number are curved.

The conclusions to be drawn from this visual examination of the result are that there is more possibility of Q being less than the most likely value rather than being greater, and that the convex nature of the flanks has the effect of increasing the possibility that the conditions will be worse than a single-valued calculation would imply. Lastly, it should be noted that the final distribution of Q and the associated conclusions are not at all obvious from an examination of the nature of the original fuzzy component parameters.

12.6.2 Use of RES (Rock Engineering Systems)

The principle behind the RES system (Hudson, 1992) is that the information obtained should match the engineering objective. The two main classification systems—those of RMR and Q—utilize six main parameters which are not the same. The developers of these systems have decided on which parameters are most important for tunnel design, and designed their classifications accordingly. Both proponents of the systems have warned users not to attempt to extrapolate the classification methods without modification and not to make predictions outside the original subjects for which the classification schemes were intended.

A more general approach is to consider for any specific project the relative importance of all rock engineering parameters, and then to concentrate on the most important, say, six or 10 parameters. One could go further and establish how many parameters contributed to, say, 95% of the design process, and allocate resources accordingly. To illustrate this point, we refer the reader to Fig. 11.2 in which the parameters associated

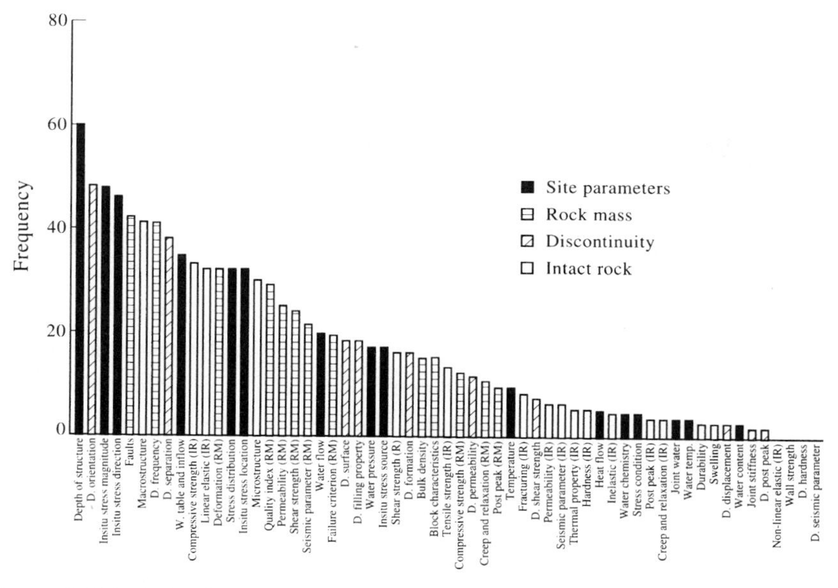

Figure 12.5 Relative importance of rock engineering parameters in the design of large rock caverns (as established from the literature and compiled by Tamai, 1990).

with pressure tunnel design are shown, and to Figs 12.5 and 12.6 which show the parameters in order of importance for large underground caverns and radioactive waste disposal, respectively.

Such histograms can be compiled on the basis of past experience, current practice and recommended practice (the latter, perhaps, arising from modelling of design requirements). The histograms in Figs 11.2, 12.5 and

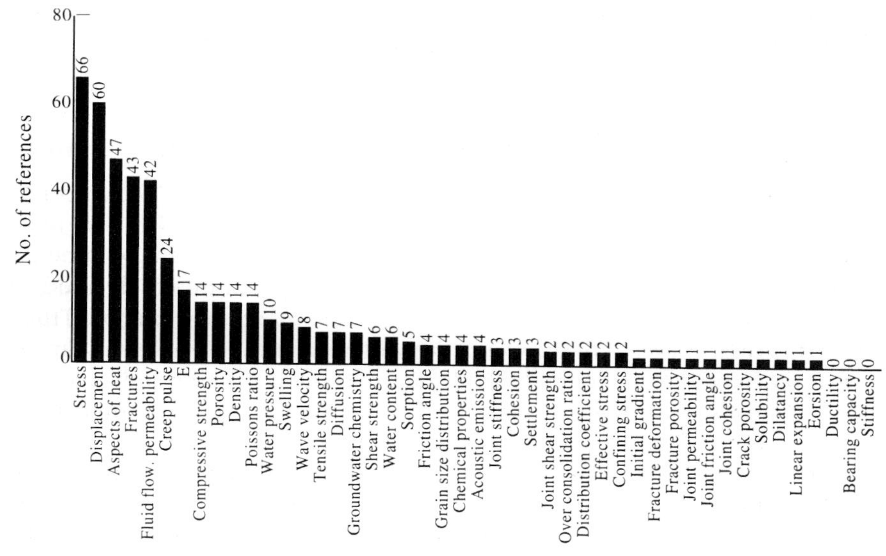

Figure 12.6 Relative importance of rock engineering parameters in the design of radioactive waste repositories (as established from the literature and complied by Arnold, 1993).

Table 12.4 Relative importance of rock engineering parameters in three engineering activities

WATER PRESSURE TUNNELS IN HYDROELECTRIC SCHEMES	LARGE UNDERGROUND CAVERNS	RADIOACTIVE WASTE REPOSITORIES
In situ stress	Depth of cavern	*In situ* stress
Discontinuity persistence	Discontinuity orientation	Induced displacements
Topographic factors	*In situ* stress	Thermal aspects
Presence of faults/folds	Presence of faults	Discontinuity geometry
Location of tunnel	Rock type	Permeability
Discontinuity aperture	Discontinuity frequency	Time dependent properties
Rock mass geometry	Discontinuity aperture	Elastic modulus
Discontinuity fill	Pre-existing water conditions	Compressive strength
Tunnel water pressure	Intact rock elastic modulus	Porosity
Pre-existing water conditions	Rock mass elastic modulus	Density

12.6 were developed from literature reviews and could be different if they were based on current practice or design-led practice.

We present Table 12.4 which lists the most important parameters for the three cases which were derived from their frequency of occurrence in the literature, reflecting the concentration of research effort. Forty-four papers were studied for pressure tunnels, 70 for large underground caverns, and 208 for radioactive waste disposal. The key point is that the list of most important parameters will not be the same for different types of engineering project, nor indeed will it coincide with one of the current classification schemes—as is clear from the table. Furthermore, we cannot divorce the rock mass properties from some of the site and operational characterisitics. The table includes not only the properties of intact rock, discontinuities and the rock mass, **but also factors relating to the prevailing boundary conditions, site circumstances and project parameters.**

12.7 Concluding remarks

In conclusion, it is clear that rock mass classification schemes have assisted engineers in the past especially in the absence of any other approach. There are pitfalls associated with the use of the schemes, primarily associated with the absence of what may be critical parameters for various projects. The classification systems can be enhanced by the use of fuzzy methods and strategic parameter selection. In the long term, perhaps beyond the lifetimes of the readers, and certainly the authors, rock mass classification systems will be superseded by direct solution of the engineering problems, although there will always be some comfort in using the classification schemes to ensure that modelling results do not contravene hard-earned practical experience. In the meantime, rock mass classification systems are useful but must be used with care.

13 Rock dynamics and time-dependent aspects

In this chapter we will be discussing a variety of effects related to the different strain rates that occur throughout the range of rock mechanics processes and rock engineering applications. Following the introduction, highlighting the wide time ranges over which these effects are manifested, we discuss the basic theory of rock dynamics, obtaining dynamic rock properties and the relevance of the ideas in engineering.

13.1 Introduction

In Chapter 6, it was noted that the compressive strength is the maximum stress that can be sustained by a specimen of rock. Let us now say that the compressive strength is reached at 0.1% strain, i.e. 0.001. If this strain is developed in 1 μs—for example, during an explosion—the strain rate is $1 \times 10^3\ s^{-1}$. If, on the other hand, this strain is developed over a period of 30 years, the average strain rate is of the order of $1 \times 10^{-12}\ s^{-1}$. Between these two extremes, there are 15 orders of magnitude of strain rate, and so, if the rock exhibits any time-dependent behaviour, we would not expect to be able to use the same rock properties for an analysis of both cases. In Fig. 13.1, we illustrate two manifestations of these extremes of strain rate. Fig. 13.1(a) shows hackle marks that develop on rock fracture surfaces formed during high strain rate failure, in this case on the surface of a blast-induced fracture. Fig. 13.1(b) shows the effect of the gradual deterioration, and subsequent failure, of the pillars in an old mine in chalk beneath a main road.

Within the rock mechanics principles, it is necessary to be able to account for time-dependent behaviour. In the development and discussion of the elastic compliance matrix in Chapter 5, when we considered the addition of the component strains caused by the component stresses, no time-dependent behaviour was incorporated. Indeed, one aspect of the theory of elasticity is that there is no time component and that all strain energy is recoverable, i.e. all strain energy introduced into a material through loading will be subsequently liberated on unloading. However, when time dependency is involved, there is always hysteresis in the

(a)

(b)

Figure 13.1 Examples of visible effects of (a) high and (b) low strain rates on rock.

loading–unloading stress–strain curve. The area under this curve represents energy, so the presence of such hysteresis loops indicates the non-recovery of energy, which causes an increase in entropy.

13.2 Stress waves

Stress waves are the manifestation of dynamic stress changes. They occur when the body is not in static equilibrium as described so far, and are essentially sound waves in solid material. The differential equations of equilibrium, represent the fact that, for any given axis, the resultant force on a body is zero when the body is in equilibrium. Considering now that an infinitesimal cube of material is accelerating, and applying Newton's Second Law of Motion, these equations become the differential equations of motion:

$$\frac{\partial \sigma_x}{\partial x} + \frac{\partial \tau_{yx}}{\partial y} + \frac{\partial \tau_{zx}}{\partial z} = -\rho \frac{\partial^2 u_x}{\partial t^2}$$

$$\frac{\partial \tau_{xy}}{\partial x} + \frac{\partial \sigma_y}{\partial y} + \frac{\partial \tau_{zy}}{\partial z} = -\rho \frac{\partial^2 u_y}{\partial t^2}$$

$$\frac{\partial \tau_{xz}}{\partial x} + \frac{\partial \tau_{yz}}{\partial y} + \frac{\partial \sigma_z}{\partial z} = -\rho \frac{\partial^2 u_z}{\partial t^2}.$$

Although these equations may appear daunting, they are quite simple to understand. The three components on the left-hand side of the equations are the increments of stress in each Cartesian direction—note that in each equation the last subscript in the numerators is the same, indicating that the stress increments are all in the same direction. ρ is the density, the u variable is for displacement and t is for time. In static equilibrium the right-hand side of the equations is zero, because the infinitesimal cube is static: in the equations above, the right-hand side is the equivalent of the *mass* × *acceleration* term associated with dynamics.

If we consider a compressive stress wave travelling in the x-direction, independent of its position in the y–z plane, then the equations of motion reduce to

$$\frac{\partial \sigma_x}{\partial x} = -\rho \frac{\partial^2 u_x}{\partial t^2}$$

$$\frac{\partial \tau_{xy}}{\partial x} = -\rho \frac{\partial^2 u_y}{\partial t^2}$$

$$\frac{\partial \tau_{xz}}{\partial x} = -\rho \frac{\partial^2 u_z}{\partial t^2}.$$

It is possible, through the differential forms of the constitutive relations, to modify the left-hand sides of these equations to give

$$-(\lambda + 2\mu)\frac{\partial^2 u_x}{\partial x^2} = -\rho \frac{\partial^2 u_x}{\partial t^2}$$

$$-\mu \frac{\partial^2 u_y}{\partial x^2} = -\rho \frac{\partial^2 u_y}{\partial t^2}$$

$$-\mu \frac{\partial^2 u_z}{\partial x^2} = -\rho \frac{\partial^2 u_z}{\partial t^2}$$

where λ and μ are the Lamé elastic constants, which are related to the customary engineering constants E and ν for isotropic materials through $\lambda = E\nu/[(1 + \nu)(1 - 2\nu)]$ and $\mu = E/2(1 + \nu)$.

We can write the one-dimensional wave equation in terms of displacement for wave propagation in each of the x-, y- and z-directions as

$$C_x^2 \frac{\partial^2 u_x}{\partial x^2} = \frac{\partial^2 u_x}{\partial t^2},$$

$$C_y^2 \frac{\partial^2 u_y}{\partial x^2} = \frac{\partial^2 u_y}{\partial t^2}$$

$$C_z^2 \frac{\partial^2 u_z}{\partial x^2} = \frac{\partial^2 u_z}{\partial t^2}$$

where C_x, C_y and C_z are the wave velocities for waves propagating in the x-direction and with particle motion in the x-, y- and z-directions, respectively. Two types of stress wave are propagated: one has particle motion in the x-direction (*longitudinal* or P-waves), the other has particle motion in the y- or z-directions (*transverse* or S-waves), with velocities given by $C_P^2 = (\lambda + 2\mu)/\rho$ and $C_S^2 = \mu/\rho$, respectively. A more complete analysis of this topic is presented in, for example, *Stress Waves in Solids* by H. Kolsky.

Expressing these velocities as v_P and v_S, and using the engineering elastic constants rather than Lamé's constants, we find that

$$v_P^2 = \left[\frac{E}{\rho}\frac{(1-\nu)}{(1+\nu)(1-2\nu)}\right]$$

$$v_S^2 = \left[\frac{E}{\rho}\frac{1}{2(1+\nu)}\right].$$

Also, with these relations, we find the ratio $v_S/v_P = [(1 - 2\nu)/2(1 - \nu)]^{1/2}$.

We are also interested in the velocities of these waves when they occur in thin bars. In this case, the longitudinal and shear wave velocities in a bar, respectively, are

$$(v_{\text{PBAR}})^2 = \left[\frac{E}{\rho}\right]$$

$$(v_{\text{SBAR}})^2 = \left[\frac{E}{\rho}\frac{1}{2(1+\nu)}\right]$$

with the velocity ratio being $v_{\text{SBAR}}/v_{\text{PBAR}} = [1/2(1 + \nu)]^{1/2}$.

The modes of transmission of the longitudinal and transverse waves are shown in Figs 13.2(a) and (b). Two other types of stress wave which are important are Rayleigh and Love waves. Both of these waves occur near interfaces and free surfaces and have elliptical particle motion which is polarized and perpendicular to the free surface: with Rayleigh waves the

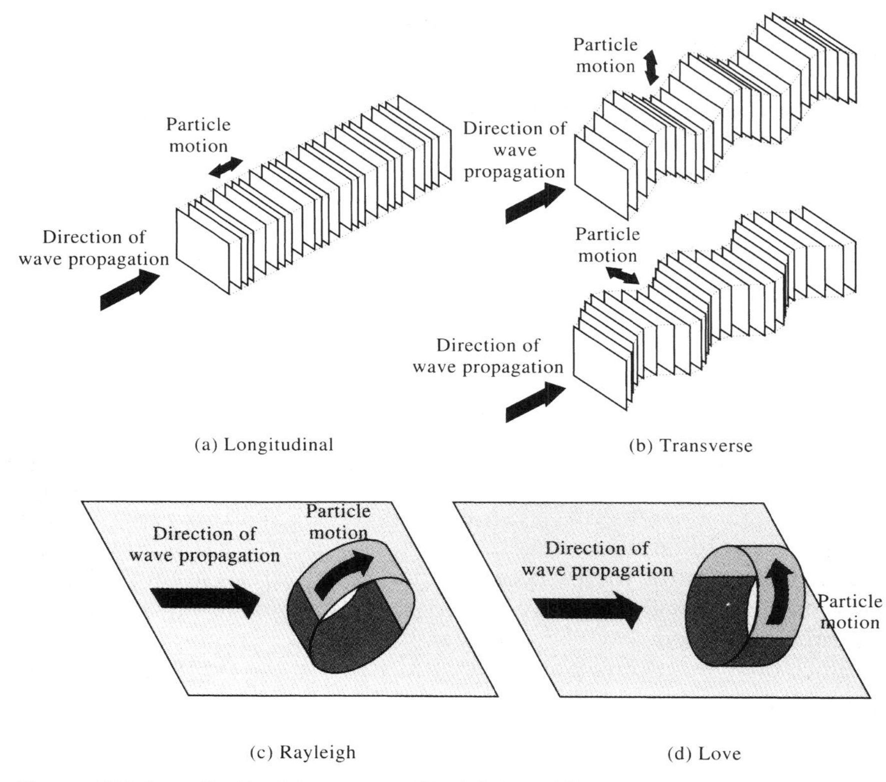

Figure 13.2 Longitudinal, transverse, Rayleigh and Love waves.

particle motion is parallel to the direction of wave propagation, as illustrated in Fig. 13.2(c); with Love waves the particle motion is perpendicular to the direction of wave propagation, as illustrated in Fig. 13.2(d). Love waves are found to occur under certain conditions in a stratified solid, depending on the relative shear wave velocities in the different strata.

It is instructive to consider the numerical value of the longitudinal and shear wave velocities and their ratios for an example rock. Taking $\rho = 25$ kN/m^3, $E = 20$ GPa and $\nu = 0.35$, the various relations presented above give $v_P = 1133$ m s^{-1}, $v_S = 544$ m s^{-1}, $v_S/v_P = 0.48$, $v_{PBAR} = 894$ m s^{-1}, $v_{SBAR} = 544$ m s^{-1} and $v_{SBAR}/v_{PBAR} = 0.61$. In a site investigation, and with the assumption of a CHILE material, we might use the *P*-and *S*-wave velocities, together with a density assumption, to estimate the *in situ* values of *E* and ν.

In the laboratory the dynamic properties of rock can be studied with the Hopkinson bar, or by directly inputting *P*- and *S*-waves via piezoelectric transducers. These two tests are illustrated in Fig. 13.3. In the Hopkinson bar, a single controlled *P*-wave pulse passes along the first steel bar, through a cylindrical rock specimen, and into the second steel bar. Using strain gauges installed on both the steel bars, the amplitude of the wave can be studied both before and after it passes through the rock specimen. By steadily increasing the amplitude of the pulse, the

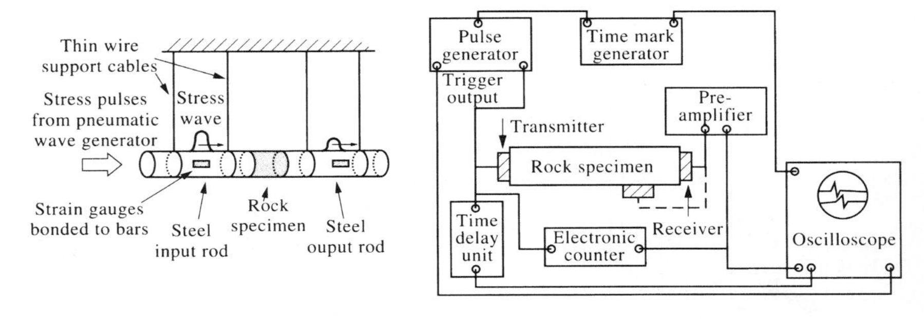

(a) Hopkinson bar (a) Piezoelectric transducers

Figure 13.3 The Hopkinson bar and piezoelectric transducer methods for dynamic properties. (a) Hopkinson bar. (b) Piezoelectric transducers.

way in which the energy is absorbed during the creation of the dynamic complete stress–strain curve can be studied. Alternatively, piezoelectric transducers can be used: these can be calibrated to directly indicate the elastic constants for the pre-peak portion of the stress–strain curve. The transducers are widely used, and apparatus is commercially available for its implementation.

Since the wave propagation velocities are a function of the elastic rock properties, it follows that the *P*- and *S*-wave velocities and associated factors such as attentuation can be used to estimate the rock properties on a literally global scale. A shear wave can only travel through a material that can sustain a shear stress: consequently, shear waves do not travel through liquids. During the early recording of seismic waves generated by earthquakes at long distances from the foci, it was found that at certain points on the Earth's surface there would be an absence of early shear wave arrivals—normally the faster *P*-wave arrives first, followed by the slower *S*-wave, followed by the surface waves and a complex mixture of reflected and refracted waves. From the absence of the early shear wave arrivals, it was realized that the Earth has a liquid core.

On a smaller scale, the manifold ways in which dynamic waves can be generated and recorded provides great potential for ground exploration techniques. Through the installation of a suitable array of geophones, and by measuring the wave transit times and hence estimating velocities, there is a powerful indicator method of establishing inhomogeneity, anisotropy and, indeed, continuity and linear elasticity—i.e. all the CHILE versus DIANE factors. Moreover, with the increasing sophistication of such techniques, e.g. the recent developments in tomography, there is the opportunity to use non-destructive testing to provide a three-dimensional mechanical characterization of a rock mass. This is one of the most exciting developments in rock characterization methods.

Another ramification of the existence of stress waves is the importance of waves being reflected at a free face. In Fig. 13.4(a), we show the general circumstances when a stress wave encounters an interface between two continua with different elastic properties. In the general case, part of the wave will be refracted as it passes into the second medium and part of the

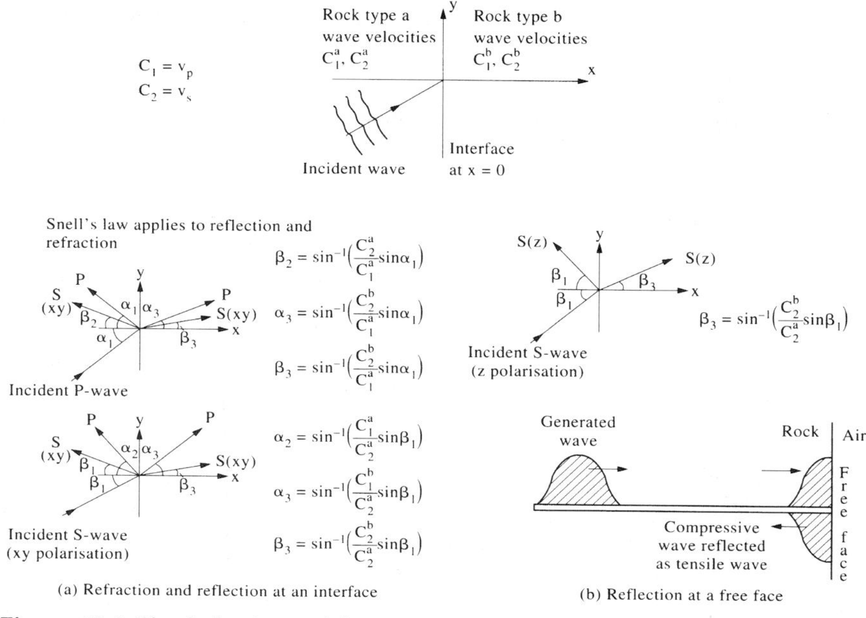

Figure 13.4 The behaviour of longitudinal stress waves at material interfaces. (a) Refraction and reflection at an interface. (b) Reflection at a free face.

wave will be reflected back into the first medium. This has implications for the measurements of such waves and the mechanisms of wave propagation and rock failure during blasting.

However, the phenomenon of paramount importance is illustrated in Fig. 13.4(b)—i.e. a compressive longitudinal stress wave is reflected as a tensile longitudinal stress wave at a 'free face'. Such free faces occur regularly in rock masses, in the form of discontinuites, but the one that has greatest significance is formed by the rock–air interface. In fact, Fig. 13.4(b) is a special case of Fig. 13.4(a) where almost all the energy is reflected, with little being refracted into the air. The concept of the 'free face' is critical in the design of all blasting rounds, and forms the basis of the specialized pre-splitting and smooth-wall blasting techniques (see Chapter 15).

In the next section, we consider the broad spectrum of material behaviour occurring over the wide range of strain rates likely to be encountered in engineering.

13.3 Time-dependency

We noted that no time component is incorporated in elasticity theory: it is assumed that the stresses and strains develop instantaneously on loading or unloading. However, we noted in Chapter 6 when discussing the complete stress–strain curve, that the exact form of the curve will depend on the strain rate at which it is determined. It is commonly observed at rock engineering sites, that the rock continues to deform after a stress change occurs—e.g. convergence of well bores and tunnels. So, it is evident that, whilst the theory of elasticity is of assistance in understanding and analysing the mechanics of rock masses, a theory is also required for time-dependent effects.

Words used to describe time-dependent behaviour are clarified in the glossary below.

Glossary of Terms

Elastic
Stresses are related to strains in a time-independent manner (i.e. $\sigma = S\varepsilon$, where S is the elastic compliance matrix). All strain energy is recoverable. It is assumed, in this context, that elastic materials remain elastic and so have infinite strength.

Plastic
Stresses are related to strains in a time-independent manner, but the material undergoes *plastic flow* when stressed (i.e. $d\sigma = Pd\varepsilon$, where P is a 6×6 plasticity matrix whose coefficients are stress- or strain-dependent). Deformation continues indefinitely without any further increase in stress. Strain energy is lost through permanent plastic straining. Generally, plastic behaviour is a function of distortional strains and deviatoric stresses.

Viscous
Stresses are related to strain rate (i.e. $\sigma = \eta\dot{\varepsilon}$, where η is a 6×6 viscosity matrix). Generally, viscous behaviour is also a function of distortional strains and deviatoric stresses.

Elastoplasticity
Time-independent theory combining elasticity and plasticity: materials behave elastically up to certain stress states and plastically thereafter.

Viscoelasticity
A generic term for a time-dependent theory in which strains are related to stresses and time. Instantaneously, viscoelastic materials have effectively infinite strength.

Viscoplasticity
Time-dependent behaviour in which the deviatoric stresses (or distortional strains) give rise to viscous behaviour, or plastic behaviour if the instantaneous strength of the material is temporarily exceeded.

Elastoviscoplasticity
This is the same as viscoplasticity, except that the instantaneous response of the material is purely elastic.

Creep
Under the action of a constant stress state, straining continues (see Fig. 6.16).

Relaxation
Under the action of a constant strain state, the stress within a material reduces (also see Fig. 6.16).

Fatigue
A generic term generally used to describe the increase in strain (or decrease in strength) due to cyclical loading.

Rheology
The study of flow.

Rheological models

These are analogues of different material behaviour, formed from assemblages of mechanical components, usually springs, dashpots and sliders. They assist in understanding the material behaviour and allow the formulation of the various constitutive relations.

Using just three rheological elements—spring (or Hookean substance), dashpot (or Newtonian substance) and slider (or St. Venant substance)—it is possible to produce a bewildering array of rheological models, depending on whether the elements are connected in series, parallel or a mixture. In Fig. 13.5, all of the two-element models are illustrated, together with their names. In Fig. 13.6, some multi-element models are shown.

In Fig. 13.6, the top left-hand rheological model is a viscoelastic material (element *2,1* in Fig. 13.5) with an additional spring in series, producing what is known as the generalized Kelvin substance. In the top right-hand model of Fig. 13.6, another dashpot has been added in series to the generalized Kelvin model, producing Burger's substance. Note that this is an elastoviscous material in series with a viscoelastic material, i.e. a Maxwell model in series with a Kelvin model (cf. Fig. 13.5). The two lower substances of Fig. 13.6 are the behavioural models associated with the leading diagonal of Fig. 13.5, first in series (i.e. formed by working down the leading diagonal) and second in parallel. Note that a clockwise convention is used when connecting the elements on the leading diagonal in this matrix form

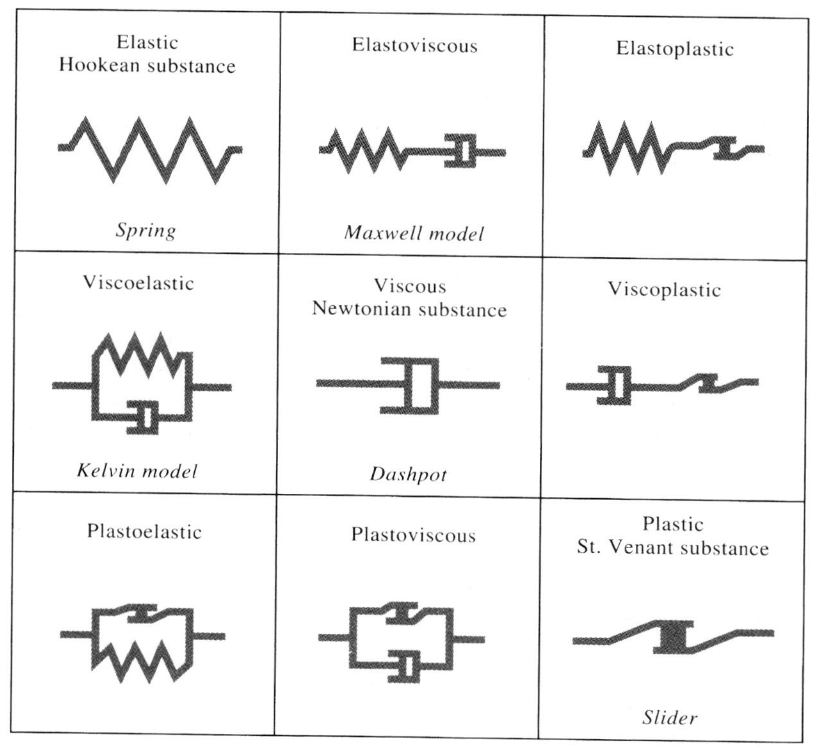

Figure 13.5 Two-component rheological models.

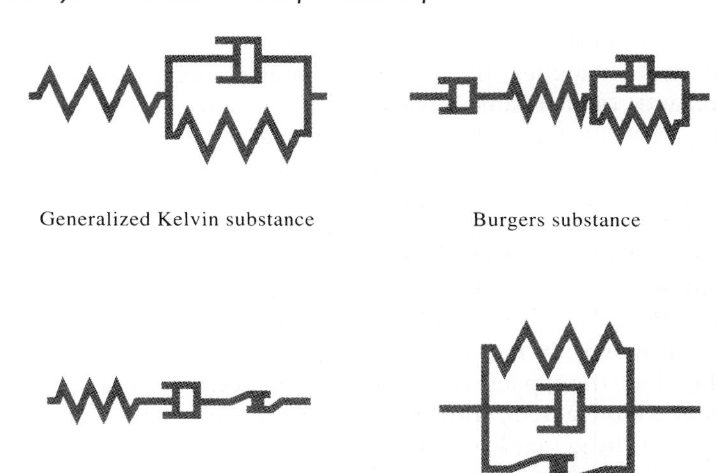

Figure 13.6 Multi-component rheological models.

of presentation. From the matrix structure, these substances are therefore termed elastoviscoplastic and plastoviscoelastic materials, respectively. Also note that the elastoviscoplastic substance's behaviour is dictated in turn by the spring, the dashpot and the slider, i.e. by its elastic, viscous and plastic elements in sequence. Conversely, when the elements are linked in parallel (the plastoviscoelastic substance), the total model behaviour is dictated in turn by the slider, dashpot and spring. The off-diagonal rheological substances above the leading diagonal in Fig. 13.5 are series models, whereas those below the leading diagonal are parallel models.

With multi-component models there are many ways in which the components can be combined in series and parallel sub-networks. In Fig. 13.6 we have shown one extension of the Maxwell model, the combination of the Maxwell and Kelvin models, and the two simplest ways of combining the three basic rheological elements. In theory, and by analogy with electrical resistors, capacitors and fuses, we could generate any n-component model and establish its global constitutive behaviour. Such time-dependency models contain a large number of terms and so are lengthy and can be difficult to assimilate. It is, therefore, instructive to consider, mathematically, the simpler Maxwell (elastoviscous) and Kelvin (viscoelastic) models (i.e. elements *1, 2* and *2, 1* of the matrix in Fig. 13.5) as examples of time-dependent behaviour.

The two fundamental elements, viscous and elastic, have basic uniaxial constitutive laws of $\sigma = F\mathrm{d}\varepsilon/\mathrm{d}t$ and $\sigma = E\varepsilon$, respectively, where F and E are the uniaxial constants of viscosity and elasticity.

The **Maxwell model** consists of viscous and elastic elements in series. Consequently, the stress is identical in each of the elements and the strain developed in the material, ε_S, is the sum of the strains developed in the elastic and viscous elements, i.e. ε_E and ε_V, respectively. Thus,

$$\varepsilon_S = \varepsilon_E + \varepsilon_V$$

which upon differentiation gives

$$\frac{d\varepsilon_S}{dt} = \frac{d\varepsilon_E}{dt} + \frac{d\varepsilon_V}{dt}.$$

Differentiating the fundamental constitutive relation for an elastic element gives $d\varepsilon/dt = (1/E)d\sigma/dt$. Substituting this, and the relation for the viscous element, into the above relation gives

$$\frac{d\varepsilon_S}{dt} = \frac{1}{E}\frac{d\sigma}{dt} + \frac{1}{F}\sigma.$$

This is the differential equation governing the behaviour of a Maxwell material. By considering two loading cases (constant stress and constant strain), it is possible to demonstrate its behaviour more clearly. For example, if we assume that from $t = 0$ to $t = t_1$, a constant stress is applied, σ_0, and then from $t = t_1$ constant strain is maintained,

$$\int d\varepsilon = \frac{1}{E}\int \frac{d\sigma}{dt} dt + \frac{\sigma}{F}\int dt + C$$

which after integration becomes, as the stress is constant,

$$\varepsilon = \frac{\sigma}{E} + \frac{\sigma}{F}t + C.$$

At $t = 0$, the material behaves instantaneously as an elastic material, with $\varepsilon = \sigma_0/E$. Hence, $C = 0$. Consequently, under the action of constant stress, the behaviour of a Maxwell material is

$$\varepsilon = \frac{\sigma}{E} + \frac{\sigma}{F}t \quad \textbf{linear creep.}$$

The strain that has accumulated at $t = t_1$ is thus

$$\varepsilon_1 = \frac{\sigma_0}{E} + \frac{\sigma_0}{F}t_1.$$

However, for constant strain, $d\varepsilon/dt = 0$, so the basic differential equation becomes

$$0 = \frac{1}{E}\frac{d\sigma}{dt} + \frac{1}{F}\sigma.$$

Rearranging and integrating gives

$$\log_e \sigma = -\frac{E}{F}t + C.$$

Now, at $t = t_1$, $\sigma = \sigma_0$ with the result that $C = \log_e\sigma + (E/F)t_1$, and hence

$$\sigma = \sigma_0 e^{-\frac{E}{F}(t-t_1)} \qquad \textbf{non-linear relaxation.}$$

These two types of behaviour are illustrated in Fig. 13.7.

The **Kelvin model** consists of viscous and elastic elements in parallel. Consequently, the strain is identical in each of the elements and the stress developed in the material, σ_S, is the sum of the stresses developed in the elastic and viscous elements, σ_E and σ_V, respectively,

$$\sigma_S = \sigma_E + \sigma_V = E\varepsilon + F\frac{d\varepsilon}{dt}.$$

Again, considering a period of constant stress followed by constant strain,

$$\sigma_0 = E\varepsilon + F\frac{d\varepsilon}{dt}$$

which, on rearranging, becomes

$$\frac{1}{F}\int dt = \int \frac{d\varepsilon}{\sigma_0 - E\varepsilon} + C.$$

Integration and substitution of $C = (1/E)\log_e\sigma_0$ (because at $t = 0$, $\varepsilon = 0$) yields

$$\varepsilon = \frac{\sigma_0}{E}\left(1 - e^{-\frac{E}{F}t}\right) \qquad \textbf{non-linear creep.}$$

The strain that has accumulated at $t = t_1$ is thus

$$\varepsilon_1 = \frac{\sigma_0}{E}\left(1 - e^{-\frac{E}{F}t_1}\right).$$

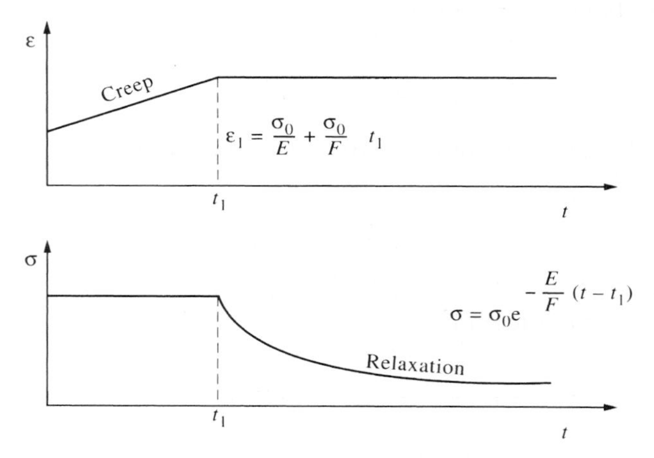

Figure 13.7 Linear creep and non-linear relaxation for the Maxwell substance.

However, for constant strain, $d\varepsilon/dt=0$, the basic differential equation reduces to $\sigma = E\varepsilon$, with the result that

$$\sigma_1 = \sigma_0\left(1 - e^{-\frac{E}{F}t_1}\right) \quad \textbf{stepped relaxation.}$$

These two types of behaviour are illustrated in Fig. 13.8.

The description which we have given of the Maxwell and Kelvin basic rheological models have only incorporated two components—in series for the Maxwell model and in parallel for the Kelvin model. One can consider three, four or indeed any number of such rheological elements connected in series and parallel networks: for example, combining the Maxwell and Kelvin models in series produces the Burger's substance illustrated in Fig. 13.6. These rheological models are models of one-dimensional behaviour and to use them to analyse the behaviour of three-dimensional continua, it is necessary to assume that the viscoelastic response is due only to the deviatoric and distortional components of stress and strain, respectively, with the spherical and dilatational components causing time-independent volume change. The fundamental differential equation for an isotropic Maxwell material in terms of the distortional and deviatoric components is

$$\varepsilon_x^* = \frac{\sigma_x^*}{2\mu} + \frac{\dot{\sigma}_x^*}{2G}$$

where the superscripted asterisk denotes deviatoric and distortional components, the overdot represents the first derivative with respect to time, and $\mu = F/3$, $G = E/2(1 + \nu)$.

In order to write this in terms of total strain, stress and stress rate, we make use of the relations between total, spherical and deviatoric components, i.e.

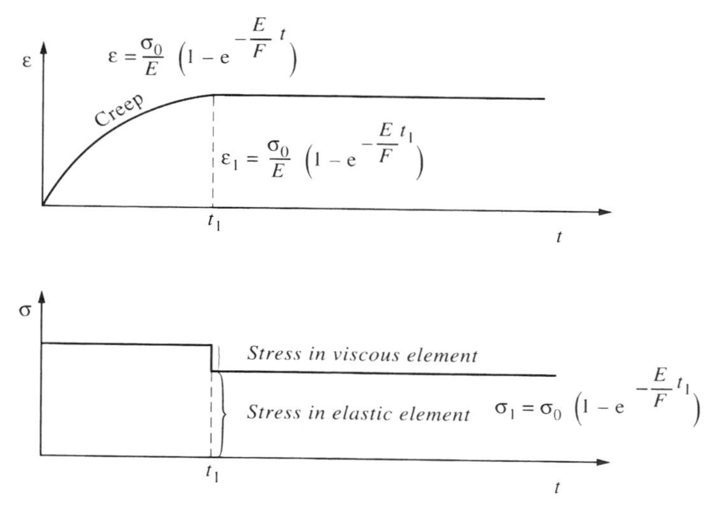

Figure 13.8 Non-linear creep and stepped relaxation for the Kelvin substance.

$$\varepsilon = \bar{\varepsilon} + \varepsilon^*,$$

$$\sigma = \bar{\sigma} + \sigma^*$$

$$\text{and } \dot{\sigma} = \dot{\bar{\sigma}} + \dot{\sigma}^*$$

where the overbar represents spherical and dilatational components. Substituting and rearranging eventually leads to three expressions of the form

$$\varepsilon_x = \frac{1}{3\mu}\left(\sigma_x - \frac{1}{2}\sigma_y - \frac{1}{2}\sigma_z\right) + \frac{1}{E}(\dot{\sigma}_x - \nu\dot{\sigma}_y - \nu\dot{\sigma}_z).$$

So the normal strain in any direction is then coupled with all three normal stresses and all three normal stress rates. There is a correspondence between the factors of ½ applied to the stress components and the factor of ν applied to the stress rate components—because $\nu = ½$ for incompressible materials, i.e. the spherical component of the stress tensor.

The multi-component rheological models and the continuum analysis shown above can lead to complex relations with many material constants. In the practical application of rock mechanics it has been found convenient to simply use empirical relations which fit observed strain versus time curves. Some of many possibilities have been collated by Mirza (1978) and are shown in Table 13.1.

For applications in rock mechanics and rock engineering, the importance of viscoelasticity has either not been fully recognized, or has been neglected owing to the difficulty of developing closed-form solutions to even basic problems. This is now being redressed in the development of numerical methods which explicitly take viscoelasticity into account, as discussed by

Table 13.1 Empirical creep laws (after Mirza, 1978)

1	$\varepsilon = At^m$
2	$\varepsilon = A + Bt^m$
3	$\varepsilon = A + Bt + Ct^n$
4	$\varepsilon = A + Bt^m + Ct^n + Dt^p$
5	$\varepsilon = At^m + Bt^n + Ct^p + Dt^q + \cdots$
6	$\varepsilon = A \log t$
7	$\varepsilon = A + B \log t$
8	$\varepsilon = A \log(B + t)$
9	$\varepsilon = A \log(B + Ct)$
10	$\varepsilon = A + B \log(C + t)$
11	$\varepsilon = A + B \log(t + Dt)$
12	$\varepsilon = At/(1 + Bt)$
13	$\varepsilon = A + B \sinh(Ct^n)$
14	$\varepsilon = A + Bt - C \exp(-Dt)$
15	$\varepsilon = At + B[1 - \exp(-Ct)]$
16	$\varepsilon = A[1 - \exp(-Bt)] + C[1 - \exp(-Dt)]$
17	$\varepsilon = A + B \log t + Ct^n$
18	$\varepsilon = A + Bt^n + Ct$
19	$\varepsilon = A + B \log t + Ct$
20	$\varepsilon = \log t + Bt^n + Ct$
21	$\varepsilon = A \log[1 + (t/B)]$
22	$\varepsilon = A[1 - \exp(B - Ct^n)]$
23	$\varepsilon = A[1 - \exp(-Bt)]$
24	$\varepsilon = A \exp(Bt)$

Pande *et al.* (1990). Because time-dependency is ubiquitous in rock mechanics problems, these developments and the ability to incorporate the techniques in design are critical to the advancement of rock engineering.

13.4 Time-dependency in rock engineering

Engineers have found it convenient to consider phenomena as either associated with very high strain rates or very low strain rates. This is because the process of rock excavation (e.g. by blasting) occurs rapidly, whereas deformation (e.g. displacement occurring throughout the life of an excavation) occurs slowly. In the high strain rate category we include blasting, vibrations and fatigue; in the low strain rate category we include creep, subsidence and long-term displacements. We noted that the strain rates can be spread over 15 orders of magnitude, with the result that it is debatable whether any generic time-dependent model can be valid over such a large range. In the newer applications of rock engineering, such as radioactive waste disposal, the specified design lives can be large, of the order of 1000 years. Thus, not only is there concern with the time-dependent behaviour but we have to consider whether **all** the rock properties and mechanisms can be considered to be uniform over such an extreme time period. This is exacerbated by the fact that we can only conduct testing procedures in the range of medium to high strain rates.

If the rock properties are determined by geophysical means, at very high strain rates, we should ask ourselves how valid is it to apply these to engineering applications of, say, a billion times greater duration than the test period? This question has profound implications for the validity of theoretical models, test results and the interpretation of field measurements. We are led to the conclusion that engineering judgement must still play a large part in determining the type of time-dependent analysis that is used.

14 Rock mechanics interactions and rock engineering systems (RES)

In this chapter, we introduce a method of structuring all the ways in which rock mechanics parameters and variables can affect one another—the rock mechanics interactions. The method is presented within the wider context of an approach to integrate all the relevant information in rock engineering design and construction, rock engineering systems, Hudson (1992). The interaction matrix is explained first with examples and a general consideration of the nature of matrix symmetry. Then larger matrices are discussed demonstrating the links between rock mechanics and rock engineering, especially the pre-construction and post-construction interactions. Further example applications are given so that readers will feel confident in generating their own matrices for any problem. The RES approach aims to identify the parameters relevant to a problem, and their interactions, thus providing overall coherency in approaching rock mechanics and rock engineering problems.

Referring back to Fig. 1.12, the three-tier approach to all rock engineering problems, the inner ring represents the analysis of individual subjects. The chapters in this book have so far followed such single-subject themes but this chapter is about interactions and coupled mechanisms (represented by the middle ring of Fig. 1.12). It is only through the understanding of these interactions that we can arrive at the outer ring of Fig. 1.12 and the corresponding solution of complete rock engineering problems using theory and experience.

The need to study the interactions has always been present. Now we have much better computational capability and are being faced with increasingly large and complex problems in which it is by no means clear what the main factors are, how they interact, and how best to build the most appropriate conceptual, mathematical, numerical or mechanical models.

14.1 Introduction to the subject

In several of the earlier chapters, we have touched on the fact that one rock mechanics parameter can affect another. This is illustrated in Fig. 14.1, which shows the six binary interactions of *in situ* stress, rock structure and

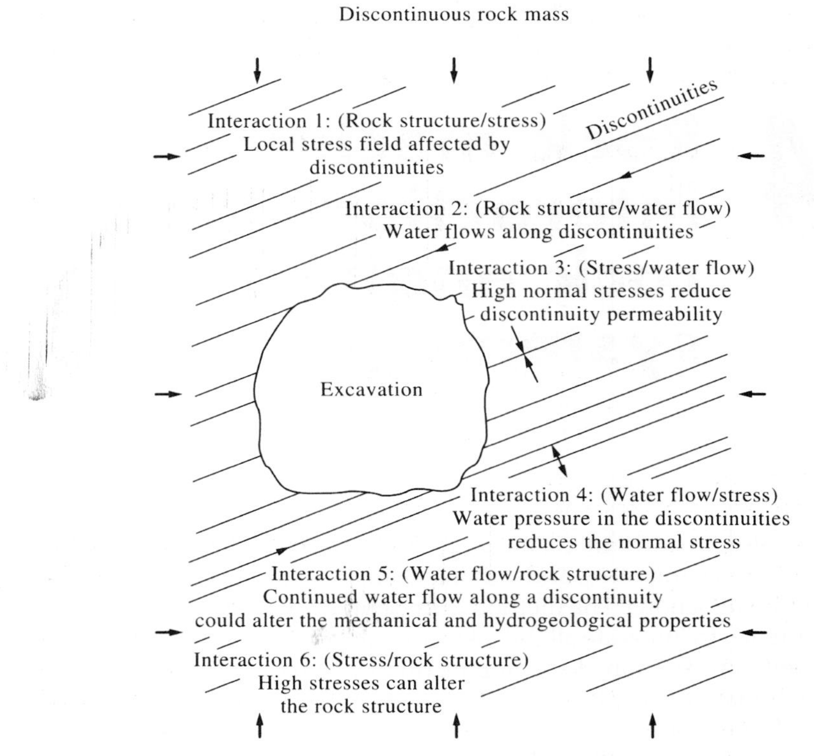

Figure 14.1 Six of the main rock mechanics interactions.

water flow. The interactions annotated (1)–(6) in the figure are listed below.

- Interaction 1: *Rock structure/stress*—stress field affected by discontinuities.
- Interaction 2: *Rock structure/water flow*—water preferentially flows along discontinuities.
- Interaction 3: *Stress/water flow*—high normal stresses reduce discontinuity permeability.
- Interaction 4: *Water flow/stress*—water pressure in discontinuities reduces effective normal stress.
- Interaction 5: *Water flow/rock structure*—water flow causes discontinuity surface alteration.
- Interaction 6: *Stress/rock structure*—high stress can alter the rock structure.

The figure shows the interactions occurring around an underground excavation but a similar diagram could be drawn for any rock engineering application. Also, these are only six of a large number of interactions which occur in rock mechanics and during rock engineering: we have chosen these because of their fundamental nature.

Interaction matrices are used to provide a systematic approach to the interactions. Generally, matrices are used as assemblages of detached coefficients, subject to agreed manipulative procedures for mathematical utility. However, the basic presentational approach is also useful if the matrix components are concepts, or subjects, *and* the interactions between these concepts or subjects. The matrix presentation is not merely a pedagogic device: it serves to identify and highlight the interactions between subjects, and forms the structure for coupled modelling.

14.2 Interaction matrices

The idea that there may be a relation between all things has been expressed by Francis Thompson, the English Victorian poet, who wrote the lines

> All things by immortal power,
> Near or far,
> Hiddenly
> To each other linkèd are,
> That thou canst not stir a flower
> Without troubling of a star.

In fact, the concept of considering the relations between quantities which have the property of perpendicularity is extremely old. The margin sketch is from Ch'ou-pei Suan-king (an ancient Chinese treatise dating from *circa* 1100 BC, which is housed in the British Museum), and is an early illustration of the proof of what is now known as Pythagoras' theorem. In the history of mathematics there has been considerable development of the mathematics associated with orthogonality (n-dimensional perpendicularity) via matrix and tensor analysis, so that the foundation for many subjects has already been laid. For example, when considering the variability of many different parameters, the subject of multivariate analysis is used, where n individual parameters are considered along n orthogonal axes in n-dimensional space, and there are other examples of subjects built on this mathematical foundation, e.g. Fourier analysis.

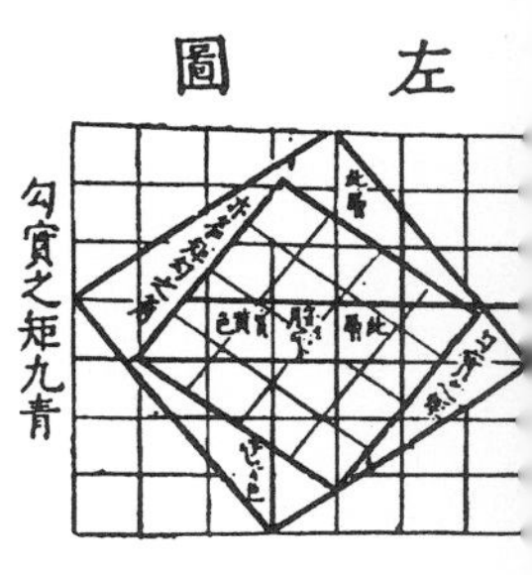

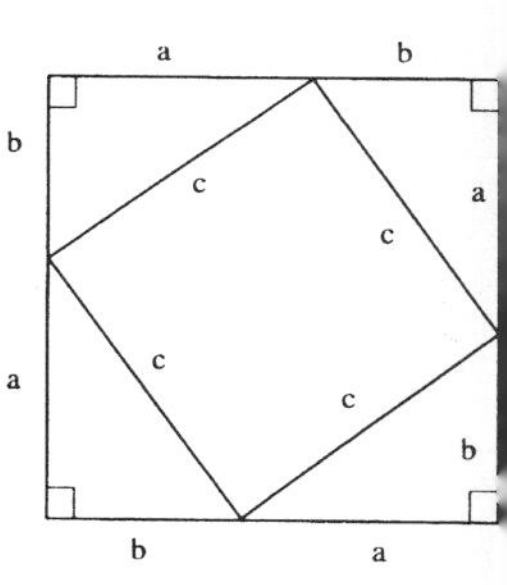

$(a + b)^2 = c^2 + 4\,(\tfrac{1}{2}ab)$
$a^2 + 2ab + b^2 = c^2 + 2ab$
$a^2 + b^2 = c^2$

The basic concept here is to study the combination, interaction or influence of one subject on another. We begin with 2×2 matrices, but it should be remembered that all the ideas can be extended to an $n \times n$ matrix. In Fig. 14.2, the main subjects, here denoted by A and B, are placed in the leading diagonal positions, i.e. from the top left to the bottom right of the matrix. A matrix is a list, and we are considering subjects, rather than the more usual numerical quantities. We are also considering the interactions—shown in the off-diagonal boxes—that are studied by clockwise rotation, as indicated by the arrows in the figure.

In the construction of such matrices, the primary parameters are always listed along the leading diagonal, as in Fig. 14.2. The off-diagonal terms could represent the combination, influence or interaction of the primary parameters, as shown in Fig. 14.3. Combination can be demonstrated simply by inserting numbers in the leading diagonal with the off-diagonal terms being, for example, their sums. Similarly, influence is demonstrated by considering discontinuity aperture and water flow and, finally, interaction by considering how, for a given stress state, normal stresses give rise to shear stresses.

In the first matrix of Fig. 14.3, the off-diagonal terms in the matrix represent the addition of the leading diagonal numbers. Because $3 + 2 = 2 + 3 = 5$, the two off-diagonal terms are the same and the matrix is therefore symmetrical about the leading diagonal.

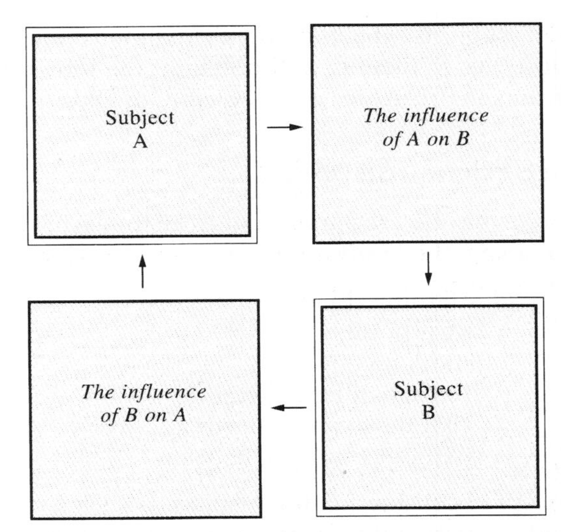

Figure 14.2 A 2×2 matrix illustrating the positioning of the primary variables and their interactions.

In the second matrix of Fig. 14.3, one example of how two rock mechanics parameters can influence each other is shown. On the one hand, a larger discontinuity aperture leads to increased water flow on the other, increased water flow can lead to mineral deposition in the discontinuity, or erosion of the discontinuity surfaces, resulting in alteration of the aperture. In this matrix, instead of numbers there are rock mechanics quantities (albeit as words) as the primary parameters. Here, the influence of *A* on *B* is not the same as the influence of *B* on *A*, which means that the matrix is **asymmetric.** We will be discussing the significance of, and reasons for, symmetry and asymmetry of matrices later in this chapter.

In the third matrix of Fig. 14.3, we have reproduced the two-dimensional stress tensor presented in Chapter 3. This is an example of interaction between the primary parameters. Given a specific stress state, defined, for example, in terms of principal stresses, the values of the normal stresses σ_{xx} and σ_{yy} uniquely define τ_{xy} and τ_{yx}. We have already noted in Section 3.6 that $\tau_{xy} = \tau_{yx}$, and so the stress matrix is symmetrical. With the analogous normal strains and shear strains, we also noted at the end of Section 5.1

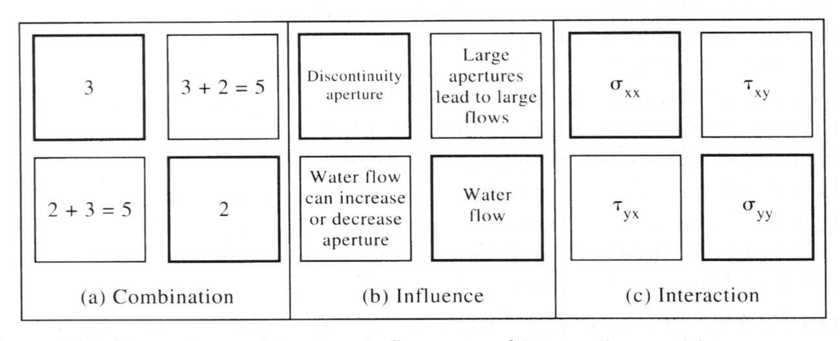

Figure 14.3 Example combination, influence and interaction matrices.

that shear strain involves an interaction between the axes because the amount of shear strain in the x-axis depends on the position along the y-axis, cf. Fig. 5.3.

Another example of the way in which interaction matrices can be used to present information is illustrated in Fig. 14.4. The leading diagonal terms are a square, a rectangle, a rhombus and a parallelogram. These are four geometrical shapes which can be converted into one another by either adding or subtracting the constraints of four equal sides or four equal angles. The condition boxes shown indicate the constraints necessary to produce the given shapes from a generic quadrilateral. The off-diagonal terms are condition icons representing the various transformations necessary to change one leading diagonal term into another—these conditions being added or subtracted. When transforming any pair of leading diagonal shapes from one to another, the same conditions are involved, whether one is in the upper half of the matrix or the lower half; only the sign of the condition is different, and hence the matrix is skew-symmetric. The link with group theory is intimated in the Venn diagram also presented, but a full discussion of this subject is not necessary here.

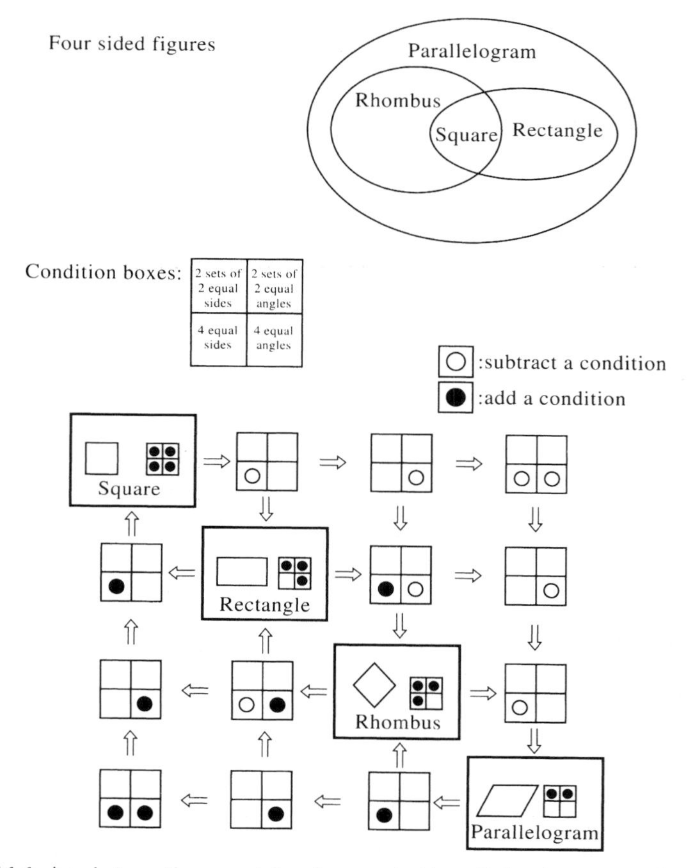

Figure 14.4 An interaction matrix demonstrating links between the leading diagonal terms.

14.3 Interaction matrices in rock mechanics

In Fig. 14.5, we show the conceptual link between stress and strain. In Section 5.5, it was recalled that stress and strain are second-order tensors, and that each component of the strain tensor can be linearly related to the six components of the stress tensor, through an elastic compliance matrix with 36 terms, of which 21 are independent. This is a result of the application of linear elasticity. In the 2×2 matrix of Fig. 14.5(a), we show this same link, **except** that we have now introduced the concept of *path dependency*, i.e. how do we compute the strains from a knowledge of the stresses, or vice versa? The top right-hand box of this matrix illustrates the term S_{11}, or $1/E$, of the elastic compliance matrix. Alternatively, we can calculate the stresses from the strains as illustrated in the bottom left-hand box. Note that the stress–strain curves are drawn with the independent variable on the horizontal axis, in accordance with the conventions of scientific presentation. Thus, this 2×2 matrix is not symmetrical in the sense that the content of the off-diagonal terms are not equal, but is symmetric in terms of functionality: that is, one can travel around the matrix with no change occurring in the stress or strain state being represented.

However, the 2×2 matrix in Fig. 14.5(b) represents constitutive behaviour which includes rock failure—and is therefore beyond simple linear elasticity—with the result that the behaviour of the rock now critically depends on whether stress or strain is the independent variable. In some cases, standardized methods for rock testing will specify values for the rate of stress increase and, as illustrated in the top right-hand box of Fig. 14.5(b), this will result in uncontrolled failure because the stress cannot be increased beyond the compressive strength. Conversely, if the strain rate is specified then strain becomes the independent variable and the complete stress–strain curve is obtainable, as explained in Section 6.3. In every sense, therefore, this matrix is asymmetric: the off-diagonal stress–strain behaviour is different; and one cannot cycle repetitively through the matrix at all stress levels.

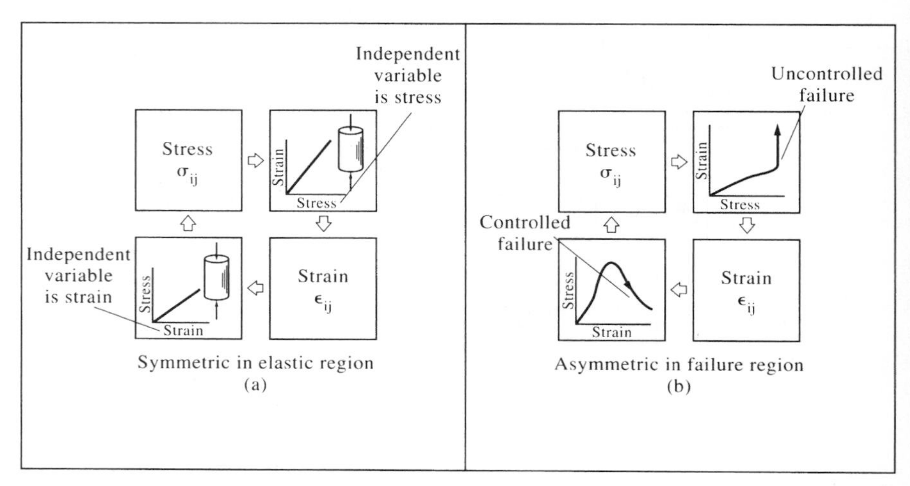

Figure 14.5 Stress–strain relations illustrated conceptually for elastic and inelastic conditions utilizing 2×2 interaction matrices.

In Fig. 14.6, we illustrate another 2 × 2 interaction matrix in which the parameters *point load strength* and *compressive strength* appear on the leading diagonal. In calculating the regression line linking these two parameters, we could consider either one as the independent variable, as is illustrated in the off-diagonal boxes. The data points are the same in each case: the axes have simply been exchanged, and a different regression line is obtained in each of the two cases. Regression analysis using these 13 data points produces

$$\sigma_c = -3.13 + 23.51 I_s$$

and

$$I_s = 0.563 + 0.0367 \sigma_c.$$

These two equations are not the same line. They are plotted in the off-diagonal boxes, with the solid line being the appropriate one in each case. The two off-diagonal boxes are not the same: the matrix is asymmetric—caused by the two estimation directions. The point is not the accuracy with which these two parameters are related, but the asymmetry of the matrix resulting from the path-dependency.

In Fig. 14.7, there is a 4 × 4 matrix illustrating the inter-relations between the orientation, spacing, persistence and roughness of a discontinuity. These are four of the primary geometrical properties of discontinuities illustrated in Fig. 7.3. On examining the schematic in Fig. 7.3, it might appear at first sight that the four parameters are independent: however, as Fig. 14.7 illustrates, all four of these parameters are interdependent. This interaction matrix demonstrates the comprehensive identification of all of the binary interactions—some of which might otherwise be overlooked.

14.4 Symmetry of interaction matrices

We have seen that the stress tensor is symmetrical about the leading diagonal of the matrix because of the equilibrium conditions, as illustrated in Fig. 3.6. Both the stress and strain tensors, and other similar second-order

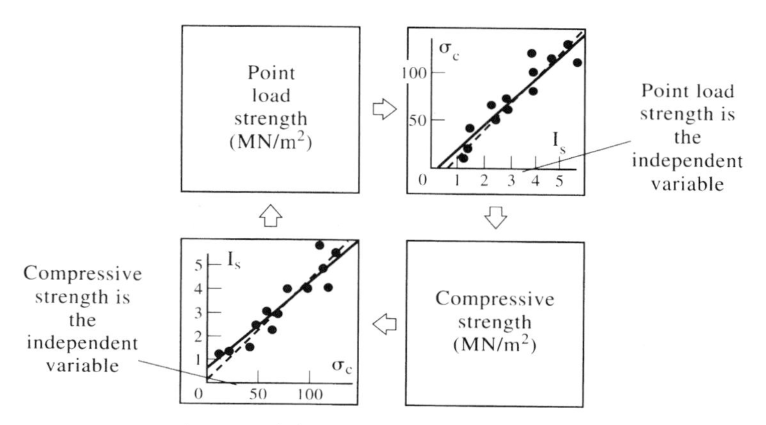

Figure 14.6 Point load strength–compressive strength relations, illustrating asymmetry of the interaction matrix induced by the estimation direction.

ORIENTATION	The frequency of intersections along a discontinuity is given by $\lambda_s = \Sigma\lambda_i \cos\theta_i$	As two interesting discontinuity sets move away from being orthogonal, so trace lengths as exhibited as block faces will either increase or reduce	If a failure plane is defined by a series of non-parallel intersecting discontinuities it will have an *effective roughness* due to its step-like form
Measurements of orientation may be imprecise and inaccurate in a highly fractured rock mass	SPACING	Intersecting discontinuity sets will tend to procude trace lengths in proportion to their spacing values	Spacing affects the absolute size of *effective roughness*
Discontinuities of small extent may be unrelated in orientation to the major rock mass structure	If discontinuities in a set have a finite trace length then spacing values will change when a discontinuity appears between two previously adjacent features	EXTENT	Extensive discontinuities tend to be planar (e.g. slickensided fault planes)
Depending on the relative scales of the roughness and the instrument used for measuring orientation, roughness may be identified as a spread of orientations	Spacing between adjacent discontinuities is not uniquely defined for rough surfaces	Extent of extremely rough surface (e.g. stylolites) may be difficult to assess due to effective departure from a plane	ROUGHNESS

Figure 14.7 Interdependence between discontinuity geometry parameters.

tensors, e.g. permeability and moment of inertia, are symmetrical: this is due to the basic equilibrium inherent in these quantities. If, now, we consider the first 2×2 matrix shown in Fig. 14.3, this is also symmetrical because of the commutative properties of addition. Note, however, that had we chosen to consider subtraction as the binary operator, the off-diagonal terms would have had the same absolute value, but with different signs (e.g. $3 - 2 = 1$, whereas $2 - 3 = -1$), resulting in a skew-symmetric matrix. The second matrix of Fig. 14.3 is quite clearly asymmetric, because the influence of discontinuity aperture on water flow is not the same as the influence of water flow on discontinuity aperture.

Considering further the other matrices we have presented, Fig. 14.4 is another skew-symmetric matrix, because the condition boxes are reversed in sign, depending on which shape conversion is being considered. The symmetry and asymmetry of the 2×2 matrices illustrating the stress–strain relations for elastic and inelastic materials, respectively, are a result of path-dependency. This is also the case illustrated in Fig. 14.6, where the regression is different when different parameters are assumed to be the independent variable. Finally, the 4×4 matrix of Fig. 14.7, which shows the interdependence between discontinuity geometry parameters, is also asymmetric.

Asymmetry of matrices is associated with path-dependency. An asymmetric matrix is shown in Fig. 14.8: this is an example of a transition probability matrix for a Markov chain of state changes. A parameter can have the states *A*, *B* or *C*. Once the parameter is in one of these states, the

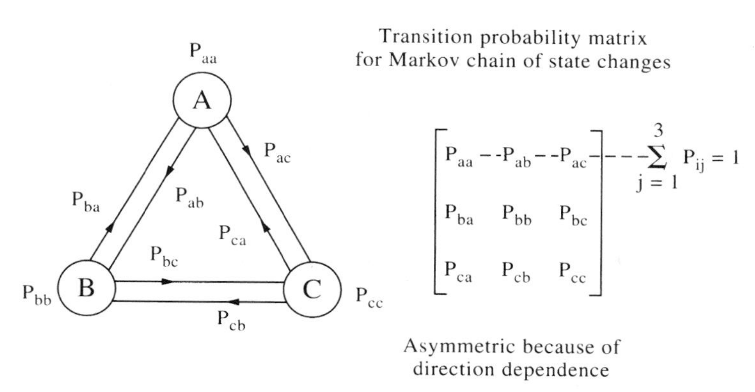

Figure 14.8 Transition probability matrix for Markov chain of state changes.

probability of it remaining in that state or moving to one of the other states is given by the transition probabilities shown in the matrix. If the parameter is in state A, the probability that it will remain in that state is P_{aa}; the probability that it will move to state B is P_{ab}; and the probability that it will move to state C is P_{ac}. These probabilities are given in the first row of the matrix, and their sum is unity. The second and third rows of the matrix represent similar transition probabilities for parameters in states B and C. Such transition probability matrices are used for generating Markov chains of events and for studying the sensitivity of the occurrence of certain states as a function of the transition probabilities. One could, for example, consider in a geological analysis the types of sedimentary sequences that will occur as the result of different depositional states.

The matrix illustrates symmetry and asymmetry conditions well. It may be that $P_{ab} = P_{ba}$ or that $P_{ab} \neq P_{ba}$. Quite clearly, any asymmetry of the matrix results from directional dependence in the state change.

Another excellent example of asymmetry is the formulae for the transformation of axes. In Fig. 14.9, the new co-ordinates x' and y' are given as a function of the old co-ordinates x and y and the angle θ through which the axes have been rotated. It can be seen that the basic operation of rotating the axes produces a cos θ term along the leading diagonal (representing the primary operation), and a sin θ term in the off-diagonal positions (representing the interaction between the axes). However, the matrix is skew-symmetric because of the rotational nature of the transformation: if we were to rotate the axes in the opposite direction, the sign of the off-diagonal terms would be reversed. It should be noted that this axis interaction is directly analogous to the note in Fig. 5.4, where simple shear also involves an interaction between the axes.

Finally, in Fig. 14.10, the axes are rotated through 45°, 90° and 180°. For the first case, the primary operation terms on the leading diagonal and the interaction terms in the off-diagonal positions have equal importance, but the matrix is still asymmetric. In the second case, the primary operation has been reduced to zero, because a rotation of 90° can be considered simply as an interchange of the axes with the signs of the off-diagonal components indicating the positive directions of the new axes relative to

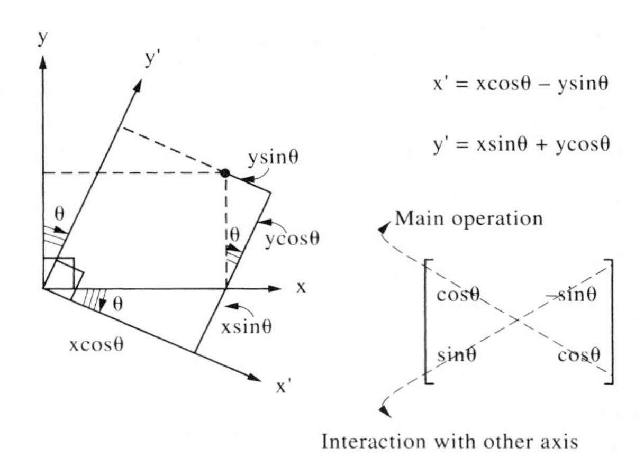

Figure 14.9 Transformation matrix for rotation of axes.

the old axes: the matrix is still asymmetric. On rotating through 180°, as shown in the third case, the axes remain in the same orientation, with only the positive directions changing. Thus, the primary operation is a multiplication by –1, and there is no interaction between the axes. **This matrix is symmetric**, because it does not matter whether we rotate clockwise or anticlockwise through 180°: i.e. there is no path-dependency in the transformation **for this specific angle of rotation**.

This example has been included because of the importance of understanding the nature of the off-diagonal terms in the rock mechanics interaction matrices. Consideration of whether path dependency is inherent in any matrix will assist in determining the symmetry or otherwise of the terms.

14.5 A rock mechanics–rock engineering interaction matrix

Referring again to Fig. 1.12 and the three-tier approach to all rock engineering problems, the analysis of coupled mechanisms is necessary in extrapolating the analysis of individual subjects to the solution of complete rock engineering problems. In Fig. 14.11, the primary parameters on the leading diagonal are rock mass structure, *in situ* stress, water flow and construction. This is a 4 × 4 matrix, with four primary variables and 12 interaction terms. The indentification and location of these interactions within the matrix is of help when interpreting the rock mechanics and rock engineering components. If we were to add two more leading diagonal terms—for example, thermal effects and time-dependency—the matrix would increase to a 6 × 6 size with six primary parameters and 30 interaction terms. It is unlikely that these interactions could be coherently identified and studied without the use of the interaction matrix or some similar approach.

There are several points to be noted about this matrix. First, the boxes have been numbered according to conventional matrix notation, with the

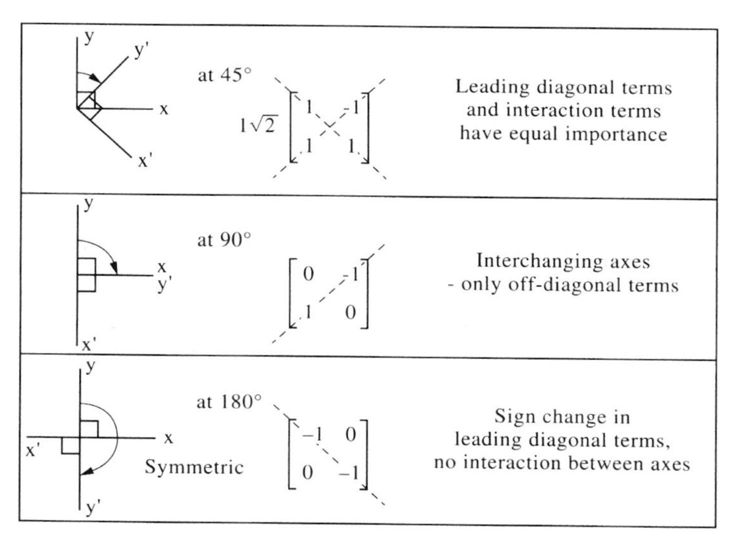

Figure 14.10 Three examples of the transformation matrix for rotation of axes.

first number representing the row and the second number the column in which the term resides. Second, the matrix is asymmetric. Third, as illustrated in Fig. 14.12, the basic rock mechanics component is contained within the uppermost 3×3 sub-matrix, with the interaction between rock mechanics and rock engineering occurring in the fourth row and column. Note that with the clockwise directionality applied to this matrix, the effect of the rock mechanics on the rock engineering is contained in the fourth column, whereas the effect of the rock engineering on the rock mechanics (or rock properties) is given in the fourth row. These matrix component sets should therefore be seen as design (or pre-construction) considerations and engineering (or during and post-construction) effects, respectively.

The interactive elements in Fig. 14.11 are self evident from the sketches. Well-known concepts arise and areas can be identified which need research. Element *32* shows the influence of the presence of water on the *in situ* stress, which is embodied in the well-known concept of effective stress: however, the complementary element, *23*, is the influence of *in situ* stress on water flow—a subject still in its infancy. Similarly, we can point to elements *13* and *31*. The first of these represents the influence of rock mass structure on water flow—i.e. the fact that discontinuities dictate permeability, a subject discussed in Chapter 9. The second of these is the influence of water flow on the rock mass structure, i.e. the weathering of discontinuities caused by water flow: almost nothing is known of the mechanics of this subject.

The reader is encouraged to interpret the diagrams representing each element within the structure of the total interaction matrix. The technique can be used for all interactions, so it is helpful not only to understand the underlying structure of such matrices, but also to be able to create new matrices oneself. For example, the technique was used to present the

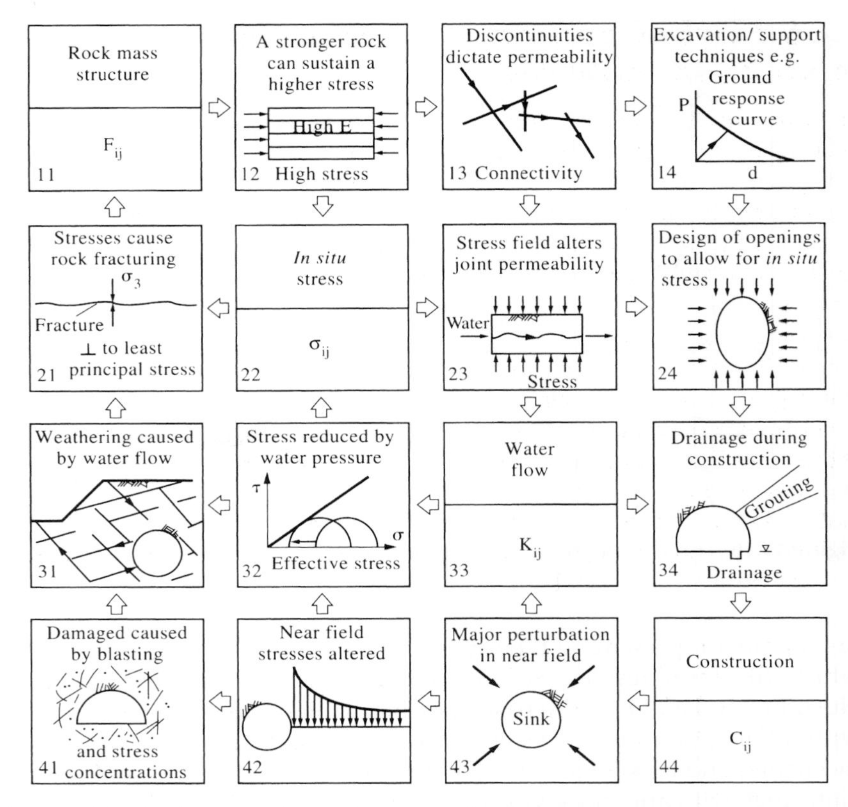

Figure 14.11 Rock mechanics–rock engineering interaction matrix.

variation of discontinuity frequency for a sampling line passing through one, two and three sets of mutually perpendicular discontinuities in Fig. 7.11. Along the leading diagonal of the three-dimensional loci in the figure are illustrated the individual variations for each set; the binary combinations for of any two sets are shown in the off-diagonal terms. This matrix is symmetrical because the frequencies are added for each set—a process which is commutative. However, it is only when adding the three sets, as shown at the bottom of the figure, that we obtain the full three-dimensional locus.

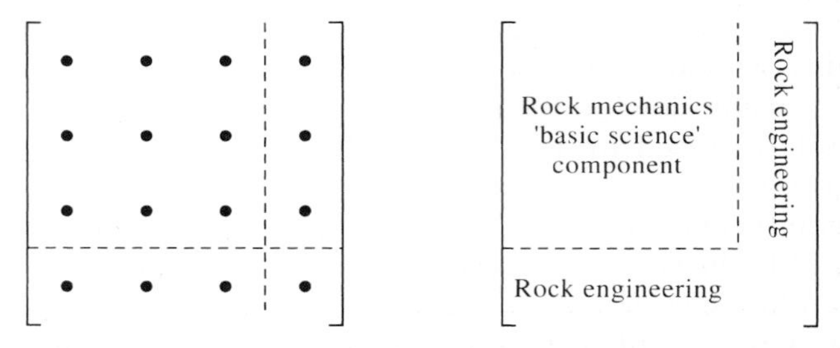

Figure 14.12 Architecture of the rock mechanics–rock engineering matrix illustrated in Fig. 14.11.

14.6 Further examples of rock mechanics interaction matrices

The three modes of fracturing rock to produce a discontinuity are shown in Figs 2.15–2.17. In these figures, the modes are shown in isolation, although it may well be that the prevailing stress state at a given location is such that the mode of failure is a hybrid of one or more of these fundamental modes. In an attempt to understand the hybrid modes, an interaction matrix can be drawn in which the fundamental modes are placed on the leading diagonal. This is shown in Fig. 14.13—in which we have assumed, for the purposes of illustration, that there is no path-dependency. Also, in the sketch shown below the matrix, all three modes occur simultaneously.

In Fig. 14.14, there is a related matrix in which the leading diagonal terms are normal, wrench and thrust faults, as dictated by the relative magnitudes of the three principal stresses causing the faulting. In this case, the presence of one type of fault will perturb the local *in situ* stress field, with the result that it could be reactivated as a different type of fault: the principal stresses will have changed and there will have been path-dependency introduced into the off-diagonal, double-faulting, elements.

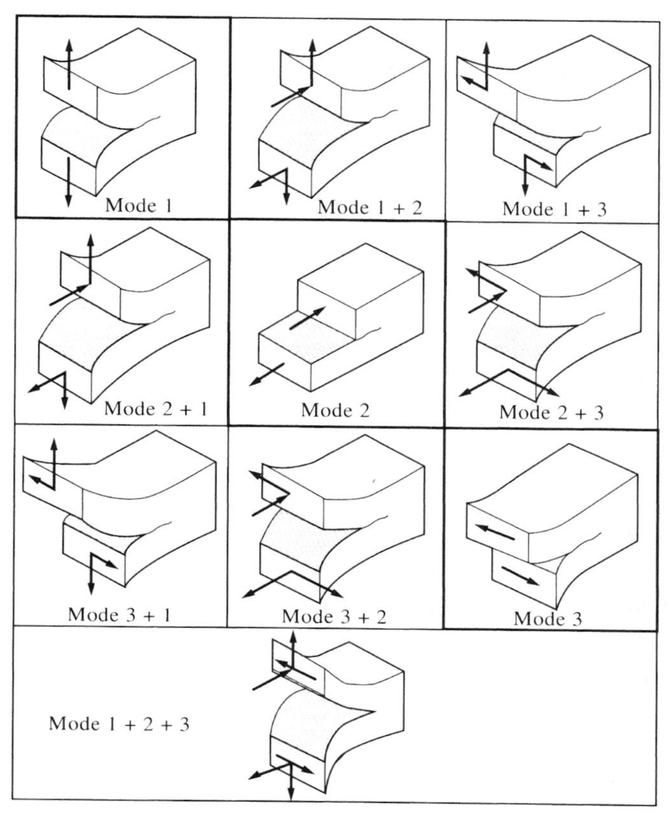

Figure 14.13 Unary, binary and ternary combinations of the fundamental modes of rock fracture.

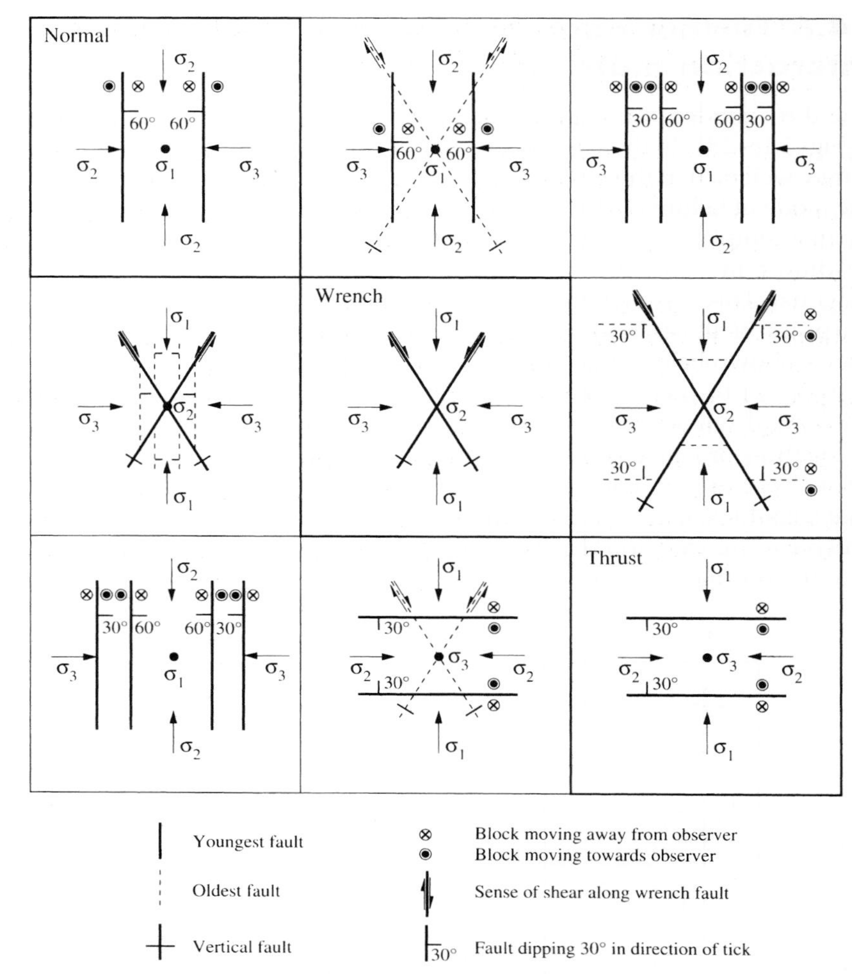

Figure 14.14 Binary combinations of normal, wrench and thrust faults (produced with help from Dr J. W. Cosgrove).

We have also presented some information in parts of the book using the interaction matrix concept, as for example Fig. 10.1 to consider the connections between the attributes of discontinuousness, inhomogeneity, anisotropy and non-elasticity.

14.7 Concluding remarks

The interaction matrix is the basic device used in rock engineering systems. For a rock engineering project, the most important step in the RES methodology is to establish the objectives of the project and the analysis. Once that has been done, the relevant 'state variables' are chosen—i.e. the terms to place along the leading diagonal of the interaction matrix. In some problems, these variables have to be more conceptual in nature; in some problems, there may be enough information to use well-defined physical properties with definite units. Then, all the interactions are established so that the problem structure is developed. Immediately, an information audit is possible considering what is known about the content of the interaction matrix boxes.

If the state variables are conceptual in nature, the off-diagonal interactions can be assessed using a semi-quantitative method of coding (Hudson, 1992). By summing the coding values in the matrix rows and columns, an estimate of each variable's interactive intensity and dominance in the system can be determined. This is known as the **soft systems approach**.

If the state variables are physical variables, a new modelling technique known as the **fully-coupled model** can be used (Hudson and Jiao, 1996). The physical mechanisms linking the variables are identified for each off-diagonal box and the relation quantified. A coupling algorithm is invoked to produce a new interaction matrix in which all the terms represent a fully-coupled system response; in other words, the response for all mechanisms operating simultaneously. The new matrix allows quantitative prediction of the effect of any applied perturbation on the state variables, whether natural or by engineering. This is known as **the hard systems approach**.

15 Excavation principles

In this chapter, we explain the fundamental principles of rock excavation. Initially, the excavation process is discussed in its most basic form, i.e. reaching the post-peak region of the complete stress–strain curve and altering the *in situ* block size distribution to the excavated fragment size distribution. Then the basic principle of blasting rounds is discussed; all blasting rounds involve a stress wave effect, stress wave reflections at free surfaces and a gas pressure effect. We highlight the pre-splitting method of near-surface excavation in which the final face is blasted **before** the bulk of the material is removed, in order to produce a boundary across which fragmentation is reduced. The complementary method for underground excavation, known as smooth-wall blasting, is also explained.

The principles of excavation using mechanical means, such as picks, discs and buttons, on partial- and full-face excavation machines are explained. The use of such machines within the excavation system is considered to demonstrate how important it is to consider the machine excavation process within the context of the overall construction strategy. Finally, we introduce the concepts associated with assessing effects of vibrations emanating from the excavation process on other structures.

15.1 The excavation process

It is instructive to consider the fundamental objective of the excavation process—which is to remove material from within the rock mass resulting in an opening (the geometry of which is set by some operational criteria). Bear in mind that there are two potential objectives in removing the rock: one is to create an opening; the other is to obtain the material for its inherent value. Examples of the first case include civil engineering works, temporary or permanent mine accesses and petroleum wellbores. In the second case, the material may contain some valuable mineral, or may be required *in toto* as, for example, concrete and road building aggregate. There will be projects where some parts of the work involve permanent openings and some parts involve temporary openings, e.g. a mine in which the access/egress through shafts and tunnnels must be by permanent openings, whereas the stopes (the openings left by mining ore) can be allowed to collapse during mining, as governed by the design process.

There are different design considerations in civil and mining engineering, related to the operational life of the openings themselves. A radioactive waste repository might have to operate satisfactorily for 5000 years, an undersea transport tunnel for 125 years, a mine shaft for 20 years, a metal mine stope for 1 year and a longwall mining opening for 1 week. These time spans relate to the discussion in Chapter 13 on the time-dependent aspects of rock mechanics and the implementation of appropriate models in rock engineering projects.

In order to remove part of a rock mass, it is necessary to introduce additional fractures over and above those occurring *in situ*. Three critical aspects of excavation are immediately introduced:

(a) the post-peak portion of the complete stress–strain curve must be reached (cf. Fig. 6.1);
(b) the *in situ* block size distribution must be changed to the required fragment size distribution; and
(c) by what means should the required energy be introduced into the rock?

We will discuss each of these aspects in turn.

15.1.1 Attaining the post-peak portion of the complete stress–strain curve

In Chapter 6, the complete stress–strain curve has a pre-peak portion of the curve which is mainly, although not completely, associated with linearly elastic behaviour. In this portion, there is little large-scale failure and little dissipation of energy on load cycling. In order to provide the necessary large-scale fragmentation, a part of the intact rock must be taken into the post-peak portion of the complete stress–strain curve. Note that subsequently we wish to remain in the pre-peak portion of the curve for rock stability. It follows that an excavation boundary is an interface between two fundamentally different engineering objectives and materials, as illustrated in Fig. 15.1.

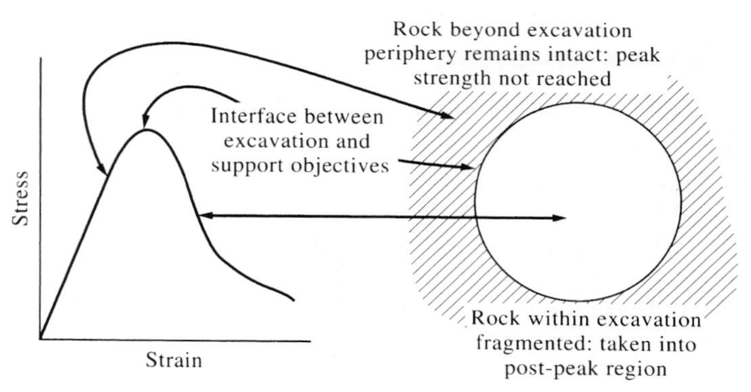

Figure 15.1 The complementary objectives of excavation and support, as related to the complete stress–strain curve.

A variety of considerations is involved, such as the brittleness of the material, as expressed via the complete stress–strain curve, and matching the explosive type and quantity to this for optimal fragmentation. In such an excavation process, is it best to break the rock in compression, tension or shear? Is there a choice? Because the tensile strength of rock is about one-tenth the compressive strength and the energy beneath the stress–strain curve is roughly related to the square of the peak strength of the rock, breaking the rock in tension requires only about one-hundredth of the energy required to break the rock in compression. So, not only do we need to match the explosive to the rock type, but we need to consider carefully how to use the energy in an optimal way to achieve the objective.

15.1.2 The *in situ* rock block and excavated fragment size distributions

Rock is naturally fractured and consists of rock blocks of certain sizes, which can be conveniently presented in an integrated way via a volumetric block size distribution, the concept of which was introduced via a cumulative block size distribution in Section 7.2.4. The fracturing of the rock during excavation changes this natural block size distribution to the fragment size distribution as illustrated in Fig. 15.2.

The engineer can consider how best to move from one curve to the other in the excavation process. There is also the connection with the basic objective mentioned earlier. If production of the resulting 'hole in the ground' is the end product, the excavation is an end in itself, although naturally coupled with the construction system. If, however, the excavated material is the desired end product, then the primary mining excavation process is the first of many stages of comminution. The first of these problems would require the minimum of change between the pre- and post-excavation block distributions, whereas the second may not involve such a constraint.

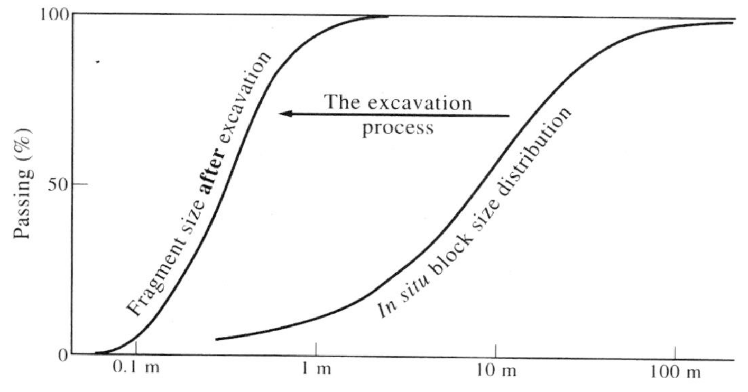

Figure 15.2 The process of excavation interpreted as changing the pre-existing natural rock block size distribution to the debris fragment size distribution.

15.1.3 Energy and the excavation process

There has been considerable discussion over the years on optimizing the use of energy during excavation, traditionally expressed as the concept of specific energy, i.e. the amount of energy required to remove a unit volume of rock (J/m^3). The design of excavation machines may well incorporate the idea of minimizing specific energy in order to maximize energy savings and increases in penetration rate, and to minimize destructive vibrations in the machine.

Seminal work, conducted many years ago at the US Army Cold Regions Laboratory, New Hampshire, involved a study of the specific energy values associated with different forms of ice fragmentation. These ran the whole gamut, from hand excavation, through conventional explosives, to machine excavation and more exotic techniques such as thermal lances. Predictably, perhaps, it was found that the more exotic forms of fragmentation were associated with higher specific energies, and hand excavation was associated with the lowest specific energy. This was because man could take advantage of the pre-existing fractures, but all other forms of excavation failed to exploit this opportunity.

In the history of mining, there have been few revolutionary advances in technology. The first of these was the application of explosives to rock excavation, with others being the ability to pump water from great depths and the development of self-guided tunnel boring machines—which are able to construct tunnels automatically to fine tolerances.

There are only two fundamental ways of inputting energy into the rock for excavation: one is by blasting, the other is by mechanical means. The two methods are illustrated in Fig. 15.3, where energy input is plotted against time. The energy is either input in large quantities over very short durations, or in smaller quantities essentially continuously. So, from purely practical considerations, excavation has to utilize a cyclical method with periodic blasting or a continuous method by machine. As far as the authors are aware, no one has yet developed a technique for combining the methods, e.g. excavating by continuous blasting which takes place immediately ahead of a tunnel boring machine.

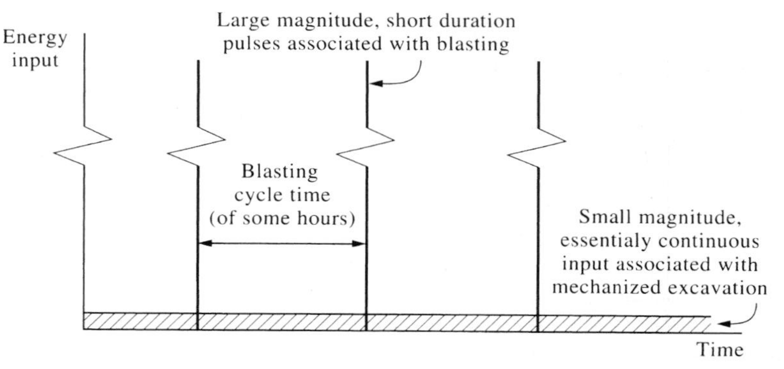

Figure 15.3 Energy input rates for blasting and mechanized excavation.

In this section, we introduced the three basic concepts of

(a) attaining the post-peak portion of the complete stress–strain curve;
(b) changing the rock fragment size distribution; and
(c) the only practical methods of inputting the required energy.

These concepts were presented in order to provide an overall conceptual background to the discussions that follow on rock blasting, specific methods of blasting and mechanized excavation. The book's objective is to provide an understanding that allows for maximal creative engineering thinking, and rock excavation is one area ripe for advancement through the application of innovative developments.

15.2 Rock blasting

In Chapter 13 rock dynamics and stress waves were discussed. Here, we consider the specific application of stress waves (and the associated gas pressure that is generated during blasting) to the explosive breakage of rock. The technique of rock breakage using explosives involves drilling blastholes by percussive or rotary-percussive means, loading the blastholes with explosive and then detonating the explosive in each hole in sequence and in accordance with a programme which depends on the type of blasting round being used.

The explosion generates a stress wave and significant gas pressure, resulting in complex reflections and refractions of stress waves at discontinuities within the rock mass. However, even with this complexity, it is possible to isolate key features of the process which allow the blasting rounds to be designed successfully.

15.2.1 The stress wave and gas pressure effects

In Fig. 15.4, an idealized development of borehole radial stress with time after detonation is shown. The stress rapidly builds up to a peak—within a few microseconds—and then reduces as the gas pressure is dissipated through discontinuities. As indicated in the figure, it is convenient to

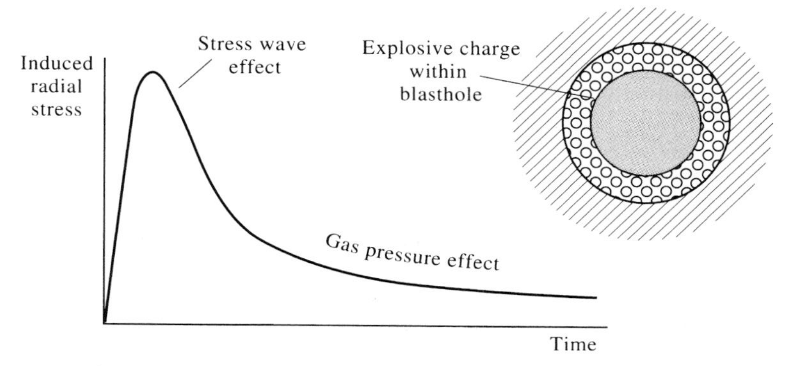

Figure 15.4 Stress wave and gas pressure effects during an explosion.

consider this phenomenon as being composed of a 'stress wave effect' and a 'gas pressure effect'. The exact form of the curve in Fig. 15.4 will depend on the type of explosive, the charge size, the coupling between the explosive and the rock and the degree of fracturing of the rock around the blasthole, among other things. Although there are many different types of proprietary explosive, by far the most widely used explosive in the civil and mining industries is '*ANFO*'—*A*mmonium *N*itrate and *F*uel *O*il. This material was found to be explosive when a ship, which had originally transported a consignment of fuel oil, was subsequently loaded with ammonium nitrate fertilizer and, unexpectedly, it exploded. An example of the maximum pressure generated in a blasthole after detonation of an *ANFO* charge density of 820 kg/m^3 is 2490 MPa.

A compressive stress wave is reflected at a free face as a tensile stress wave, see Fig. 13.4. Thus, after detonation of a borehole charge, there tends to be local pulverization due to the high compressive stresses caused by the blasthole pressure, followed by a dissipation of the stress wave. However, if there is a proximate free face, as illustrated in Fig. 15.5, the compressive stress wave is reflected at such a free face as a tensile stress wave, and surface spalling occurs due to the low tensile strength of the intact rock. This is because the rock is able to sustain a compressive wave at a given stress magnitude, but unable to sustain a tensile wave with stresses of the same magnitude. The energy remaining in the spalled fragment of rock is then liberated as kinetic energy—the fragment flies off the rock face.

Following the local fracturing at the blasthole wall and the spalling at the free face, the subsequent gas pressure then provides the necessary energy to disaggregate the broken rock. It is emphasized that no one understands the precise way in which the stress wave effect and the gas pressure effect combine to break the rock, but our understanding of the phenomenon associated with the proximity of a free face is sufficient to allow us to design all blasting rounds. This has been well proven in practice, despite the difficulties incurred by the presence of discontinuities, illustrated in Fig. 15.6.

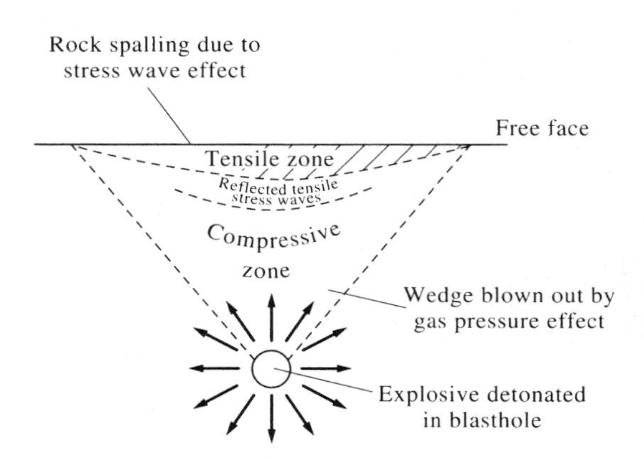

Figure 15.5 Effect of stress waves and gas pressure adjacent to a free face.

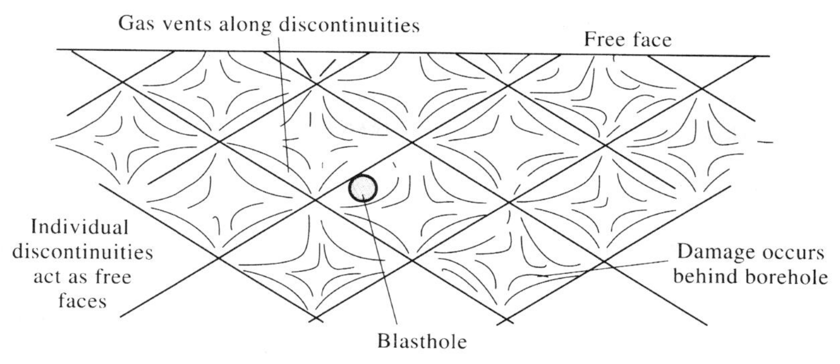

Figure 15.6 Complex effect of discontinuities on stress waves and gas pressure.

15.2.2 Blasting rounds

One of the basic principles of designing all blasting rounds, i.e. the configuration and sequential detonation of blastholes in one blast, is the presence of a free face parallel or sub-parallel to the blast holes, *as detonation occurs*. In some cases, these free faces may be automatically present (benches in a quarry), but in other cases may need to be created by the blast itself (a tunnel face).

It is interesting to consider what might be the 'most ideal' blasting round. When excavating a circular tunnel, a circular free face is required in the form of the final cylindrical tunnel outline, as illustrated in Fig. 15.7. Such a free face, or kerf, could be cut—in a weak rock—by a long, tungsten carbide-tipped chainsaw (Fig. 15.7(a)). Given the basic process of fragmentation which we have described, it might then be sufficient to detonate an explosive charge in a single blasthole at the centre of the free cylinder created by the kerfing. The compressive stress wave would radiate outwards, in a cylindrical form, and be reflected back towards the centre

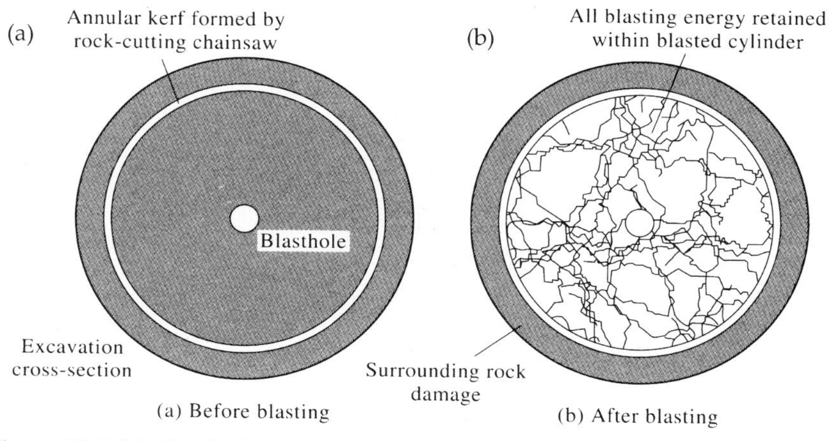

Figure 15.7 Idealized blasting geometry using a circular kerf. (a) Before blasting. (b) After blasting.

at the kerfed free face as a tensile stress wave, resulting in complete fragmentation (Fig. 15.7(b)). This is the ideal form of the free face for the excavation geometry. Although this system has been used in chalk in the United States, it is not generally practical because of the difficulty of cutting the kerf and also because of the 'bulking factor'. Broken rock occupies a significantly greater volume than intact rock, and hence allowance must be made for dilation.

Another way of generating the free face would be to drill a large diameter 'relieving' borehole at the centre of the face, and leave it uncharged. As shown in Fig. 15.8, the geometry of the free face could then spiral out from this initial small free face as successive blastholes are initiated, using delay detonators. With reference to Fig. 15.8, blasthole 1 is detonated with the initial uncharged borehole being its free face, and the process of fragmentation occurs as shown in Fig. 15.5, but on a smaller scale. This creates a larger free face, at a different orientation, which can be used by blasthole 2. The same process continues with blastholes 3, 4 and 5, demonstrating the important concept of progressively increasing the free face by the use of successive detonations.

From these fundamental concepts it is clear that part of the practical optimization of blasting rounds should include strict control of the drilling, to ensure correct geometry of the blastholes, and the use of precision delay detonators, to ensure the correct initiation sequence.

The face of a tunnel is a free face, but it is not parallel to any practically obtainable blasthole orientations because of drilling constraints: note from Fig. 15.5 that the free face should be parallel, and not perpendicular, to the blasthole.

However, as illustrated in Fig. 15.9, one can compromise by having a sequence of inclined boreholes which are successively detonated. This is known as the wedge cut, detonated in concentric cones from the centre of the face to the periphery, gradually increasing the area and changing the orientation of the free face as detonation proceeds. Note again, that the use of delay detonators is critical: the fragmentation process would not operate if the blastholes were detonated simultaneously.

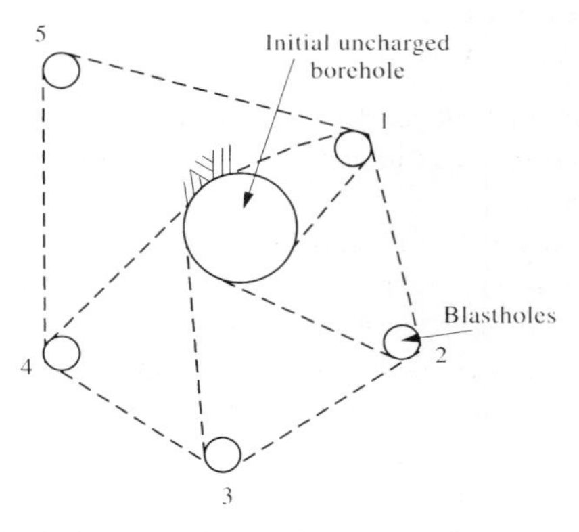

Figure 15.8 Practical application of the free-face concept using one form of the burn cut.

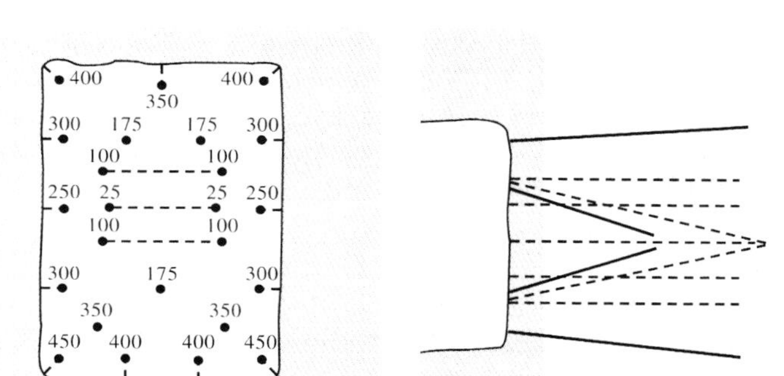

Figure 15.9 Use of the free-face concept with delay detonators and the wedge cut.

15.2.3 Explosives

It is also important to tailor the type of explosive to the overall objective and the type of blasting round. In considering the type of explosive to be used for a particular purpose, we should take into account their various characteristics. These include:

(a) *strength:* a function of the energy content (J/g) and the energy release rate;
(b) *density:* varies from 800 to 1500 kg/m^3;
(c) *velocity of detonation*: speed at which the detonation wave passes through a column of explosive (the higher the velocity the greater the shattering effect), typically in the range of 3000–5500 m/s with extremes of 1500 m/s for *ANFO* and 6700 m/s for detonating cord;
(d) *sensitivity*: ease of initiation;
(e) *sensitiveness*: ability to maintain the detonation wave;
(f) *water resistance*: capacity for detonation in wet conditions;
(g) *fume characteristics*: applicability in zones of poor ventilation;
(h) *gas volume*: gas liberated on detonation (l/kg); and
(i) *stability:* includes chemical and storage stability.

Some of these properties are related to one another and we take advantage of the best combination for our particular circumstances.

There are many types of explosive now available, the main types being:

(a) *gelatine explosives*: nitroglycerine thickened with nitrocellulose to give a gelatinous consistency. These are the most widely used nitroglycerine-based explosives;
(b) *semi-gelatine explosives*: which have a consistency between a gelatine and a powder. These are special-purpose explosives for use in wet conditions and small-diameter boreholes;
(c) *nitroglycerine powder explosives*: these are made from ammonium nitrate with nitroglycerine as a sensitizer, have a moderate bulk strength and are economical;
(d) *ammonium nitrate–fuel oil mixtures*: these are effective because ammonium nitrate is a cheap chemical source of oxygen for the explosive reaction. They have a low bulk strength;

(e) *detonators:* electric detonators consist of an aluminium tube with an electrically activated fusehead which initiates a priming charge and then a base charge of high explosive. Generally, delay detonators are produced as either 'short delay', measured in milliseconds, or 'half-second delay', measured in seconds.

These are the basic principles of blasting and we have illustrated, with idealized and practical cases, how blasting rounds are designed. There has been much practical experience accumulated in this subject area, and all major manufacturers of explosives have produced *Blasters' Handbooks* which give excellent guidance on details of all technical matters, including safety, associated with the use of explosives.

With the knowledge and understanding of the principles explained here, all this information is readily assimilated, and the reader can create a new type of blasting round for a new purpose. Without such understanding, these handbooks are simply series of instructions for technicians.

In the military area, explosives have been developed for a variety of specialized techniques. The power of these explosives is specified by the heat generated in the explosion (Q-value, units of J/g) and the volume of gas produced (P-value, units of cm^3/g). The standard, taken as 100, is that of picric acid, which has a Q-value of 3745 J/g and a P-value of 790 cm^3/g. On this scale, the power of other military explosives is:

Picric acid	100
Gunpowder	20
TNT (trinitrotoluene)	110
RDX (Research Department Explosive)	160
Thermonuclear explosives	>1,000,000.

The explosives currently available for civil purposes are adequate for rock fragmentation, and it is more important to consider the blasting technique itself than the explosive for optimal engineering.

15.3 Specialized blasting techniques

As illustrated in Fig. 15.6, the explosive damage may not only occur according to the blasting round design, but there may also be extra rock damage *behind* the borehole wall, particularly if there are major discontinuities present which reflect and refract the stress waves and provide paths for the gas pressure. When blasting to produce a final rock surface, such damage is malignant because it is out of sight and alters the rock in the very region where we require optimal quality (cf. Fig. 15.1).

For permanent rock faces at or near the ground surface, an elegant form of blasting has been devised which takes advantage of the principles we have outlined to minimize the damage to the rock. This technique is known as pre-splitting, and its fundamental function is to 'create the final plane first'. For permanent rock faces at depth, pre-splitting cannot generally be used, and because of the alteration of the local stress field during initial blasting, another technique known as smooth-wall blasting is used for the final surface. Both of these techniques tend to be used only where it is essential to produce a high-quality, high-strength and low-maintenance final surface.

There is considerable history in the modern development of blasting techniques, and indeed the cover of the 1963 book *Rock Blasting* by Langefors and Kihlström shows a perfect example of pre-splitting at a conduit wall in the Niagara project. Their book provides many examples of the application of blasting technology and, as we will be emphasizing later, the on-site problem is more one of blasting management rather than the requirement for new technology.

15.3.1 Pre-split blasting

In Fig. 15.10, we illustrate the excavation of a cutting through a rock outcrop. The primary purpose of the blasting is to remove the rock to form the cutting, but there are several operational reasons why a high-quality final slope profile may be required. This is the perfect circumstance for utilizing pre-split blasting.

As shown in the figure, a series of small-diameter, parallel boreholes are first drilled in the planes of the required final slopes. The principle is then to tailor the explosive parameters such that detonation of the explosive in these initial holes will primarily create a plane intersecting the holes. Under these circumstances, no provision has been made for dilation, but a complete new fracture is formed in the rock. When, subsequently, the main body of rock is bulk blasted to form the cutting, the pre-split plane reflects the stress waves back into the rock being excavated and dissipates excess gas pressure, such that the bulk blast has little effect on the rock behind the pre-split plane.

As Fig. 15.10 indicates, we have now used engineering knowledge to separate the two concepts, shown in Fig. 15.1, of excavation and support:

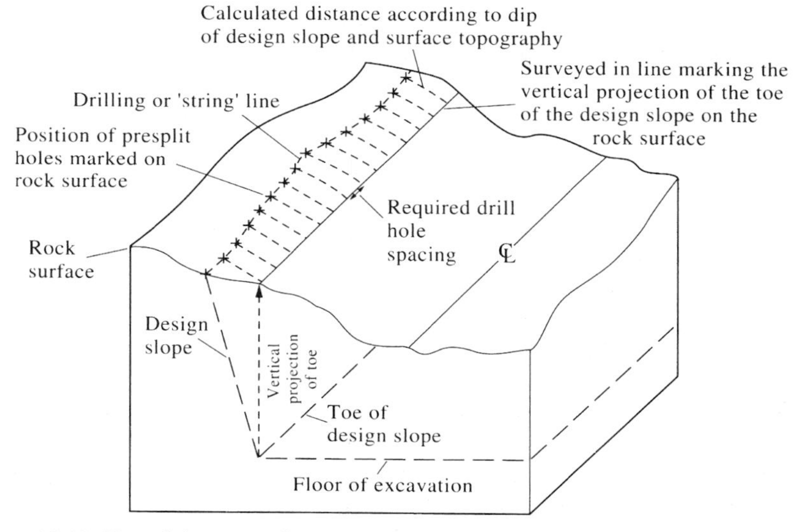

Figure 15.10 Use of the pre-splitting technique to create high-quality final surfaces (from Matheson, 1983).

the bulk blasting will be optimized through the use of blasting rounds; and the natural stability of the rock slopes will be optimized by minimizing the damage at and behind the slope surface through use of the pre-splitting technique. This latter point is particularly significant in a highway or railway cutting, where the lack of either any dilation of pre-existing fractures in the rock face or the creation of new fractures leads to a maintenance free slope. Bulk blasting alone could leave a highly damaged surface leading to potentially expensive post-excavation maintenance. This concept also applies to the final slopes of quarries and open pit mines.

Mechanism of pre-split blasting. As with regular blasting, the mechanisms by which the pre-split plane is created and the way in which the stress wave and gas pressure individually contribute to the process are not completely understood. Application of the principles of blasting does, however, enable us to produce an outline design of blasthole geometry and detonation sequence, such that the generation of a single pre-split plane is favoured.

In Fig. 15.11, there are two stages in the detonation sequence of a series of coplanar blastholes: a few microseconds after detonation; and a few milliseconds after detonation.

Initially, the stress wave effect generates radial fractures with some bias towards the plane of the blastholes, due to reflection from the nearest point

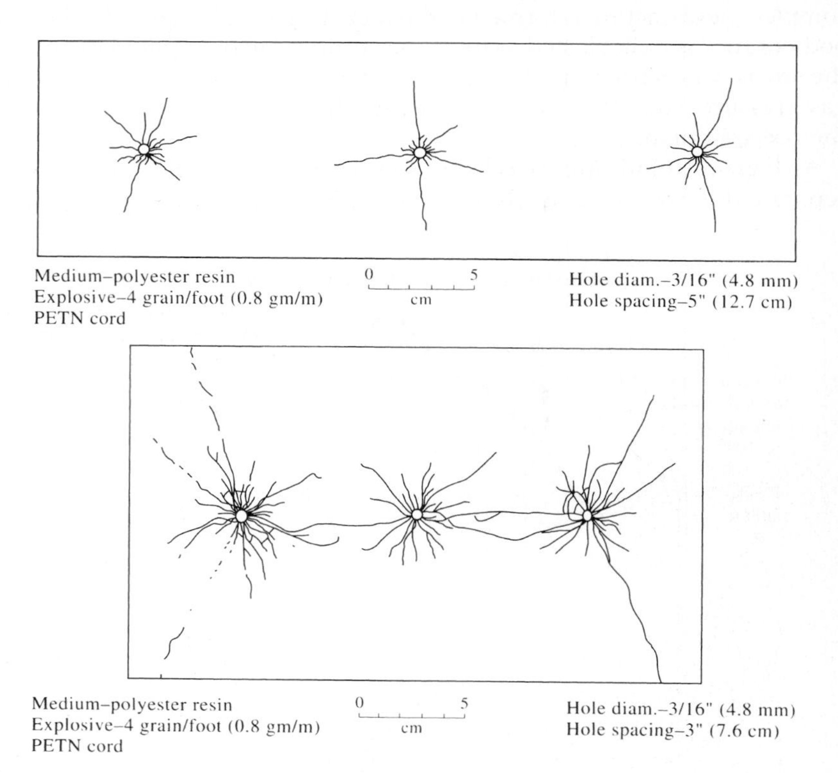

Figure 15.11 The progressive creation of a fracture plane during pre-split blasting (from Matheson, 1983).

of the adjacent blastholes. Second, there are dynamic and quasi-static effects of the gas pressure reinforcing this tendency. These are:

(a) the gas travels along the fractures and causes further cracking in a manner similar to hydraulic fracturing; and
(b) the pressurized blastholes and radial fractures induce high 'elastic' stress concentrations which contribute to failure of the rock along the plane.

Thus, as illustrated in the lower diagram of Fig. 15.11, the fractures lying in the plane of the blastholes coalesce to form a continuous plane. Because the resultant plane is formed by the coalescence of these radial fractures, it will tend to be rather rough on the scale of the blastholes (a few centimetres) but will be smooth on the engineering scale (a few metres). Note that if the creation of the pre-split plane is successful, the half cylinders of the blastholes should be visible on the final slope, as shown in Fig. 15.12.

Guidelines for successful pre-split blasting. From the discussion so far, it is possible to directly deduce the guidelines that govern the successful management of pre-split blasting. These are as follows.

(a) The blastholes must be parallel and locally coplanar—so that the blasting mechanisms occurring in each blasthole can interact to successfully form the desired final surface.
(b) Use closely spaced boreholes—the distance between the blastholes should not exceed about 10 times the blasthole diameter. This is important to ensure that blastholes are, indeed, close enough to allow interaction.
(c) The blastholes should be lightly charged and the charges decoupled from the blasthole wall—this is to ensure that local pulverization is minimized whilst maximizing the gas pressure effect (in the margin sketch we show the effect of decoupling an explosive charge on the stress–time curve).
(d) Detonate simultaneously—this maximizes the interaction between adjacent holes, such that the preferred plane for fracturing is the pre-split plane.

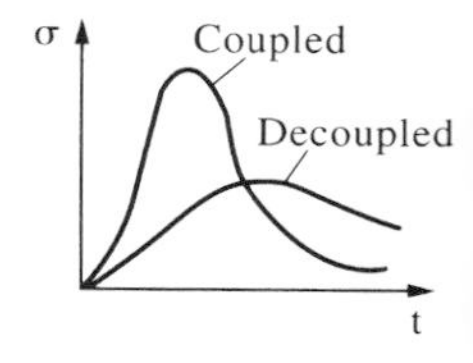

Figure 15.12 A successful pre-split face.

(e) Ensure good site management—accurate surveying and setting out of blastholes followed by accurate drilling is essential if truly parallel and coplanar blastholes are to be achieved.

If these guidelines are followed, successful pre-splitting will generally occur. The method is forgiving and robust. The authors have seen examples of successful pre-splitting under remarkably adverse conditions.

There are three main factors which can mitigate against pre-splitting.

1. To avoid the pre-split blasting degenerating into bulk blasting (as shown in Fig. 15.5), it is important that the blast occurs sufficiently far from a free face parallel to the blasthole plane, so that the blasting energy is preferentially guided into forming the pre-split plane.
2. Discontinuities within the rock mass can act as free faces such that, if they are at a low enough angle, as shown in Fig. 15.13(a), a ragged pre-split could result. Conversely, discontinuities which are almost perpendicular to the pre-split plane have little effect on the outcome, as shown in Fig. 15.13(b).
3. *In situ* stresses can induce an effect analogous to the proximity of discontinuities, because the rock will tend to fracture perpendicular to the least principal stress, as illustrated in Fig. 15.13(c). If the principal stresses are adversely orientated and of sufficient magnitudes, the pre-splitting mechanisms may be rendered ineffective.

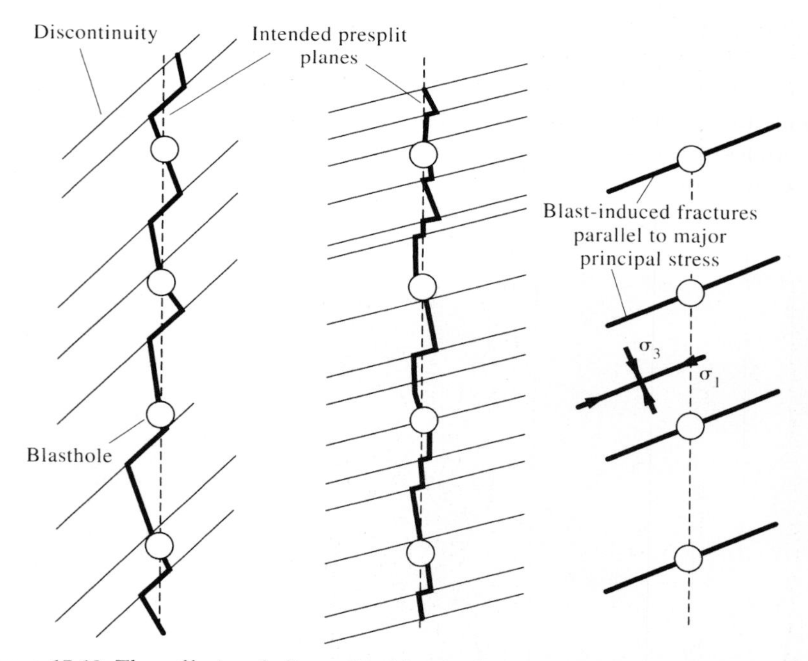

Figure 15.13 The effects of discontinuities and *in situ* stress on the creation of the pre-split plane. (a) Low-angle discontinuities. (b) High-angle discontinuities. (c) *In situ* stress.

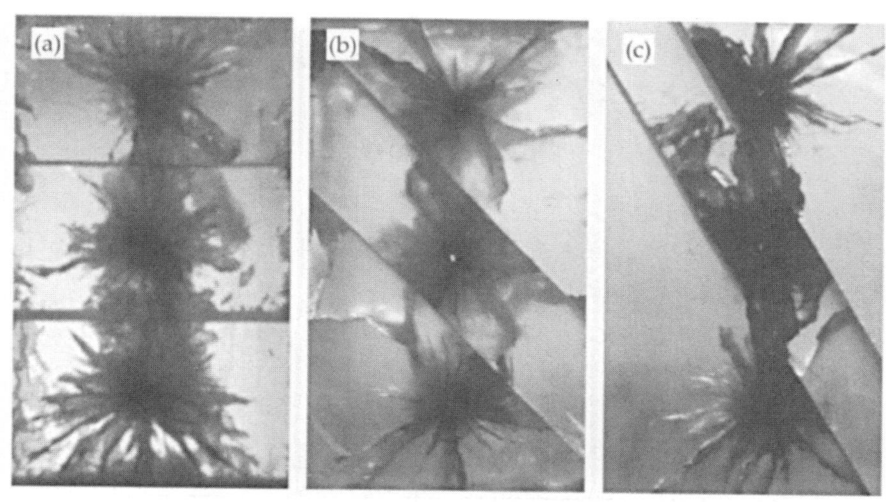

Figure 15.14 Illustration of pre-splitting effect in model tests in Perspex with introduced discontinuities (from Worsey, 1981).

The pre-split mechanism we have discussed, and the associated factors affecting the outcome, have been investigated and reported by Worsey (1981). In Figs 15.14 and 15.15 there are sets of three photographs illustrating the creation of pre-split planes in model tests utilizing Perspex (polymethyl-methacrylate) and rock. These photographs illustrate many of the points discussed, and the reader should note the clear evidence, in the tests conducted using Perspex, of the reflected stress wave, and the robustness of the pre-splitting technique almost regardless of the discontinuity orientation.

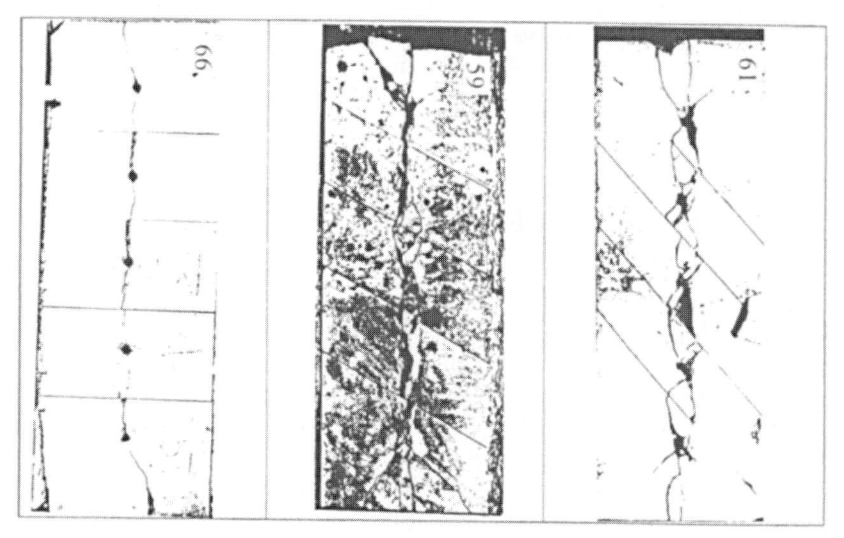

Figure 15.15 Illustration of pre-splitting effect in model tests in rock with introduced discontinuities (from Worsey, 1981).

Figure 15.16 A static pre-split formed using expanding grout (Isle of Lewis, Scotland).

An extreme example of the robustness of the technique is illustrated by the rock face shown in Fig. 15.16, where pre-splitting has been successful using an expanding grout instead of explosive—in this case it took several weeks for the pre-split plane to be developed. This technique was used to avoid vibration damage to an immediately adjacent building. Note that there was a complete absence of the dynamic stress wave effect in this situation, i.e. one of the fundamental mechanisms was not invoked, and yet the pre-split plane formed.

15.3.2 Smooth-wall blasting

We have seen the advantages and effectiveness of the pre-split method of blasting as it applies to excavations at or near the ground surface. Can pre-splitting be successful in an underground excavation? The answer is generally no, because of the existence of the *in situ* stress field. The stress field is all pervasive, and hence encourages any fracturing to form linear features, thus not being conducive to assisting in the creation of a circular tunnel. There is, however, an elegant method of utilizing the stress field to assist in producing the geometry required—by post-splitting or smoothwall blasting.

As shown in Fig. 15.17, once an initial opening has been created, the principal stress directions at the excavation boundary become radial and tangential. The radial principal stress is reduced to zero, and the tangential

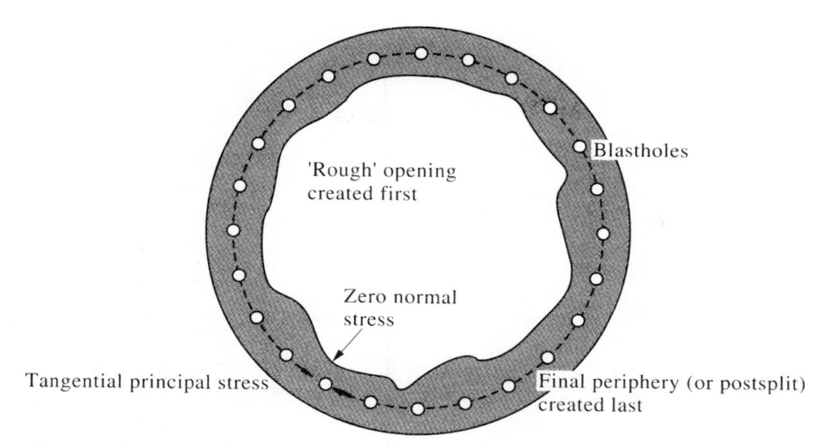

Figure 15.17 Smooth-wall (or 'post-split') blasting.

principal stress will be concentrated to become the major principal stress. So, we use this principle—by blasting twice (as with pre-split blasting but in the reverse order)—to produce the desired high-quality excavation boundary.

First, an opening is created 'roughly' which is close to the desired size and shape, and hence the stress field is changed all around the opening. Second, using similar blasting principles to those required for pre-split blasting (i.e. closely spaced and lightly charged parallel holes, decoupled charges and simultaneous detonation), a smooth-wall blast follows. The fracturing will be perpendicular to the minor principal stress and reinforced by the adjacent blastholes, thus forming a 'perfect' excavation geometry.

The techniques of, and distinction between, pre-splitting and post-splitting illustrate clearly the application of rock mechanics principles to the enhancment of rock engineering. They show how, through the application of engineering principles, blasting can be a sophisticated excavation technique, rather than a series of uncontrolled explosions.

15.4 Mechanical excavation

When this book was written, the Channel Tunnel, providing a land link between Britain and Continental Europe, was under construction using many types of tunnel boring machines (TBMs). The project is the final realization of a long-standing dream of civil engineers: indeed, in 1882 Col. F. Beaumont, a director of the Submarine Continental Railway Company, drove 1.5 km of a tunnel towards France using a steam-driven full-face tunnel boring machine. In a similar way to blasting, the development of mechanized excavation has an interesting history.

Mechanized excavation is considered first through an outline of the machines involved and then there is a discussion of the rock mechanics mechanisms involved.

15.4.1 Tunnel boring machines

There are two basic types of machine for underground excavation: partial- and full-face machines, as illustrated in Fig. 15.18. Partial-face machines

Figure 15.18 Partial-face (top) and full-face (bottom) tunnel boring machines.

have a cutting head on the end of a movable boom, the whole normally being track mounted. Full-face machines have a rotating head, armed with cutters, which fills the tunnel cross-section completely and hence excavates, almost invariably, circular tunnels. As the figure shows, the main components of both machines are some form of cutters on the rotating head, the rotating head itself, the power unit (usually electro-hydraulic), steering and mucking mechanisms, and supply lines.

The two machine types have different advantages: partial-face machines are cheaper, smaller and much more flexible in operation, whereas full-face machines—when used for relatively straight and long tunnels (radii exceeding about 500 m and length exceeding about 2 km)—permit high rates of advance in a smooth, automated construction operation. With respect to the rate of constructing tunnels, it is important to distinguish between

(a) cutting rate, C: the rate at which the tunnel is constructed when the tunnelling machine is operating; and
(b) advance rate, A: the rate at which the tunnel is constructed.

These are related through the utilization factor, U, which is the amount of time (expressed as a proportion) that the machine is cutting rock, as $A = U \times C$.

Machine manufacturers may specify the cutting rate, whereas the actual tunnel construction rate will depend on the utilization factor, which will itself depend on site conditions. There have been examples of such machines operating with $U = 10\%$ in bad ground conditions and even $U = 0\%$, i.e. the machine has become stuck. It is unlikely that machines will ever run at $U = 100\%$ because of scheduled maintenance, but values around 90% are possible when the total tunnelling system has been carefully designed. Factors contributing to low utilization rates are difficulties with ground support and steering, the need to frequently replace cutters, blocked chutes and conveyors, and a wide variety of other problems.

Over the years, while these machines have been developed, there has been continuing discussion on the best philosophy of design. Should a robust machine be developed that will overcome almost any ground conditions encountered, or should the machine be tailored to the ground conditions at a specific site? The advantage of the former is that 'off-the-shelf' machines and components will be available; the disadvantage is that considerable over-expenditure may be incurred through the provision of unnecessary quality and/or components. The advantage of bespoke machines is that they represent optimal ground engineering; they are, however, susceptible to unexpected ground conditions (there have been examples of soft ground machines designed to excavate sand encountering granite boulders, and of hard rock machines encountering clay).

The two main factors that will stop tunnel boring machines are either that the rock is too hard to cut or that the rock is too soft to sustain the reaction necessary to push the machine forward. Tunnel boring machines will operate within certain ranges of rock deformability and strength: the more the machine is tailored to the ground, the narrower these ranges (or suites of ranges) will be. All tunnelling engineers say that the one certain thing about tunnelling is the uncertain ground conditions.

15.4.2 The mechanics of rock cutting

One of the primary elements of the rock excavation system is the mechanism of rock cutting at the front of the machine. The four main types of device that are used as cutters, and their requirements in terms of applied forces, are shown in Table 15.1. The symbols F_n and F_t represent the forces acting normal and tangential to the rock face. In tunnelling terms, the F_n is related to the thrust and F_t is related to the torque applied by the TBM. The figures above are only intended as a guide, and indeed, on most machines, there will be a combination of two or more cutter types. For example, on a machine with discs, there may be button cutters on the rim of the cutting head to ensure a smooth sidewall; also, water jets can be used to assist all three of the directly mechanical devices. There are many exotic

Table 15.1 Rock cutting devices and associated parameters

Cutter type	Rock strength	Applied forces
drag picks *(conical or flat–bladed)*	<70MPa	$F_n \ll F_t$
discs *(mounted singly or multiply)*	70-275MPa	$F_n > F_t$
button bits *(mounted on discs or cylinders)*	275-415MPa	$F_n \gg F_t$
water jets	all strengths	not applicable

types of device used for rock excavation, including the use of flame cutters, but we are only considering the conventional and widely used devices here.

The action of all four devices is shown in Fig. 15.19. The top left-hand diagram illustrates the cutting action of a drag pick. As can be seen from the diagram, the cutting mode is likely to be a complex mixture of tensile, shear and compressive modes of failure. We noted the basic Mode I, Mode II and Mode III types of failure, together with their binary combinations, in Fig. 14.13. The actual failure mode will be far more complex than this, and it is questionable whether any directly practical modelling of the process can currently be undertaken. There are many extra factors such as the vibration of the cutting head, the stiffness of the cutting tools and holders, and the irregular nature of the cut face (which is comparable to the dimension of the cutting edge of the tools).

In Fig. 15.20, we illustrate the tangential (i.e. cutting) forces recorded by a dynamometer, in which a drag pick was mounted, over a cutting distance of about 250 mm. It is tempting to assume that there will be a direct correlation between the peaks in this graph and the liberation of individual chips during the cutting process. However, deeper investigation reveals that this is not the case; nor do any of the suggested failure criteria or proposed cutting mechanisms adequately model the variation of forces illustrated.

Because of the complex nature of the cutting process, with all the attendent micro-structural effects and external complicating factors, the pragmatic approach has been to use familiar strength parameters, e.g. compressive strength (as demonstrated in Table 15.1), to empirically predict cutting rates. Since most geomechanical rock properties are inter-related, this approach has achieved some success. Current research is exploring the possibility of modelling the cutting process as a chaotic system.

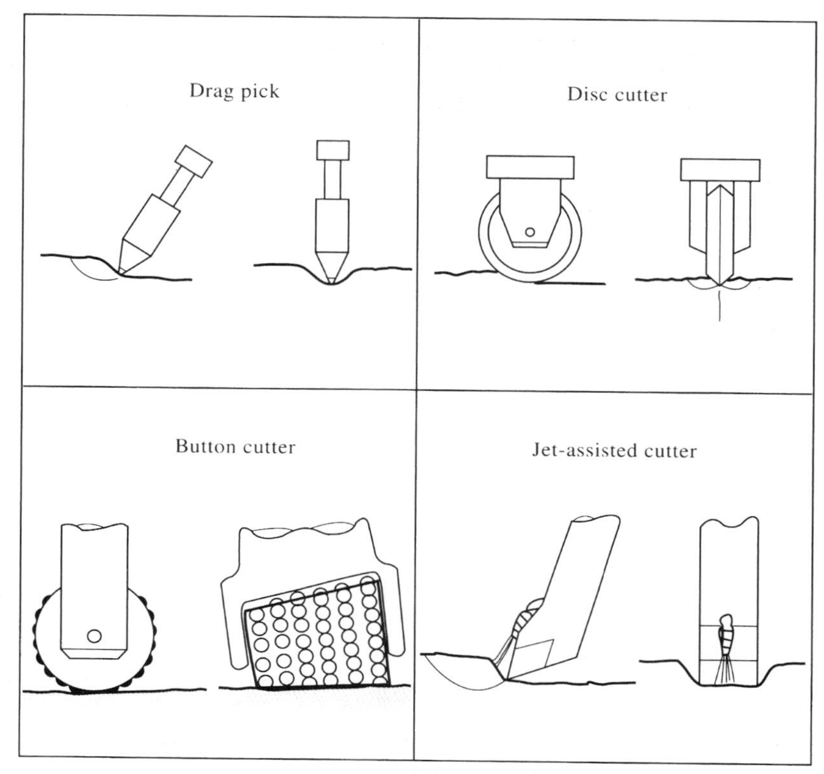

Figure 15.19 The action of four primary cutting devices.

The engineer wishes to know which cutting tool is the best, in what way the tools should be configured on a machine cutting head, how to minimize the need to replace the cutters, how to avoid damaging the cutter mounts, and how to minimize vibration. This involves the use of experience, empirical criteria as we have described, and varying the operating characteristics (such as torque and thrust) during tunnel construction.

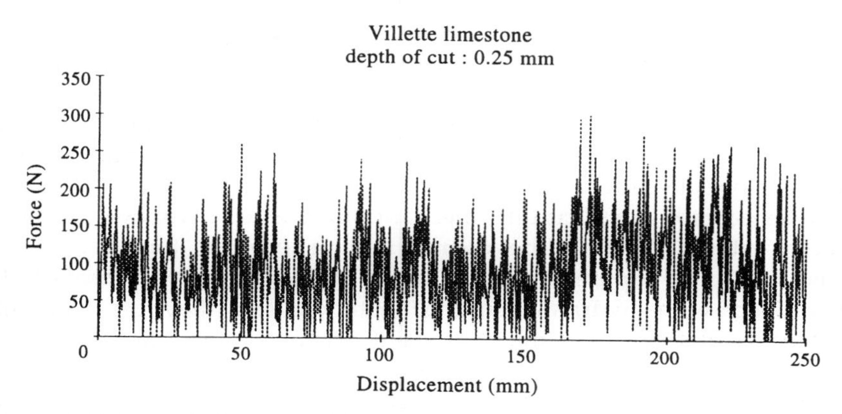

Figure 15.20 Record of tangential force during 250 mm cutting traverse using a drag pick (after Almenara, 1992).

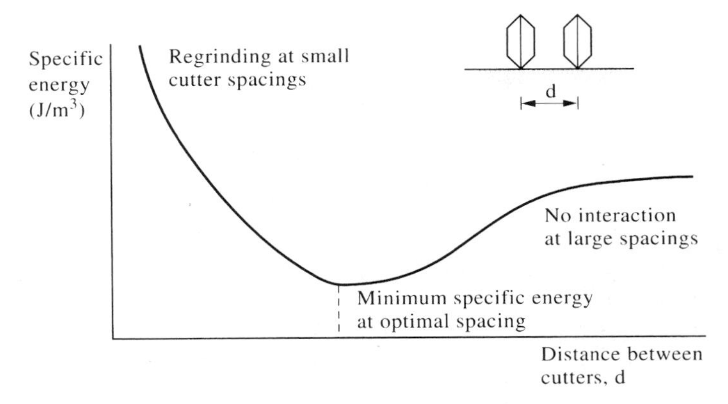

Figure 15.21 Generic curve of specific energy versus cutter spacing.

Considering the other three cutting mechanisms illustrated in Fig. 15.19, the same remarks and applications apply in practice. There has always been debate about the failure mechanisms operating beneath freely rolling disc cutters, and whether they should be mounted singly or in sets to take advantage of interaction during the cutting process. In Fig. 15.21 we illustrate a generic curve of distance between the cutters on the horizontal axis and specific energy (the energy required to remove a unit volume of rock) on the vertical axis. When the distance between the cutters is low, considerable overcutting or accessory grinding takes place; when the minimum specific energy is reached the cutters are positioned such that optimal interaction occurs during the breakage process; when the distance between the cutters is large, the cutters are acting independently and hence there is no benefit gained from cutter interaction. One approach to the application of rock mechanics in TBM design has been to minimize specific energy requirements. In particular, this has the effect of reducing vibrations and increasing the life of the cutters and other machine components. However, the construction process using mechanized excavation is a complete system, and the optimization of the system (either in terms of advance rate or cost) may or may not involve minimizing the specific energy of cutting.

The action of water jets, shown in the bottom right-hand diagram of Fig. 15.19, is a specialized subject beyond the scope of this book and the reader is referred to the publication by Hood *et al.* (1990) for further details. High pressures, of the order of 70 MPa, are used. We also mentioned earlier that combinations of the cutter types shown in Fig. 15.19 could be used. This is especially so for water jets, because they can either be used for water jet assisted mechanical cutting, mechanically assisted hydraulic cutting or as an integral part of the overall cutting process, for example, in dust and spark suppression. Very high specific energies are likely to be associated with hydraulic cutting methods, but they do have the advantage that there is no mechanical link between the rock and the cutting machine.

More rock is excavated by large tracked machines (flywheel power up to 0.5 MW) fitted with massive ripping tines (see Fig. 15.22) than by other mechanical means. The main method by which the appropriate machine

Figure 15.22 Large tracked-type tractor fitted with ripping attachment.

is chosen to match the soil and rock conditions is using seismic wave velocity. The theory relating the different seismic wave velocities and rock properties was outlined in Section 13.2: these equations show how the seismic velocities are related to the elastic properties of the rock mass. There is a correlation between rock mass moduli and rock mass strength, which is why the method is so effective.

In Fig. 15.23, there is a ripper performance chart for a Caterpillar D10 tractor with multi- or single-shank rippers. The chart illustrates the consistency of the rippability estimation through the rock spectrum, based on seismic velocity.

15.5 Vibrations due to excavation

All rock excavation induces vibrations in the ground and surrounding structures: the vibrations may be very large when blasting is used; or they may be relatively small when mechanized techniques are used. It is of engineering benefit to understand the generation of these vibrations, how they travel through the rock mass, and their possible effect on adjacent structures. In Chapter 13, various types of stress wave were discussed. Here we concentrate on the engineering implications of the vibrations caused by excavation, in particular those due to blasting.

In order to evaluate rock blasting effects, it is helpful to

(a) estimate ground displacements resulting from the blast;
(b) evaluate the response of engineered structures to the blast; and, hence,
(c) establish tolerable limits to prevent damage.

These three concepts are covered in Sections 15.5.1–15.5.3.

15.5.1 Estimating ground displacements

First, the parameters involved in estimating ground displacements must be established. These fall into two categories: independent and dependent

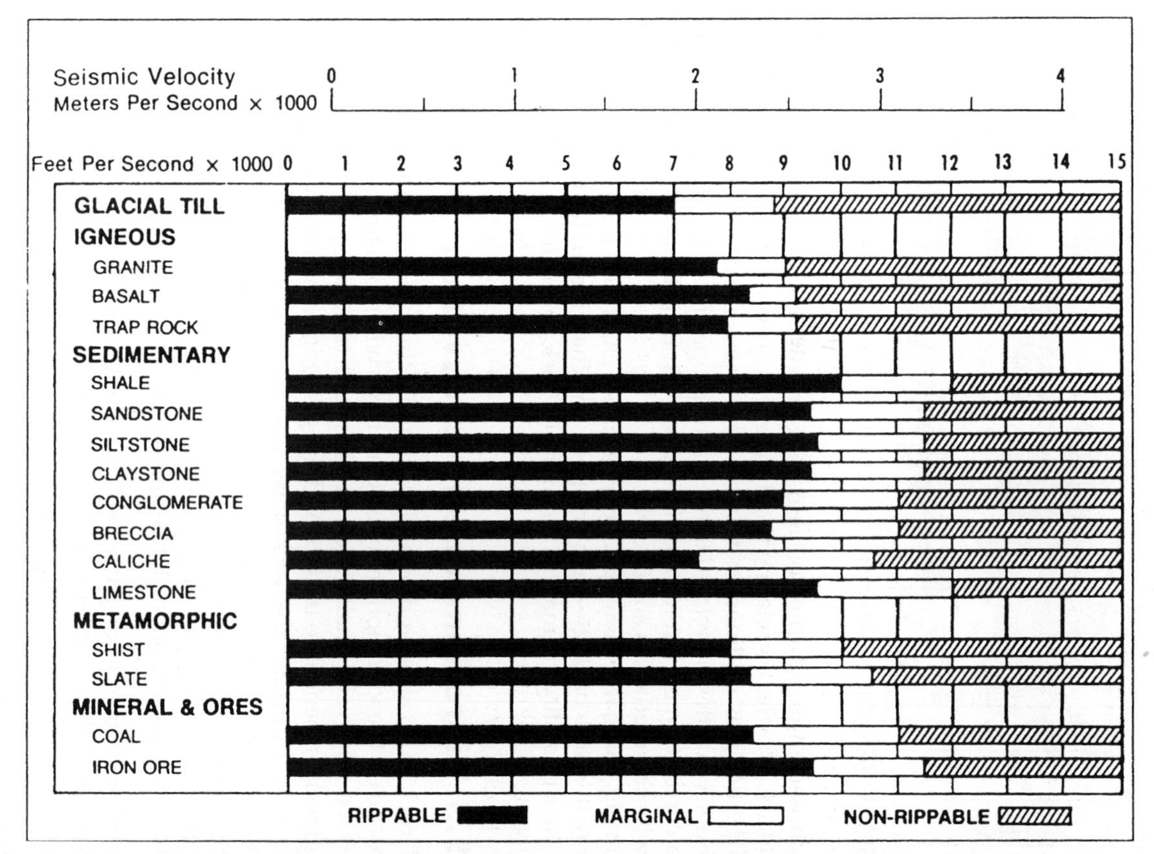

Figure 15.23 Rippability in terms of seismic velocity. From Caterpillar Performance Handbook (1983).

parameters. Independent parameters in some way control the blast; the dependent parameters relate to the ground response.

The main independent parameters are: blasting energy (kg/delay), W; distance from the blast (m), R; wave propagation velocity in the rock mass (m/s), c; rock bulk density (kg/m^3), ρ; and time (s), t.

The dependent parameters are: maximum ground displacement (m), u; maximum ground velocity (m/s), v; maximum ground acceleration (m/s^2), a; and frequency (Hz), f.

Dimensional analysis of these parameters results in the following six dimensionless variables: tc/R, $W/\rho c^2R^3$, u/r, v/c, aR/c^2, ft. The first two are independent variables; the final four are dependent variables. It is helpful to graphically present the ground displacement information using the dimensionless variables.

One of the most important variables is v, the velocity of ground displacement (we note that this is a vector and should be considered as the resultant velocity, i.e. $v = \{v_x^2 + v_y^2 + v_z^2\}^{1/2}$).To determine v, the maximum component of velocity, the maximum resultant velocity, or the vector sum of the maximum components (which may be temporally separated) can be used. The first of these formulations is, historically, the most used. In Fig. 15.24(a), this velocity is plotted against $R/W^{1/3}$, which is the inverse cube root of the dimensionless variable $W/\rho c^2R^3$, assuming that ρ and c are sufficiently constant to be neglected. The graph shows the advantage of the dimensionless approach, because of the coherency of the results from many different sites and blasting operations.

An alternative approach is to plot the maximum value of v (the peak particle velocity, PPV) versus different distances from the source for various vibration inducing operations. In this case, as illustrated in Fig. 15.24(b), there is a suite of straight lines for the different operations.

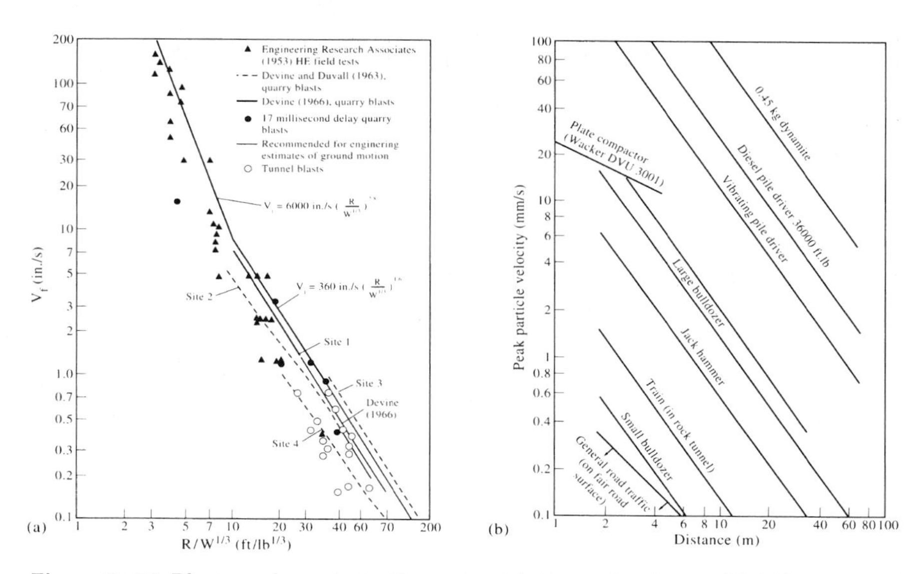

Figure 15.24 Blasting characterization using (a) dimensionless and (b) dimensional methods (from Hendron, 1977).

15.5.2 Evaluation of structural response

The next step in establishing the effects of vibrations due to excavation is to consider the tolerable limits of structures to the various wave characteristics. These limits depend on the type of structure, the construction materials, the history of the structure and the use of the structure. Thus, factors such as the type of foundation, existence of any finishings and claddings, whether other types of strain have already occurred and individuals' perceptions are all important. Each structure will be susceptible in different ways, but general guidelines have been incorporated into design codes to assist engineers and to provide a basis for constructional operations. As an example, we list below some tolerable limits based on PPV, emphasizing that these are examples and not specific guidelines.

Type of structure	Tolerable PPV limits (mm/s)
Residential masonry buildings	12–50
Retaining walls, bridge abutments, industrial buildings	100
Lined and unlined rock tunnels	500–600

Further detailed discussion of this topic is beyond the scope of this book, but we refer the reader to Hendron (1977), New (1984) and Dowding (1985) for more information.

Another factor determining the response of a structure is the range of frequencies present in the vibration with respect to the frequency response of the structure itself. The frequencies can be presented either as predominant frequency histograms or complete frequency spectra. Similarly, the response of a structure can be illustrated through a response spectrum. In Fig. 15.25(a), we show how the predominant frequencies, measured at a 'structure of concern', can change with the type of blasting operation. Figure 15.25(b) demonstrates the response of low-rise residential structures in terms of their natural frequency.

The response of structures to ground vibrations can be quantified through the use of mathematical models, such as single degree of freedom models (as shown in the margin sketch). Such an approach can be difficult, given the difficulty of adequately determining values for the various components in such a model. Consequently, a pragmatic approach is usually adopted.

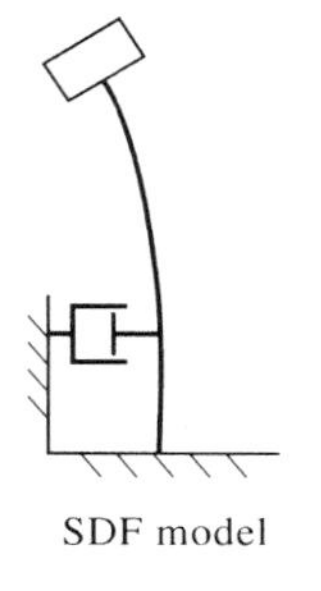
SDF model

15.5.3 Engineering approach to blast-induced vibrations

Faced with the complexity of information which has been indicated in Section 15.5.2, some form of pragmatic guidance is required for engineering. The four main steps in considering the effect of blasting vibrations on structures are:

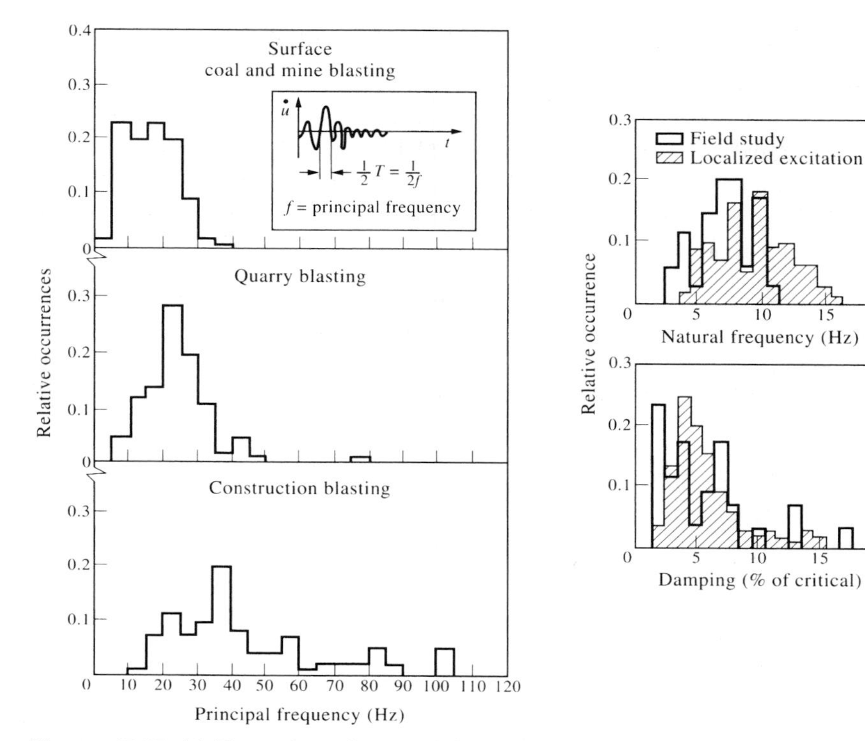

Figure 15.25 (a) Examples of principal frequencies induced by blasting (from Siskind *et al.*, 1980). (b) Examples of fundamental frequencies for low-rise residential structures (from Dowding, 1985).

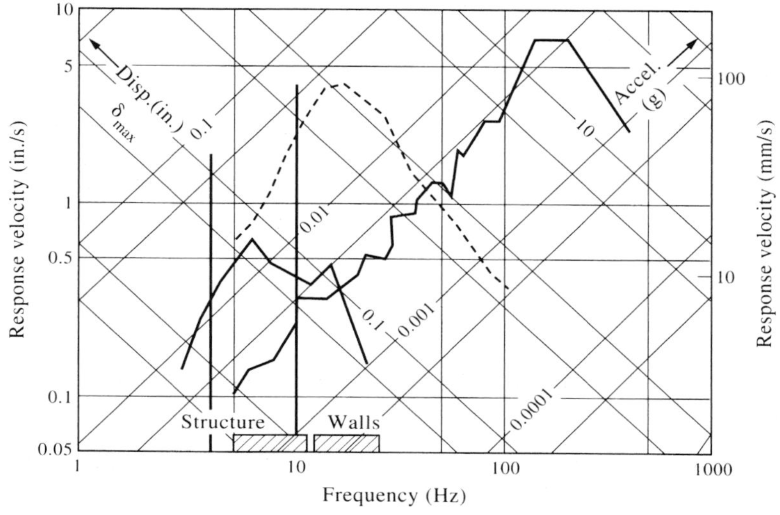

Figure 15.26 Example ground motion and structure response spectra (from Dowding, 1985).

(a) establish the relation between the ground motions due to the blast and scaled distances (cf. Fig. 15.24);
(b) determine the structural amplification of ground motions with mathematical models such as a single degree of freedom model which incorporates attentuation due to the structure itself;
(c) estimate the reponse spectrum from the two steps above; and
(d) compare the natural frequency of structures with the response spectra in order to evaluate the structural response.

Figure 15.26 demonstrates this procedure, with a predicted response spectrum and predicted peak ground motions plotted on four-axis tripartite paper. Tripartite paper is developed from a consideration of the sinusoidal approximation to ground wave motions, and is useful for the rapid assessment of blast-induced vibrations. At any given site, it is always advantageous to calibrate this approach using site trials, for various blast parameters and utilizing ground motion recording apparatus.

The complementary subject to excavation (or rock removal) is reinforcement and support (i.e. maintaining the engineering quality of the rock immediately adjacent to the excavation periphery). We will adopt an identical approach in describing the principles of reinforcement and support in the next chapter.

16 Stabilization principles

In the previous chapter on excavation principles, we discussed the idea of taking the rock into the post-peak region of the complete stress–strain curve for excavation. Here, we discuss the principles of stabilization—whether for staying in the pre-peak region, or for allowing the rock to pass into the post-peak region and utilizing its residual strength.

Following the introduction on the form of the disturbances caused by the excavation process, we discuss the two fundamental methods of stabilization: **rock reinforcement** and **rock support**. In each case, the behaviour of essentially continuous and discontinuous rock masses is discussed separately. Finally, we consider the principles of stabilization when the rock mass behaviour has attributes of both a continuum and a discontinuum (e.g. slip on planes of weakness). These principles are amplified and their practical application illustrated in the later chapters dealing with rock engineering.

16.1 The effect of excavation on the rock mass environment

There are two aspects of rock excavation that we will concentrate on here. The first is that 'one cannot prevent all displacements at the excavation boundary'. The second is that 'a mistake in excavation design can be a major problem'. In order to understand the displacements and avoid problems, we will consider the three primary effects of excavation and then decide on the ramifications for stabilizing excavations of all kinds.

The three primary effects of excavation are:

(a) displacements occur because stressed rock has been removed, allowing the remaining rock to move (due to unloading);
(b) there are no normal and shear stresses on an unsupported excavation surface and hence the excavation boundary must be a principal stress plane with one of the principal stresses (of magnitude zero) being normal to the surface. Generally, this will involve a major perturbation of the pre-existing stress field, both in the principal stress magnitudes and their orientations; and

(c) at the boundary of an excavation open to the atmosphere, any previous fluid pressure existing in the rock mass will be reduced to zero (or more strictly, to atmospheric pressure). This causes the excavation to act as a 'sink', and any fluid within the rock mass will tend to flow into the excavation.

These three primary affects are illustrated in Fig. 16.1.

With respect to the likely *displacements* (top right-hand diagram in Fig. 16.1), there is the choice of allowing them to occur or providing some method of stabilization to resist them. The engineering objective dictates the significance of any rock displacement and its maximum tolerable magnitude. It is important to know whether the displacements are

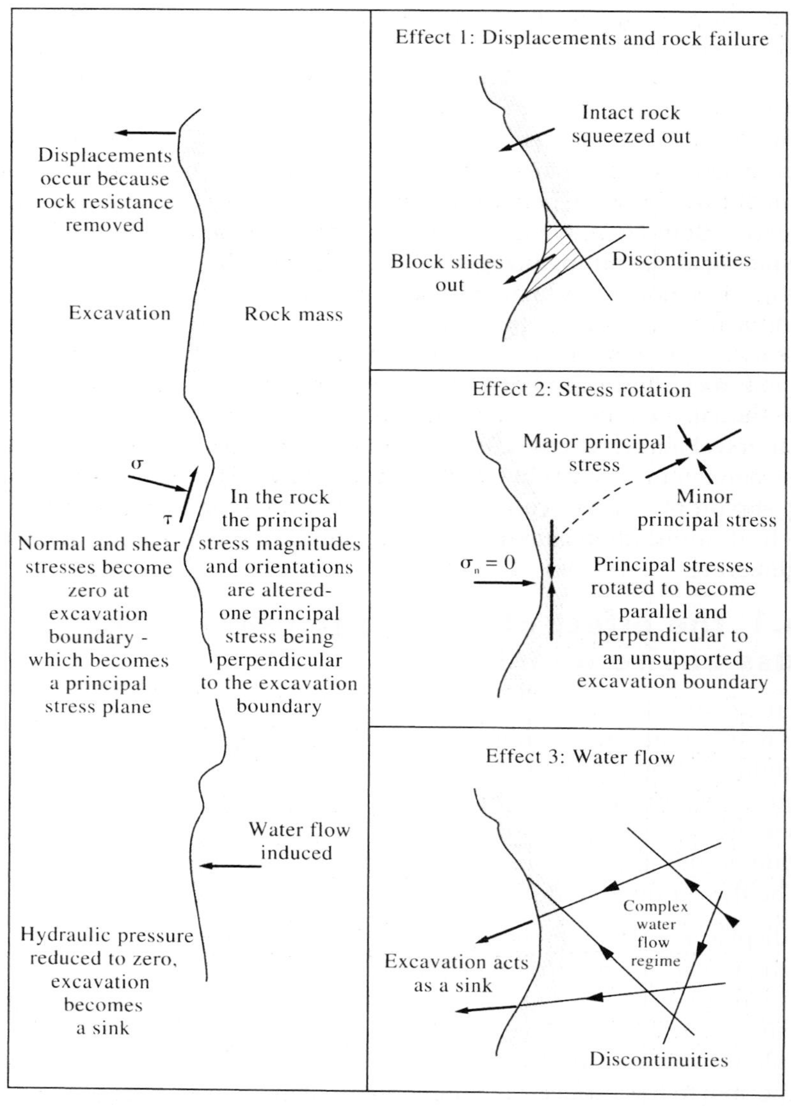

Figure 16.1 The three primary effects of excavation on the rock mass environment.

associated with entire rock blocks moving into the excavation, or whether the rock mass is deforming as a whole, or whether failure is occurring in the rock. It is possible for all three of these mechanisms to be operating simultaneously, and it is necessary to understand the mechanisms in order to decide on the stabilization strategy.

The most significant consequence of the second effect—*disturbance of the stress field* (middle right-hand diagram of Fig. 16.1)—is that the rock is more likely to fail, owing to the increased magnitude of the deviatoric stresses. The increase in the deviatoric stresses arises from the change in the magnitude of the major principal stress, together with the fact that an arbitrary polyaxial stress state has become an effectively uniaxial or biaxial stress state.

The third effect, that of *increased water flow* (bottom right-hand diagram of Fig. 16.1), is significant because there will be higher differential heads within the rock mass which will tend to push rock blocks into the excavation, with the attendant possibility of increased weathering and time dependent deterioration as the water flow increases.

These three primary effects, and the optimal way in which the rock engineering strategy is developed to account for them, have one thing in common: we should not blindly attempt to maintain the original conditions (e.g. by installing massive support or reinforcement and hydraulically sealing the entire excavation); rather, we should treat these effects as providing the opportunity to understand the rock behaviour and to develop the engineering sympathetically.

As the displacements occur, engineering judgement may determine that they can be allowed to develop fully, or be controlled later. Similarly, perhaps the engineer can utilize the alteration of the stress field to advantage when designing the shape of the excavation to minimize the induced deviatoric stresses. Knowing that water will flow into the excavation means that the flow can be controlled in accordance with the engineering objective: some tunnels may be excavated as drainage or water-gathering systems; others may have to be totally dry.

16.2 The stabilization strategy

A categorization of rock reinforcement and rock support in continuous and discontinuous rock is required because rock reinforcement and rock support are not the same. In Fig. 16.2 we show this categorization based on the engineering viewpoint.

If failure around an excavation, whether at the surface or underground, is due to blocks of rock moving into the excavation, two approach philosophies can be considered for stabilization:

(a) the block displacements are occurring because the rock mass is a discontinuum, and hence the rock is reinforced so that it behaves like a continuum; or
(b) direct support elements are introduced into the excavation in order to maintain block displacements at tolerable levels.

The first option is known as rock reinforcement; the second is known as rock support, as presented in Fig. 16.2. Note that with rock reinforcement

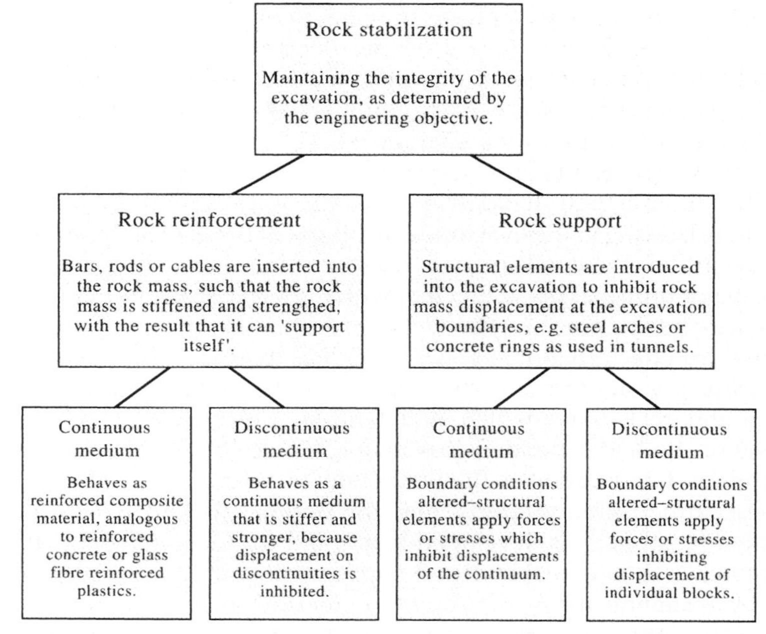

Figure 16.2 Basic categorization of rock reinforcement and support.

the engineering elements are inserted within the rock mass and with support they are inserted within the excavation.

The hierarchical system shown in Fig. 16.2 is for the purpose of separating the stabilization concepts: in practice, two or more of the conditions shown in the bottom row may be operating concurrently.

In the case of reinforcement, as illustrated in Fig. 16.3(a), steel cables or bars grouted within boreholes are used in an attempt to minimize displacement occurring along the pre-existing discontinuities—**so that the rock supports itself.** An associated preventative measure is the spraying of concrete or cement mortar onto the rock surface immediately on exposure, not as a direct structural support, but in conjunction with the bolting in order to protect the surface and inhibit minor block movements. Other materials can be considered as reinforcing elements for particular purposes, e.g. wood or glass fibre, if it is necessary to subsequently excavate through the reinforced rock mass.

In the case of support, structural elements—such as steel arches or concrete rings, as illustrated in Fig. 16.3(b)—are introduced to inhibit rock displacements at the boundary of the excavation. These elements, which are external to the rock mass, provide load carrying capability, with the result that the rock does not totally support itself: **the rock is supported.** Support does not directly improve the intrinsic strength of the rock mass, but does alter the boundary conditions. We will be discussing later how the utility of such support can be assessed by means of the 'ground response curve', and how developments such as using knuckle joints with pre-cast concrete segments (so that the applied loads are resisted as compressive hoop forces rather than bending moments) are helpful. At this stage, we note that the

Figure 16.3 Illustrations of (a) rock reinforcement and (b) rock support.

basic principle of rock support is fundamentally different from rock reinforcement, as illustrated in Fig. 16.4.

In Fig. 16.2, the concept of rock stabilization was defined in the top box, the two primary methods of achieving this objective in the middle boxes, and the way in which the two separate concepts of rock reinforcement and rock support apply to continuous and discontinuous rock masses in the bottom boxes. Each of the four elements shown in the bottom row of Fig. 16.2 will now be described in detail.

16.3 Rock reinforcement

Rock reinforcement as applied to essentially continuous rock masses differs from that used in discontinuous rock masses because of the mode of action of the reinforcing elements. Sections 16.3.1 and 16.3.2 describe this difference.

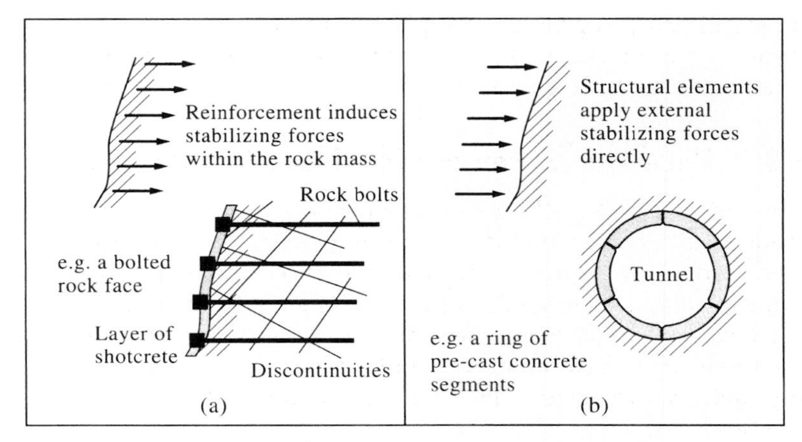

Figure 16.4 Principles of (a) rock reinforcement and (b) rock support.

16.3.1 Rock reinforcement in continuous rock

It may be thought that the use of rock reinforcement, e.g. rock bolts, is only of use in discontinuous rock masses in order to prevent discrete block displacements. However, the use of rock reinforcement in a continuous medium can also be of benefit because of the reinforcement effect on the overall rock properties and hence rock behaviour. If a continuous rock is strong, it may well be capable of withstanding the induced stresses without further assistance. Conversely, if a continuous rock is weak, heavy direct support may be required, such as segmental pre-cast concrete rings. The case we are discussing here is where improvement to the intrinsic strength by the rock reinforcement is all that is required for rock stabilization. The mechanics of this problem are similar to those of reinforced concrete.

Consider an element of reinforced rock adjacent to the excavation boundary (see margin sketch). The effect of the reinforcing elements is to produce an effective confining stress of

$$\sigma_{\mathrm{r}} = A'E'\nu\sigma_{\theta}$$

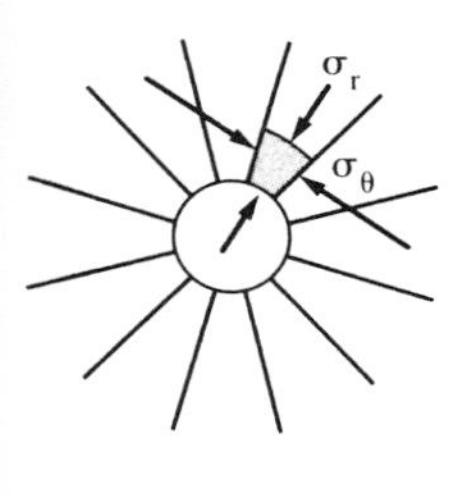

Sketch of radially reinforced annulus

where A' and E' are the ratios of the cross-sectional areas and the Young's moduli of the reinforcing element to that of the rock being reinforced, respectively,
ν is Poisson's ratio for the rock, and
σ_{θ} is the tangential stress.

Note that the larger the ratios A' and E', i.e. for an increased rock bolt density and lower stiffness rock, respectively, the larger the effective confining pressure will be. As an example, consider a chalk (E = 1 GPa) being reinforced with 25 mm diameter steel bars at a density of four bars per square metre of rock face, $\sigma_{\mathrm{r}} = 0.12\sigma_{\theta}$. Although the induced confining stress is only 12% of the tangential stress in this example, it will have a profound effect on the strength and failure properties of the chalk. Figure 6.15 shows the marked effect of a small confining stress on the strength and shape of the complete stress–strain curve of rock in compression. Use of this type of analysis provides a rapid means of determining the value of reinforcing continuous rock—which will clearly be most effective in low-stiffness, low-strength, brittle rocks.

16.3.2 Rock reinforcement in discontinuous rock

The mode of action of the reinforcement in a discontinuous medium is somewhat different to that described in Section 16.3.1 because, not only are we considering improvement of the rock structure properties, but also the avoidance of large displacements of complete blocks. The method of analysis for the kinematic feasibility of rock blocks will be discussed in Chapters 18 and 20 for surface and underground excavations, respectively: here we will highlight the factors relating to rock reinforcement principles.

Two of the most important factors are whether the blocks are free to move, given the geometry of the rock mass and excavation, and the character (quantity, length and orientation) of the reinforcement. In Fig. 16.5 we show the simplest case of reinforcing a discontinuous material: a single block on a rock surface is reinforced by a tension anchor. The tension anchor should be installed such that the block and the rock beneath act as a continuum, and block movement is inhibited. We may wish to know the optimal bolt length, orientation and tension, and indeed whether the reinforcement is required at all, cf. the engineering objective referred to in the uppermost box in Fig. 16.2.

For the simple geometry shown in Fig. 16.5 without the rock bolt, basic mechanics indicates that the block will slide if the angle of the slope exceeds the angle of friction of the rock surfaces for a cohesionless interface. This is therefore the first criterion for indicating the potential for failure. Considering now the length and diameter of the bolt, these have to be sufficient to ensure that the strength of the bonds across the anchor–grout and grout–rock interfaces are capable of sustaining the necessary tension in the anchor, which in turn will depend on the fracturing of the rock mass. Finally, the anchor diameter may also be determined on the basis of the tensile strength of the anchor material.

With respect to the bolt orientation and tension, it is not obvious at what angle the anchor should be orientated for optimal effect, taking into account the basic mechanics and the rock structure. If we regard the

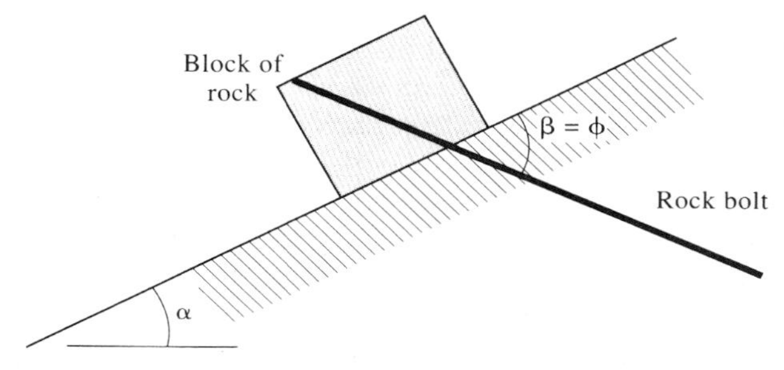

Figure 16.5 Optimized rock reinforcement for the case of a block on a rock surface.

optimal orientation for the anchor as that which enables the anchor tension to be a minimum, then the angle between the anchor and the slope surface is equal to the friction angle between the block and the slope. Many other factors may be involved in this analysis: these will be covered in Chapter 17. The intention here is to indicate the fundamental philosophy.

The key point to be made is that, *if the reinforcement inhibits block movement, and sufficient stress can be transmitted across the interface, then in principle the rock reinforcement has changed the rock discontinuum to a rock continuum.*

In practice, when rock anchors are installed in a discontinuous rock mass, the rock surface is often covered with wire mesh and then covered in shotcrete (sprayed concrete). It is emphasized that the wire mesh and shotcrete are part of the rock reinforcement system: the purpose of the shotcrete is to provide a stiff coating to inhibit local block rotation and movement. Before rotation, forces may be being transmitted across complete block-to-block interfaces; after even very small rotation, these forces become concentrated at the edges or vertices of the blocks, with high local stresses being developed. It is this sequence of block rotations that leads to the progressive failure of a discontinuous rock mass and subsequent loss of integrity of the engineered structure.

16.4 Rock support

The term 'rock support' is used for the introduction of structural elements into a rock excavation in order to inhibit displacements at the excavation boundary. As in the case of rock reinforcement, rock support is considered separately for continuous and discontinuous media. In reality, the distinction between continuous and discontinuous rock masses may not be quite as clear as implied; the transitional case is discussed in Section 16.5.

16.4.1 Rock support in continuous rock

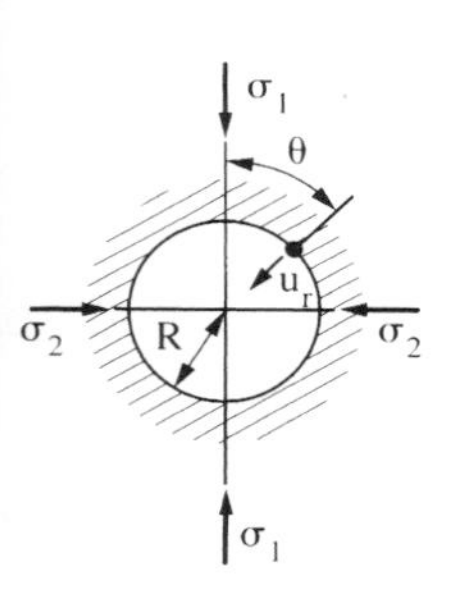

Consider the stresses and displacements induced by excavating in a CHILE material. For example, the radial boundary displacements around a circular hole in a stressed CHILE rock in plane strain are

$$u_r = (R/E)[\sigma_1 + \sigma_2 + 2(1 - \nu^2)(\sigma_1 - \sigma_2)\cos 2\theta - \nu\sigma_3]$$

where R is the radius of the opening,
σ_1 and σ_2 are the far-field in-plane principal stresses,
σ_3 is the far-field anti-plane stress,
θ is indicated in the margin sketch, and
E and ν are the elastic constants.

Recall that the *stress concentrations* around an opening in similar circumstances are *independent* of both R and the elastic constants—the stress concentrations around circular openings of different diameters and in different CHILE materials are the same. However, the magnitude of the radial *displacement* must *depend* on both the radius of the opening and the values of the elastic constants, as indicated in the equation above:

displacements are proportional to the radius of the opening and inversely proportional to Young's modulus. Moreover, any deviation from CHILE behaviour towards DIANE characteristics results in increased displacement values.

The rock stabilization strategy can be based on the need to restrict the displacements as governed by the engineering objective. The *ground response curve* is a graph of the support pressure required to maintain equilibrium of the boundary at a given displacement value versus the displacement value. The ground response curves shown in Fig. 16.6 illustrate this relation for the cases of linearly elastic, 'stable' non-elastic and 'unstable' non-elastic behaviour.

Where the elastic ground response curve intersects the boundary displacement axis in Fig. 16.6(a), the u_r-value is found from the expression above: this point represents total elastic deformation of the boundary of the excavation and no support pressure is required, providing that the magnitude of this displacement is acceptable. For most rock engineering situations, such an elastic displacement will be less than 0.1% of the radius and will be acceptable.

Considering the 'stable' non-elastic curve of Fig. 16.6(a), the intersection of the curve with the boundary displacement axis occurs at a higher displacement value, say up to 10% of the radius. Whether such a displacement is acceptable or not depends on the engineering objective: for example, in a high-speed rail tunnel it may be unacceptable, whereas in a temporary mine opening it may be tolerable.

Finally, the curve in Fig. 16.6(a) corresponding to 'unstable' non-elasticity definitely indicates the need for support, because the curve does not intersect the boundary displacement axis, i.e. the opening will collapse without support. Because of the general nature of the ground response curve concept and the ability to study a variety of associated factors, it has become a widely used semi-empirical tool in the design of support for excavations.

As an example of the utility of the ground response curve method, consider the curves in Fig. 16.6(b), which are similar to those in Fig. 16.6(a) but occur when *the same* rock mass is excavated by *different* methods.

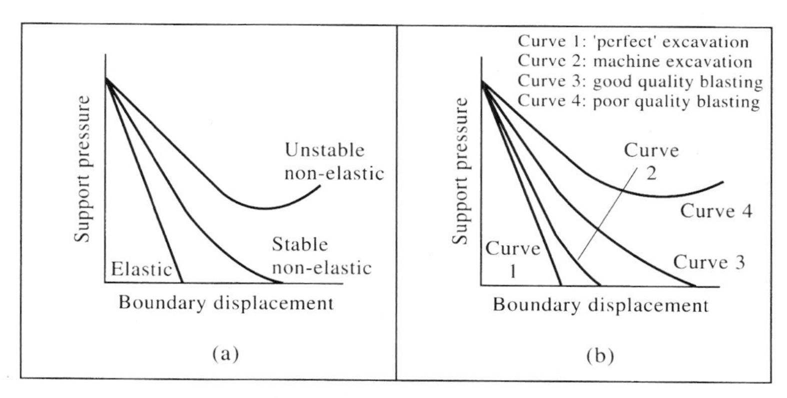

Figure 16.6 Ground response curves in (a) different types of rock and (b) in the same rock type but excavated by different methods.

- **Curve 1** represents the 'perfect' excavation case, in which there is no disturbance of the remaining rock and there is no deviation from CHILE behaviour.
- **Curve 2** may occur in a machine-driven tunnel, in which there is a slight disturbance to the remaining rock with the result of increased final displacement.
- **Curve 3** could represent high-quality blasting, where more disturbance is inevitable and the resulting displacements are increased yet further. All three of these curves intersect the boundary displacement axis which indicates that no support is required, providing the displacements are tolerable.
- **Curve 4**, representing the response following poor-quality blasting, indicates that support is essential if stability of the excavation is to be maintained. This is another example of interaction within rock engineering—in this case, the direct link between excavation technique and stabilization requirement.

If support is required, we can gain an indication of the efficacy of particular support systems by plotting the elastic behaviour of the support, the available support line, on the same axes as the ground response curve. We have plotted a soft support and a stiff support (considered as a radial stiffness) together with ground response curves in Fig. 16.7. The points of interest are where the available support lines intersect the ground response curves: at these points, equilibrium has been achieved.

There are other aspects of practical significance which can also be noted in relation to the ground response curve; two of these are also illustrated in Fig. 16.7. First, we remember that the support cannot be installed contemporaneously with excavation, and so some initial displacement must occur before the support is installed (and, strictly speaking, displacements will have occurred even before the excavation reaches the point in question—because of the alteration of the complete stress, and hence strain, states within the rock mass). Thus, the available support line starts, as shown in Fig. 16.7, with a displacement offset.

Another useful aspect of this approach is illustrated in Fig. 16.7, i.e. the concept of a yielding support. As shown in the diagram, the available support line for a yielding support has a maximum strength, which cannot

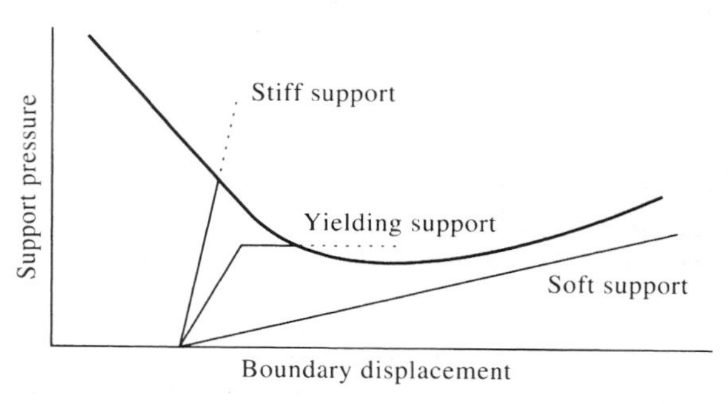

Figure 16.7 Ground response curves and available support lines.

be exceeded. This strength may be reached before equilibrium with the ground response curve is attained. Such an effect can be achieved by the use of compressible inserts placed between the knuckle joints of pre-cast concrete segments, or clamped joints in steel arch supports. In a more precise way, the yielding of the support can be determined by the control of the fluid pressure in hydraulic rams, a technique used effectively in longwall mining practice.

There are other engineering conclusions to be drawn from a ground response curve diagram. No support is necessary to achieve equilibrium in the elastic and stable non-elastic cases: if support is used in these cases, the displacement at equilibrium is simply reduced. In the case of the unstable, non-elastic curve, support must be used. For the case illustrated in Fig. 16.7, the stiff support (e.g. pre-cast concrete segments) will be successful, but the soft support (e.g. steel arches at 1 m centres) will not bring the system to equilibrium. Another point that is demonstrated by this diagram is that the engineer should never attempt to achieve zero displacement by introducing as stiff a support as possible—this is never possible, and will also induce unnecessarily high support pressures. The support should be in harmony with the ground conditions, with the result that an optimal equilibrium position is achieved.

Through a knowledge of the mechanics of support as illustrated by the ground response curve and associated available support lines in Fig. 16.7, a purely observational approach to providing support can be utilized. It has been seen that it is unnecessary to install stiff support elements at an early stage: it is better to allow the rock to displace to some extent and then ensure equilibrium is achieved before any deleterious displacement of the rock occurs. In practice, it may not be possible to establish the exact form of the ground response curve, but we can measure the displacement that occurs, usually in terms of the convergence across an excavation. The ground response curve and and convergence curves are linked because they are different manifestations of a single phenomenon.

The three curves in Fig. 16.8 serve as an aid to understanding this linkage. Commencing with a ground response curve and an available support line, the information is redrawn as a single curve of the pressure 'difference'

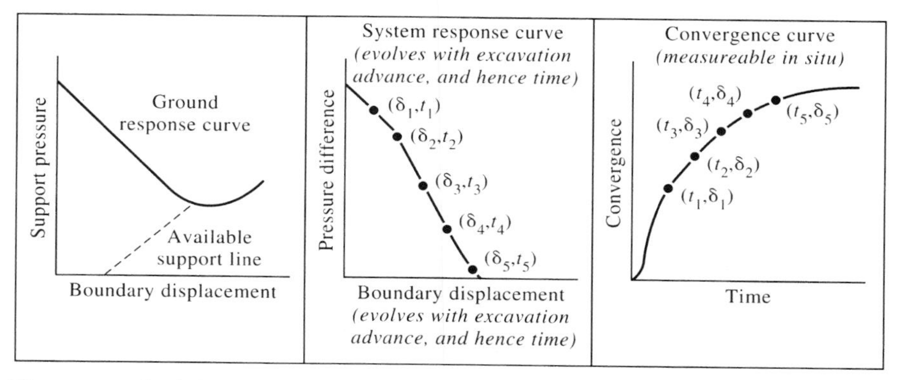

Figure 16.8 Link between ground response curves and observed convergence.

between the two (i.e. the pressure remaining to be equilibriated) versus displacement. Equilibrium is reached when the pressure difference is zero. Accepting that the pressure difference is a function of the displacement of the lining, and that the displacement of the tunnel wall is a function of time (because the displacement results from advancing the tunnel face), then a convergence–time curve can be drawn. This final curve is the one that is measureable in practice.

Convergence occurs rapidly as excavation proceeds; subsequently the convergence rate decreases as equilibrium is approached. This leads directly to the observational method (sometimes referred to as the New Austrian Tunnelling Method), in which sufficient support is installed, after the period of initial high displacement rate. The disadvantage of this approach is that the rock–support mechanics is not precisely known; the advantage of the approach is that it is based on sound engineering principles and can be tailored on site to the idiosyncrasies of the project. The technique was used in the Bochum metro (Fig. 16.9) and the Channel Tunnel sub-sea running tunnel crossover chambers.

An extensive review of excavation support techniques with tunnel linings in the United Kingdom has been conducted by Craig and Muir Wood (1978).

16.4.2 Rock support in discontinuous rock

A directly analogous ground response curve approach can be considered for the use of rock support in discontinuous rock. As the rock becomes more and more fractured with the attendant loss of strength, the ground response curve would be expected to become progressively flatter, as illustrated in Fig. 16.10. This effect is similar to the reduction in rock mass modulus with increasing discontinuity frequency, as illustrated in Fig. 8.2.

Figure 16.9 Construction of the Bochum metro in Germany.

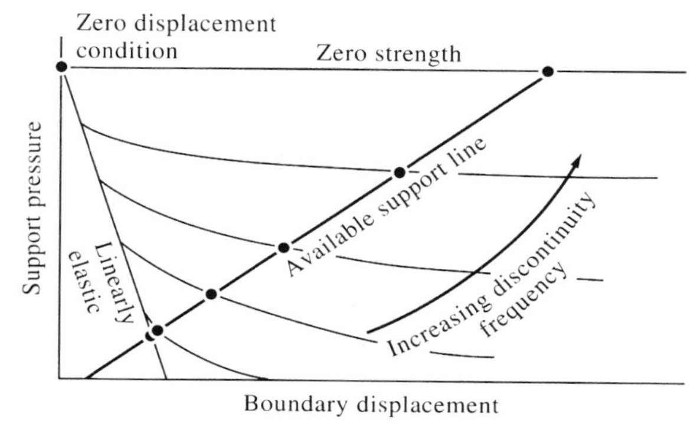

Figure 16.10 Ground response curve in discontinuous rock masses.

The two limiting cases of the suite of ground response curves in Fig. 16.10 are the linearly elastic behaviour at the left part of the figure and the zero strength behaviour represented by the uppermost horizontal curve. Note that in order to achieve a *zero displacement* condition in either case, it is necessary for the introduced support pressure to equal the *in situ* rock pressure. However, equilibrium is reached when the available support line intersects the ground response curve, so that in most cases for a continuous rock it is not necessary to replace the *in situ* rock pressure with an equivalent support pressure. Nevertheless, considering Fig. 16.10, it can be seen that increasingly higher support pressures are required for equilibrium as the introduction of more and more discontinuities into the rock mass flattens the ground response curve. So, at the other limit, there is a zero-strength material in which it is always necessary to replace the *in situ* rock pressure with an equivalent support pressure.

The circumstances are not only affected by the overall discontinuous nature of the rock mass, but are exacerbated by the existence of discrete rock blocks which will create point loads on the support elements. Moreover, there is the basic danger to personnel of rock blocks falling from the roof during construction and the difficulty of localized high water flows.

16.5 Stabilization of 'transitional' rock masses

The previous four sub-sections have followed the lower boxes in Fig. 16.2, and have concentrated on the major features of the subjects. In practice, there will be a wide spectrum of rock media and associated rock behaviour. The term 'transitional' in the heading to this section is used to indicate that the rock mass around an excavation may have attributes associated with continuous *and* discontinuous rocks. There is a wide range of such attributes and consequential behaviour; here, we highlight one transitional case—slip on discontinuities in a layered rock. In such a case, the stress distribution around the opening can be found from a continuum analysis, but the mode of failure is due to the discontinuous nature of the rock. The 'ϕ_j theory' described here was presented by Goodman (1989). The parameter ϕ_j is the angle of friction between two discontinuity surfaces.

There are three basic concepts that permit consideration of the potential for interlayer slip and establishing the extent of the regions thus affected:

(a) when rock is excavated and an opening formed, the excavation surface becomes a principal stress plane, with the result that the components of normal stress in the rock rotate to become parallel to the excavation surface (see Fig. 16.11(a));
(b) slip on a discontinuity can occur when the major principal stress (in the plane of the excavation surface) applied across the discontinuity acts at an angle greater than ϕ_j to the discontinuity normal (see Fig. 16.11(b));
(c) a convenient geometrical construction, utilizing (a) and (b) above, can be used for establishing the extent of the boundary of the opening over which the potential for interlayer slip exists (see Fig. 16.11(c)).

In Fig. 16.11(c), the geometrical construction used involves drawing a construction line normal to the single discontinuity set under consideration, followed by two further construction lines inclined at an angle ϕ_j to the normal. Tangents to the excavation boundary, which are parallel to these two construction lines, are then drawn. The key to this construction is that between these two tangent points, criterion (b) above is fulfilled, and hence this is the region for potential inter-layer slip. This construction is valid for the entire excavation boundary.

The construction applies to any concave shape of excavation. In Fig. 16.12, the construction is shown for a much more irregularly shaped excavation. It is not necessary to know the geometrical centre of the excavation shape; keeping the normal line of the construction and the discontinuity normals parallel, all three lines of the geometrical construction are moved until the outer lines touch the excavation boundary on both sides. In Fig. 16.12(a), the construction is shown for $\phi_j=20°$, whereas in Fig. 16.12(b) the construction is shown for $\phi_j=50°$. This not only demonstrates the construction method, but also shows that there is potential for inter-layer slip in the walls for the lower friction angle, but not for the higher friction angle. Such a reduction in the friction angle could result from deterioration of the discontinuity surfaces with time, with the result that previously stable zones of an excavation boundary become liable to instability.

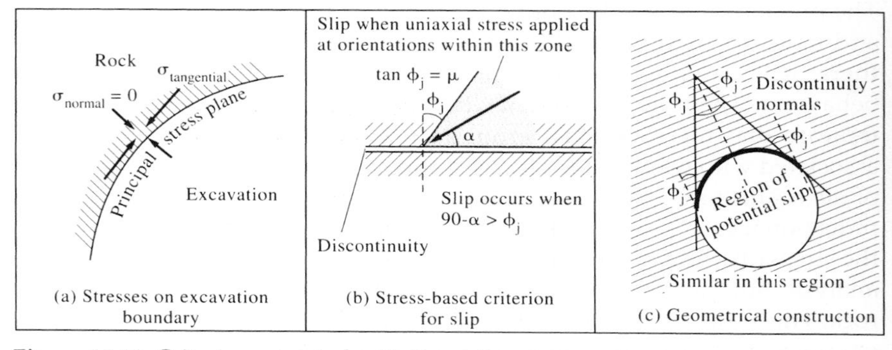

Figure 16.11 Criteria associated with the ϕ_j theory.

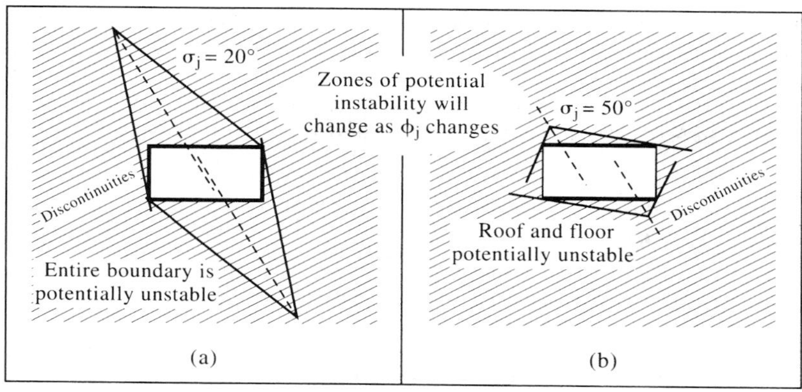

Figure 16.12 The use of the ϕ_j construction for a non-circular opening with (a) $\phi_j=20°$ and (b) $\phi_j=50°$.

The analysis as presented assumes that the normal to the discontinuities is in the plane of the cross-section. If this is not the case then the angle projected onto the plane of the cross-section will be an 'apparent ϕ_j''. This angle can conveniently be determined using hemispherical projection methods.

Considering the case when a support pressure is applied normal to the excavation boundary, it is possible to calculate the pressure required to inhibit inter-layer slip. In the plane of the cross-section, the stresses acting on an element of material immediately adjacent to the boundary are σ_θ and p, as shown in Fig. 16.13. Application of the stress transformation equations and utilization of the Mohr–Coulomb failure criterion results in the following expression for the required support pressure:

$$p = \sigma_\theta \frac{\tan(\alpha - \phi_j)}{\tan\alpha} \quad \phi_j < \alpha < 90.$$

In general, the tangential stress σ_θ will vary around the boundary. For simple geometries, closed-form solutions exist for the tangential stress, but for more complex shapes a numerical procedure is adopted.

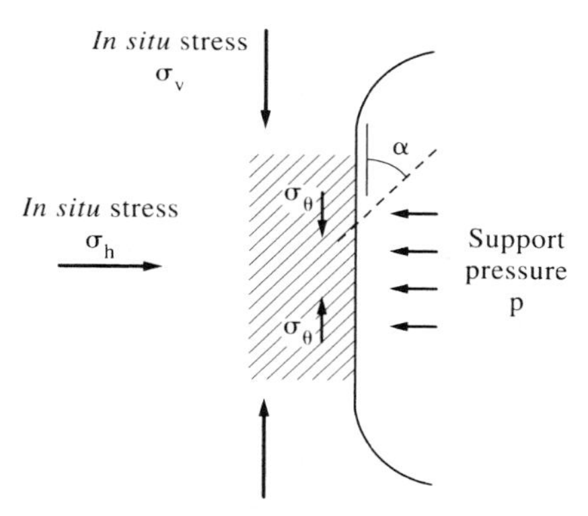

Figure 16.13 The ϕ_j theory applied to the calculation of support pressure.

Regardless of the technique used for determining the tangential stress, the formula above shows that the optimal support pressure varies around the excavation boundary.

Although we have used only one discontinuity set in this example, it is clear that the same approach can be adopted for all the discontinuity sets present and the solutions superposed. If a particular region of excavation boundary is found to have the potential for inter-layer slip with respect to multiple discontinuity sets, then it will be particularly prone to failure. From the point of view of provision of support at any point on the boundary, then the highest calculated support pressure is the one required to inhibit slip on all sets.

16.6 Further comments on rock stabilization methods

Following the explanations earlier in this chapter, it follows that the simplest way to stabilize a rock mass containing discontinuities is to install rock bolts in order to reduce the mechanical effect of the discontinuities. In the extreme case, were we able to eliminate these effects completely, the rock discontinuum would have been changed to a rock continuum. The excavation surface is a special case, requiring extra reinforcement and so, as mentioned earlier, a combined system can be used which would include shotcrete and wire mesh. This can be either a primary operation used for temporary protection, or it could be the final stabilizing operation.

In Fig. 16.14, there is an example of a rock mass which has been reinforced by rock bolts alone, and an example in which rock bolts and shotcrete have been used together. The bolts can be either tensioned or untensioned: there are advantages and disadvantages to each.

The advantage of a tensioned bolt is that it can provide extra force across the discontinuity surfaces and hence inhibit further block displacement. The disadvantage of such a bolt is that the tension may not be sustained, due to relaxation, over the design life. Hence, unless there is a continual monitoring programme, the engineer may not have sufficient continuing confidence that the bolting system is operating as designed and originally installed. When rock bolt heads have to remain accessible for monitoring following the installation of secondary reinforcement or support (in particular, further layers of shotcrete or cast *in situ* concrete), the extra complication will incur higher costs.

The advantage of untensioned rock bolts is that block displacements induce the necessary tension within them, due to dilation of the shearing discontinuity: thus, they respond directly as required. The disadvantage of these bolts is that they do not possess the small amount of pre-tension which could inhibit the initial displacements.

There are many proprietary types of rock bolt available, and many ways in which the tension is applied through the rock bolt–rock bond to create compression in the rock. There are also operational factors to be considered: tensioned rock bolts require the necessary equipment to be available, and time for its use to be scheduled; with untensioned bolts the engineer can never be certain that the rock bolt–rock bond has sufficient strength to allow the required tension to be induced.

Figure 16.14 Underground excavation stabilized by (a) rock bolts and (b) rock bolts and shotcrete.

This form of rock stabilization, where the rock mass 'supports itself', is now generally accepted by all aspects of the industry. When one is underground, however, it is an act of intellectual faith to believe that the rock is, in fact, being stabilized by the rock bolt technique—because there is little visual evidence of any engineering work. It is a good idea to explain the principles to all personnel in order to improve confidence and instill the necessary discipline to ensure correct installation.

In recent years, reinforcement by the installation of long lengths of steel wire rope—cable bolting—has been introduced, thus enabling the reinforcement to be of the same dimension as the structure. For example, installing cable anchors of sufficient length in a mine stope hanging wall prevents smaller blocks, reinforced by the shorter rock bolts, from becoming

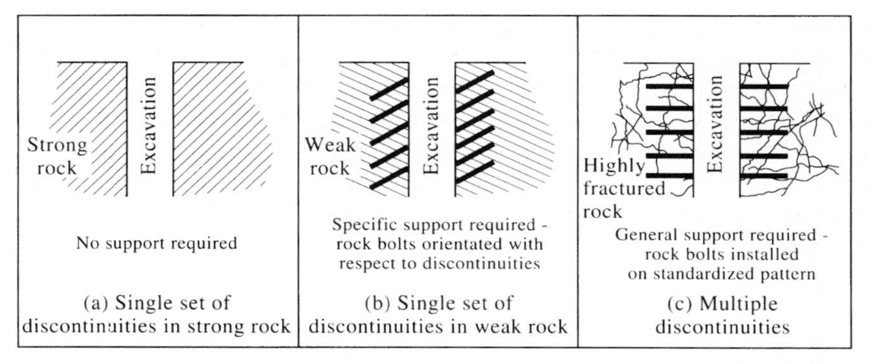

Figure 16.15 Rock bolting in discontinuous rock masses. (a) Single set of discontinuities in strong rock. (b) Single set of discontinuities in weak rock. (c) Multiple discontinuities.

detached.

Rock bolts can be installed on a fixed production pattern (designed according to the rock structure), or the pattern can be varied according to local conditions, or both. One extreme in the rock structure spectrum is a single set of parallel, persistent and planar discontinuities: the other end of the spectrum is the case of discontinuities at all orientations with many idiosyncratic features. Much can be established about the optimal rock bolting pattern from engineering judgement.

If there were only one set of discontinuites, and the rock were strong, failure would not occur at all. The minimum number of faces required to form a discrete rock block is four: thus, a single set of discontinuities intersecting an excavation does not result in any rock blocks being formed, as illustrated in Fig. 16.15(a). Conversely, if the rock mass were weak, reinforcement would be needed and the optimal direction for the bolts could be determined, which would not be consistently radial to the excavation, as illustrated in Fig. 16.15(b).

If the numbers and properties of the discontinuities tend to form a homogenous and isotropic rock mass structure, a standardized pattern can be adopted. This case is illustrated in Fig. 16.15(c).

Shotcrete can be regarded as a rock reinforcement method, because it inhibits block movement rather than providing structural support. This method of operation may be understood if we consider that a thin (say 100 mm) membrane of shotcrete may be sufficient to stabilize a 5 m diameter tunnel. Such a thin unreinforced concrete element is incapable of supporting large radial pressures without failing. The pressures do not develop because the slight lateral restraint offered to the rock around the excavation by the shotcrete is sufficient to inhibit the block movement, and effectively increase the rock mass strength. It follows that the shotcreting is most effective when a complete ring is formed, and drainage holes should be provided if significant water pressures are likely to develop.

The structural operation of a reinforcement system consisting of tensioned rock bolts and a shotcrete membrane is conceptually analogous to that of a spoked bicycle wheel. In such wheels the thin spokes are tensioned before the wheel carries any load, with the effect that the weight of the rider is transmitted by a reduction in tension through the spokes rather than compression. This allows use of thin spokes, which are capable of withstanding significant tensile forces whilst being weak in compression. The alternative is to provide spokes with significant compressive strength, as in cumbersome timber cartwheels—in the context of rock excavations this is analogous to the engineer having to provide cumbersome concrete linings.

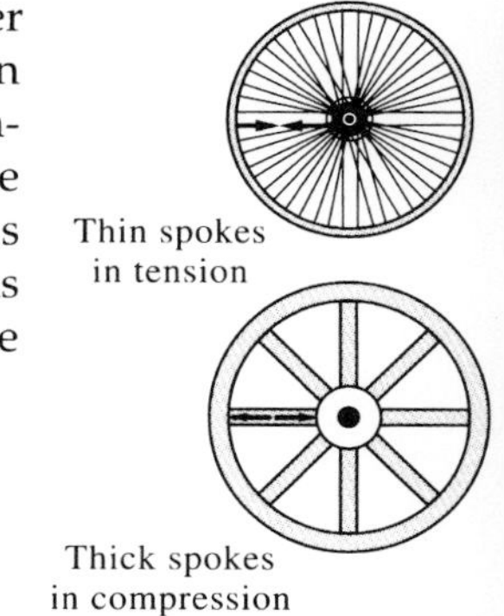

17 Surface excavation instability mechanisms

In this chapter, we discuss instability mechanisms in relation to slopes and foundations. It is possible, in an underground excavation, for the excavation walls and floor to fail in a similar way to surface slopes and foundations, but these topics are covered in the next chapter.

17.1 Slope instability

In the history of rock mechanics and rock engineering, more attention has been paid to slope instability considerations than any other topic, and this topic remains one of the most important today. In Fig. 17.1, there are two slope failure mechanisms. Figure 17.1(a) illustrates slope instability when the rock is behaving as an equivalent continuum; Figure 17.1(b) illustrates slope instability when the rock is behaving as a discontinuum. One of our first considerations must be to identify the basic mechanisms of slope instability. The sketches in Fig. 17.1 also highlight the CHILE versus DIANE nature of the rock which was discussed in Chapter 10.

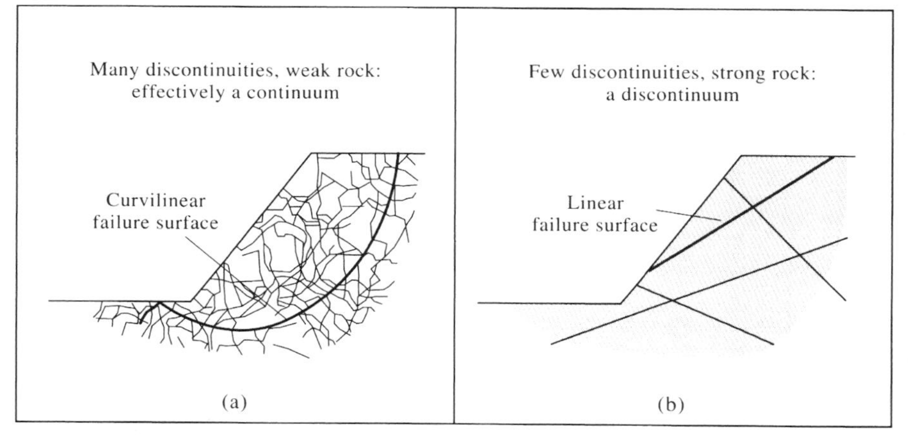

Figure 17.1 Slope failure mechanisms in (a) a continuum and (b) a discontinuum.

In Fig. 17.1(a), the failure surface has been created through the rock mass (behaving as a continuum), whereas, in Fig. 17.1(b), the failure surface is dictated more directly by the presence of specific pre-existing discontinuities. It is also possible to have intermediate cases where the failure occurs partly along the discontinuities and partly through bridges of intact rock, but we are concentrating here on the essential differences between continuous and discontinuous behaviour. Although most soil slope instability is of the continuous nature, the majority of rock slope instability is caused by individual discontinuities. This is because the strength of the intact rock can be high, with the result that the pre-existing discontinuities are the weakest link.

As an amplification of the sketches in Fig. 17.1, the four diagrams in Fig. 17.2 and associated photographs in Fig. 17.3 illustrate the mechanisms which are traditionally regarded as the four basic instability mechanisms for rock slopes. The geometry of the slip in Fig. 17.2(a) is a function of the geometry of the slope and the strength of the material forming the slope, but the slope instabilities in Figs 17.2(b)–(d) show how the boundaries of the instability are governed by the discontinuities, giving essentially planar faces to the sliding and toppling blocks. Each of these mechanisms is discussed separately in Sections 17.1.1–17.1.4.

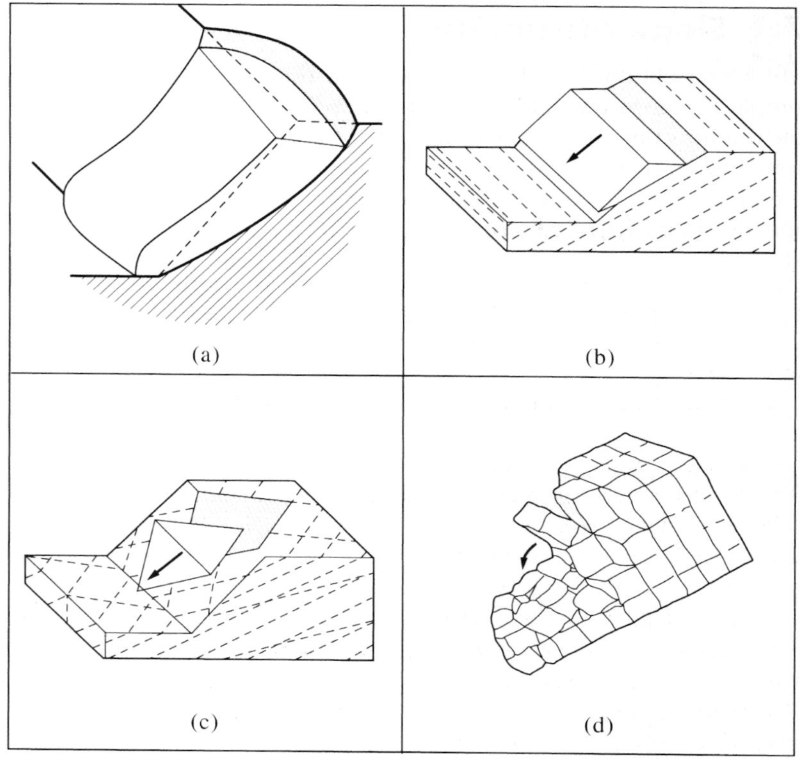

Figure 17.2 The four basic mechanisms of rock slope instability: (a) circular slip; (b) plane sliding; (c) wedge sliding; and (d) toppling ((b), (c) and (d) from Matheson, 1983).

Figure 17.3 Photographs illustrating the four basic mechanisms of rock slope instability shown in Fig. 17.2: (a) circular slip; (b) plane sliding; (c) wedge sliding; and (d) toppling.

17.1.1 Curvilinear slip

The term 'curvilinear slip' is used to describe the group of instabilities shown in Fig. 17.4. This term should be regarded as synonymous with the more usual one 'circular slip', which is normally understood to also include non-circular slips. Only in exceptional circumstances will instabilities occurring in a continuum have truly circular slip surfaces; they will usually be curvilinear. Hence, we will analyse general curvilinear slips, and present truly circular forms as a special case.

The text in the following sub-section is based on lecture notes produced by Dr J.W. Bray (formerly of Imperial College and co-author with Professor E. Hoek of the seminal book *Rock Slope Engineering*), to whom we should like to express our gratitude here.

In Fig. 17.4 there are five diagrams of geological circumstances under which curvilinear slips may develop. Experience indicates that with these materials, the slip surface is curved and usually terminates at a tension crack at the upper ground surface. The shape and location of the slip surface depends on the strength characteristics of the ground mass, which in turn depend on the structure—as indicated in Fig. 17.4.

In analysing the potential for slip, one has to consider (a) the location of the slip surface and (b) determination of the factor of safety for a given slip

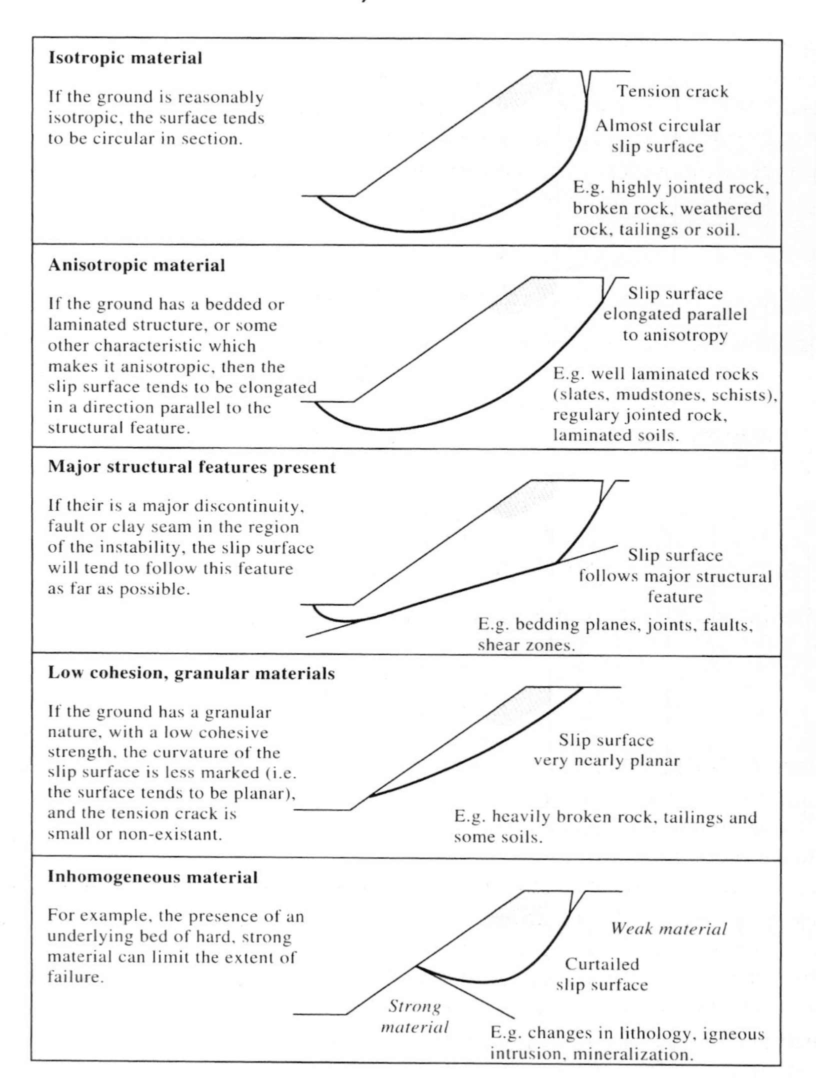

Figure 17.4 Development of curvilinear slips.

surface. In practice, the factor of safety is determined for assumed slip surface locations. In the margin sketch, the slip surface is shown discretized into four elements, each of which has normal and shear forces applied to it. Each element has three unknowns associated with it: the normal (N) and shear (S) forces, and the location of the line of action of the normal force relative to the element itself (n).

For the case shown, therefore, there is a total of 12 unknown parameters in the problem. However, there are only three equations of static equilibrium available to solve this problem: $\Sigma F_x = 0$, $\Sigma F_y = 0$ and $\Sigma M = 0$, where F_x are components of forces in the x-direction, F_y are components of forces in the y-direction and M are moments in the x–y plane. There are insufficient equations to determine the unknowns: i.e. the problem is statically indeterminate.

To solve the problem, we have to make assumptions which reduce the number of unknowns. The usual method of doing this is by sub-dividing the mass under consideration into 'slices', and analysing each slice on the basis that it is in limiting equilibrium, i.e. each N and S is linked through the strength criterion of the slip surface.

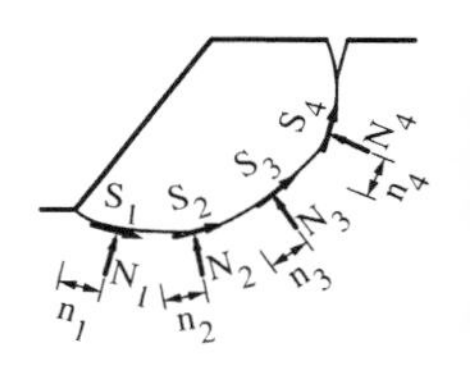

The margin sketch shows a typical slice with the various forces applied to it, and equilbrium analysis leads to

$$S = \frac{[(W - ub)\tan\phi + bc]\sec\alpha}{F[1 + (\tan\alpha\tan\phi)/F]}$$

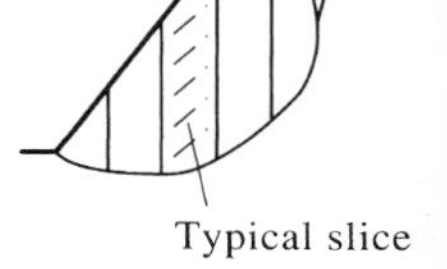

which enables S to be expressed in terms of the other slice parameters. The analysis of the factor of safety, F, of the entire mass then depends on whether the slip surface is generally non-circular, or specifically truly circular.

In the former case, resolving horizontally and vertically for all of the slices leads to

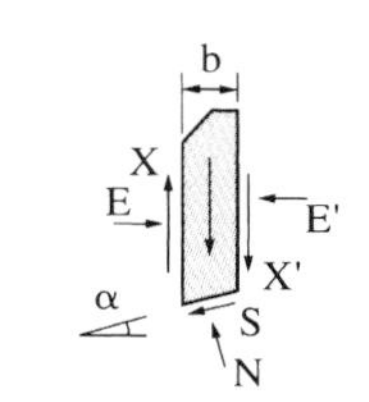

$$F = \frac{\sum(FS\sec\alpha)}{\sum(W\tan\alpha) + \sum(S\sec\alpha - W\tan\alpha)}.$$

In the latter case, the equation is simpler, and reduces to

$$F = \frac{\sum(FS)}{\sum(W\sin\alpha) + Hz/R}$$

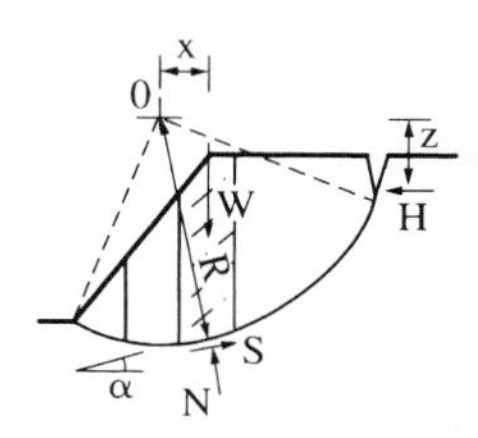

where H is the hydrostatic thrust from the tension crack and the other parameters are illustrated in the margin sketch.

The anticipated location of the slip surface can now be found from analysis of the whole range of possible surfaces, and taking the actual surface to be that which gives the lowest factor of safety. Curvilinear slips are, in general, truly three-dimensional in that they resemble the bowl of a spoon, and hence the analysis here is an approximation. The seminal references for this type of two-dimensional analysis are Bishop (1955) for the circular slip surface and Janbu (1954) for the non-circular slip surface, with further explanation specifically related to rock slopes in Hoek and Bray (1977).

17.1.2 Plane sliding

In Fig. 17.4, we showed the variety of curvilinear slips that can occur for different geological circumstances. In this section, we concentrate on the type of failure illustrated in the central diagram of Fig. 17.4, where major structural features are present which are much weaker than the rock on either side. Because the slip generally occurs on a major discontinuity, it will usually have a planar form—owing to the planar nature of the pre-exisiting discontinuity. In fact, when the instability is dictated by the presence of pre-exisiting discontinuities, the instability takes the form of plane sliding, wedge sliding or toppling, as illustrated in elements (b)–(d) of both Figs 17.2 and 17.3. In this and the two following sub-sections, we deal with these in turn.

The case of plane sliding is unlike that of curvilinear slip, in that it is statistically determinate. We can calculate the factor of safety for plane sliding directly, and by making suitable assumptions to render the problem two dimensional, the solution is straightforward.

The right-hand side of Fig. 17.5 shows an idealized form of the plane instability condition. This demonstrates two of the underlying assumptions in the analysis: the strikes of the plane of sliding and the slope face are parallel, and there are no end restraints caused by adjacent blocks of rock. The free body diagram shows the forces acting on the unstable block of rock. In the case shown, a partially water-filled tension crack has been included, with the result that there are water pressure distributions along the tension crack and the plane of sliding. The usual assumption for these distributions is that they are linear, and the water pressure on the plane of sliding is zero at the plane's intersection with the slope face.

Making the assumptions that the rock mass is impermeable, the sliding block is rigid, the strength of the sliding plane is given by the Mohr–Coulomb criterion and that all forces pass through the centroid of the sliding block (so that moment equilibrium is automatically maintained), then by defining the factor of safety as the ratio between the forces resisting sliding and the forces driving the sliding, we have

$$F = \frac{c'(H - z)\operatorname{cosec}\psi_p + (W\cos\psi_p - U - V\sin\psi_p)\tan\phi'}{V\cos\psi_p + W\sin\psi_p}.$$

Similar formulations can be derived for other cases, such as a horizontal sliding plane, no tension crack, a sloping upper surface or dry conditions. The last of these cases can over-estimate the stability of the slope and should only be used when there is confidence in the knowledge of the hydraulic regime.

The effective stress parameters c' and ϕ' have been used in the analysis above. It is by no means clear without further information whether, in fact, the most appropriate parameters are the traditionally used total stress parameters c and ϕ of rock engineering which imply drained conditions, or the traditionally used effective stress parameters c' and ϕ' of soil engineering which incorporate the effect of water pressure resulting from undrained conditions. This is a complex subject, and a thorough knowl-

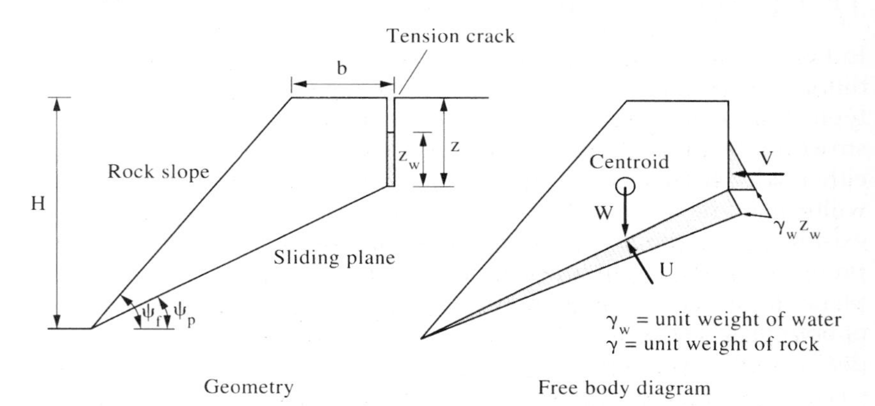

Figure 17.5 Geometry of static analysis of plane instability.

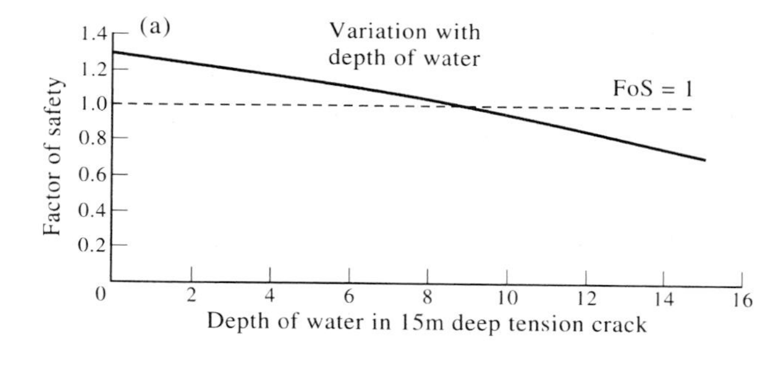

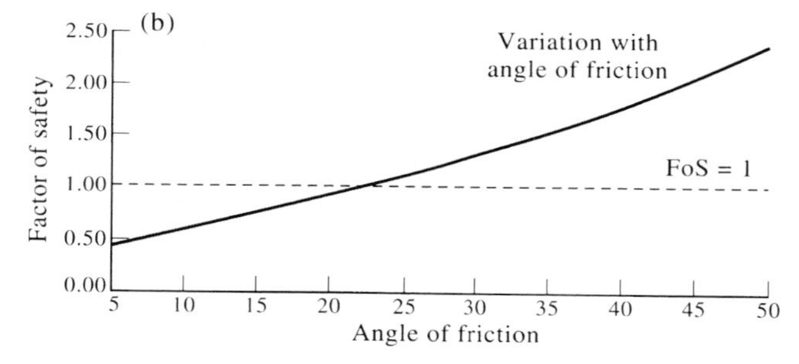

Figure 17.6 Simplified analysis of plane failure demonstrating variation in factor of safety with (a) depth of water in tension crack varying and (b) angle of friction of sliding plane varying.

edge of the history of the site, the nature of any infilling and the hydraulic conditions are required in order to determine whether total stress or effective stress parameters are to be used.

To illustrate the utility of the equation presented above, Fig. 17.6(a) shows how the factor of safety may vary for different depths of water in the tension crack, indicating a possible significant effect of heavy and prolonged rainfall. It can be seen from this graph that, as the depth of water in the tension crack varies from 0 to 15 m (the overall depth of the tension crack itself) and the angle of friction of the sliding plane remains constant at 30°, the factor of safety reduces from 1.30 to 0.72.

In Fig. 17.6(b), we show the complementary case of variation of the effective angle of friction along the plane of sliding, for the instance of a dry slope and all other parameters remaining constant. In this case, the factor of safety reduces from 2.36 to 0.45 as the angle of friction varies from 50° to 5° for the dry slope.

The curves in Fig. 17.6 show how, for even a simple model, the factor of safety varies dramatically with just two critical parameters. A more realistic analysis would have to include the manifold aspects of a real plane instability, such as the end restraints, the roughness and possible partial impersistence of the sliding plane, water pressures in the discontinuity network, the nature of any filling material in the discontinuities, and so on. It is unlikely though that the general thrust of the factor of safety variation trends shown in Fig. 17.6 would be altered by the adoption of a more realistic model. In the following chapter we will present more thorough methods of analysing the instability of plane slides, both kinematically and statically.

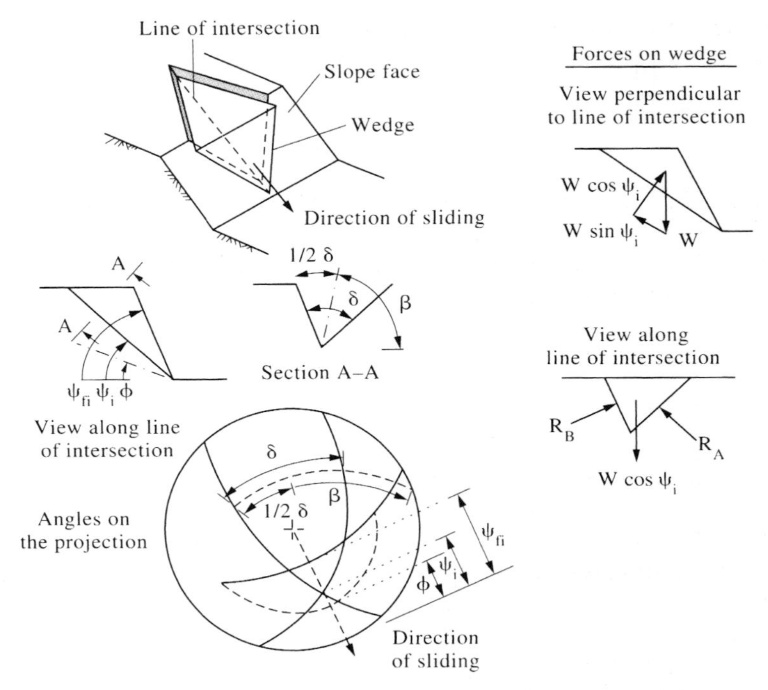

Figure 17.7 Geometry of static analysis of friction-only wedge instability.

17.1.3 Wedge sliding

The previously presented method of analysing the basic mechanism of plane sliding can be adapted to the case of wedge sliding. Wedge sliding is illustrated in Figs 17.2(c) and 17.3(c), and the extension from plane sliding is to consider sliding on the two sliding planes simultaneously. In Fig. 17.7, the geometry of the wedge instability and the primary forces acting on the system are shown. The problem has been simplified to one in which there is no cohesion on either sliding plane, and both of the planes possess the same angle of friction. A solution to the comprehensive problem, in which both planes possess differing cohesion and angles of friction, as well as the existence of a water-filled tension crack, is presented by Hoek and Bray (1977).

Assuming that the direction of sliding is parallel to the line of intersection of the two sliding planes, forces parallel to this line and perpendicular to the two sliding planes can be resolved in order to determine the factor of safety. This analysis leads to

$$F = \frac{(R_A + R_B)\tan\phi}{W \sin\psi_i}$$

$$\text{and } R_A + R_B = \frac{W \cos\psi_i \sin\beta}{\sin\frac{1}{2}\delta}.$$

The various forces and angles used in these formulae are shown in the individual parts of Fig. 17.7. Consolidating these formulae results in

$$F = \frac{\sin \beta}{\sin \frac{1}{2}\delta} \times \frac{\tan \phi}{\tan \psi_i}$$

which provides a simple method of evaluating the effect of the main parameters on the factor of safety for wedge sliding.

A direct insight into the fundamental mechanism of wedge instability is achieved by abbreviating the equation to

$$F = k_W \times F_P$$

i.e.

wedge factor of safety = **wedge factor** × **plane factor of safety.**

In Fig. 17.6, the factor of safety varied with two of the main parameters. For wedge sliding, we can study the effect of k_W, the wedge factor. This is a purely geometrical parameter, concerning how upright and how sharp the wedge is.

In Fig. 17.8, we show how the factor of safety varies with the parameter δ, the sharpness of the wedge, and β, the verticality of the wedge. Again, the utility of the application of a simple model to a complex problem is clearly demonstrated. Considering the suite of curves in Fig. 17.8, it is not obvious that thin, upright wedges would have a higher factor of safety than thin, inclined wedges; nor, indeed, that the verticality of the wedge will be more critical for thin wedges than for thick wedges (remembering that in the diagram constant angles of friction and intersection line plunge have been used).

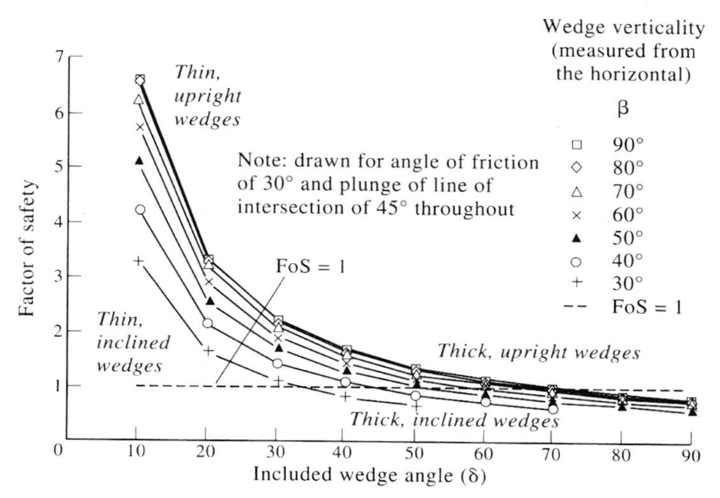

Figure 17.8 Simplified analysis of wedge failure demonstrating variation in factor of safety with included wedge angle and wedge verticality.

17.1.4 Toppling

To complete the set of fundamental mechanical modes of structurally controlled instability, toppling failure is considered. Toppling failure has traditionally been regarded as occurring in two modes: direct toppling and flexural toppling. The former occurs when the centre of gravity of a block of rock lies outside the outline of the base of the block, with the result that a critical overturning moment develops. The latter occurs under certain circumstances when a layered rock mass outcrops at a rock slope, and the principal stress parallel to the slope face induces inter-layer slip which causes the intact rock to fracture and the resulting blocks to overturn. The distinction between these two toppling modes of instability is illustrated in Fig. 17.9.

Direct toppling instability. Will a block resting on an inclined rock surface be stable, or slide, or topple, or simultaneously slide and topple? The nature of the instability, if any, is determined from considerations of the block geometry and the angle of friction between the block and the surface on which it is resting. The four possibilities are shown in Fig. 17.10, as the various regions in a graph of block aspect ratio versus friction angle.

Sliding will only occur when the dip of a plane exceeds the angle of friction. This results in the vertical line dividing Fig. 17.10 into regions—with no sliding on the left and sliding on the right.

To establish the equilibrium due to toppling, consider the location of the line of action of the force due to gravity. This passes through the centre of gravity of the block and will coincide with the lower apex of the block if $b/h = \tan \psi$, which is the limiting equilibrium condition. Thus, toppling will not occur if $b/h > \tan \psi$, and will occur if $b/h < \tan \psi$.

The resulting four categories of equilibrium are

(a) no sliding and no toppling: $\psi < \phi$ and $b/h > \tan \psi$;
(b) sliding but no toppling: $\psi > \phi$ and $b/h > \tan \psi$;
(c) no sliding but toppling: $\psi < \phi$ and $b/h < \tan \psi$;
(d) sliding and toppling: $\psi > \phi$ and $b/h < \tan \psi$.

These four fundamental categories represent the basic circumstances of toppling and related sliding, and enable a rapid initial analysis of whether direct toppling could take place and hence whether further analysis is necessary.

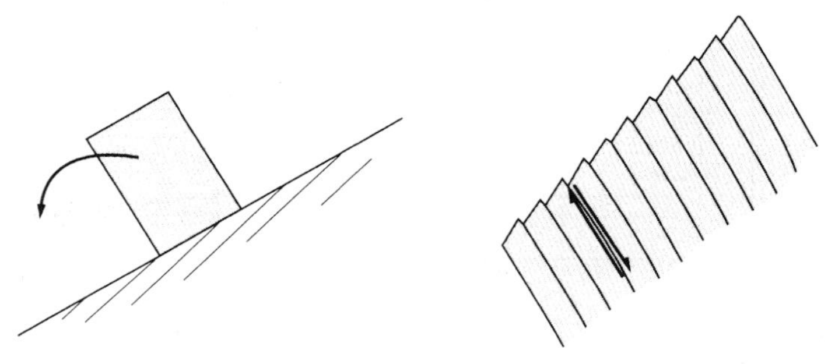

Figure 17.9 Direct and flexural modes of toppling instability.

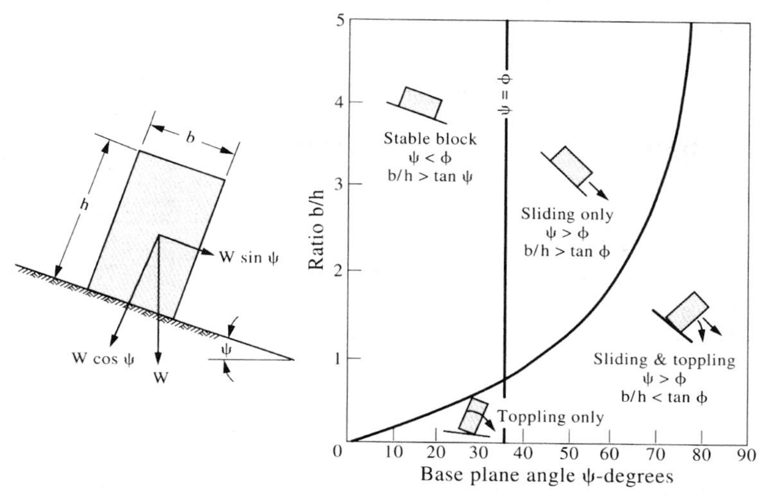

Figure 17.10 Sliding and toppling instability of a block on an inclined plane (from Hoek and Bray, 1977).

Flexural toppling instability. In Chapter 16, the stability of underground excavations was discussed in relation to the potential for inter-layer slip, the ϕ_j theory (see Figs 16.11 and 16.12). Here we adopt an analogous approach to the potential for slope instablity.

Remembering that the creation of a new excavation surface results in the principal stresses being parallel and perpendicular to the excavated face, we consider the potential for inter-layer slip given the geometry illustrated in Fig. 17.11(a). An analysis of instability will include these geometrical parameters as well as the angle of friction. In Fig. 17.11(b), the ϕ_j theory is applied directly to inter-layer slip along the slope surface. The geometrical construction, which includes the normal to the discontinuities

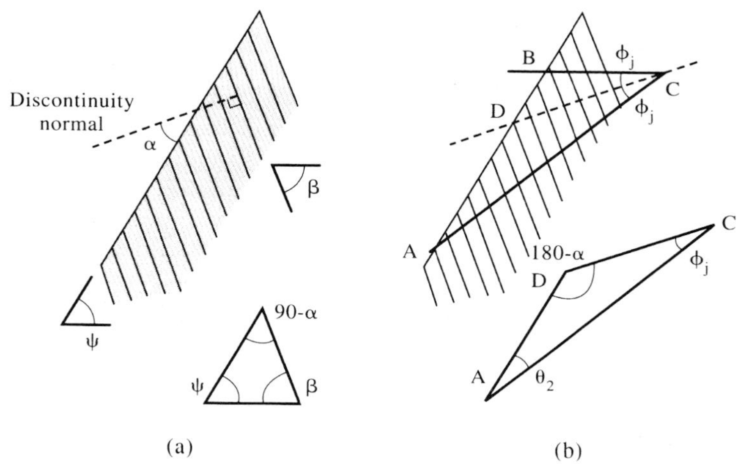

Figure 17.11 Flexural toppling: (a) geometry and (b) ϕ_j analysis.

and the limiting lines at an angle of ϕ_j on either side of this normal, is shown. By analysis of the geometry of this figure, the basic criterion for inter-layer slip potential can be established.

Fig. 17.11(b) shows that for inter-layer slip to take place, the geometry of the system must be such that the triangle ABC will be formed: if the orientation of the discontinuities relative to the slope surface is such that AC and AB are parallel or diverging downwards, the conditions for inter-layer slip will not be met. The inset diagram of Fig. 17.11(b) shows the geometry of triangle ACD, from which it can be seen that $\alpha - \phi > 0$. The basic geometry of the system shows that $\alpha = \psi + \beta - 90$, with the result by inspection that, for inter-layer slip to take place,

$$\psi \geq 90 + \phi - \beta.$$

Using these angles, we can utilize a 'geometrical factor of safety' to provide some indication of how close the slope conditions are to this criterion. If the factor of safety is defined as that factor by which $\tan \phi$ must be divided to bring the slope to limiting equilibrium,

$$F = \frac{\tan \phi}{\tan(\psi + \beta - 90)}.$$

As an example, if we require $F = 1.3$ when $\phi = 30°$ and $\beta = 70°$, then the limiting angle for ψ is 44°. For steeper slopes the factor of safety is reduced; for shallower slopes it is increased.

This concludes the descriptions of the basic mechanics of rock slope instabilities. In Section 17.2, foundation instability is discussed, this being the other manifestation of surface excavation instability. The application of these basic analyses to the design of surface excavations, with additional techniques, is described in Chapter 18.

17.2 Foundation instability

Instabilities in slopes are caused by alteration of the rock mass geometry, whereas foundation instabilities are caused by the direct application of load. In Fig. 17.12, this fundamental difference between the two mechanisms is illustrated, with the distinction being reduced to one of gravitational versus applied load instability. Also shown in Fig. 17.12 is the fact that the foundation instability may result from the creation of new slip surfaces or from movement on a pre-exisiting discontinuity. Since the load is being applied by a structure, the rock–structure interaction has to be considered. This is summarized in the flow chart in Fig. 17.13.

17.2.1 Equlibrium analysis of foundations

As an illustration of the equilibrium analysis approach to foundation instability, consider the plane two-dimensional case of a uniformly distributed line load inducing instability. Two different approaches exist to the solution of this problem:

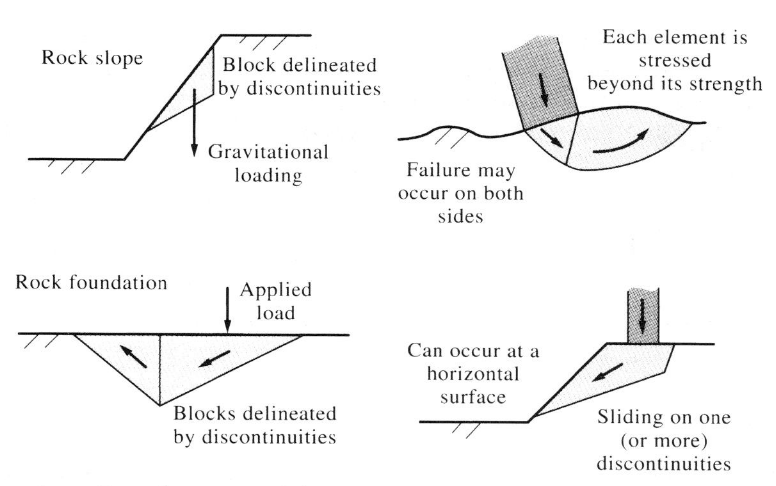

Figure 17.12 Foundation instability.

(a) to postulate a geometry of discrete blocks and evaluate the associated forces and instability; and
(b) to consider the sustainability of a postulated stress distribution beneath the loaded region.

To demonstrate the fundamentals of the methods of solution, only loaded areas are being considered, and not loads applied through structures. In the latter case, the strength and stiffness of the structure must be taken into account and these have a marked effect on the results.

These approaches have been used extensively in the study of plasticity. Two fundamental theorems exist for plastic analysis, and Brown (1987), with reference to the theory of plasticity, quotes these as:

1. **Upper bound theorem.** If an estimate of the plastic collapse load of a body is made by equating the internal rate of dissipation of energy to the rate at which external forces do work in *any* postulated mechanism of deformation of the body, the estimate will be either high or correct.

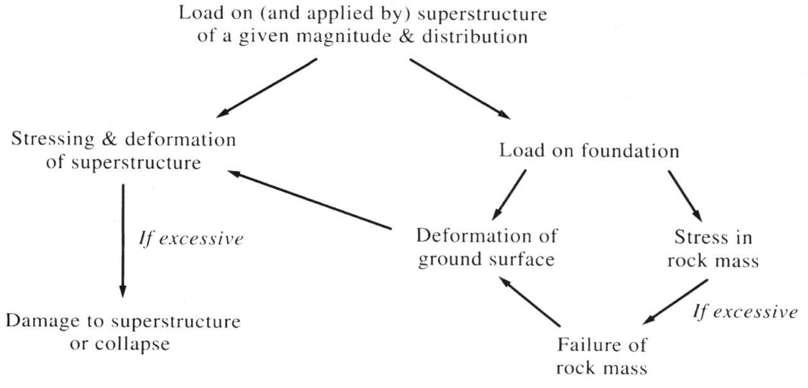

Figure 17.13 Simplified rock–structure interaction flow chart for foundation instability (from lecture notes by S. D. Priest).

2. **Lower bound theorem.** If *any* stress distribution throughout the structure can be found which is everywhere in equilibrium internally and balances certain external loads and at the same time does not violate the yield condition, those loads will be carried safely by the structure.

An *upper bound solution* results from an analysis in which a geometry of discrete blocks is postulated and the associated forces then determined, and a *lower bound solution* results from an analysis in which the sustainability of a stress distribution is analysed.

At the surface of a rock mass, the applied and *in situ* stresses are generally so low as to prevent ductile behaviour and plasticity theorems will be inapplicable. However, the concepts can be applied usefully to rock foundations by:

(a) using the upper bound analysis in the study of foundations where the instability is governed by the movement of rigid blocks along pre-existing discontinuities; and
(b) using lower bound analysis in the study of foundations where the instability is governed by a general yielding of the rock material, which could occur for highly loaded weak rocks.

Discontinuum analysis. In Fig. 17.14, there is a cross-section through a uniform line loading of width D on a rock foundation containing three discontinuities. For simplicity in this analysis, the discontinuities are assumed to have some cohesion but zero angle of friction, although the analysis can easily take account of a non-zero angle of friction. Application of the equations of static equilibrium to the forces shown acting on the free-body diagrams of the two wedges (also shown in the figure) permits calculation of the applied load which will cause instability of the system and, for the geometry and discontinuity strength shown, this is $p = 6c$.

The analysis refers to the problem of a discontinuous rock and the solution is mechanically correct. However, if the analysis were being considered as part of a plastic analysis of a continuum, then this solution

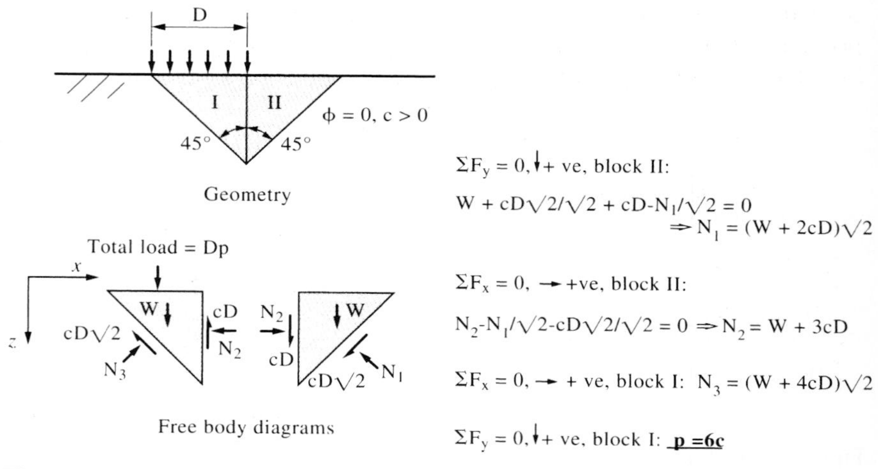

Figure 17.14 Equilibrium analysis of a foundation on discontinuous rock.

would be one of many upper bound solutions to the actual collapse load. The geometry of the assumed plastic wedges would then require variation in an attempt to produce increasingly lower values of collapse load, with the result that each one would be closer to the actual plastic collapse load.

An alternative approach which is more concise and less prone to error is to apply the concept of virtual work, allowing the equilibrium to be established by considering a small amount of work done by the forces involved. For example, in Fig. 17.15, we show three forces acting at a point. Considering the imposition of an imaginary displacement of magnitude u in the direction shown in Fig. 17.15, then

work done by force = (force magnitude) × (component of displacement in direction of force)

and

$$\text{virtual work} = \Sigma \text{ (work done by all forces).}$$

The magnitude of the virtual work will be zero if the system is in equilibrium—because the work done by the resultant force (which is zero for a system in equilibrium) must be zero. For the forces shown in Fig. 17.15, the inset table gives the calculation of the virtual work.

The application of the concept of virtual work to a more complex foundation problem is illustrated in Fig. 17.16. Although this is intended to represent a system of discrete blocks formed by discontinuities, it may also be regarded as a refinement to the upper bound plastic problem shown in Fig. 17.14. In this case the angle of friction is non-zero.

As a first stage in the analysis, the directions of the virtual displacements associated with the forces arising from the strength of the discontinuities are drawn on the diagram. These directions, shown by the vectors v_1, v_2, v_3, v_{12} and v_{23} in Fig. 17.16(a), are drawn inclined at an angle ϕ, the angle of friction, to the discontinuity. This results in each virtual displacement being orthogonal to the resultant force on each discontinuity. To evaluate the compatibility relations between the various virtual displacements, the polygon of displacements shown in Fig. 17.16(b) is constructed. This is initiated by assuming a unit magnitude for the virtual displacement v_1, and

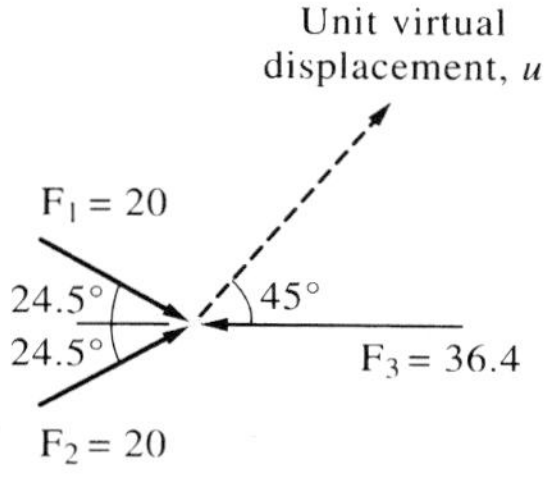

Unit virtual displacement, u

$F_1 = 20$

24.5°

24.5°

45°

$F_3 = 36.4$

$F_2 = 20$

Force	Angle with line of virtual displacement, u	Component of u in direction of force	Work done
$F_1 = 20.0$	69.5°	cos 69.5 = 0.3502	7.00
$F_2 = 20.0$	20.5°	cos 20.5 = 0.9367	18.73
$F_3 = 36.4$	-135°	cos -135 = -0.7071	-25.73
		Sum of virtual work components	**0.00**

Figure 17.15 The principle of virtual work applied to the analysis of equilibrium.

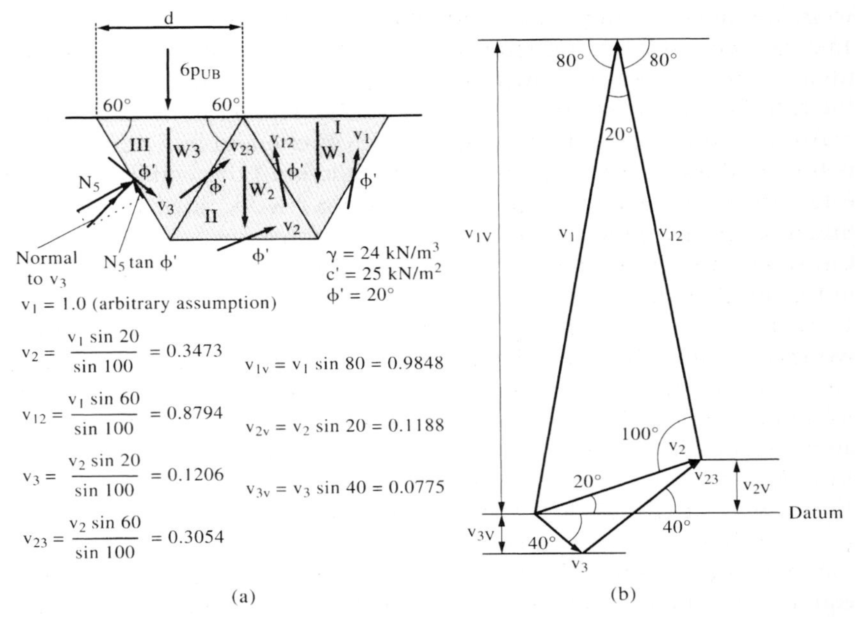

Figure 17.16 Virtual work applied to rock foundation instability: (a) foundation geometry; and (b) associated virtual displacements.

then vectorially adding the other virtual displacements triangle by triangle. Thus, following from v_1, the triangle $v_1 - v_{12} - v_2$ is completed, followed by the triangle $v_2 - v_{23} - v_3$. The displacements may be determined either by trigonometrical calculation or by drawing the polygon accurately to scale and measuring directly.

From these various virtual displacements, the virtual work can be calculated. As an aid in considering the various components of virtual work, they can usefully be assigned to one of two categories: external virtual work, EVW (due to the applied forces and the weight of the rock blocks); and internal virtual work, IVW (due to the work done by forces arising from the strength of the discontinuities). Hence we have

$$\text{EVW} = (dp_{\text{UB}} + W_3)v_{3\text{v}} + W_2 v_{2\text{V}} + W_1 v_{1\text{V}} \text{ and}$$

$$\text{IVW} = (v_1 + v_2 + v_3 + v_{12} + v_{23})c'L\cos\phi'$$

with the result that, because EVW + IVW = 0 and solving for p_{UB} with $d = 6$ and the values indicated in Fig. 17.16, $p_{\text{UB}} = 1629 \text{ kN/m}^2$.

When conducting this type of analysis one must ensure that the correct signs are maintained for the virtual displacements associated with the external forces: this applies for both the virtual displacement polygon *and* the calculation of virtual work. An inspection of Fig. 17.16 reveals that $v_{3\text{V}}$ is the only negative displacement in this example, resulting from an initial inherent assumption that displacements upwards are positive.

Continuum analysis. Studying the sustainability of a stress distribution in terms of the lower bound theorem of plasticity is mostly applicable to rock foundations under high loads and where the rock mass is effectively continuous and weak. Such circumstances will be rare, and so the simplified analysis presented here is included mainly for completeness.

With the same loading geometry as for the discontinuum analysis illustrated in Fig. 17.14, but for a mesh of square elements, consider the stresses acting on the sides of the elements to determine whether and, if so, where local plastic failure occurs according to a suitable yield criterion. In the interests of simplicity, a Mohr–Coulomb criterion with $\phi = 0$ has been used here, with the added assumption that the sides of the elements have zero cohesion.

Figure 17.17 illustrates the basic problem. The stresses acting on elements I and II can be estimated by considering the stresses resulting from the overburden and the applied load in conjunction with the yield criterion. Analysing elements I shows that the overburden stress acting at these locations—remote from the loaded area—is γz. It follows from inspection of the yield criterion that the horizontal stress cannot exceed $\gamma z + 2c$ (see inset Mohr's circle in Fig. 17.17). By inspection, we see that at element II the vertical stress due to the applied load and the overburden is greater than the horizontal stress. However, the horizontal stress has the same magnitude throughout, i.e. $\gamma z + 2c$, and hence the vertical stress acting on element II cannot exceed $\gamma z + 2c + 2c$, that is, $\gamma z + 4c$. But, because we can approximate the vertical stress acting on element II as $p + \gamma z$, it follows that $p = 4c$—which is a lower bound solution and should be compared to the result of $p = 6c$ as an upper bound solution found earlier.

In the case of a more realistic yield criterion and stress distribution, the analysis becomes much more complex. Closed form solutions exist for the simpler cases experienced in soil mechanics but, in general, numerical methods are required to produce solutions.

17.2.2 Stress distributions beneath applied loads

Two of the classic closed form solutions in stress analysis are for normal and shear line loads applied to the surface of a CHILE half-space. These are commonly attributed to Boussinesq (1883) and Cerruti (1882), respectively. We illustrate these problem geometries and key aspects of the solutions in Figs 17.18(a) and (b).

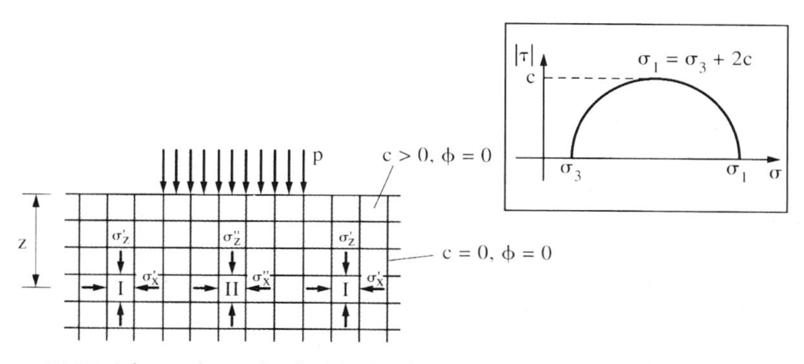

Figure 17.17 A lower bound solution for foundation collapse load with associated Mohr's circle.

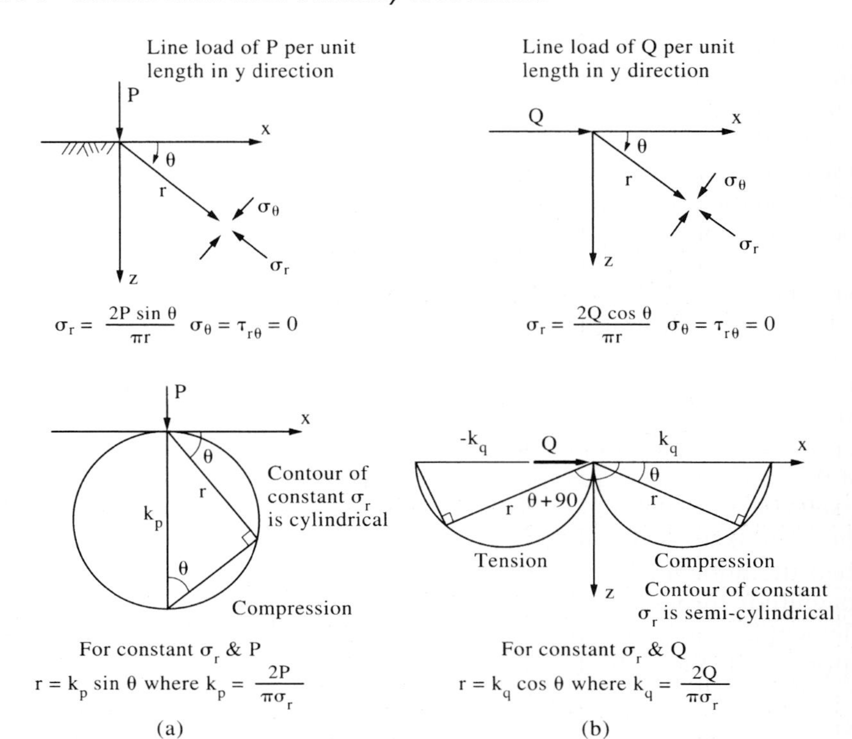

Figure 17.18 Boussinesq and Cerruti solutions for line loads on the surface of a CHILE half-space.

The interest is in the application of a line load at an arbitrary angle to the surface. This can be obtained by resolving the force into its normal and shear components and then superposing the Boussinesq and Cerruti solutions, respectively. After some algebraic manipulation, the radial stress induced in the solid can be expressed with reference to the line of action of the inclined line load as

$$\sigma_r = \frac{2R\cos\beta}{\pi r}.$$

For values of $-\pi/2 < \beta < \pi/2$, $\cos\beta$ is positive and hence the radial stress is compressive, whereas, for angles outside this range, $\cos\beta$ is negative—giving tensile radial stress.

The resulting locus of radial stress for an inclined load is shown in Fig. 17.19. The reader should verify that, in the extreme cases of $Q = 0$ or $P = 0$, the locus would be that of the Boussinesq and Cerruti solutions, respectively. This interpretation assists in the understanding of the contribution made by the normal and shear components to the inclined load solution. Note that the left lobe of the locus represents a tensile radial stress and the right lobe represents a compressive radial stress.

In applying this solution to a real rock, it would be necessary to be able to sustain the induced tensile stress in order for the solution as shown to

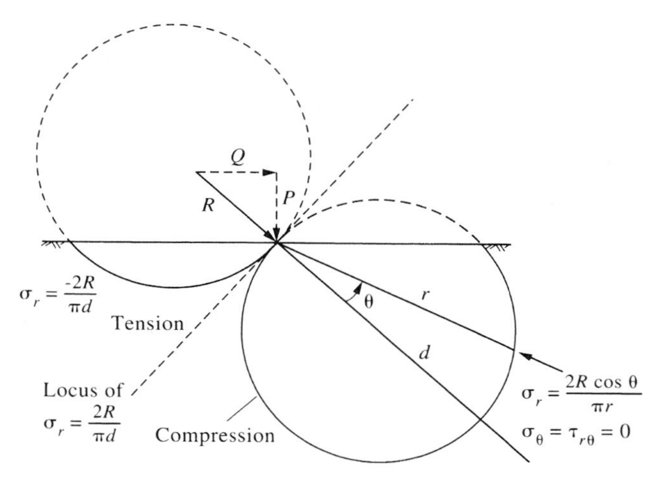

Figure 17.19 Contours of radial stress induced by line loading inclined at an arbitrary angle to the half-space surface (from Goodman, 1989).

be valid. If the rock is laminated and inter-layer slip is possible (as has been discussed previously in connection with the ϕ_j theory), then the rock may not be able to sustain the compressive and shear stresses either, even at very low magnitudes of applied loads. An idea of the regions in which inter-layer slip could occur can be obtained by applying the ϕ_j theory to a contour of constant radial stress computed using the Boussinesq solution, as developed by Goodman (1989) and as shown in Fig. 17.20.

The shape of these contours of radial stress, commonly known as 'bulbs of pressure', is then seen to be affected by the occurrence of inter-layer slip, with the result that the applied foundation load affects a greater depth of rock than in the case of a CHILE material. The modified contour is only

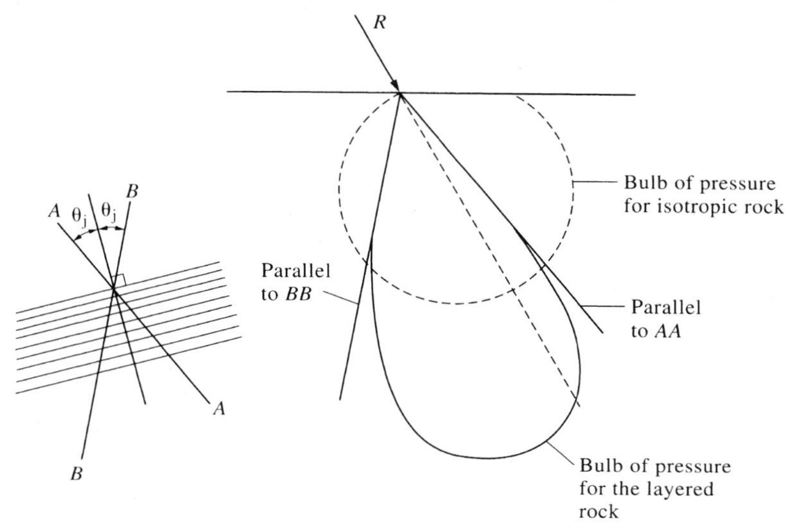

Figure 17.20 Modification of radial stress contours due to inter-layer slip (from Goodman, 1989).

approximate, because once an inter-layer slip failure criterion has been applied, the elastic Boussinesq solution itself is not valid.

In applying these ideas in practice, it is prudent to study the influence of rock anisotropy. Dr Bray developed a solution for an 'equivalent isotropic medium' for a line load inclined at an arbitrary angle to the surface. The solution is developed by considering the effect of a single set of discontintuies which have been subsumed into an equivalent transversely isotropic rock—but the solution does explicitly include the normal and shear stiffnesses and mean spacing of the discontinuities. The solution is given below and the geometry illustrated in Fig. 17.21:

$$\sigma_\theta = 0, \quad \tau_{r\theta} = 0, \quad \sigma_r = \frac{h}{\pi r}\left[\frac{X\cos\beta + Yg\sin\beta}{(\cos^2\beta - g\sin^2\beta)^2 + h^2\sin^2\beta\cos^2\beta}\right]$$

where

$$g = \left[1 + \frac{E}{(1-\nu^2)k_n\bar{x}}\right]^{1/2} \text{ and } h = \left\{\left(\frac{E}{1-\nu^2}\right)\left[\frac{2(1+\nu)}{E} + \frac{1}{k_s\bar{x}}\right] + 2\left(g - \frac{\nu}{1-\nu}\right)\right\}^{1/2}$$

and where k_n and k_s are the normal and shear discontinuity stiffnesses, respectively, and $\bar{x}$ is the mean discontinuity spacing.

The resulting contours of radial stress for an equivalent isotropic medium with the plane of anisotropy at various angles to the surface of the half-space are shown in Fig. 17.22 (note that the forms of these contours will vary with the exact values of all the elastic constants, including the normal and shear discontinuity stiffnesses). Experimental data produced by Gaziev and Erlikhman (1971) are shown in Fig. 17.23 for comparative purposes.

The significance of Figs 17.22 and 17.23 is clear: contours of radial stress can be deeper than those predicted with a CHILE solution; and they can be severely distorted, so that they are not only extended downwards

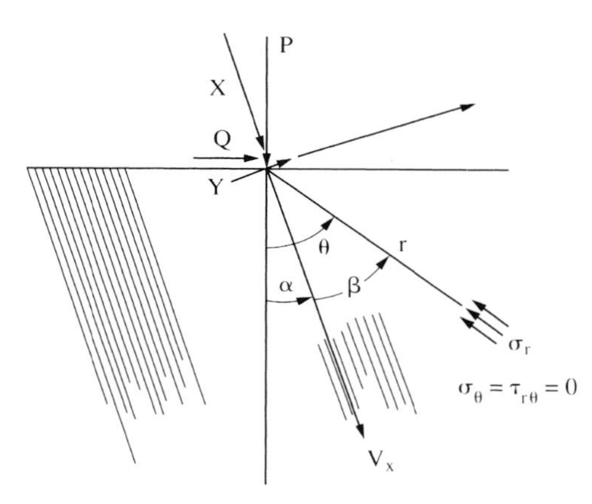

Figure 17.21 Geometry of Bray's equivalent continuum solution (from Goodman, 1989).

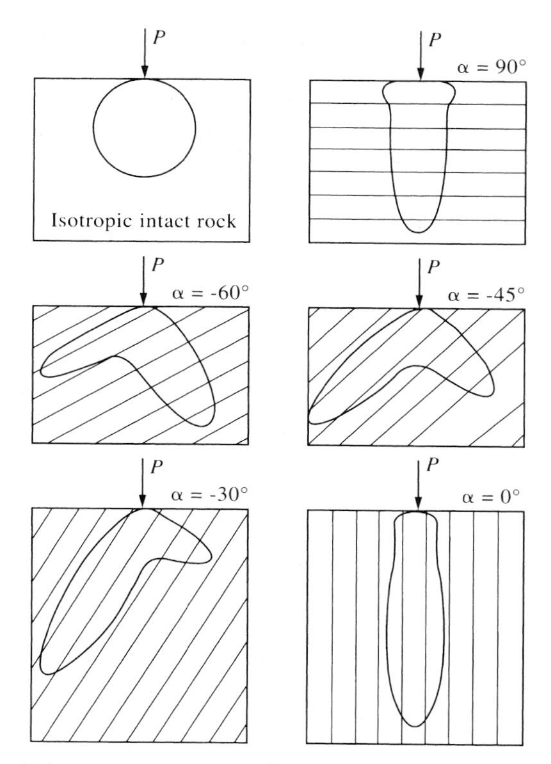

Figure 17.22 Radial stress contours produced using Bray's solution for an equivalent anisotropic medium, with the plane of anisotropy at angles as shown (from Goodman, 1989).

but also significantly sideways. Thus an understanding, even only in terms of qualitative trends, of the extent of the stress distribution within the rock will lead to the design of more appropriate site investigation procedures—because the effect of any proximate ground weaknesses must be evaluated.

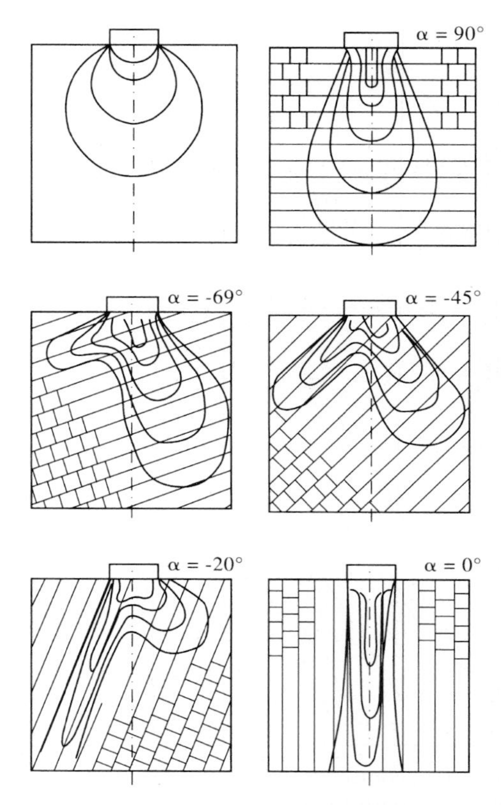

Figure 17.23 Model test data for loading on the surface of an artificial anisotropic material (after Gaziev and Erlikhman, 1971).

18 Design and analysis of surface excavations

In the last chapter, some of the idealized models were presented that have been developed to provide a basic grasp of the instability mechanisms associated with slopes and foundations in rock. Although the models are highly simplified, they do assist understanding and indicate the important parameters, together with their sensitivity. When faced with the design and analysis of an actual surface excavation, one has to go beyond these primary models and match the model with the site reality and the nature of the rock.

Thus we should ask what is the exact design objective, what mechanisms are likely to be operating, which data are required as a consequence, and does the model incorporate the discontinuous, inhomogeneous, anisotropic and non-elastic behaviour of the ground, together with factors such as the effects of blasting, rainfall, seismic risk and so on? The approach should therefore be to examine the potential for instability mechanisms and to gradually refine the design and analysis, from an initial skeletal approach through to a comprehensive finely-tuned design. In this book we are discussing principles and hence the techniques which may be applied during the initial approach to all projects. Any further development should be tailored to the rock, site and project circumstances—in the knowledge that the initial analyses have been performed thoroughly.

18.1 Kinematic analysis of slope instability mechanisms

One of the best examples of an initial approach is establishing the possibility of instability by the method of kinematic analysis of slopes. 'Kinematics' refers to the study of movement, without reference to the forces that produce it. For some geometries of slope and discontinuities, movement is possible (i.e. the system is kinematically feasible). For other geometries, movement is not possible (i.e. the system is kinematically infeasible).

A method based on checking the kinematic feasibility of a rock slope–discontinuity system will provide a 'first pass' analysis, although kinematic feasibility checks are but the first in a long line of design and analysis tools.

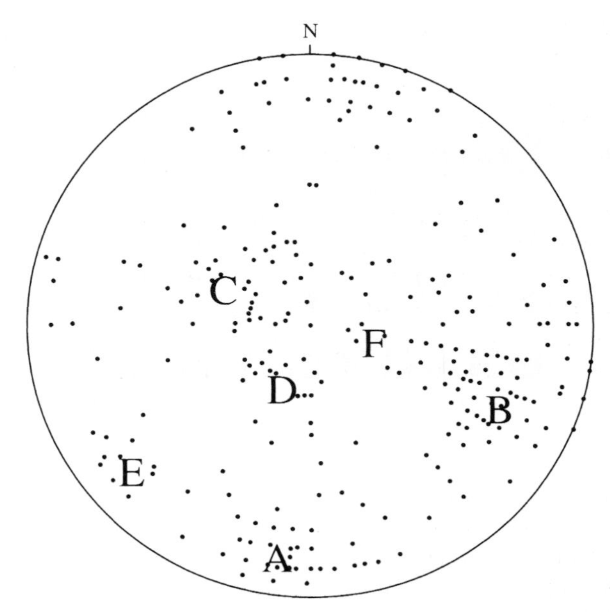

Figure 18.1 Pole plot of discontinuities in the rock mass under consideration (case example data from Matheson, 1983).

They do not provide a numerical measure of the degree of safety of the slope, but whether or not instability is feasible in the first instance. If the system is kinematically infeasible, a great deal has been established rapidly.

The kinematic analysis of plane, wedge and toppling instabilities for rock slopes is explained next—in which the instability is governed by the geometry of the slope and discontinuities. The method follows that presented by Hoek and Bray (1977), and refined by Matheson (1983) and Goodman (1989). The examples presented in the text use a data set based on field records, recorded by Matheson.

In Fig. 18.1, the lower-hemispherical projection of the poles to the discontinuities in the rock mass is shown. The second Appendix covers the basics of hemispherical projection. The initial impression is that there are two main sets of sub-vertical discontinuities, one (set A) striking approximately E–W, another (set B) striking approximately N–S. There are four minor sets, some (sets C, D and F) being sub-horizontal, one (set E) being sub-vertical striking NW–SE.

If necessary, we can return to these data to consider the dispersion of the poles within each set and the different strength parameters associated with each set. Firstly, however, consider the kinematic feasibility associated with constructing a proposed slope of dip direction 295° and dip angle 75°, assuming that all discontinuity sets follow a Mohr–Coulomb strength criterion with $\phi = 30°$ and $c = 0$ kPa.

18.1.1 Plane instability

To consider the kinematic feasibility of plane instability, four necessary but simple criteria are introduced, as listed below.

(a) **The dip of the slope must exceed the dip of the potential slip plane** in order that the appropriate conditions for the formation of discrete rock blocks exist.
(b) **The potential slip plane must daylight on the slope plane.** This is necessary for a discrete rock block formed by criterion (a) to be capable of movement.
(c) **The dip of the potential slip plane must be such that the strength of the plane is reached.** In the case of a friction-only plane, this means that the dip of the plane must exceed the angle of friction.
(d) **The dip direction of the sliding plane should lie within approximately ±20° of the dip direction of the slope.** This is an empirical criterion and results from the observation that plane slides tend to occur when the released blocks slide more-or-less directly out of the face, rather than very obliquely.

In Figs 18.2(a) and (b), the generation of hemispherical projection instability overlays based on the criteria above is shown. These will be used over a plot such as Fig. 18.1. There can be uncertainty about the directions on these overlays, and so it is important to understand the location of a slope in plan and the associated directions of the slope in these overlays, together with the kinematic criteria.

Each family of lines or curves in the overlay of Fig. 18.2(a) represents one of the criteria listed above. The radial solid line pointing to the left is taken to be the slope direction. (*Note that if the perimeter of the projection represented the plan of a circular projection, then the* **location** *on the crest of a slope dipping in this direction would be diametrically opposite, i.e. on the right-hand side of the perimeter*). The two radial dashed lines to the right represent criterion (d), and serve to concentrate the search for instability within a region ±20° of the slope direction. Note that this overlay is to be used with *pole plots*. Therefore, concentric circular arcs within the sector—which represent criteria (a)–(c), the dips of the slope and the potential plane of sliding—are *numbered away from the centre of the overlay* and so provide the remaining bounding lines of the region of instability.

Figure 18.2(b) shows the completed specific overlay for a slope dip of 75° and an angle of friction of 30°. The innermost bounding arc is the friction angle (criterion (c)) and the outermost bounding arc is the slope angle (criterion (a)). Because pole plots are being used, the region of instability on the overlay is on the opposite side to the dip direction of the slope.

The final step in assessing the kinematic feasibility for plane instability is to superimpose the specific overlay (in this case Fig. 18.2(b)) onto the projection representing the rock mass discontinuity data (in this case Fig. 18.1). The result for this example is shown in Fig. 18.3.

The advantage of the overlay technique is immediately apparent. We can say directly that there is a severe potential for plane instability associated with discontinuity set B. Plane instability cannot occur on any other discontinuity set. The exact value of the innermost bound of the instability region, i.e. the friction angle, is not critical in the analysis—any variation between, say, 30° and 50° will not prevent instability. The dip of the slope

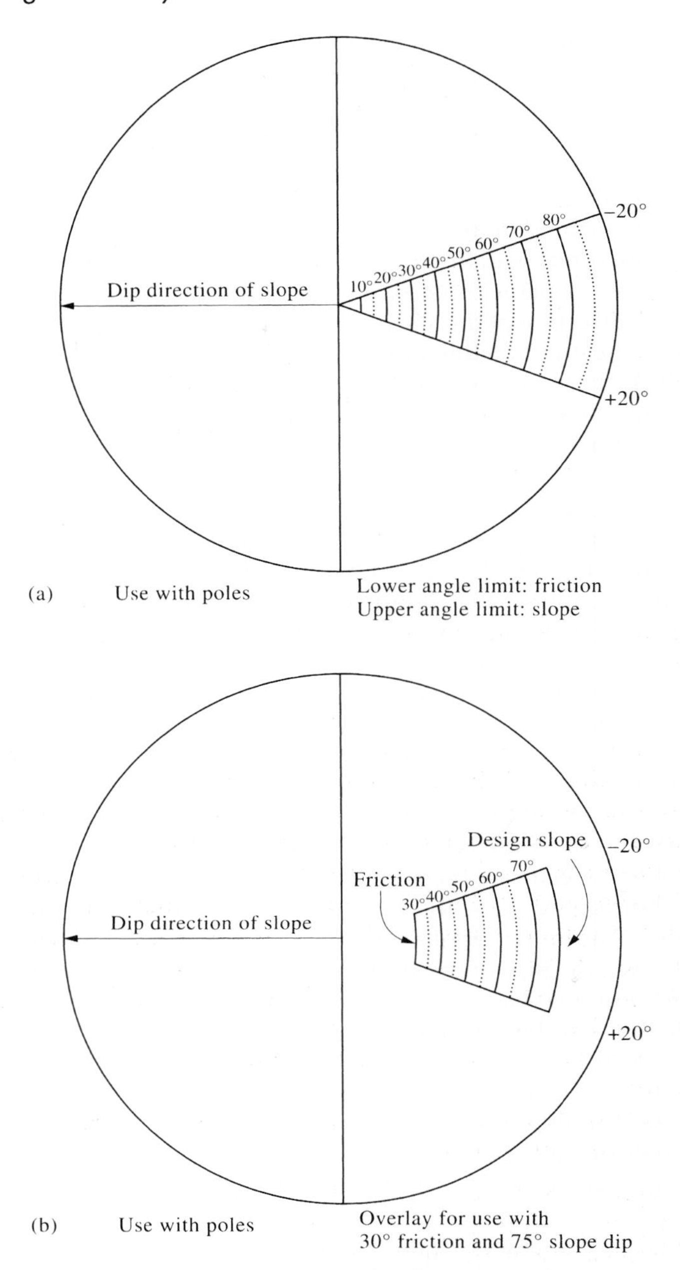

Figure 18.2 Construction of instability overlay for plane sliding.

is critical: any major increase or decrease in the dip will significantly alter the degree of instability, because slope dip angles around this value coincide with the dip of the majority of discontinuities in set B. Finally, the orientation of the slope itself is critical: were we to be able to alter the dip direction of the slope by ±30°, the potential for plane instability would be reduced considerably.

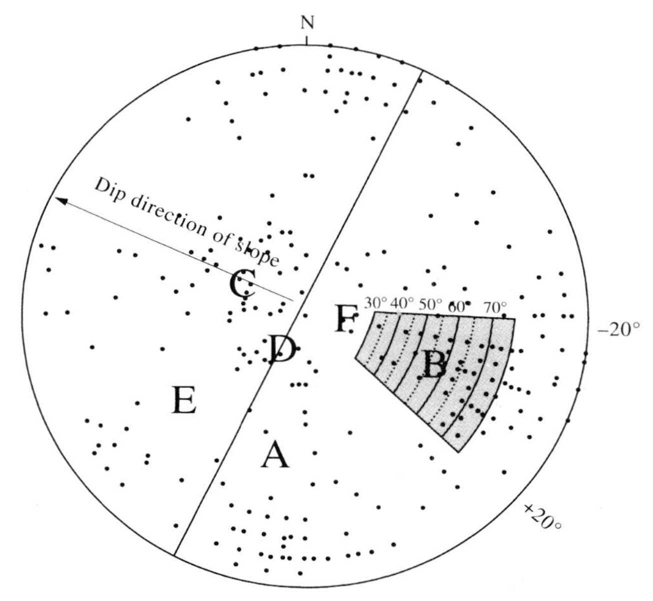

Figure 18.3 Example assessment for a slope of orientation 295°/75°—plane instability.

Armed with this information, it is necessary to check that the data for discontinuity set B have been correctly recorded. Then, can the dip direction and dip angle of the slope be altered? If so, the potential for instability can be eliminated by re-orientating the slope. If not, further analysis has to be conducted to decide on the optimal excavation and support techniques. In the event of detailed numerical analyses being required, the variation associated with the orientation of the individual discontinuities within a set must always be considered. One might be tempted to try to produce one factor of safety using a single discontinuity orientation but, as the overlay method clearly demonstrates, the variation within a set means that such an approach is meaningless without qualification.

There is also the interaction between the design process and the rock mass. There are three options:

- the design can be altered to account for the rock mass;
- a fixed design can be applied to the fixed rock mass and the structure engineered accordingly; and
- the rock mass can be altered to account for the design.

The power of the kinematic feasibility overlays lies in an 'immediate appreciation' of the primary parameters and design possibilities.

18.1.2 Wedge instability

An overlay for wedge instability potential can be constructed by the same method used for plane instability. Wedge instability can be considered as a variation of plane instability, in that the sliding takes place on two discontinuity surfaces, as shown in Figs 17.2(c) and 17.3(c). The resultant sliding direction is assumed to be in a direction common to both surfaces, i.e. along their line of intersection.

To consider the kinematic feasibility of wedge instability, we therefore need to consider only three criteria relating to the line of intersection, as listed below. The plane instability criterion relating to the ±20° variation in sliding direction is no longer required, as the sliding direction is uniquely defined by the line of intersection.

(a) **The dip of the slope must exceed the dip of the line of intersection of the two discontinuity planes associated with the potentially unstable wedge** in order that the appropriate conditions for the formation of discrete rock wedges exist, in a similar fashion to criterion (a) of plane instability.
(b) **The line of intersection of the two discontinuity planes associated with the potentially unstable wedge must daylight on the slope plane.** This is necessary for a discrete rock wedge formed by the first criterion to be capable of movement.
(c) **The dip of the line of intersection of the two discontinuity planes associated with the potentially unstable wedge must be such that the strengths of the two planes are reached.** In the case of friction-only planes, each possessing the same angle of friction, the dip of the line of intersection must exceed the angle of friction.

In an analogous fashion to the analysis of plane instability, in Figs 18.4(a) and (b), the generation of the hemispherical projection instability overlays based on the criteria above is shown.

The radial solid line at the right but pointing to the left is taken to be the slope direction. (*Note that, as before, if the perimeter of the projection represented the plan of a circular projection, then the* **location** *of the crest of a slope dipping in this direction would be on the right-hand side of the perimeter.*) However, because we are analysing lines of intersection, this overlay is to be used with *intersection plots* and consequently the construction to locate the region of instability will be on the same side of the projection as the slope dip.

Thus, criterion (a) is implemented using the series of great circles (because the slope is a plane, and planes are plotted as great circles) and criterion (c) is implemented by the series of concentric circles (because lines of equal dip form a concentric circle). Because this is a direct plot of dips and dip directions, the dips of the slope and the line of intersection are *numbered towards the centre of the overlay*. Because **intersection** plots are being used, the region of instability on the overlay is on the **same** side as the considered dip direction of the slope.

Note the large size of the region of instability developed on the projection—often covering a range of dip directions as large as 150°. This means that attempting to vary the slope orientation as a means of reducing instability is not likely to be as effective as in the case of plane instability.

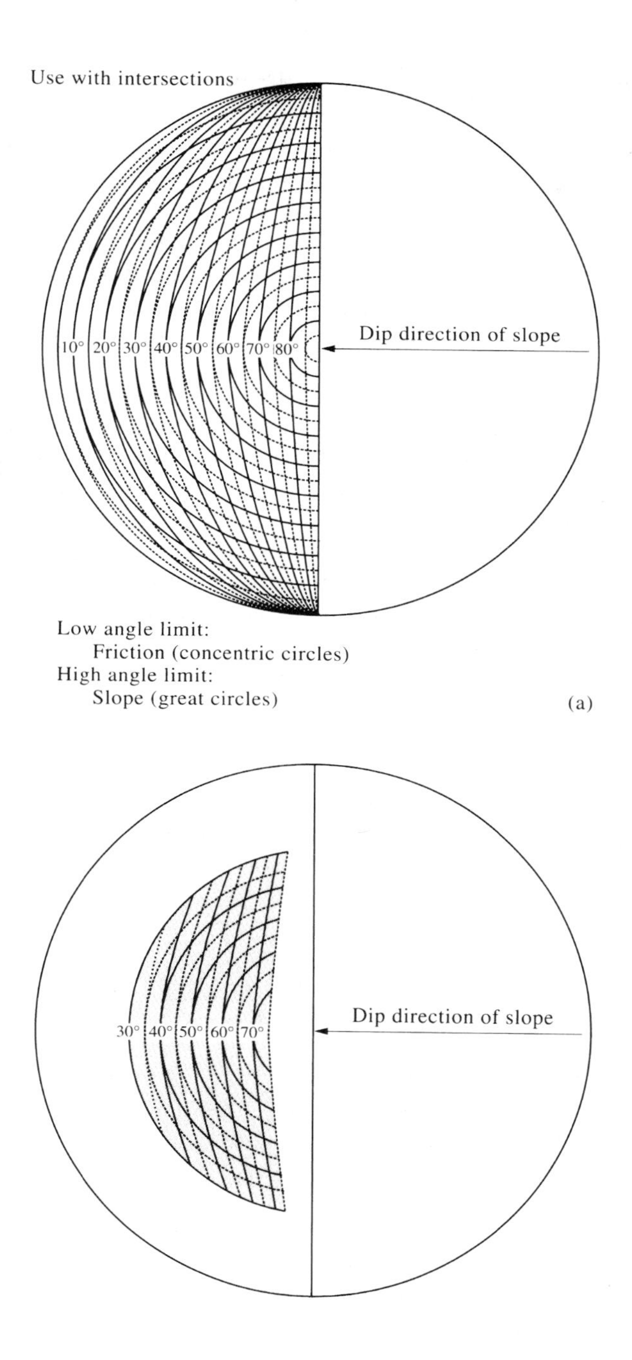

Figure 18.4 Construction of instability overlay for wedge sliding.

Figure 18.4(b) shows the completed specific crescent-shaped overlay for a slope dip of 75° and an angle of friction of 30°. In the design process, it will probably be the innermost boundary of the crescent which is the most variable—i.e. how steep the slope can be without wedge instability occurring.

The final step in assessing the kinematic feasibility for wedge instability is to superimpose the specific overlay (in this case Fig. 18.4(b)) onto a projection representing all the intersection possibilities for the rock mass discontinuity data. This is usually achieved by considering one representative plane from each discontinuity set and determining the set intersections. A more accurate method would be to determine the intersections resulting from all inter-set combinations of discontinuities and treat these as a set of intersections. The result for this example, using the former method, is shown in Fig. 18.5.

Once more, the advantages of the overlay technique are apparent. First, there are only two lines of intersection along which wedges are potentially unstable—these are formed by the intersection between discontinuity set B and sets A and E. Again, the exact value of the angle of friction (i.e. the position of the outermost boundary of the crescent) is unimportant, but the slope angle itself is paramount. By reducing the slope angle, and hence moving the innermost boundary of the crescent away from the centre of the projection, wedge instability can be minimized. Returning to the field, one can visually assess the nature of the lines of intersection I_{AB} and I_{BE} to establish the shape and size of the wedges.

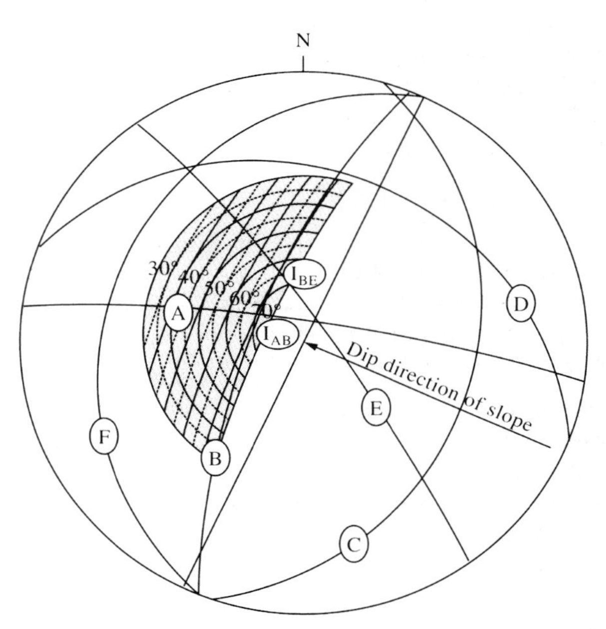

Figure 18.5 Example assessment for a slope of orientation 295°/75°—wedge instability.

From the pole plot of the discontinuities shown in Fig. 18.1, a basically orthogonally fractured rock mass was indicated, with the result that the intersections would be expected to be sub-vertical and sub-horizontal. Thus, wedge instability problems are only likely to arise for steep slopes or low angles of friction—as indicated in Fig. 18.5. Note, though, that an essentially orthogonally fractured rock mass which has undergone a tilting of only 30° or so will give rise to problems of wedge instability.

What are the implications of having a different friction angle on the two discontinuity planes forming the wedge? Utilizing a 'generalized friction circle', Goodman (1989) presents a method of analysing wedge instability with different friction angles. He notes that "in view of the uncertainty with which friction angles are assigned in practice, it is more useful to express the degree of stability in terms of such a sensitivity study [referring to his stereographic method] than to force it to respect the factor of safety concept". So, although using a different method, Goodman is also of the opinion that an appreciation of the problem is more important than a rigid adherence to the factor of safety concept, as stated at the end of Section 18.1.1.

18.1.3 Toppling instability

For the third mode of instability, toppling, both direct toppling and flexural toppling as illustrated in Fig. 17.9, will be studied. The same overlay technique that was presented for plane and wedge instability can be used, except that there is the need to analyse intersections (defining the edges of toppling blocks) and poles (defining the basal plane about which toppling takes place). An overlay is required which makes use of both pole and intersection plots, as a result of the feasibility criteria associated with toppling. It is also important to note (with reference to Fig. 17.10) that toppling instability is being considered in isolation. Plane and wedge instability, which may or may not be occurring contemporaneously, can be established from the instability analyses already presented.

Direct toppling instability. In the case of direct toppling instability, the kinematic feasibility criteria will only relate to the geometry of the rock mass, rather than the geometry plus the strength parameters—although the latter can be used to establish the cut-off between toppling only and sliding plus toppling illustrated in Fig. 17.10. Therefore, the only two criteria required are as follows (see Fig. 17.2(d)).

(a) **There are two sets of discontinuity planes whose intersections dip into the slope,** in order to provide the appropriate conditions for the formation of the faces of rock blocks.
(b) **There is a set of discontinuity planes to form the bases of the toppling blocks,** so that, in association with criterion (a), complete rock blocks may be formed.

Naturally, toppling is more likely if the basal planes dip out of the slope, but such a condition is not necessary. If the dip of the basal planes is less than the friction angle, then sliding will not occur in association with toppling.

From criterion (a) an overlay for an intersection plot is required; from criterion (b) an overlay for a pole plot is required. For this the intersection and pole plots are superimposed and a composite overlay is used.

In an analogous fashion to the previous analyses of plane and wedge instability, in Figs 18.6(a) and (b) the generation of the hemispherical projection instability overlays based on the criteria above is shown.

In Fig. 18.6(a), the radial solid line pointing to the left is again taken to be the slope direction. Because the interest is in the angles between the vertical and both the plunge of the lines of intersection (criterion (a) above) and the dip of the basal plane (criterion (b) above), the overlay will consist only of concentric circles. The concentric circles are numbered from the perimeter inwards for the intersections, and from the centre outwards for the poles. (*Because the intersection lines are dipping* **into** *the slope, whereas the basal planes are dipping* **out of** *the slope, the overlay criteria are both on the same side of the composite generic overlay shown in Fig. 18.6(a)—and on the opposite side to the direction of the slope dip.*) The two dashed radial lines represent a 'sub-criterion', in that observations have indicated that toppling tends to occur within a ±20° sector of the slope dip, except for very steep slopes where the sector can be considerably enlarged.

Given the criteria, the necessary bounds can be drawn and the overlay produced. Figure 18.6(b) shows the overlay for this example. There are many instability regions associated with a direct toppling overlay, depending on the combinations of the occurrences of overlaid poles and intersections. Figure 18.7 clarifies these possibilities. The upper suite of sketches refers to the basal plane occurrences; the lower suite of sketches refers to the intersection occurrences. In this sense, the occurrence of direct toppling instability is not so sharply focused as with the previous two overlays, but again illustrates the value of this approach.

The many modes of toppling instability can be established from the sketches in Fig. 18.7 and any specific example can be interpreted with the aid of the overlay technique. Moreover, once a potential mode has been established from the analysis, the engineer can return to the field and consider the mechanism *in situ*. This provides a powerful technique for establishing the real likelihood of instability: attempting to establish the direct toppling modes without such a visual and integrating analysis would be most unsatisfactory.

To assess the kinematic feasibility for direct toppling instability, the specific overlay (in this case, Fig. 18.6(b)) is superimposed onto a composite projection representing all the inter-set intersections and all the poles for the rock mass discontinuity data (in this case, the data shown in Fig. 18.1). This is shown in Fig. 18.8.

It can be seen that the potential for toppling is not high. The main possibility is for set F to form the basal plane, and the block edges to be formed by intersection I_{AE}—a typical example of the need to return to the field and assess the mechanism visually. The circumstances are akin to a combination of the two left-hand sketches in Fig. 18.7, with oblique toppling occurring due to the intersections not falling within the main region of instability. In the specific example shown in Fig. 18.8, the potential toppling direction is diagonally southwards across the slope.

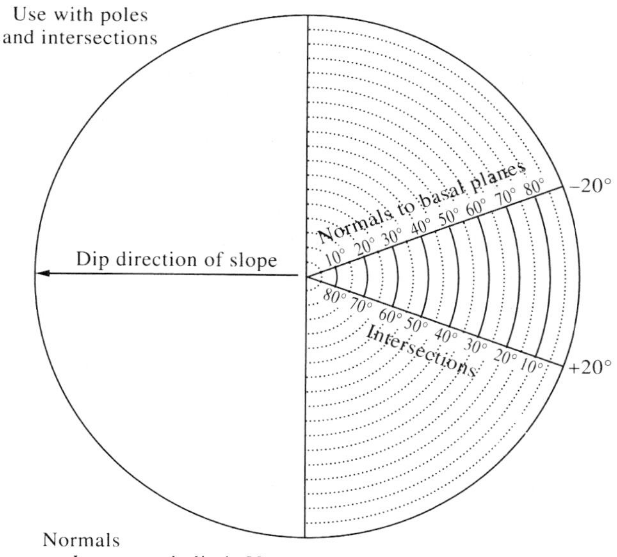

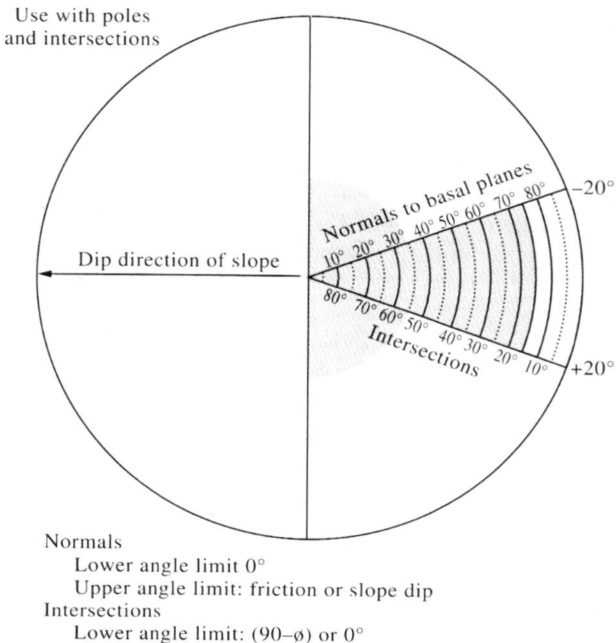

Figure 18.6 Construction of instability overlay for direct toppling.

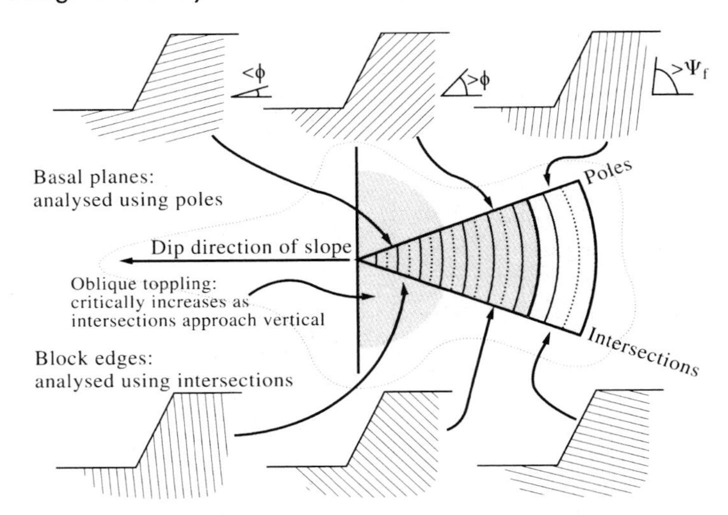

Figure 18.7 Illustration of the direct toppling instability modes.

An interesting aspect of this analysis is that the slope angle is not explicitly used. The direction of the slope serves to indicate in which area of the projection one needs to look for poles and intersections as potential candidates for inducing instability. Similarly, although not shown in Fig. 18.8, it must be remembered that those basal planes whose dips lie outside the main region of instability must also be considered as candidates for defining toppling blocks. It is quite likely that oblique blocks formed by basal planes dipping *into* the slope will topple.

Flexural toppling instability. In Figure 17.11, the analysis of flexural toppling was illustrated using the ϕ_j theory. The geometrical analysis and the associated criteria for inter-layer slip can be used to form the basis of a kinematic feasibility analysis for flexural toppling using the overlay method. In fact, the angles and the criterion are clarified by the use of such a method. From Section 17.1.4, the geometrical criterion for inter-layer slip to occur is $\psi \geq 90 + \phi - \beta$, where ψ is the dip of the slope, ϕ is the friction angle associated with the discontinuities and β is the dip of these discontinuities.

The criterion was expressed in this way because we were interested in the slope angle at which inter-layer slip could occur. In terms of the projection overlay, we wish to know the positions of the discontinuity poles on the projection which would indicate the potential for inter-layer slip. Thus, the criterion above can be rewritten as $\beta \geq \phi + (90 - \psi)$. This allows not only the creation of the instability overlay for flexural toppling, but also identification of the various components of the criterion on the projection.

The criteria are as follows.

(a) **There is one set of discontinuity planes dipping into the slope,** at a sufficiently high angle to generate inter-layer slip, following the criterion above.

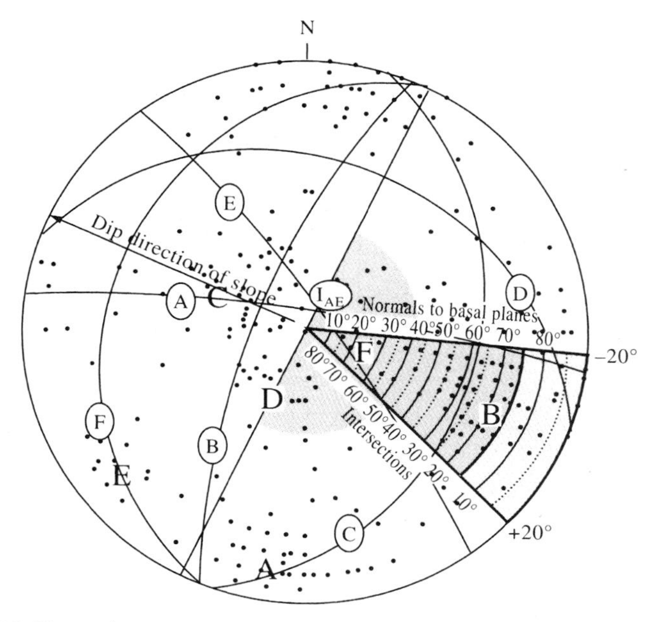

Figure 18.8 Example assessment for a slope of orientation 295°/75°—direct toppling instability.

(b) **The dip direction of the slip planes should lie within approximately ±20° of the slope.** As with plane instability, this is an empirical criterion and results from the observation that inter-layer slip tends not to occur when discontinuities occur obliquely to the slope.

From criterion (a), an overlay is required which is constructed from great circles (representing the plane of the slope) and yet is used with a pole plot projection (to establish regions of instability associated with the dip of the discontinuity planes).

In Figs 18.9(a) and (b), we illustrate the construction of the generic flexural toppling overlay, together with the specific overlay for these example data. In Fig. 18.9(a), the radial solid line directed to the left is again taken to be the slope direction and the great circles represent planes corresponding to both the slope and the friction angle of the slipping discontinuity planes. Location of the region of instability is best understood from an analysis of Fig. 18.9(b). The dip angle of the dotted great circle in Fig. 18.9(b)—representing the slope—is ψ, and the complement of this angle (i.e. the angle to the vertical) is $90 - \psi$. Inter-layer slip will only occur for discontinuities dipping at an angle of ϕ greater than this (the geometrical criterion above), giving a region of instability outside the solid great circle. Finally, using the second criterion above, we produce the shaded instability region—for superimposition on pole plots.

Compare the construction of this overlay with that of Fig. 18.2 (plane instability), and note that, although both overlays are to be superimposed on pole plots and the direction of the slope dip relative to the overlay

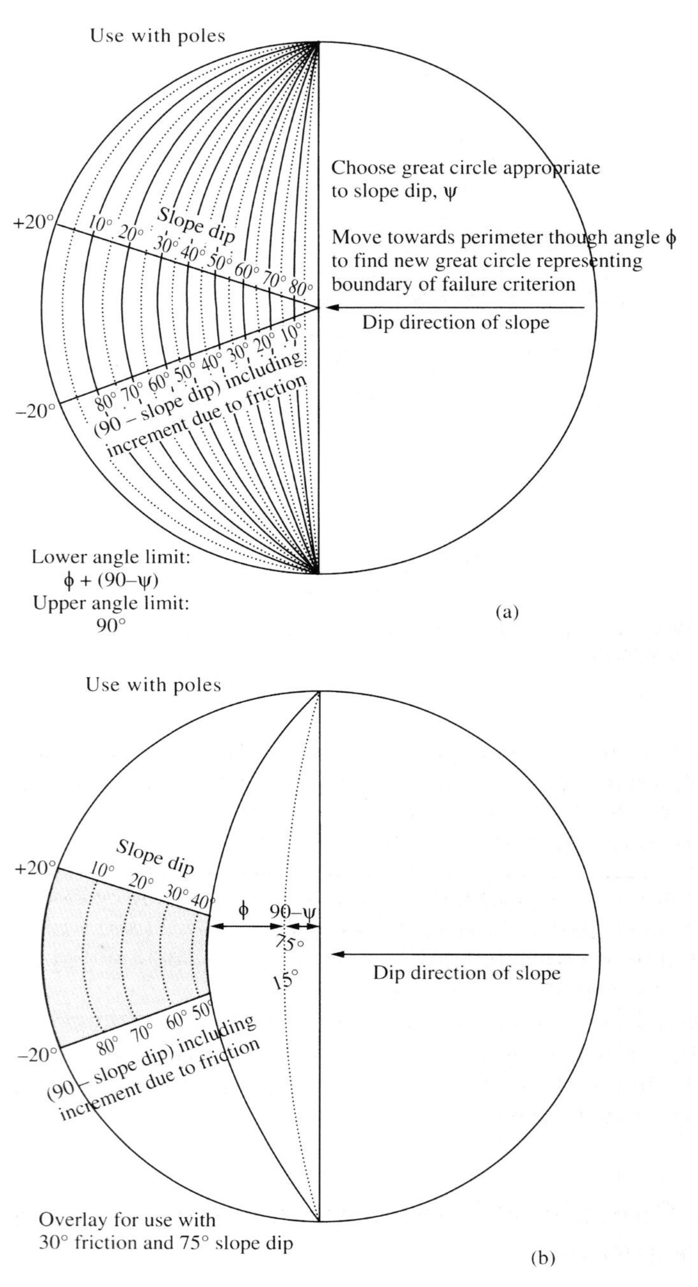

Figure 18.9 Construction of instability overlay for flexural toppling.

construction is the same, the location of the overlay is different in the two cases. This is because the discontinuities dip in the same direction as the slope for plane instability, but into the slope for flexural toppling instability.

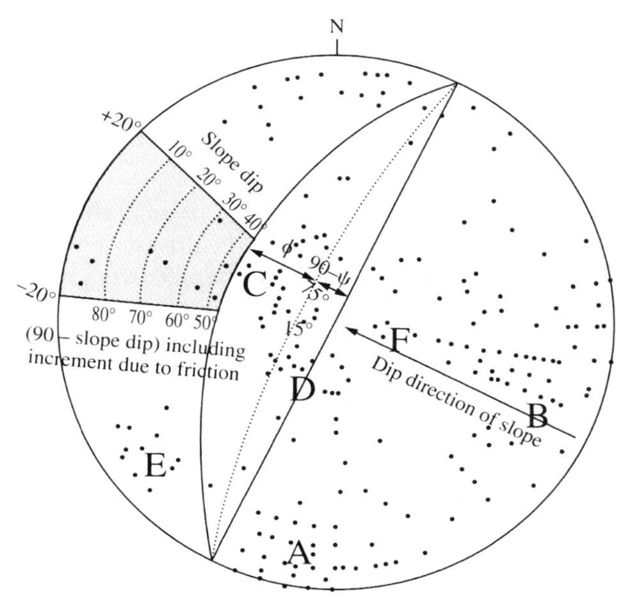

Figure 18.10 Example assessment for a slope of orientation 295°/75°—flexural toppling instability.

Thus, to assess the kinematic feasibility for direct toppling instability, we superimpose the specific overlay (in this case, Fig. 18.9(b)) onto a projection of the poles for the rock mass discontinuity data (in this case, the data shown in Fig. 18.1), with the result shown in Fig. 18.10.

It can be seen that the potential for flexural toppling is low, as the region of instability coincides with the limit of the cluster associated with discontinuity sets C (and B, bearing in mind that some of the discontinuities associated with this set appear within the region of instability). As before with the other instability mechanisms, however, we would wish to identify the precise nature of this geometry in the field to ensure that, indeed, the possibility of such an instability mechanism was low—e.g. are the relevant discontinuities sufficiently persistent, or are they a minor impersistent set with short trace lengths?

18.2 Combined kinematic analysis of complete excavations

When considering a proposed surface excavation in a rock mass, the kinematic feasibility of all of the four mechanisms described in Section 18.1 must be established, and for all potential slope orientations. In some projects, the slope dip direction may be dictated by considerations other than rock mechanics, e.g. a fixed highway route requiring cuttings. Even the slope dip may be fixed, but the rock engineer will be able to make a contribution to optimizing the stability of the slope. In other projects, such as an open-pit mine or quarry, *all* slope dip directions may have to be evaluated by an engineer who is expected to recommend the various slope dip values to ensure stability. In this case, we adopt an approach whereby all of the overlays in turn are rotated around the complete perimeter of the projection, indicating the regions of kinematic feasibility associated with the different mechanisms.

Because discontinuities occur in sets, the analysis leads to identification of slope orientations which are kinematically **infeasible** and other orientations which are kinematically **feasible for the different mechanisms.** The orientations then have to be considered together. We usually find that there are ranges of slope dip directions where steep (or even vertical) slopes are safe and other ranges which are susceptible to one or more mechanisms of instability. It must be remembered that the hemispherical projection technique *only utilizes orientations and not locations.* It follows that if the rock mass shows any degree of inhomogeneity, it may be necessary to consider the rock mass in 'structurally homogeneous' domains, each of which is analysed separately.

The results of this type of complete analysis—performed using the data in Figure 18.1—are shown in Fig. 18.11, in which the steepest slope dips preventing the development of the relevant instability mechanism are tabulated, together with the net result which is the steepest safe slope. The associated diagram shows a plan of how the crest of an excavation with a circular floor would appear.

Slope of direction	Plane sliding		Wedge sliding		Flexural toppling		Direct toppling		Overall
	maximum slope angle	discontinuity set	maximum slope angle	intersection of sets	maximum slope angle	discontinuity set	maximum slope angle	intersection of sets	maximum slope angle
000	25	D	62	BE	–	–	–	–	25
015	25	D	65	BE	–	–	–	–	25
030	25	D	68	BE	–	–	–	–	25
045	25	D	75	BE	–	–	–	–	25
060	80	E	80	AE	–	–	–	–	80
075	90	–	80	AE	–	–	–	–	80
090	27	C	83	AE	–	–	–	–	27
105	27	C	83	AE	52	B	68	AB	27
120	27	C	90	–	52	B	68	AB	27
135	90	–	90	–	72	F	–	–	72
150	90	–	90	–	–		–	–	90
165	90	–	90	–	–		–	–	90
180	90	–	90	–	38	A	–	–	38
195	90	–	90	–	38	A	–	–	38
210	90	–	90	–	75	D	–	–	75
225	90	–	85	AB	40	E	–	–	40
240	90	–	80	AB	40	E	–	–	40
255	90	–	76	AB	–	–	–	–	76
270	20	F	73	AB	64	C	–	–	20
285	20	F	71	AB	64	C	–	–	20
300	20	F	68	BE	64	C	–	–	20
315	20	F	63	BE	–	–	–	–	20
330	90	–	62	BE	–	–	–	–	62
345	90	–	61	BE	–	–	–	–	61

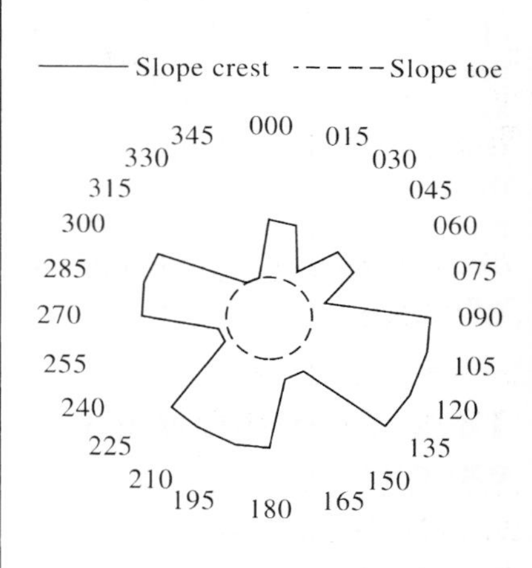

Figure 18.11 Example assessment for a circular floored excavation—plane, wedge and direct and flexural toppling instabilities.

The analysis indicates that, were a circular (in plan) quarry to be operating in this rock mass, then the absence of potential instability mechanisms for slopes with dip directions in the range 150–165° indicates that they could be steep, but slopes with dip directions in the ranges 90–120° and 180–240° would be vulnerable to instability unless cut to shallow dip angles. Is this acceptable, or is an alternative solution required? One such alternative is to avoid creating slopes within these ranges of dip directions. A generalized corollary of the example is that circular excavations can never be optimal in terms of maximizing slope dip: an elliptical or irregular polygonal geometry will always be better, in that these allow the flexibility necessary to harmonize the engineering geometry with the rock structure geometry. A correctly orientated elliptical floor plan will always be better than a circular plan for a quarry based on these slope instability considerations.

The entire analysis is based on the simple criteria established for each of the instability mechanisms. Further analysis is required to confirm that the failure mechanisms are likely to be operative. The strength of the technique lies in its underlying philosophy utilizing primary instability criteria. With the technique, it is possible to design a stable excavation without recourse to mathematical analysis and subsequent interpretation of factors of safety.

18.3 Foundations: stress distributions beneath variably loaded areas

We have extended the slope mechanisms approach of Chapter 17 to the study of the kinematic feasibility of four different potential slope instability mechanisms. By analogy, we now extend the earlier consideration of the stress distributions beneath point loads for foundations to the stress distributions that occur beneath variably loaded areas, i.e. considering the more realistic circumstances. In the next section of this chapter, we consider other factors as they relate to both slopes and foundations.

18.3.1 Cartesian form of the Boussinesq and Cerruti solutions

In Section 17.2.2, the cylindrical polar form of the solutions was given for the stress distributions associated with single normal and shear point loads on the surface of an infinite CHILE half-space, due respectively to Boussinesq and Cerruti. In order to give these solutions greater utility in the case of loaded areas and varying loads, it is helpful to first express them in Cartesian form so that loaded areas can be discretized as elemental components, each of a given magnitude, and then compute the total solution by integration of the components over the area in question.

Poulos and Davis (1974) provide the solutions for various Cartesian components of stress and displacement in a form similar to those tabulated in Fig. 18.12. Given that these are available, and knowing from the theory of elasticity that the solutions for two or more separate loadings

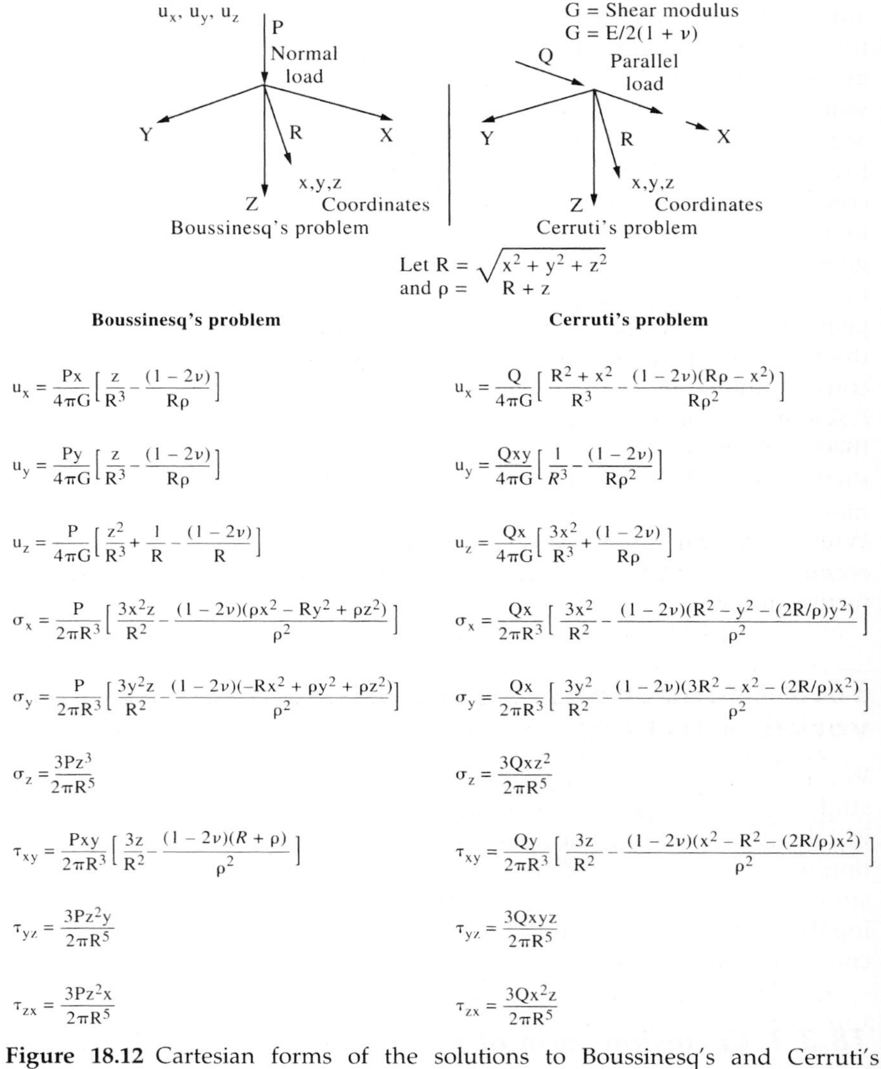

Figure 18.12 Cartesian forms of the solutions to Boussinesq's and Cerruti's problems.

can be superimposed, the stresses and displacements associated with any loading of the surface can be estimated. It is only necessary to be able to discretize the load into suitable component areas over which any normal and shear stresses acting can be considered to be uniform, as illustrated in Fig. 18.13.

18.3.2 Analytical integration over loaded areas

Here we consider only the cases of stresses and displacements in the z-direction, for the Boussinesq solution, to demonstrate the principle of

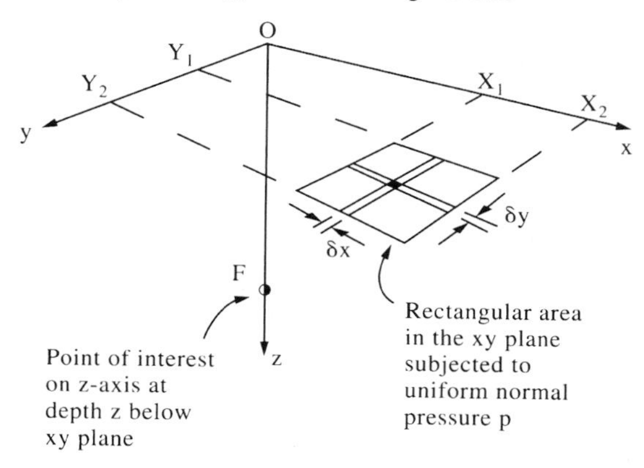

Figure 18.13 Integration of Boussinesq and Cerruti solutions over each component element of a loaded area.

determining **influence functions** for the component areas. This technique applies, with suitable variation, to all of the components of stress and displacement.

In Figure 18.13, the area bounded by X_1, X_2, Y_1 and Y_2 is assumed to be loaded with a uniform normal stress, p, and we wish to consider the consequential stress component σ_z and displacement component u_z at the point F at depth z below the surface of the half-space. These are found by integrating the relevant expressions given in Fig. 18.12 over the loaded area. Considering a small element $dx - dy$, as shown in Fig. 18.13, the equivalent point load is $P = p\delta x \times \delta y$ and thus the relevant expression for the stress component induced by this infinitesimal element is:

$$\sigma_z = \frac{3pz^3}{2\pi R^5} dxdy.$$

To calculate the total stress component at F, integrate between the appropriate limits in the x–y plane as follows:

$$\sigma_z = \int_{Y_1}^{Y_2} \int_{X_1}^{X_2} \frac{3pz^3}{2\pi R^5} dxdy.$$

Although z is independent of x and y, $R = (x^2 + y^2 + z^2)$, with the result that evaluating the integral is not straightforward. However, a standard form exists for the integral (ref. *Handbook of Mathematical Functions*, Abromovitch and Stegun, 1965) as

$$\iint \frac{3z^3}{R^5} dxdy = \tan^{-1}\left(\frac{xy}{Rz}\right) + \frac{xyz}{R}\left(\frac{1}{x^2+z^2} + \frac{1}{y^2+z^2}\right) = I_s(x,y).$$

The term $I_s(x,y)$ is referred to as a *stress influence function* and the stress itself is therefore given as

$$s_z = (p/2p)[I_s(x_2,y_2) + I_s(x_1,y_1) - I_s(x_2,y_1) - I_s(x_1,y_2)].$$

Once all the influence functions have been evaluated (these being geometrical functions), the stresses and displacements at any point can be found as the result of any loading distribution—by discretizing the surface loading appropriately and applying the relevant influence functions to summate the individual contributions made by each element.

Exactly the same approach is used for the displacements. The total vertical displacement induced by the element $\delta x - \delta y$ is given by

$$u_z = \frac{4}{4\pi G}\int_{Y_1}^{Y_2}\int_{X_1}^{X_2}\left[\frac{z^2}{R^3} + \frac{2(1-\nu)}{R}\right]dxdy$$

from which the *displacement influence function* is evaluated as

$$I_d(x,y) = 2(1-\nu)[x\log_e(R+y) + y\log_e(R+x)] - (1-2\nu)z\tan^{-1}(xy/Rz).$$

Again, the total displacement induced by the loading over the particular element is calculated as

$$u_z = (p/4\pi G)[I_d(x_2,y_2) + I_d(x_1,y_1) - I_d(x_2,y_1) - I_d(x_1,y_2)].$$

As with the stresses, the displacement contributions from each of the individual uniformly loaded discrete component elements are added to give the total displacement at the point F.

18.3.3 The sector method

In the circumstances of an irregular boundary of a uniformly loaded area, analytical integration of the Boussinesq and Cerruti solutions may be either intractable or impossible, but a simplified form of the stress or displacement

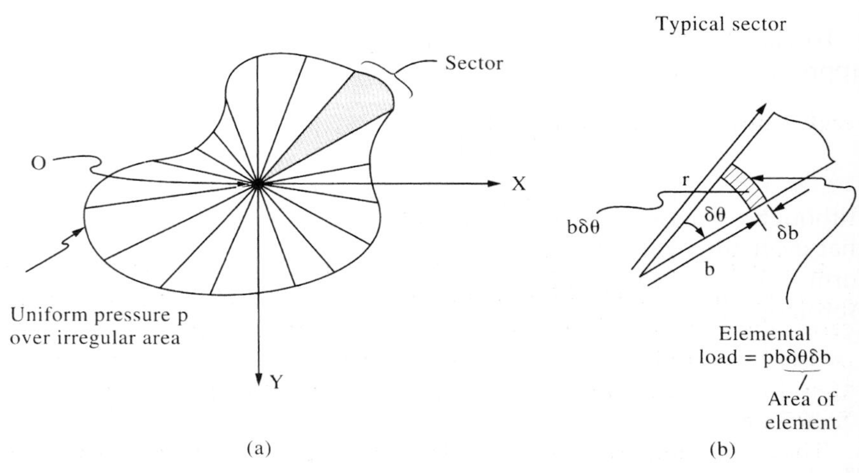

Figure 18.14 The sector method for loaded irregularly-shaped areas: (a) irregularly-shaped areas are divided into sectors; and (b) geometry of a typical sector.

influence function may be developed and used through implementation of the sector method. The principle is that the uniformly loaded area is divided into sectors around the point of interest, analytical integration performed over each sector and the total effect found by summation of the sectorial contributions. The technique can be conducted in graphical, semi-graphical or numerical fashion. Figure 18.14 demonstrates the basic principle.

A loaded area with an irregular boundary is shown in Fig. 18.14(a). Around some arbitrarily chosen point of interest a number of sectors have been drawn. Figure 18.14(b) shows a typical sector in detail, indicating an element over which the analytical integration will be performed. The subtended angle at the origin of the sector is assumed to be sufficiently small to enable adequate representation of the irregularity of the boundary.

Considering the element shown in Fig. 18.14(b),

$$\text{elemental load} = p\, b\, \delta\theta\, \delta b.$$

As an example, consider the formula for the vertical displacement due to a normal point load given in Fig. 18.12, i.e.

$$(u_z)_{\text{point}} = \frac{P}{4\pi G}\left[\frac{z^2}{R^3} + \frac{1}{R} - \frac{(1-2\nu)}{R}\right].$$

We then substitute the elemental load for the point load P at the elemental position $z = 0$ and $R = b$. This reduces to

$$(u_z)_{\text{element}} = \frac{p(1-\nu)}{2\pi G}\delta\theta\delta b.$$

To obtain the displacement induced by loading over the complete sector, the above expression is integrated for $b = 0$ to $b = r$, giving

$$(u_z)_{\text{sector}} = \frac{p(1-\nu)}{2\pi G} r\delta\theta.$$

and finally, for the total loaded area,

$$(u_z)_{\text{area}} = \frac{p(1-\nu)}{2\pi G} \Sigma r\delta\theta.$$

Evaluation of the term $\Sigma r\delta\theta$ involves either graphical, semi-graphical or numerical techniques to determine a value of r for each value of θ. In most cases, the number of sectors required to produce a result of acceptable accuracy is modest—as the reader can verify for the case of a circular area using the formula above, knowing that the analytical result to $\Sigma r\delta\theta$ is $2\pi r$.

The sector method is a simplified version of the stress influence function method, where the loading is uniform over the entire area and polar co-ordinates have been used. Given the conditions of a uniform load, the sector method could be used to develop expressions for any of the displacement or stress components that may be required through the use of the expressions listed in Fig. 18.12.

Within the overall context of the design and analysis of surface excavations, discussions of the 'first pass' studies of slope design and foundation design have followed different approaches to slopes and foundations. With slopes, kinematic feasibility was used and it was found that a great deal of information could be obtained from the preliminary overview analysis. With foundations, we considered how to estimate the stresses and displacements beneath a non-uniformly loaded area. Again, this type of analysis would immediately highlight areas for concern and, if discontinuities were present, indicate the regions of potential instability.

To extend both these approaches, one would need to consider the effects of stress in very high slopes and the effects of discontinuities on the stress distributions beneath foundations. Moreover, there is a range of site factors that should be taken into account in more extended analyses. These include overall rock anisotropy and inhomogeneity, local variation in material properties, the effects of natural factors such as groundwater, rainfall, seismic risk and the effects of engineering factors such as blasting. We have chosen, therefore, to present in the next section techniques for considering these together, rather than extend each of the analyses separately to account for the wide variety of deviations from the assumed CHILE circumstances.

18.4 Techniques for incorporating variations in rock and site factors into the analyses

There is a range of factors that can influence the instability of surface excavations. The techniques presented so far do not explicitly allow these factors to be taken into account, nor indeed can they all be taken into account quantitatively in a direct way—because there will never be complete knowledge of the rock mass. Also, considering the total energy balance of a rock mass system via thermodynamics, it can be shown that any excavation must result in an alteration of the rock properties (Hudson, 1992). This means that even with complete knowledge of the rock properties before excavation as a result of a 'perfect' site investigation, the post-construction properties would still not be known—because the pre-construction properties will have been altered by the excavation process. It can be predicted with certainty that some form of analysis that deals with uncertainty will be required.

So far, the approach has been deterministic (i.e. estimated single values of each of the parameters involved have been used in order to produce a single result). This is because it is essential to understand the basic mechanics before superimposing methods which account for variability. There have been attempts during the development of rock mechanics to substitute probabilistic approaches for an understanding of the mechanisms, i.e. to relegate the mechanics and solve the problem by treating some of the factors as random variables.

We believe that this probabilistic approach is fundamentally flawed: *the basic mechanics must be understood first and then any variations in any properties, or any lack in our knowledge, dealt with* via *appropriate mathematical techniques*. Such techniques are still being developed and are extensive. They range from the simple application of probability theory through to the development of new systems utilizing cognitive processes and neural networks.

18.4.1 Sensitivity analysis

In Figs 17.6 and 17.8, the variation in the factor of safety is shown for plane instability and wedge instability with the depth of water in a tension crack, the angle of friction and the included wedge angle. In all of the examples, as the factor of safety varied, it passed through unity—the interface between stability and instability. Qualitative examination of the graphs in these figures shows that, for certain ranges of the independent variables, the factor of safety is more *sensitive* to these variations than for other ranges (e.g. in Fig. 17.8 the range 10–20° for the included wedge angle). This behaviour can be expressed quantitatively through a formal definition of sensitivity, as follows:

$$\text{sensitivity} = \frac{\mathrm{d}F}{\mathrm{d}(p/p_1)} = \text{slope of the factor of safety vs normalized parameter curve}$$

where p is any parameter involved in the analysis, and
p_1 is the value of this parameter which produces a factor of safety of unity (all other parameters remaining constant for this analysis).

The use of a normalized parameter is a simple device whereby the curve is scaled around the region of interest.

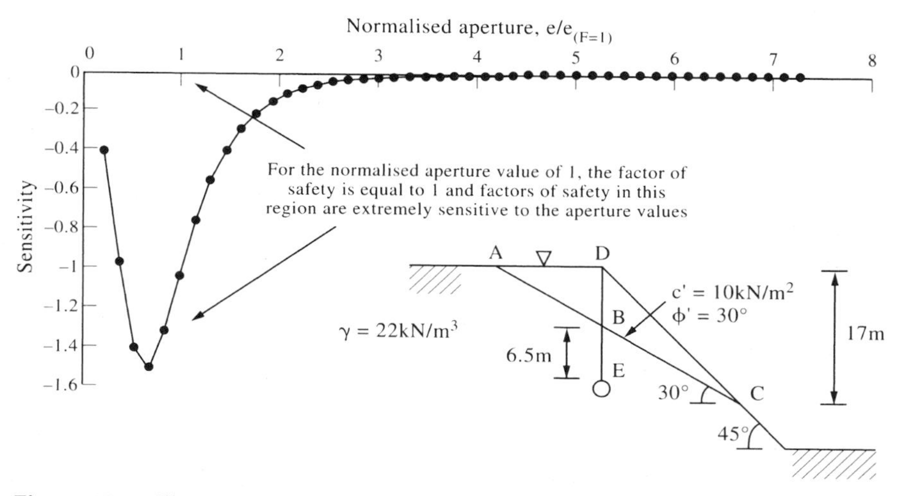

Figure 18.15 Illustration of sensitivity analysis applied to plane sliding.

In the example in Fig. 18.15, we show the sensitivity of the factor of safety for plane sliding of a rock slope to the aperture of a discontinuity. The inset sketch shows the geometry of the slope under consideration, together with the material properties. All of the discontinuities are assumed to be full of water, a drainage gallery is provided at E, and the block BCD is unstable and may slide along BC. Using the techniques described in Sections 9.4 and 17.1.2, the hydraulic pressures in the discontinuity network can be determined and thence the factor of safety against plane sliding found, in this case, as a function of the aperture of BD. Note that in Fig. 18.15, this aperture has been normalized, as described above, and it is the sensitivity that has been plotted on the vertical axis and not the factor of safety.

This illustrates that, for factors of safety around unity, the system is sensitive to changes in aperture of discontinuity BD. For this illustration we have taken a normalizing aperture related to a factor of safety of 1; any other factor of safety could equally well have been taken and produced similar curves to establish the different sensitivities under these other conditions.

Sensitivity analysis is useful (not least, in the significance for site investigation), but is not the most convenient method for either analysing or communicating the effects of variation in, what could be, a large number of relevant parameters. For this, one must turn to other techniques, as described next.

18.4.2 Probabilistic methods

A traditional method of describing the many values a parameter may take is through the use of probability theory. The key difference between the deterministic and probabilistic approaches is that in the latter we do not actually know, or even assume, a specific value for the parameter in

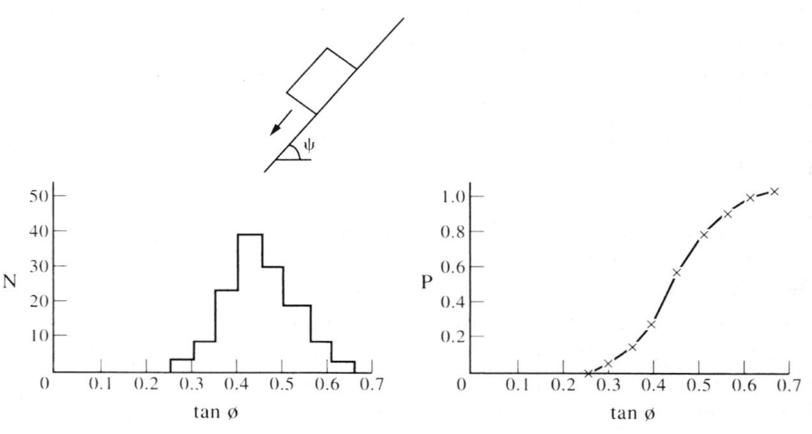

Figure 18.16 Direct probabilistic approach, illustrated by sliding of a block on a plane.

question. Instead, it is said that the parameter can take on a range of values defined by a probability density function, with the result that statements can be made about the probability that the parameter will take on values within a certain range. Thus, given any mechanical model, the effect of the various parameters in the model can be considered as random variables taken from probability density functions.

In those cases where only one or two parameters are considered as random variables, it is possible to use probabilistic statements to examine the system, and the method of solution may be by so-called direct methods. In those cases where a large number of parameters are considered to be random variables from different types of distribution, the mathematics associated with the direct probabilistic analysis becomes intractable, and so a numerical technique, e.g. the Monte Carlo method, must be used. Monte Carlo simulation involves repeatedly substituting generated random variables into a deterministic model and collation of the results into a histogram.

Direct approach. The direct approach is demonstrated in Fig. 18.16 for the simple case of a block sliding on a plane, where the angle of friction is considered as a random variable. Considering the left-hand histogram to represent the results of 133 shear box tests to determine the angle of friction, the histogram can either be used directly, or a normal distribution (for example) can be fitted to the results. In the former case, the probability density histogram is defined by the class intervals; in the latter case, it is defined by the mathematical expression for the function in question, e.g. the normal distribution with particular values of mean and standard deviation.

The important distinction between the deterministic and probabilistic methods is illustrated by the fact that the class intervals in the probability density histogram are used and not the actual separate 133 test results. Inserting the mean value of each of the class intervals into the deterministic model in turn allows a cumulative distribution function to be generated, as illustrated in the right-hand graph of Fig. 18.16. Probabilistic statements can then be made about the factor of safety, e.g. what is the probability that the factor of safety will be above 1.25 for a case when the angle of friction is a random variable from the same population as that determined by the sample tests. The probabilistic analysis can be initiated by assuming a continuous probability density function, with or without reference to test data.

Monte Carlo simulation. Monte Carlo simulation is a procedure which permits the variation in many parameters to be considered simultaneously. The calculation is performed many times for repeatedly generated sets of input data. Each calculation produces one value of the factor of safety, from which a histogram or cumulative distribution of factors of safety is generated. Figure 18.17 demonstrates the principle of the simulation, and Fig. 18.18 shows how it may be applied to the analysis of a curvilinear slip in a poor-quality rock mass, using the method of slices and following the procedure outlined by Priest and Brown (1983).

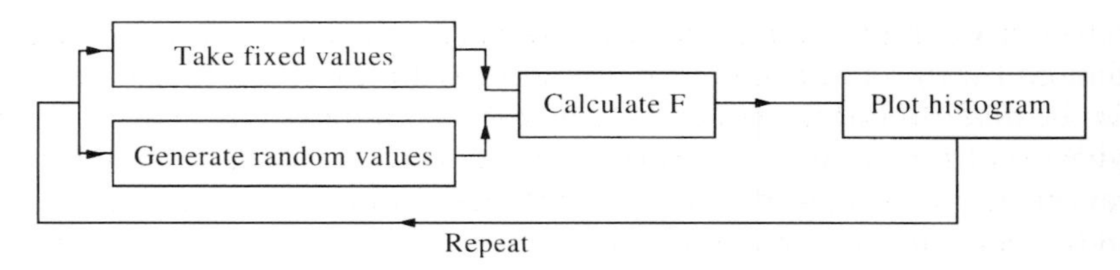

Figure 18.17 Mechanism of the Monte Carlo simulation procedure.

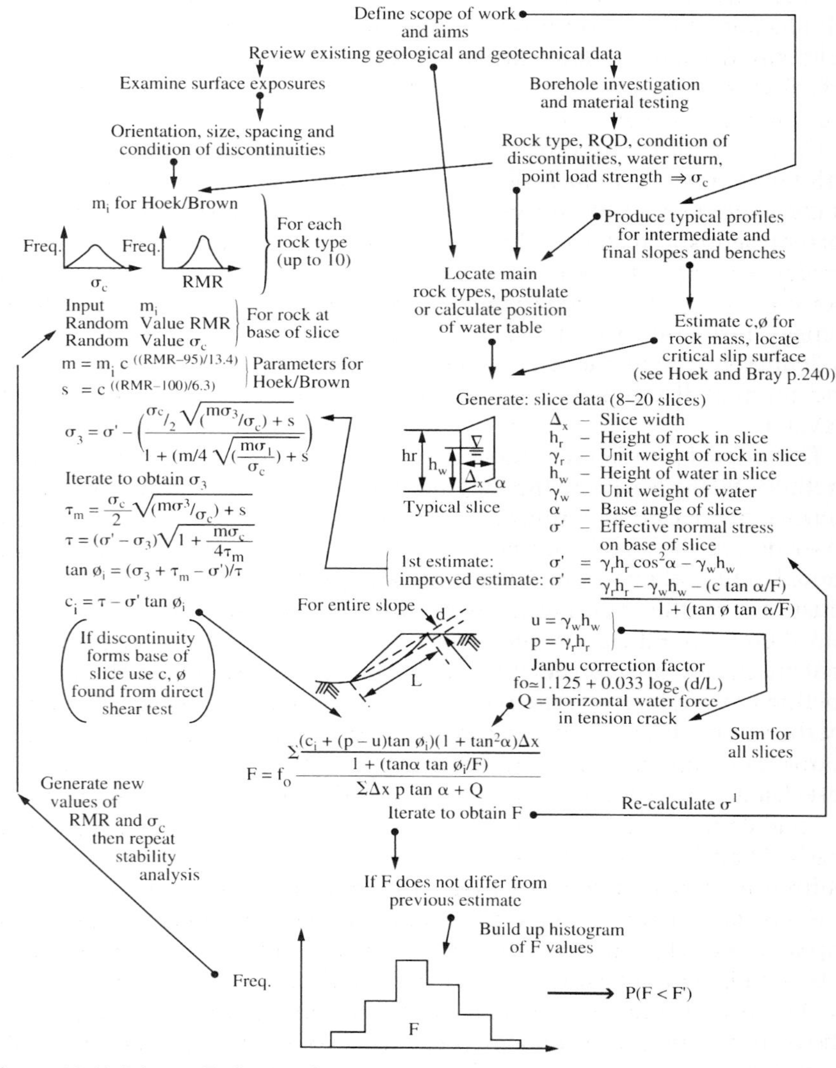

Figure 18.18 Monte Carlo simulation applied to slope instability in poor rock masses (from lecture notes by S. D. Priest).

Modern computational facilities are such that large numbers of simulations can be conducted in a short period of time on a desktop computer. To perform Monte Carlo simulation, generation of random variables for the given probability density distribution is required. These are elegantly generated by consideration of a cumulative distribution plot. Every point on the vertical axis of a cumulative distribution plot has equal probability of occurring: thus, to generate the random variable, the equation for the cumulative distribution function is inverted so that the random variable is expressed as a function of the cumulative probability. Then, inserting a random number, taken from a uniform distribution, with a value between 0 and 1 for P will give a random variable for the distribution desired. For example, with the negative exponential distribution, the cumulative probability, P, is given by $P = 1 - e^{-\lambda x}$, which upon inversion produces $x = -(1/\lambda)\log_e(1 - P)$. Uniformly distributed random variables may now be substituted for P to provide x variables from a negative exponential distribution. This technique is valid for all probability density functions, although the inversion is not always as easy as the one demonstrated here.

Interpretation of probabilistic analyses. Having conducted a probabilistic analysis in the manner just described, the resulting histogram of factor of safety values has to be interpreted for engineering purposes. The interpretation must take into account both the mean factor of safety and the spread of values about the mean. In Fig. 18.19, two tables are shown which can be used to assist in this interpretation.

The first table categorizes slopes in terms of the mean factor of safety and the probability of the actual factor of safety being less than a specific value, in this case 1.0 and 1.5. These last two conditions are used to take into account the spread of the histogram about the mean. The second table in Fig. 18.19 considers the engineering consequences of the various combinations in which the three probabilistic criteria might be satisfied. There is a degree of subjectivity in the levels at which the various probabilistic criteria are set and the associated interpretation. In practice, an engineer would have to consider the site-specific circumstances.

To use these tables, the engineer initially assesses the consequences of failure of the slope and hence establishes the slope category (the first two columns of the upper table in Fig. 18.19). This sets values for the minimal mean factor of safety and the maximal probabilities of not exceeding a factor of safety of 1.0 or 1.5 (the three right-hand columns). Having established these criteria, and the degree to which they are satisfied for a specific slope (through the use of Monte Carlo analysis and compared with the left-hand column of the lower table in Fig. 18.19), the engineer can utilize the interpretation provided (the right-hand column).

There are many potential variations on this probabilistic theme and many design techniques that can be based on alternative approaches to assessing instability. However, the basic methodology has been explained in this section and thus, by extrapolation, the reader can conceive how similar probabilistic approaches can be developed and adopted. We concentrate next on an alternative technique for assessing variations in rock and site factors using fuzzy mathematics.

Category of slope	Consequences of failure	Examples	Acceptable values		
			Minimum	Maximum	
			Mean F	P(F<1.0)	P(F<1.5)
1	Not serious	Individual benches, small (height<50m) temporary slopes not adjacent to haulage roads.	1.3	0.1	0.2
2	Moderately serious	Any slopes of permanent or semi-permanent nature.	1.6	0.01	0.1
3	Very serious	Medium sized (50m<height <150m) and high (height>150m) slopes carrying major haulage roads or underlying permanent mine installations.	2.0	0.0003	0.05

Probabilistic slope design criteria

Satisfaction of above criteria	Interpretation
Satisfies all three criteria	Stable slope
Exceeds minimum mean F but violates one or both probabilistic criteria	Operation of the slope presents a risk which may or may not be acceptable. The level of risk can be assessed by a comprehensive monitoring programme.
Falls below minimum mean F but satisfies both probabilistic criteria	Marginal slope. Minor modifications of slope geometry are required to raise the mean F to a satisfactory level.
Falls below minimum mean F and does not satisfy one or both probablistic criteria	Unstable slope. Major modifications of slope geometry are required. Rock improvement and slope monitoring may be necessary.

Slope performance interpretation

Figure 18.19 Interpretation of probabilistic design criteria (after Priest and Brown, 1983).

18.4.3 Fuzzy mathematics

It may be that the parameters influencing the instability of a slope do not conform to any known probabilistic distribution, or that the resources necessary to determine the relevant distributions are unavailable. In such circumstances, the application of probabilistic methods is inappropriate. However, the analysis of 'uncertainty' (rather than probability) may be performed using fuzzy mathematics, as described in Section 12.6.1.

The application of fuzzy mathematics to the analysis of slope instability through the use of standard equilibrium analysis is straightforward, but the interpretation of the resulting fuzzy factor of safety needs care. A procedure for this interpretation has been outlined by Sakurai and Shimizu (1987), who considered fuzzy cohesion and angle of internal friction in the analysis of plane sliding. The analysis is mechanically identical to that presented in Section 17.1.2, but in order to interpret the resulting

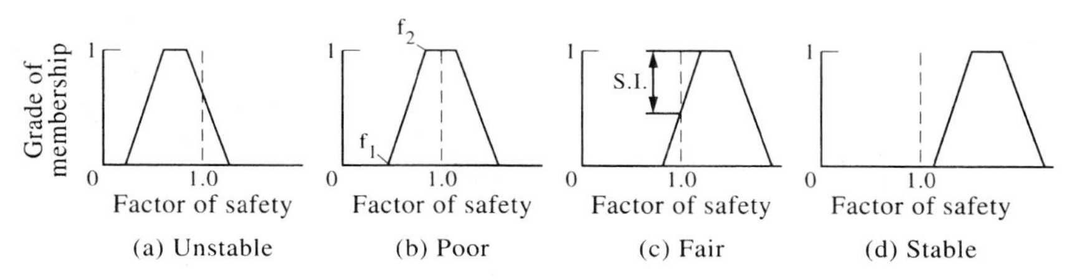

Figure 18.20 Class of slope stability based on fuzzy analysis (after Sakurai and Shimizu, 1987).

trapezoidal fuzzy number representing factor of safety, Sakurai and Shimizu defined a 'Stability Index', S.I., as

$$\text{S.I.} = (f_2 - 1)/(f_2 - f_1)$$

where f_1 and f_2 are as indicated in Fig. 18.20, which shows four classes of slope stability based on the stability index.

Using fuzzy mathematics to accommodate uncertainty in slope analysis presents no additional difficulty in an ordinary instability analysis, and so it is likely that its use will increase in the future together with experience of assessing instability in terms of the stability index or some similar measures.

19 Underground excavation instability mechanisms

Chapters 19 and 20 of this book, on instability around underground excavations, are direct analogues of Chapters 17 and 18, which were concerned with surface excavation instability. In this chapter, the underground instability mechanisms are presented. In Chapter 20, the design and analysis of underground excavations are discussed in the context of these mechanisms.

In Section 16.2, the distinction between structurally-controlled and stress-controlled instability mechanisms is explained. Accordingly, this chapter has been devoted to these two primary instability modes, considering also composite instability modes and the effect of time and weathering on excavation stability.

19.1 Structurally-controlled instability mechanisms

Structurally-controlled instability means that blocks formed by discontinuities either fall or slide from the excavation periphery as a result of an applied force (usually the force due to gravity) or stresses induced by the process of excavation. Hence, we include the kinematic feasibility of tetrahedral blocks and, later, instability in stratified rock.

19.1.1 Kinematic feasibility analysis

The minimum requirement to define a discrete block is four non-parallel planes, which give rise to a tetrahedral block. In terms of the analysis of instability around an excavation, such a block can be formed by three discontinuity planes *and one plane representing the excavation periphery*. Because we are limiting the analysis to the simplest case—that of tetrahedral blocks—the blocks themselves may be identified as spherical triangles on the hemispherical projection. This is because the plane of the projection represents a plane which is one face of the tetrahedral block and the remaining three planes which form block faces are given by the great circles representing the discontinuities. Thus, a study of instability mechanisms in terms of kinematics may be conveniently conducted through the use of hemispherical projection techniques. Initially the analysis will be limited to horizontal roofs (so that the plane of the projection is parallel to the excavation surface); later the projection will be inclined to account for any orientation of the excavation periphery.

Given that a tetrahedral block exists, there are three kinematic possibilities to be examined: the block falls from the roof; the block slides (either along the line of maximum dip of a discontinuity, or along the line of intersection of two discontinuities); or the block is stable.

Falling. Falling occurs when a block detaches from the roof of an excavation without sliding on any of the bounding discontinuity planes. In the case of gravitational loading, the direction of movement is vertically downwards. This is represented on the projection as a line with a dip of 90°, i.e. at the centre of the projection. Thus, if this point falls within the spherical triangle formed by the bounding discontinuities, falling is kinematically feasible, as illustrated in Fig. 19.1.

Sliding. In Sections 18.1.1 and 18.1.2, plane and wedge instability analyses for a surface slope were discussed. A similar method is used here to consider blocks sliding from the roof, either on one discontinuity plane (as plane failure) or on a line of intersection (as wedge failure), as illustrated in Fig. 19.2 by consideration of the spherical triangle and whether any part of it has a dip greater than the angle of friction.

Assuming that both discontinuity planes have the same friction angle, there are only two candidates for the sliding direction: either the line of maximum dip of one plane, or the line of intersection of two planes. No other part of the spherical triangle represents a line of steeper dip than these candidates.

Not all lines of maximum dip can be candidates for the sliding direction. An example is afforded by the line of maximum dip, β_3, of plane 3 in

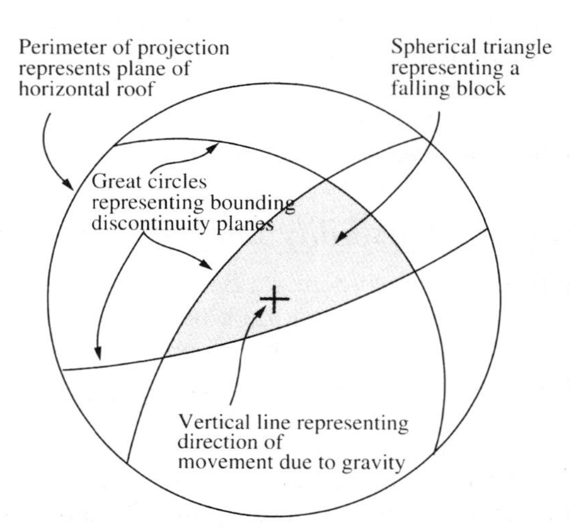

Figure 19.1 Kinematic identification of a falling block.

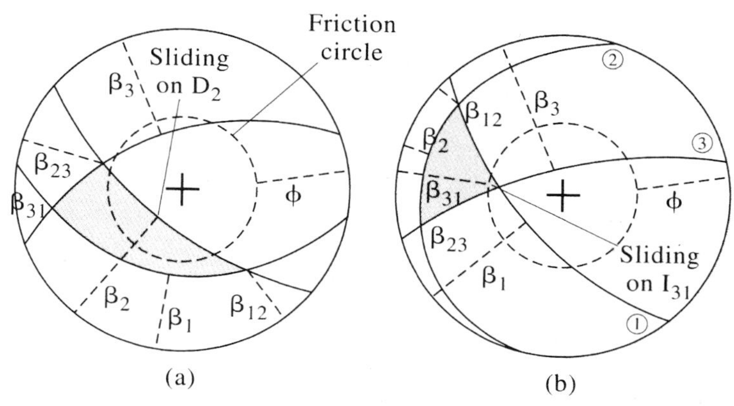

Figure 19.2 Kinematic identification of sliding blocks.

Fig. 19.2(a) and the lines of maximum dip, β_1 and β_3, of planes 1 and 3 in Fig. 19.2(b). In each of these cases, the planes, although dipping at angles greater than the friction angle, are not candidates for the sliding direction *because the line of maximum dip is not included within the block.* The spherical triangle represents the region of kinematically admissible directions of movement and any other direction represents directions directed into the rock surrounding the block. There are no restraints on the azimuth of sliding direction from the horizontal roof: there will naturally be constraints on the azimuth when sliding is considered from the side walls.

Hence, the shaded blocks in Figs 19.2(a) and 19.2(b) represent plane sliding along β_2 and wedge sliding along β_{31}, respectively. Only one friction circle has been used because all discontinuities are assumed to have the same friction angle, but in the plane sliding case the analysis could easily be extended to cater for different friction angles with associated friction circles on the projection. In the case of wedge sliding, which may be regarded as plane sliding on two planes simultaneously, the direction of movement is parallel to the direction of the line of intersection of the planes themselves. If the two planes have different angles of friction, the line of intersection must lie within both friction circles for wedge sliding to occur.

Stable. The final possibility, that the block is stable, is shown in Fig. 19.3. This occurs when a spherical triangle lies completely outside the friction circle. Again, a line of maximum dip which exceeds the angle of friction is not in itself sufficient to cause instability: it must lie on the perimeter of the spherical triangle under consideration to be part of a kinematically feasible block.

19.1.2 Use of inclined hemispherical projection methods

In Section 19.1.1, it was assumed that the blocks would move from the horizontal roof of an excavation. In order to use the simplicity and clarity of these graphical methods for any blocks—which may be moving from

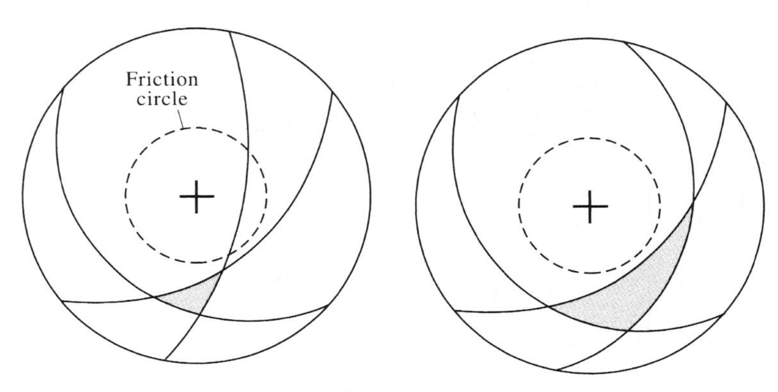

Figure 19.3 Kinematic identification of stable blocks.

surfaces which are not horizontal—the concept of inclined hemispherical projection can be used.

The purpose of inclining the projection is to be able to identify tetrahedral blocks formed by three discontinuities and the excavation surface, the latter being retained as the perimeter of the projection. This is achieved by ensuring that the plane of projection is coincident with the plane of the excavation surface. Naturally, the position of the discontinuity planes and the point representing the vertical direction on the projection will change, and friction circles will no longer be concentric with the perimeter of the net but must be constructed as circles representing friction cones around the vertical.

With the extended technique of the inclined projection, the block identification procedures can be retained and similar techniques for establishing kinematic feasibility used as for horizontal roofs. The details of the procedure required to construct an inclined hemispherical projection are given in the second Appendix, and so here we concentrate on the required angle of inclination and interpretation of the resulting diagram.

Angle of inclination of the hemispherical projection. In Fig. 19.4, there is a generic excavation with five main excavation surfaces: the roof, the shoulder, the sidewall, the knee and the floor. The inclined hemisphere is shown as it relates to each of the excavation surfaces. The lower-hemispherical projection at the roof is inverted to an upper-hemispherical projection at the floor: between these extremes the hemisphere has been inclined by 45° at the shoulder, 90° at the vertical sidewall and 135° at the knee. The hemisphere can be inclined in this way to accommodate any required excavation surface.

The key to the procedure is to incline the projection such that it becomes coincident with the outward directed normal to the excavation surface. This is achieved by performing the following steps.

- First, plot on an overlay, the normal to the excavation surface in question (N_f in Fig. 19.5), the normals to the various discontinuity surfaces (N_1, N_2 in Fig. 19.5) and the normal to the horizontal plane (N_h in Fig. 19.5, which is coincident with the centre of the projection, i.e. vertical).
- Then rotate the overlay such that N_f lies on the E–W line.

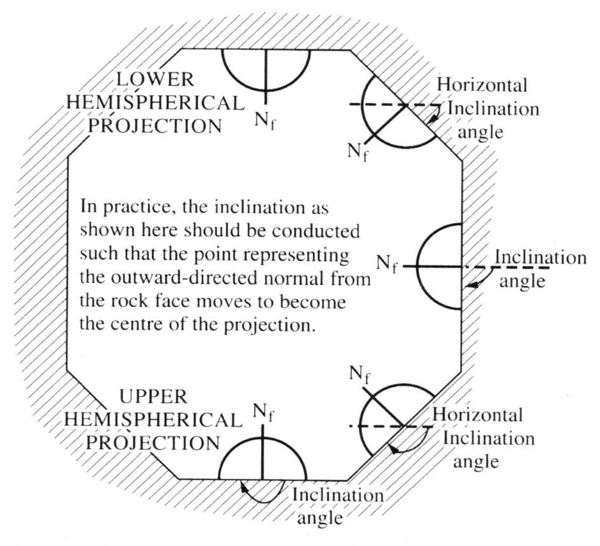

Figure 19.4 Inclination of the hemispherical projection to be coincident with the excavation surface.

- The inclination is then applied (taking care to ensure that both the direction and magnitude are correct) to all normals by moving along small circles and replotting the points.

The procedure is shown in Fig. 19.5 and explained in detail in the Appendix.

The inclined normals are labelled N_{fI}, N_{hI}, N_{1I} and N_{2I} in Fig. 19.5. N_{fI} is, by definition, coincident with the centre of the net. From these normals, the associated great circles representing the inclined projection of the various discontinuities are constructed, as shown by the solid great circles in Fig. 19.5. Similarly, the horizontal plane is derived from N_{hI} and is shown

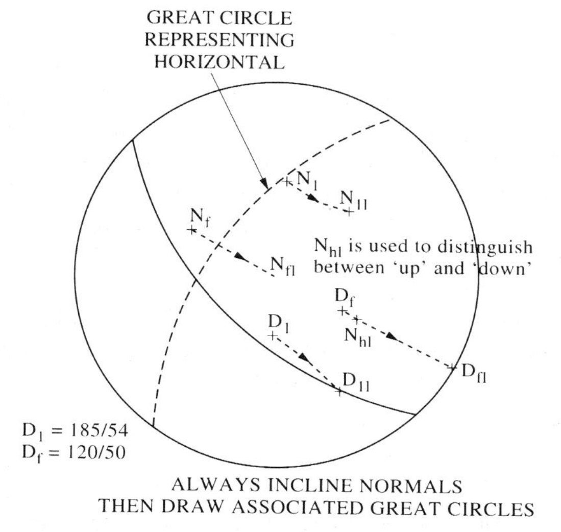

Figure 19.5 Construction of an inclined hemispherical projection.

on the inclined projection by the dashed great circle. The relative positions of the inclined horizontal plane and N_{hI} are used to distinguish between 'up' and 'down': any line which appears on the N_{hI}-side of the inclined horizontal plane is directed downwards (because we started with a lower-hemispherical projection and N_h was initially coincident with the downwards directed vertical line). The requirement to be able to distinguish between up and down is essential in interpreting potential gravitationally induced instability!

The elegance of this technique lies in the simple graphical transformation (illustrated by the curved lines representing, for example, the N_1 to N_{1I} inclination) which is the representation of an equivalent 3×3 matrix multiplication. Also, considering the associated D_1 to D_{1I} inclination, we see immediately that D_{1I} does not correspond to the mid-point of the great circle, which is expected, in terms of a lower-hemispherical projection, to be the line of maximum dip. Relative to the global frame of reference, D_{1I} remains the line of maximum dip. Relative to the local frame of reference (the inclined projection), the mid-point of the great circle is no more than the line on the plane which makes the maximum angle to the plane of the projection (which is the rock surface) and has no general engineering utility.

Thus, the inclined hemispherical projection technique retains the interpretive character of representing the three-dimensional rock structure geometry, whilst enabling rapid study equivalent to lengthy mathematical operations.

In the following paragraphs, we demonstrate the method of identifying falling, sliding and stable blocks utilizing the inclined hemispherical projection technique.

Identification of falling blocks. In Section 19.1.1, procedures were presented for identifying the kinematic feasibility of a falling block using the lower-hemispherical projection to represent a horizontal roof. These same basic procedures, shown in Fig. 19.1, can be used but with the inclined hemispherical projection accounting for excavation surfaces at any orientation.

Figure 19.6 illustrates the identification of a block falling from an inclined surface. The various great circles and poles on this diagram have been constructed using the procedures shown in Fig. 19.5. Note particularly the great circle, H, representing the horizontal plane and the associated pole, N_{hI}, representing the vertical. This vertical line is also shown in the accompanying sketch of such a block.

By comparison with Fig. 19.1, the highlighted spherical triangle in both cases contains the pole representing the vertical direction—and hence the block will fall from an overhanging surface, because the spherical triangle surrounds the downward directed vertical. The latter point is related to the following discussion on stable blocks, which cannot fall from non-overhanging surfaces because they have upward directed verticals.

Identification of sliding blocks. By comparison with Fig. 19.2, a similar procedure can be used for inclined projections to identify blocks which can

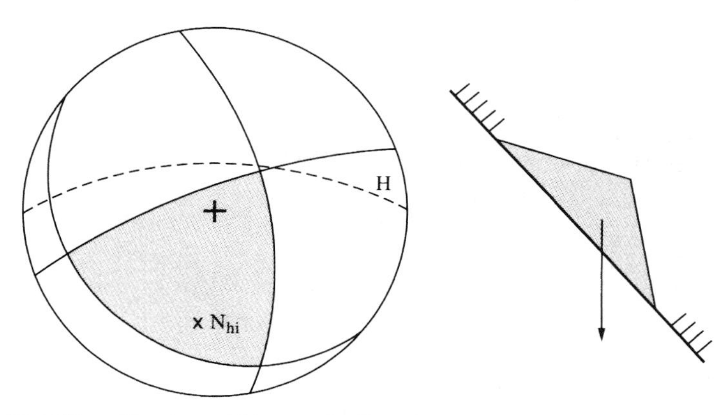

Figure 19.6 Identification of kinematically admissible falling blocks at an inclined surface.

slide—either from an overhanging or a non-overhanging surface. Figures 19.7 and 19.8 illustrate the identification of blocks sliding from inclined overhanging and non-overhanging surfaces, respectively. For the case of a falling block, note the great circle, H, representing the horizontal plane and the associated pole, N_{hI}, representing the vertical. The blocks themselves and the vertical lines are also shown in the accompanying sketches of the geometry. For overhanging surfaces, N_{hI} is directed downwards and for non-overhanging surfaces, N_{hI} is directed upwards.

In order to use the method illustrated in Fig. 19.2, the friction circle has to be included on the inclined projections. This circle is easily drawn, as it represents a cone of semi-angle $(90 - \phi)^\circ$ around N_{hI} for overhanging surfaces and $(90 + \phi)^\circ$ for non-overhanging surfaces—as implied in Figs 19.7 and 19.8 by the arrowed lines.

Thus, for an overhanging surface, as shown in Fig. 19.7, if any point on the perimeter of the spherical triangle lies between N_{HI} (the downward directed vertical) and the friction circle, sliding is kinematically feasible. Similarly, for a non-overhanging surface, as shown in Fig. 19.8, if any point

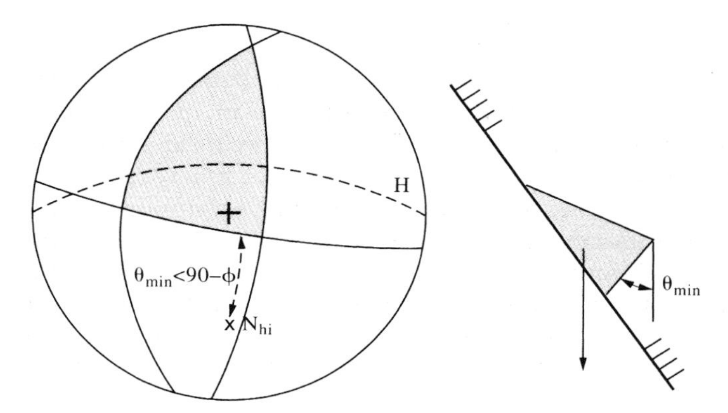

Figure 19.7 Identification of kinematically admissible sliding blocks at an overhanging inclined surface.

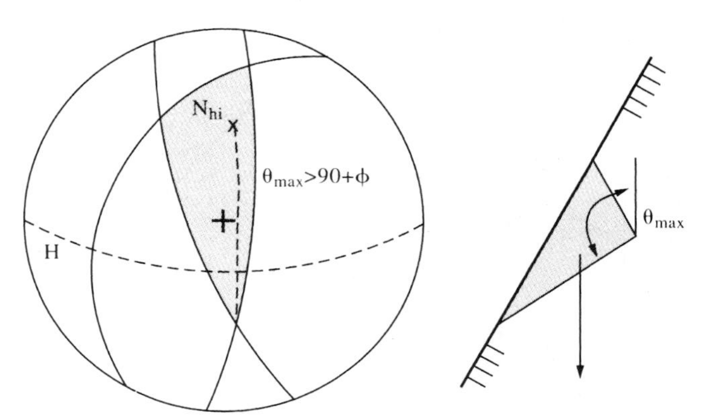

Figure 19.8 Identification of kinematically admissible sliding blocks at a non-overhanging inclined surface.

on the perimeter of the spherical triangle lies beyond the friction circle drawn below the horizontal (i.e. has a dip steeper than ϕ), sliding is kinematically feasible. In essence, the two cases are the same: it has just been necessary to account comprehensively for the excavation surface orientation.

For the case of a horizontal roof, the possible modes of sliding (parallel to the line of maximum dip of a plane or parallel to the line of intersection of two planes) can be identified from the projection geometry. The analogous interpretation for a horizontal excavation surface has been discussed in Section 19.1.1, and the same techniques are used in these cases.

Identification of stable blocks. The stable blocks will be those that do not satisfy any of the instability criteria described above. In particular, in Fig. 19.9, blocks are shown at both overhanging and non-overhanging surfaces which are stable because the friction angle is not exceeded on any relevant plane or block edge. The accompanying sketches show that the shape of the blocks is such that they can be removed from the excavation surface, but the orientation of the block faces relative to the friction angle is such that sliding cannot occur.

The analysis refers only to the instability of a single block. We have not studied the potential instability of a suite of blocks, nor whether the instability of a single block will lead to one or more blocks also being unstable. It is important to know, for support considerations, whether an unstable block is an isolated event or whether it may trigger catastrophic ravelling of the rock mass. In Chapter 20, the support requirements of individual blocks are considered, but the analysis of a ravelling rock mass is beyond the scope of this book. However, the principles and techniques presented here do form the basis for an understanding of such an analysis.

19.2 Stress-controlled instability mechanisms

The instability mechanisms described in Section 19.1 are all driven by forces, and in particular the force due to gravity. Such a force is known as

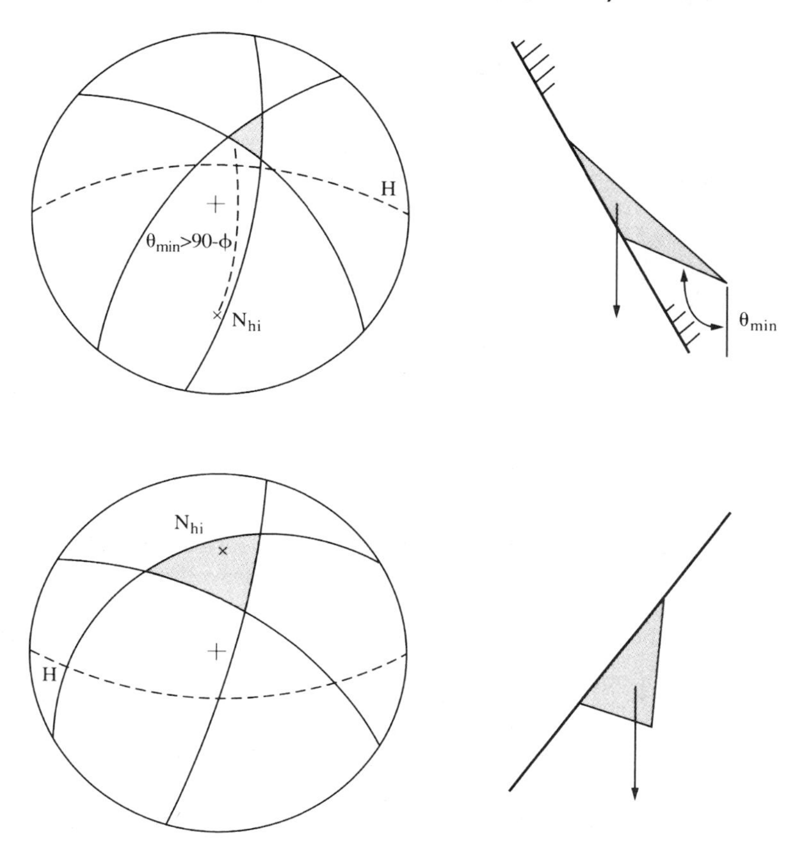

Figure 19.9 Identification of kinematically stable blocks at overhanging and non-overhanging inclined surfaces.

a body force and is unidirectional. Consequently, the blocks either move vertically downwards or along some preferred sliding direction. In this section, however, instability mechanisms are discussed which are stress controlled. Thus, the activating quantity is not a single force, but a tensor with six independent components and hence the manifestations of stress-controlled instability are more variable and complex than those of structurally-controlled instability.

For example, considering the stresses around a circular opening, the same stress concentrations can theoretically be experienced at opposite ends of any diameter of the opening, which could be the floor and the roof—with the result that **both** of these locations experience the same manifestation of instability. So it is not surprising that not only may the roof have to be supported, but **the floor may also have to be supported.** This illustrates one of the fundamental differences between structurally- and stress-controlled instability mechanisms.

Although in the case of structurally-controlled instability there is simplicity in the mechanisms, it is necessary to consider the complexity of the three-dimensional geometry of the rock mass. On the other hand, in the case of stress-controlled instability, the fundamental 'complexity' of the nature of stress has to be considered, but in CHILE materials there is a relative simplicity in the associated stress-controlled instability mechanisms.

The analysis of stress-controlled instability must begin with a knowledge of the magnitudes and directions of the *in situ* stresses in the region of the excavation. The induced stresses can then be determined, i.e. the *in situ* stresses after perturbation by engineering. There exist closed form solutions for the induced stresses around circular and elliptical openings (and complex variable techniques extend these to many smooth, symmetrical geometries), and with numerical analysis techniques the values of the induced stresses can be determined accurately for any three-dimensional excavation geometry. Finally, a rock failure criterion expressed in terms of stresses is required; failure has already been discussed in Chapter 6 for intact rock, in Chapter 7 for discontinuities and in Chapter 8 for rock masses.

It is now appropriate to consider stress distributions around underground openings in order to determine the extent of stress-controlled instability mechanisms. In the text that follows a series of elastic solutions for various geometries will be presented, but the derivation of each solution is not included.

All analytical closed form solutions must satisfy the following criteria.

(a) **Equations of equilibrium**—three equations of the form

$$\frac{\partial \tau_{zx}}{\partial x} + \frac{\partial \tau_{zy}}{\partial y} + \frac{\partial \sigma_z}{\partial z} + Z = 0.$$

(b) **Strain compatibility equations**—three equations of the form

$$\frac{\partial^2 \varepsilon_x}{\partial y^2} + \frac{\partial^2 \varepsilon_y}{\partial x^2} = \frac{\partial^2 \gamma_{xy}}{\partial x \partial y}$$

and three of the form

$$2\frac{\partial^2 \varepsilon_x}{\partial y \partial z} = \frac{\partial}{\partial x} - \left(\frac{\partial \gamma_{yz}}{\partial x} + \frac{\partial \gamma_{zx}}{\partial y} + \frac{\partial \gamma_{xy}}{\partial z} \right)$$

where the symbols are as defined in Chapters 3 and 5.

(c) **Boundary conditions**—e.g. zero traction or uniform pressure on the excavation boundary.

(d) **Conditions at infinity**—e.g. field stresses.

It is from these conditions that the solutions for the circular and elliptical openings that follow have been derived. As the conditions require that the derivatives of various functions exist, openings with sharp corners cannot be exactly modelled, although solutions for openings with small radii corners have been developed using the theory of complex variables. The solutions can be cumbersome, inhibiting simple analysis of parameter variation, and it is for this reason that we do not present them here—concentrating instead on the simpler instructive solutions, of which the Kirsch equations are perhaps the paradigm.

19.2.1 Stresses and displacements around a circular excavation

In rock mechanics and rock engineering, the Kirsch equations are the most widely used suite of equations from the theory of elasticity. They allow determination of the stresses and displacements around a circular excavation, and are given in Fig. 19.10. The pre-eminent nature of these equations is due to the requirements of stress determination techniques in circular boreholes and consideration of the stability of circular tunnels. The equations apply to openings made in previously stressed CHILE materials, rather than the case of openings made in unstressed materials. The authors had considerable difficulty in reconciling the many different expressions given in the literature for u_r and u_θ, but are confident that the expressions given in Fig. 19.10 are correct. The angle θ is measured anticlockwise positive from the horizontal axis in the figure.

Some special cases of interest are now given in which the Kirsch equations are used to demonstrate a number of important points. These occur at specific locations (i.e. the boundary of the excavation) and with specific stress fields (i.e. uniaxial and hydrostatic).

Stresses at the boundary of a circular opening. We see from Fig. 19.10 that the stresses on the boundary (i.e. when $r = a$) are given by

$$\sigma_r = 0$$

$$\sigma_\theta = p_z\{(1 + k) + 2(1 - k)\cos 2\theta\}$$

$$\text{and } \tau_{r\theta} = 0.$$

Note that the first of the stresses is zero because there is no internal pressure, and the last of the stresses must be zero at a traction-free boundary (the excavation boundary is a principal stress plane). The variation in boundary tangential stress at the end points of horizontal and vertical diameters for $0 \le k \le 1$ is shown in Fig. 19.11.

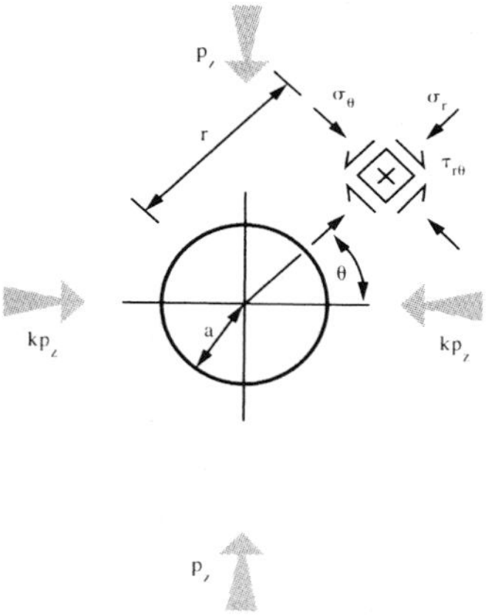

$$\sigma_r = \tfrac{1}{2}p_z\left\{(1+k)\left(1-\frac{a^2}{r^2}\right) - (1-k)\left(1-4\frac{a^2}{r^2}+3\frac{a^4}{r^4}\right)\cos 2\theta\right\}$$

$$\sigma_\theta = \tfrac{1}{2}p_z\left\{(1+k)\left(1+\frac{a^2}{r^2}\right) + (1-k)\left(1+3\frac{a^4}{r^4}\right)\cos 2\theta\right\}$$

$$\tau_{r\theta} = \tfrac{1}{2}p_z\left\{(1-k)\left(1+2\frac{a^2}{r^2}-3\frac{a^4}{r^4}\right)\sin 2\theta\right\}$$

$$u_r = -\frac{p_z a}{4G}\left\{(1+k) - (1-k)\left(4(1-\nu)-\frac{a^2}{r^2}\right)\cos 2\theta\right\}$$

$$u_\theta = -\frac{p_z a}{4G}\left\{(1-k)\left(2(1-2\nu)+\frac{a^2}{r^2}\right)\sin 2\theta\right\}$$

Figure 19.10 Stresses and displacements induced around a circular excavation in plane strain (for a CHILE material).

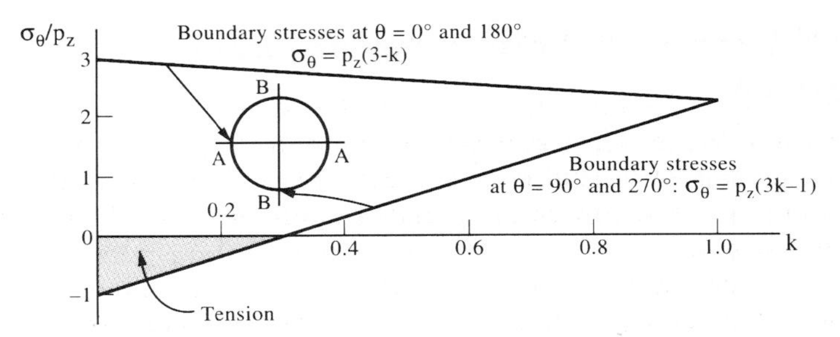

Figure 19.11 Stress concentration factors due to a circular opening.

The highlights of this diagram are:

(a) under all stress fields, the opening alters the pre-existing state of stress, i.e. the opening produces stress concentrations;
(b) there is a linear variation with k of the stress concentrations at the points A and B (and, indeed, everywhere on the boundary);
(c) in a uniaxial stress field ($k = 0$), the maximum stress concentration is 3 (i.e. *compressive*), and the minimum stress concentration is -1 (i.e. *tensile*);
(d) for a hydrostatic stress field ($k = 1$), the stress concentration is 2 everywhere on the boundary (note that this may be demonstrated using the information in (c) above, where the superposition of two orthogonal and equal uniaxial stress fields results in the stress concentration of $3 + -1 = 2$);
(e) tension on the boundary can only occur if $k < 1/3$.

In the hydrostatic case ($k = 1$), the stress concentration around the excavation boundary is always $2p_z$. The solution for stresses anywhere within the rock mass for this stress state is similarly simplified because there are no shear stresses: the terms $(1 - k)$ are all zero. Hence the equations for radial and tangential stress reduce to

$$\sigma_r = p_z\{1 - (a^2/r^2)\} \text{ and } \sigma_r = p_z\{1 + (a^2/r^2)\}.$$

For many practical applications, it is useful to superpose the solution for the stresses induced in the rock by a uniform internal pressure, p, with such a pressure being due to either fluid pressure (water or mud for boreholes) or support pressure (for tunnels and shafts). The contributions made by an internal pressure to the radial stress, tangential stress, radial displacement and tangential displacement are, respectively,

$$\sigma_r = p(a^2/r^2),\ \sigma_\theta = -p(a^2/r^2),\ u_r = pa^2/2Gr \text{ and } u_\theta=0.$$

If, again, we consider the case when $k = 1$, but now the opening is internally pressurized, the superposition of the above solutions gives

$$\sigma_r = p_z - (p_z - p)(a^2/r^2) \text{ and } \sigma_\theta = p_z + (p_z - p)(a^2/r^2).$$

From these equations we can see that:

(a) when $p = p_z$, the internal pressure replaces the hydrostatic stress field present in the rock before excavation and then $\sigma_r = \sigma_\theta = p$;
(b) considering p as a support pressure in a tunnel, the magnitude of p is typically very low compared to that of p_z, and so has little influence on either σ_r or σ_θ;
(c) by pressurizing the fluid in a borehole, it is possible to produce conditions where $p > p_z$, and if $p > 2p_z$, then σ_θ will become negative, i.e. tensile, and, depending on the tensile strength of the rock, hydraulic fracturing may occur as shown in Fig. 4.5.

Several special cases have been given here, and by extension the ideas developed can be considered for more complex situations. One concept that can be elegantly demonstrated from the Kirsch equations is the principle of the conservation of load.

Conservation of load. Figure 19.12 shows, by means of sketches representing different stages in a hypothetical excavation process, how the distribution of vertical stress across a horizontal plane changes. The argument can be used to analyse the stress distribution on any plane—we have chosen the horizontal plane coincident with the centre of the excavation for the sake of convenience. Fig. 19.12(a) indicates this cross-section through a CHILE rock mass, with the future excavation shown as a dotted line, and the horizontal plane shown as a dashed line. Fig. 19.12(b) shows a free body diagram of the rock above this horizontal plane. In this case, the effect of

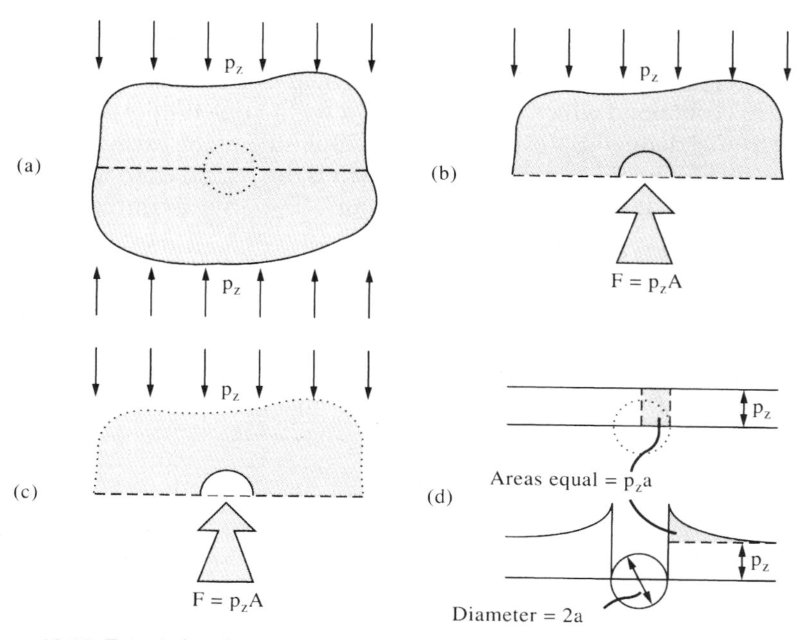

Figure 19.12 Principle of conservation of load before and after excavation.

the lower mass of rock is replaced with a statically equivalent force p_zA, where A is the area of the imaginary plane under consideration. If one now considers the case after the tunnel has been excavated, as shown in Fig. 19.12(c), it can be seen that the portion of rock mass excavated no longer transmits any stress, but the statically equivalent force must remain the same as in Fig. 19.12(b) to preserve equilibrium. However, this force now acts over a smaller area than before (because the rock which is in the location of the tunnel has been removed) and so the stress across the horizontal plane must be increased.

The load no longer carried by the rock removed from the tunnel is $p_z \times 2a$, per unit length of tunnel, and, in order to preserve equilibrium, this must equal the load re-distributed into the surrounding rock (Fig. 19.12(d)). This can be demonstrated by using the Kirsch equations for a uniaxial stress state (i.e. $k = 0$), and determining the vertical stress across the horizontal plane in question by taking the expression for tangential stress with $\theta = 0$,

$$\sigma_\theta = p_z\{1 + (a^2/2r^2) + (3a^4/2r^4)\}.$$

To obtain the total redistributed load, subtract p_z from this expression and integrate between the limits of a and ∞ (the Kirsch equations are defined for an infinite volume of rock, which implies the area A of the horizontal plane is also infinite) for both sides of the tunnel,

$$\text{redistributed load} = 2p_z\int_a^\infty \left(\frac{a^2}{2r^2} + \frac{3a^4}{2r^4}\right)\mathrm{d}r$$

which reduces to $2p_za$, the load bearing capability lost by tunnel excavation.

By integrating between limits of a and $3a$, or a and $5a$, the percentage of the load is obtained which is re-distributed within one tunnel diameter and three tunnel diameters, respectively, on either side of the excavated tunnel. These are 81.5% and 89.6%, showing that it is the region close to the tunnel in which the load re-distribution occurs. For this example, the load

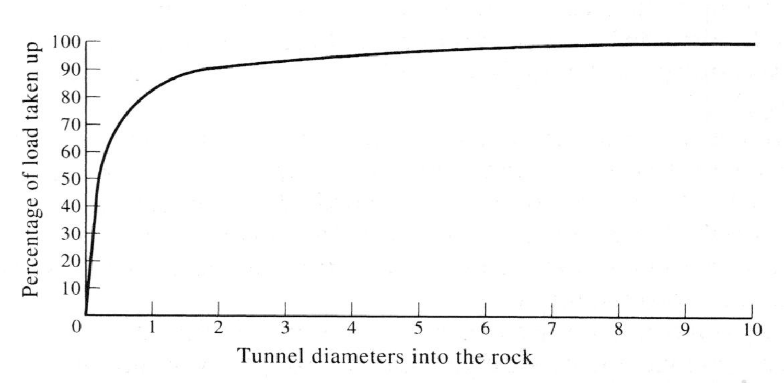

Figure 19.13 Redistribution of vertical load adjacent to a circular tunnel.

redistribution over 10 tunnel diameters is shown in Fig. 19.13. Other interesting statistics related to this re-distribution are that 50% of the load is redistributed between the tunnel boundary and 0.23 tunnel diameters into the rock, and that 95% of the load is re-distributed between the tunnel boundary and 4.5 tunnel diameters into the rock.

Note that the curve in Fig. 19.13 applies to the load re-distribution rather than the stress re-distribution. The curve shown is the cumulative load re-distribution, i.e. the integration of the stress distribution represented by the shaded area in Fig. 19.12(d).

19.2.2 Stresses around elliptical openings

The stresses around elliptical openings can be treated in an analogous way to that just presented for circular openings. There is much greater utility associated with the solution for elliptical openings than circular openings, because these can provide a first approximation to a wide range of engineering geometries, especially openings with high width/height ratios (e.g. mine stopes and civil engineering caverns). From a design point of view, the effects of changing either the orientation within the stress field or the aspect ratio of such elliptical openings can be studied to optimize stability.

Elliptical openings in isotropic rock. An elliptical opening is completely characterized by two parameters: aspect ratio (the ratio of the major axis to the minor axis) which is the eccentricity of the ellipse; and orientation with respect to the principal stresses (measured, for example, in terms of the angle between the major axis and the major principal stress). Bray (1977) derived a suite of equations for the state of stress around an elliptical opening in terms of these parameters and the Cartesian co-ordinates of the location of the point in question. These equations are given in Fig. 19.14, with reduced forms in Figs 19.15 and Fig. 19.16 for the cases of tangential stress on the boundary of an arbitrarily orientated excavation and the tangential stress on the boundary of an excavation orientated with its axes aligned with the principal stress directions, respectively. The diagram in Fig. 19.14 shows how the angle θ defines the orientation of the local reference axes l, m relative to the ellipse local axes x_1, z_1. In Fig. 19.15, the position on the boundary, with reference to the x-axis, is given by the angle χ, and in Fig. 19.16 the ellipse is aligned by taking $\beta = 0$.

It is instructive to consider the maximum and minimum values of the stress concentrations around the ellipse for the geometry of the ellipse in Fig. 19.16. It can easily be established that the extremes of stress concentration occur at the ends of the major and minor axes—points A and B in Fig. 19.16—and the corresponding stress magnitudes are as given by the equations in the figure.

In a given engineering context, k cannot generally be altered and so any design optimization must be performed through a variation in q, which is usually possible. An optimal design can be defined as one in which the maximum stress concentration is minimized. Fig. 19.17 shows how the stress concentrations at A and B vary with q, and demonstrates that the two concentrations are equal when $q = k$. Thus, an elliptical excavation has an optimal shape when the eccentricity of the ellipse is harmonized with the ratio of field stresses—an elegant result.

Elliptical openings in anisotropic rock. In order to allow for the real nature of rock, the solution for the stresses around an elliptical opening can be extended to take into account the case of transversely isotropic rock. The extension is realistic for many rock types, because transverse isotropy is a good representation of sedimentary or metamorphic rocks. There are now

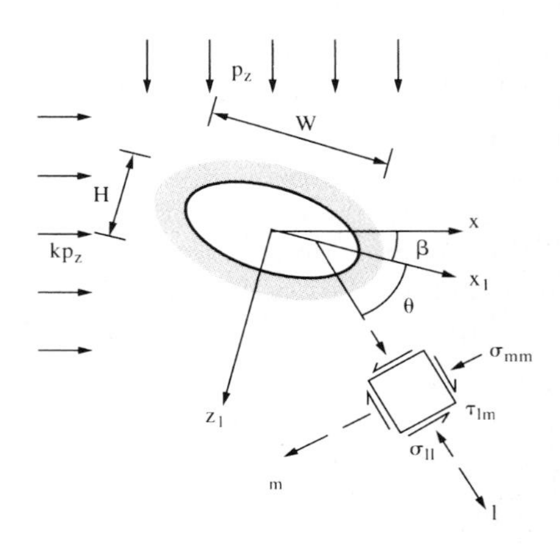

$$\sigma_{ll} = \frac{p(e_o - e)}{J^2}\left\{(1+k)(e^2-1)\frac{C}{2e_o} + (1-k)\left[\left\langle\frac{J}{2}(e-e_o) + Ce\right\rangle \cos 2(\psi+\beta) - C\cos 2\beta\right]\right\}$$

$$\sigma_{mm} = \frac{p}{J}\left\{(1+k)(e^2-1) + 2(1-k)e_o\left[e\cos 2(\psi+\beta) - \cos 2\beta\right]\right\} - \sigma_{ll}$$

$$\tau_{lm} = \frac{p(e_o - e)}{J^2}\left\{(1+k)\frac{Ce}{e_o}\sin 2\psi + (1-k)\left[e(e_o+e)\sin 2\beta + e\sin 2(\psi-\beta) - \left\langle\frac{J}{2}(e_o - e) + e^2 e_o\right\rangle \sin 2(\psi+\beta)\right]\right\}$$

where the following geometrical parameters are defined

$$e_o = \frac{(W+H)}{(W-H)},\quad b = \frac{4(x_1{}^2 - z_1{}^2)}{(W^2 - H^2)},$$

$$d = \frac{8(x_1{}^2 - z_1{}^2)}{(W^2 - H^2)} - 1,\quad u = b + \frac{e_o}{|e_o|}\sqrt{(b^2 - d)},$$

$$e = u + \frac{e_o}{|e_o|}\sqrt{(u^2 - 1)},\quad \psi = \arctan\left[\left(\frac{e+1}{e-1}\right)\frac{z_1}{x_1}\right],$$

$$\theta = \arctan\left[\left(\frac{e+1}{e-1}\right)^2 \frac{z_1}{x_1}\right],$$

Figure 19.14 Stresses induced around an elliptical excavation in plane strain for a CHILE material (after Bray, 1977; from Brady and Brown, 1985).

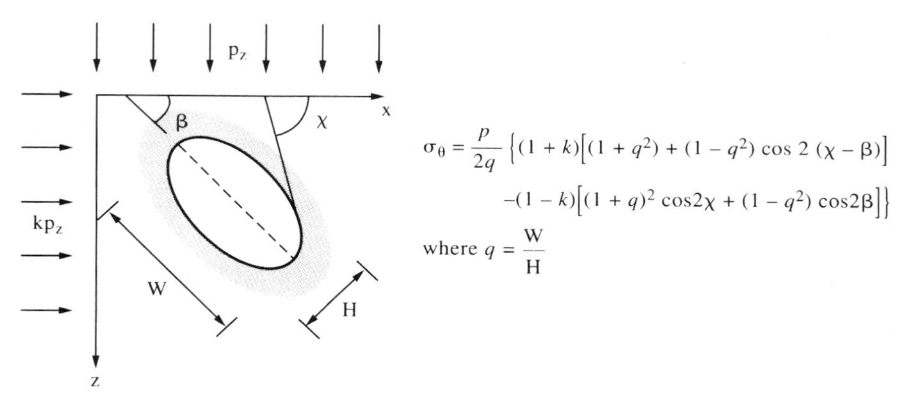

Figure 19.15 Stresses induced on the boundary of an elliptical excavation in plane strain for a CHILE material (after Bray, 1977; from Brady and Brown, 1985).

three parameter groups describing the problem—these relate to the aspect ratio of the opening, the ratio of the *in situ* principal stresses, and the five elastic moduli for a transversely isotropic material.

The cross-section through the elliptical excavation together with the salient geometrical parameters and the field stresses are shown in Fig. 19.18(a). The general three-dimensional stress field and the model chosen to represent the transversely isotropic rock are shown in Fig. 19.18(b). Note that the element shown in Fig. 19.18(b) represents the state of stress at a point, and the stress components indicated represent local stresses; this is in contrast to Fig. 19.18(a) where the field stresses are indicated.

Very often, long excavations have their longitudinal axis aligned with the strike of the plane of isotropy and therefore the problem can be simplified by assuming plane strain and hence only having to take into account four material properties—this is shown in Fig. 19.18(c).

These ideas are used with associated equations in connection with discussion on zones of influence presented in Chapter 20.

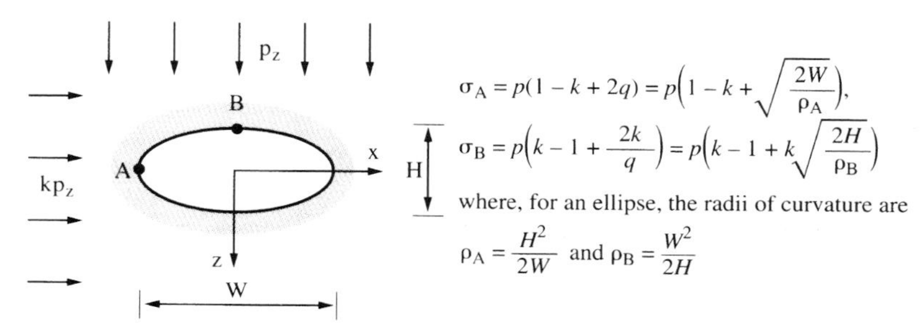

Figure 19.16 Stresses induced on the boundary of an elliptical excavation (aligned with axes parallel and perpendicular to the principal stresses) in plane strain for a CHILE material (after Bray, 1977; from Brady and Brown, 1985).

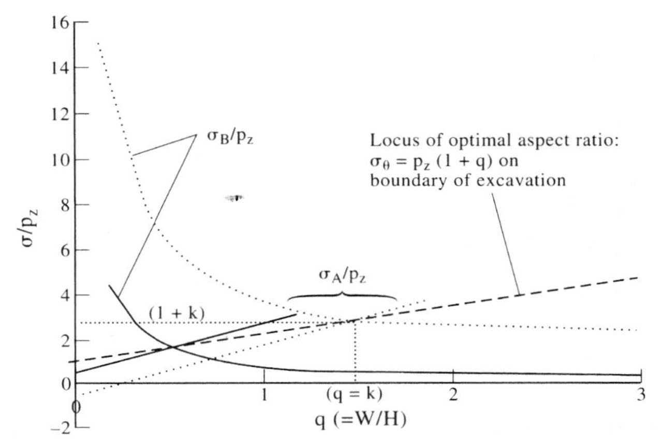

Figure 19.17 Optimal aspect ratio for an elliptical excavation.

19.2.3 Analysis of instability around underground openings

There are three main modes which will be addressed here:

(a) fracture zones around the excavations caused by stress induced failure of the intact rock;
(b) the possibility of slip on pre-existing discontinuities because of the induced stress field; and
(c) the special case of slip within a stratified rock.

Development of fracture zones. The discussion of fracture zones is illustrated with reference to circular excavations in plane strain, but the ideas apply to all excavations. In Fig. 19.19, there is a zone shown around the opening where the Mohr–Coulomb criterion for the intact rock has been satisfied. For the conditions of a hydrostatic field stress, as shown, this zone is circular and concentric with the centre of the opening.

Closed form solutions for the radial extent of the fracture zone, the stresses within it, and the stresses within the remaining elastic zone can be derived from first principles for this geometry and stress state—with the usual CHILE assumptions. The solutions are given in Fig. 19.19.

Although these equations apply for an idealized case, they can provide guidance to intact rock failure potential and to what extent the rock might be damaged. The expressions for stresses within the fractured zone and the radius of the fractured zone both contain the parameter p_i, the internal pressure. This pressure may be a fluid pressure (water or drilling mud, for example) or may be produced by the installation of mechanical support. In the latter case, the equations enable one to examine the effect of support on the stability of an excavation. This theme will be continued in connection with the ground response curve in Chapter 20.

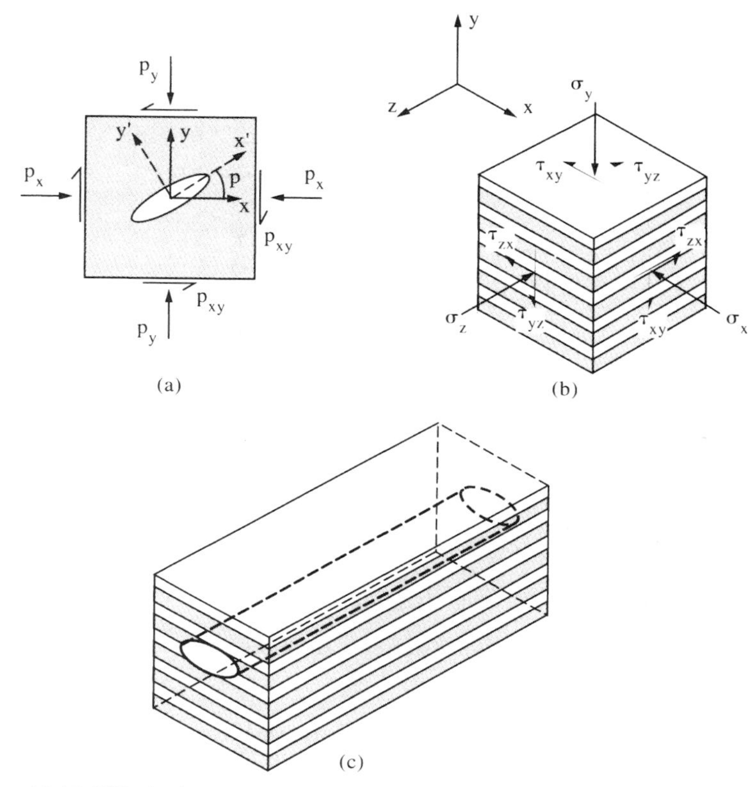

Figure 19.18 Elliptical openings in transversely isotropic rock.

Slip on pre-existing discontinuities. Another possibility is that the rock has been weakened by the presence of a pre-existing discontinuity. Assume that this discontinuity does not affect the elastic constants in any way, and so the usual CHILE assumption is valid, but the strength of the rock is reduced on the discontinuity. The extent of any potential zone of instability can be established by considering whether the induced stresses locally satisfy the discontinuity shear strength criterion.

In Fig. 19.20, there is a discontinuity in the vicinity of a circular opening. The specific procedure is then to take a point on the discontinuity to start (for computational convenience, we have chosen the closest point on the discontinuity to the centre of the opening), evaluate the stress components at the point using the Kirsch equations, transform these components into normal and shear stress components acting on the discontinuity, and finally substitute them into the Mohr–Coulomb (or any other suitable) criterion. This procedure enables a graph of the ratio of actual strength to required strength versus the parameter d to be drawn, and an example (for a cohesionless discontinuity) is also shown in Fig. 19.20.

From this curve, there is an indication of the location and intensity of the damage the discontinuity might sustain as a result of the engineering. In the graph, the line representing tan ϕ is drawn and hence the extent of the zone of potential slip is studied. The length of the zone will depend on

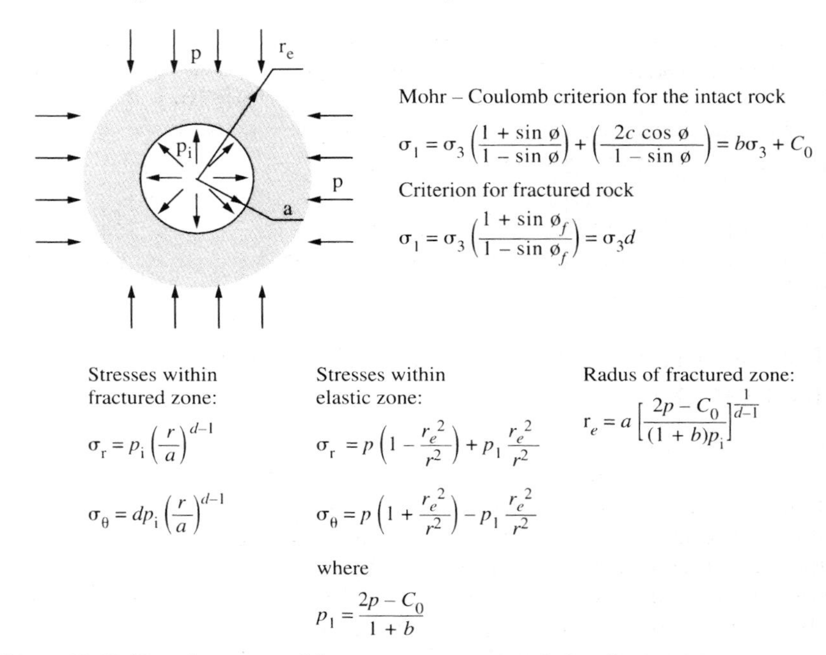

p
r_e
p_i
p
a
Mohr – Coulomb criterion for the intact rock
$\sigma_1 = \sigma_3 \left(\frac{1 + \sin \phi}{1 - \sin \phi}\right) + \left(\frac{2c \cos \phi}{1 - \sin \phi}\right) = b\sigma_3 + C_0$
Criterion for fractured rock
$\sigma_1 = \sigma_3 \left(\frac{1 + \sin \phi_f}{1 - \sin \phi_f}\right) = \sigma_3 d$
Stresses within fractured zone:
$\sigma_r = p_i \left(\frac{r}{a}\right)^{d-1}$
$\sigma_\theta = dp_i \left(\frac{r}{a}\right)^{d-1}$
Stresses within elastic zone:
$\sigma_r = p\left(1 - \frac{r_e^2}{r^2}\right) + p_1 \frac{r_e^2}{r^2}$
$\sigma_\theta = p\left(1 + \frac{r_e^2}{r^2}\right) - p_1 \frac{r_e^2}{r^2}$
where
$p_1 = \frac{2p - C_0}{1 + b}$
Radus of fractured zone:
$r_e = a \left[\frac{2p - C_0}{(1 + b)p_i}\right]^{\frac{1}{d-1}}$

Figure 19.19 Development of fracture zones around circular openings.

the orientation of the discontinuity with respect to the field stresses, the proximity of the discontinuity to the excavation, and the strength parameters of the discontinuity itself.

The analysis, although idealized, is useful for evaluating the likely influence of the parameters on construction.

Special case of stratified rock. There can be cases of composite instability in which the stress induces slip on pre-existing planes of weakness, with the

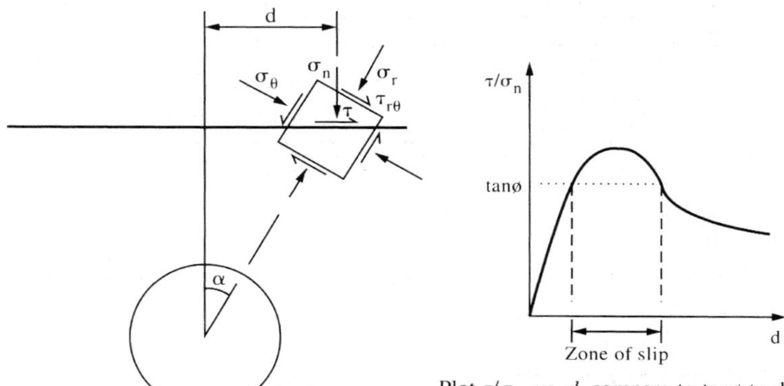

Figure 19.20 Slip on pre-existing discontinuities.

possibility of discrete blocks developing as a consequence. This is especially likely in the case of excavations in stratified rock, when the perturbed stress field causes interlayer slip, which in turn could lead to joints opening and the resulting possibility of blocks falling. This is illustrated in the suite of diagrams in Fig. 19.21. The slip is similar to the case described immediately above, except that the discontinuities may be regarded as ubiquitous, with the location of the slip being at the corners of such an excavation, where the angle between stress trajectories and the discontinuities is at its most adverse.

19.3 A note on time-dependency and weathering

We have concentrated on the mechanisms and associated solutions for simple cases in order to demonstrate the value of understanding the basic principles. We have utilized idealized cases for a CHILE rock. The rock is actually discontinuous, but in the last two mechanisms we did consider the influence of introducing discontinuities. The effects of inhomogeneity and anisotropy will have to be explicitly studied through the use of numerical analysis, although in many cases the trends would be similar.

The last of the differences between CHILE and DIANE rocks is that the latter do have mechanisms which are time-dependent. We would expect that the influence of time, creep and stress relaxation might assist engineering because stress concentrations would be reduced and displacements are not instantaneous. However, the more insidious aspect of time-dependency is that the material itself might lose strength: whilst the stress concentrations might reduce asymptotically to a given level, the strength of the rock might continue to decrease over the months and years through many processes, collectively termed weathering.

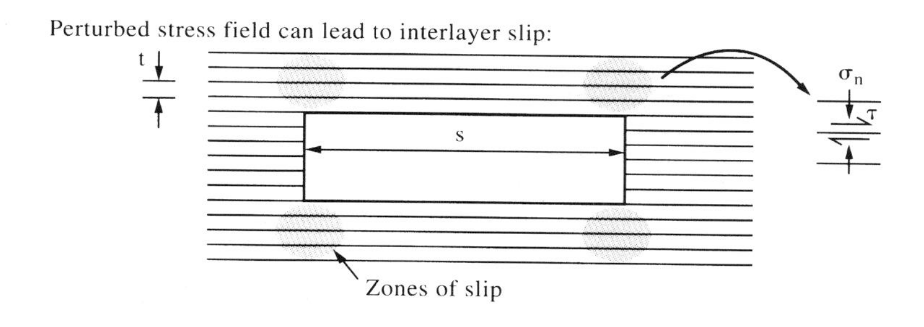

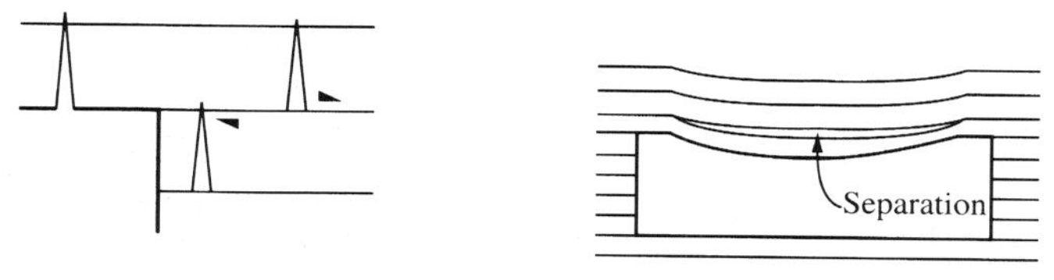

Figure 19.21 Composite instability around excavations in stratified rock.

There are index tests which assist in characterizing aspects of weathering, such as the slake durability test, although it is difficult to present the information with the same impact as the purely mechanical concepts we have dealt with above. Many aspects of the subject of weathering can be raised, especially in connection with the interaction matrices introduced in Chapter 14. At this juncture, we seek only to remind the reader that time-dependent mechanisms do exist and need to be modelled.

20 Design and analysis of underground excavations

In Chapter 19, we concentrated on the specific instability mechanisms relating to underground excavations. These were instability due to block movement, instability due to stress effects, and circumstances where both occur. In this chapter, we consider these mechanisms within the overall context of the design and analysis of underground excavations. One does not know, *a priori*, which mechanisms are the ones that will be operating, and hence the ones to defend the excavation against.

There are two essential precursors to such design and analysis. The first is a mechanical understanding of the rock mass and the requirements of the project being undertaken; the second is the consideration of which mechanisms are relevant, including factors such as the presence of a nearby fault, which may or may not have been detected by site investigation. With these provisos, the subjects described in this chapter will provide guidance for the design and analysis of all underground excavations.

20.1 Design against structurally-controlled instability

20.1.1 Background—pragmatic design, orientation and size effects

In this section, the foundation laid in Section 19.1 is used to consider design against structurally-controlled instability. Even without detailed analysis of the blocks formed by discontinuities, a great deal can be achieved by pragmatic design of roofs and other excavation boundaries. The procedure is the intelligent general use of the principles already described. In Fig. 20.1, there are two cases where instability in the roof is apparent from underground observations. One can make useful decisions on excavation location, orientation and support without detailed mechanical analysis—and one can 'design with the rock'.

In the left-hand diagram of Fig. 20.1, the roof area is likely to be unstable and one option for improving stability would, if viable, be to ensure that the excavation were located in the massive rock. Similarly, in the right-hand

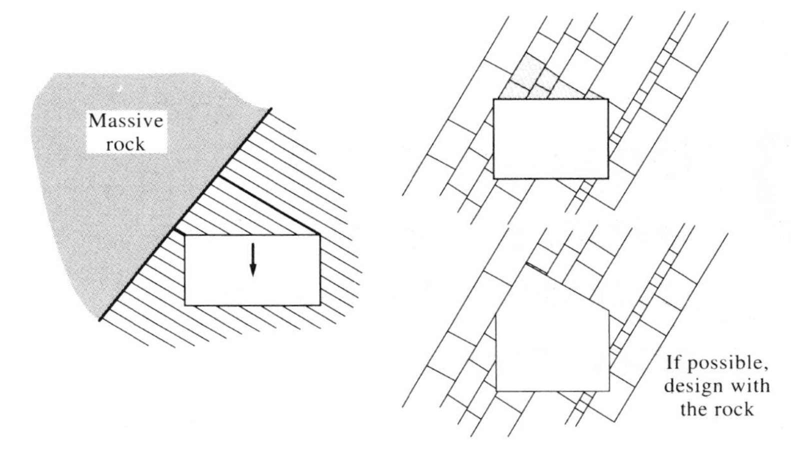

Figure 20.1 Assessing roof and sidewall structurally-controlled instability from visual observations.

part of Fig. 20.1, the stippled area of rock blocks is likely to be unstable, and, if acceptable, changing the designed outline of the excavation to that shown would harmonize the structure with the rock and reduce the degree of instability.

Another example is shown in Fig. 20.2, in which the excavation orientation is considered. As a general principle, and as is clear from the figure, tunnels parallel to the strike of discontinuity sets are likely to have more instability problems. Similarly, with reference to Fig. 20.3, a large excavation will be more unstable than a small excavation in the same rock mass.

Much can be readily estimated by careful observation. It is apparent by direct observation whether a rock type is susceptible to weathering. Similarly, visual assessment of rock discontinuity surfaces can provide guidance on their mechanical behaviour. This is not to say that we advocate abandoning more sophisticated forms of analysis: it is just that a great deal can be learnt from observation, especially once the rock mechanics principles are understood.

20.1.2 Elastic analysis applied to stratified rock

Beam analysis. Following the pragmatic design approach, more analytical approaches are now considered. One of the oldest problems that has been considered in rock engineering is that of roof beam flexure—because the main design factor for many excavations in stratified rock is the maximum acceptable unsupported roof span. In Fig. 20.4, there is an example of the flexure of a beam above an excavation. With this geometry we can calculate both the maximum tensile stress induced in such beams by their self-weight, and also the maximum deflection of the beams. In the case where the beams thin upwards, the strata will not separate directly above the opening—providing the elastic properties of the beams are equal. Conversely, if the beams thin downwards, strata separation will

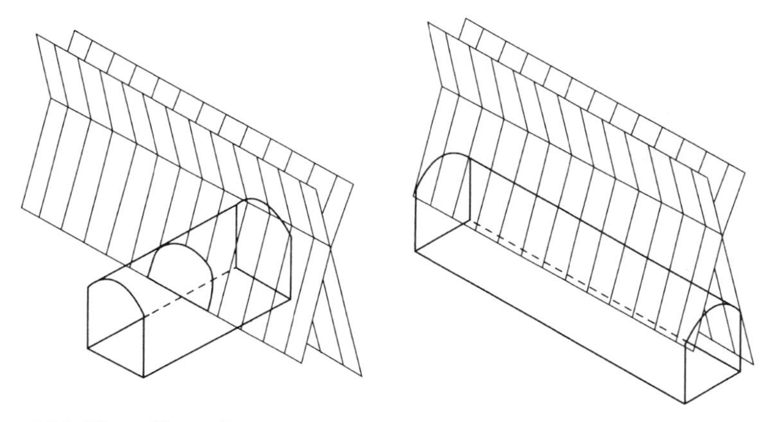

Figure 20.2 The effect of excavation orientation in relation to discontinuity set orientation.

occur immediately above the roof. A simple elastic analysis of beam flexure provides helpful indications of the stability of, and type of support that may be required for, a stratified roof.

The equations resulting from this elastic analysis are shown in the lower part of Fig. 20.4. The analysis is two-dimensional and thus is most realistic for the cross-section of a long excavation. It must be remembered that these equations apply to a CHILE material with the only discontinuities being the planes of stratification. Furthermore, a high tensile strength for the rock is required, not only in small zones of the beams, but on the scale of the excavation itself. Despite these shortcomings, the equations provide useful insight into the mechanical behaviour of stratified rock structures.

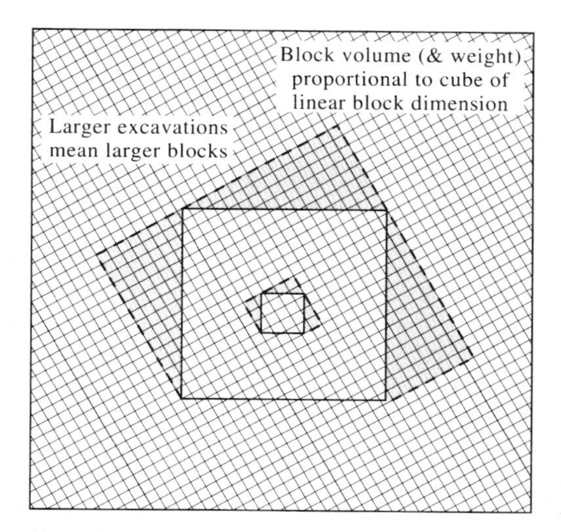

Figure 20.3 The effect of excavation dimensions on the size of potentially unstable blocks.

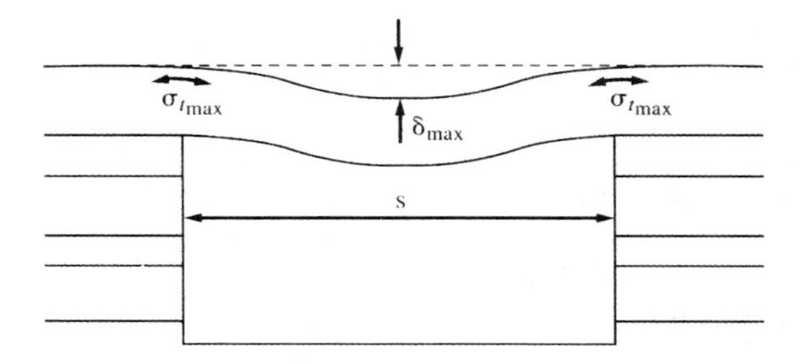

$$\sigma_{t\max} = \frac{\gamma s^2}{2t} \qquad \delta_{\max} = \frac{\gamma s^4}{32Et^2}$$

When the strata thin upwards, replace γ in these equations with γ_a:

$$\gamma_a = \frac{E_1 t_1^2 (g_1 t_1 + g_2 t_2 + \cdots + g_n t_n)}{E_1 t_1^3 + E_2 t_2^3 + \cdots + E_n t_n^3}$$

When a uniform pressure, p, acts in the zone of separation:

$$\sigma_{t\max} = \frac{\gamma s^2}{2t} + \frac{ps^2}{2t} \qquad \delta_{\max} = \frac{\gamma s^4}{32Et^2} + \frac{ps^4}{32Et^3}$$

Figure 20.4 Deflection of roof strata assumed to act as elastic beams.

Voussoir arch model. The analysis in Fig. 20.4 is only valid for a CHILE material. However, it is possible to have a stable roof in a discontinuous medium. This is the reason why many of the ancient constructions using rock blocks have been possible. Figure 20.5 shows a photograph of an ancient bridge over a river in northern Turkey. The span of the bridge is comprised of a *single* set of rock blocks with no binding agent; this type of arch is called a voussoir arch—and examples are common in masonry structures.

Figure 20.5 Masonry bridge demonstrating a voussoir arch comprised of a single set of rock blocks (northern Turkey).

The principle of the voussoir arch is open to some discussion, but a block will not fall from the arch providing the vertically acting resisting forces are greater than the sum of the weight of the block and any superimposed forces. These resisting forces are, in turn, generated by the frictional and dilational properties of the discontinuities between the blocks, which result from horizontal or sub-horizontal forces induced in the arch by the weight of the blocks themselves. Viewing the arch in Fig. 20.5, one can imagine how the weight of the central block of the arch is transmitted via the inter-block frictional shear forces to the abutments of the arch.

It is possible, as shown in the top of Fig. 20.6, to have an arch where the radius of curvature is large, with the result that the arch becomes essentially flat. A voussoir arch will have compressive stresses between all the blocks and between the abutments. Although the arch is stable, and examples such as that in Fig. 20.5 have survived for hundreds of years, the whole arch is vulnerable to any block perturbation, which will affect the total arch integrity: the voussoir arch is like a chain—all the elements have to be present in order for it to function. Thus, although the voussoir arch principle is elegant, the arch is not a robust design for underground excavation roofs. We can use the principle, but the inherent vulnerability of the arch needs to be overcome.

To do this, it is necessary to have an understanding of the precise mechanics of the voussoir arch and its potential modes of instability which can involve either insufficient or excessive inter-block compressive stress. Underground, the voussoir arch is not built from masonry blocks: the arch comes into existence once the empty space beneath it has been created by excavation, with the result that there are thus two stability aspects to consider—is the arch stable in the first place, and will it remain so?

In the event that the geometry of the arch is such that sufficient inter-block compressive stress is not generated to mobilize sufficient frictional shear forces, a block will become unstable, leading to complete collapse of the arch. Such circumstances can arise from lateral movement of the abutments, highly compliant elements in the arch (for example, low modulus of elasticity of the intact rock and low stiffness discontinuities) or simply an inappropriate rock block geometry. Conversely, if the geometry

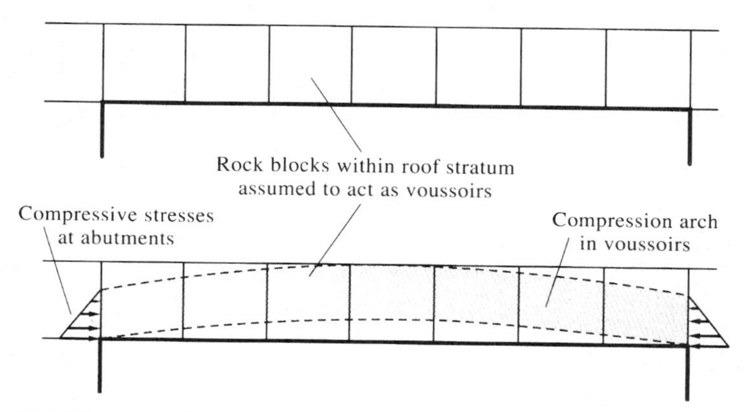

Figure 20.6 The voussoir arch in underground excavations.

of the arch is such that excessive inter-block compressive stress is generated, then instability will arise because the strength of the rock blocks forming the arch elements is reached, the block integrity is prejudiced and the arch collapses. Finally, buckling instability can occur if the ratio of the induced compressive stress to the slenderness of the arch becomes too large—for example, if a thin, highly competent, and thus highly stressed stratum forms the rock structure.

Rock bolting. Elastic materials do not fail (because the definition of elasticity is that all strain energy is recoverable). If voussoir arch mechanics is assisting in the stability of the roof but the roof is vulnerable to perturbations of stresses and strains, rock bolts can be installed in the roof strata, thus connecting the suite of potential voussoir arches and maintaining essentially elastic behaviour. This is illustrated in Fig. 20.7. Not only do the rock bolts reinforce the strata, but any block movement could lead to more stable conditions, due to the bolt forces being increased.

A first estimate with a simple model of the induced rock bolt tension is obtained by assuming that each bolt supports the representative prism of rock surrounding it as shown in Fig. 20.7. If the bolts are arranged on a square grid with a spacing of s metres and the depth of the rock prism is D metres, then the tension, T, required is simply the rock prism weight, i.e. $T = \gamma D s^2$ kN, where γ is the unit weight of the rock (kN/m^3). For example, to support strata with a unit weight of 23 kN/m^3 for a depth of 3 m using a rock bolt spacing of 1 m, a rock bolt tension of 69 kN (i.e. close to 7 tonnes) is indicated.

The calculation tacitly assumes that the bolts are ungrouted and anchored solely at the end embedded in the rock. With bolts that are grouted along their length, the support mechanics are more complicated, but in terms of stabilizing the voussoir arches they are more effective.

20.1.3 Support of falling and sliding blocks

The idea of providing the necessary force to retain blocks in the roof can be extended to the falling and sliding blocks which were discussed in Chapter 19. The calculation is achieved by determination of the block weight for the case of a falling block, with a modification to account for the angle of sliding and the effect of frictional resistance in the case of a sliding block. The calculation can be made more rigorous by accounting for the effect of the stresses present in the rock adjacent to the boundary of the excavation.

Simple falling and sliding analysis. The left-hand diagram of Fig. 20.8 shows a lower-hemispherical projection of three discontinuity planes which, together with the excavation surface, form a tetrahedral block in the roof of an excavation. In addition, diametral lines indicating the trend of a long excavation and the strikes of the three discontinuity planes are included. In the right-hand diagram of Fig. 20.8, there is a plan view of the associated largest block that can be formed by these four surfaces, in the roof of an excavation with a specific width. Note that the dashed lines representing

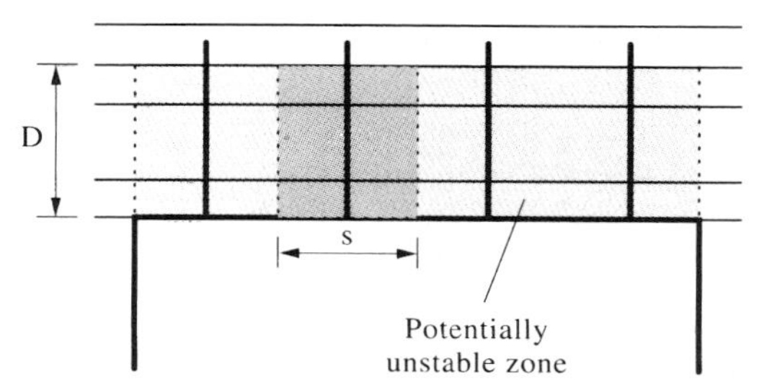

Figure 20.7 Simple model of rock bolts in stratified roofs.

the excavation side walls in the right-hand diagram correspond to the trend of the excavation in the left-hand diagram. The dotted lines in both parts of Fig. 20.8 represent the base edges and upper edges of the block, as these are formed by the intersections of the discontinuity planes. This demonstrates the geometric congruence of the lower-hemispherical projection and the plan view.

In a similar way to the calculation of the rock bolting requirements in a stratified roof (Fig. 20.7), a bolting configuration can now be established for any discontinuity geometry, assuming that the objective is to support the deadweight of individual blocks as calculated in Fig. 20.8. The circumstances may not be as simple as accounting for individual blocks because of the possibility of smaller blocks being formed behind the largest block identified and, hence, it is necessary to determine the bolt lengths by establishing the optimal anchorage positions.

If the block is not falling but sliding then there is an element of constraint, and the calculations can be modified to account for this frictional resistance,

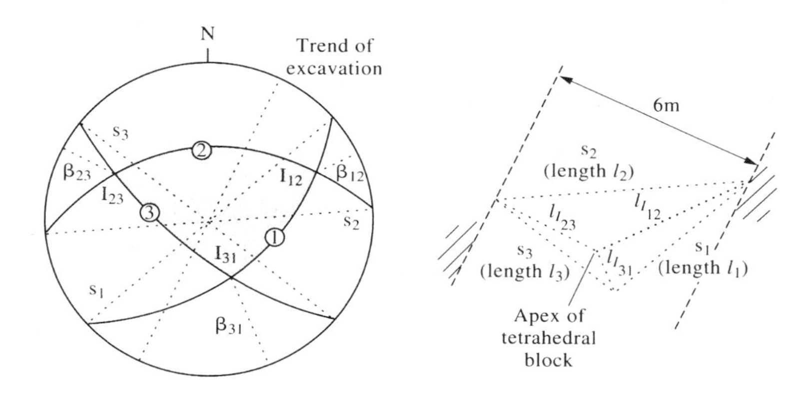

$A_f = \frac{1}{2} l_1 l_2 \sin\theta_{12} = \frac{1}{2} l_2 l_3 \sin\theta_{23} = \frac{1}{2} l_3 l_1 \sin\theta_{31}$ and $h = l_{l_{12}} \tan\beta_{12} = l_{l_{23}} \tan\beta_{23} = l_{l_{31}} \tan\beta_{31}$

Block volume, $V = \frac{1}{3} h A_f$ block weight, $W = \gamma V$ support pressure, $p = \frac{W}{A_f} = \frac{1}{3}\gamma h$

thus $A_f = 10.07\text{m}^2$, $h = 1.48\text{m}$, $V = 4.97\text{m}^3$, $W = 114.3\text{kN}$ and $p = 11.35\text{kPa}$ for the example here.

Figure 20.8 Lower-hemispherical projection of three discontinuity planes, and the associated maximal tetrahedral block in the roof of an excavation with given width and orientation.

A = area of sliding surface

$\left.\begin{array}{l} c \\ \phi \end{array}\right\}$ = Coulomb strength parameters

$$F = \frac{cA + (W \cos \psi + T \cos \theta) \tan \phi}{W \sin \psi - T \sin \theta}$$

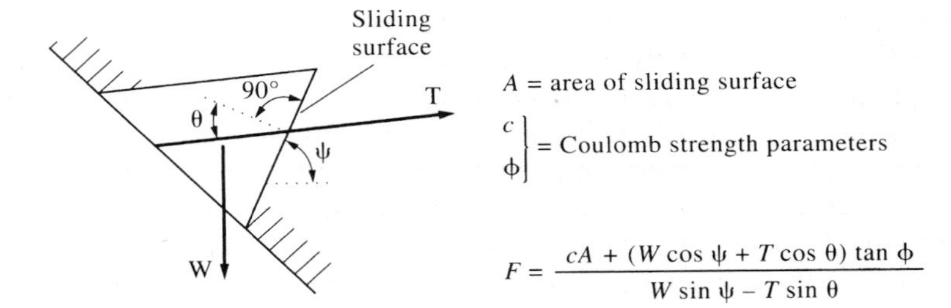

The effect of stresses in the rock has been ignored, and this analysis is for sliding on one discontinuity surface only.

Figure 20.9 Analysis of a tetrahedral block sliding on one face.

as illustrated in Fig. 20.9. This calculation is for sliding on a discontinuity surface, and hence is analogous to the plane sliding instability for slopes discussed in Chapters 17 and 18. The formula for the rock bolt force, *F*, in Fig. 20.9 is a modification to that presented in Section 17.1.2 and, from Section 16.3.2, this is minimized for these idealized circumstances when $\theta = 90 - \phi$.

Symmetric and asymmetric triangular roof prisms. To incorporate the effect of the stresses in the rock surrounding the excavation, Fig. 20.10 shows an extended analysis considering the forces in a symmetric triangular roof prism. The purpose of this analysis is to establish the influence of a horizontal force H_o on the block, and its influence on the support force necessary to hold the block in place. The presence of any such compressive stress in the rock (and hence compressive force on the block) will affect the support force required: depending on the wedge apical angle and other mechanical parameters, the block will either tend to be squeezed out or be constrained. The support force, *R*, is given by *W–P*, where *W* is the block weight and *P* is the resultant of the frictional forces acting on the block, expressed as positive downwards. The expressions for *P* are given in Fig. 20.10.

Assuming $S = N \tan \phi$ (i.e. no cohesion) and resolving forces vertically

$$P = \frac{2N \sin (\phi\text{-}\alpha)}{\cos \phi}$$

If $N > 0$, then $P > 0$ only if $\alpha < \phi$; if $\alpha > \phi$, then $R > W$

Introducing discontinuity stiffnesses and rock stress:

$$P = 2H_o \frac{(k_s \cos^2 \alpha + k_n \sin^2 \alpha)}{(k_s \cos \alpha \cos \phi + k_n \sin \alpha \sin \phi)} \sin (\phi\text{-}\alpha)$$

If $k_n >> k_s$

$$P = 2H_o \frac{\sin \alpha}{\sin \phi} \sin (\phi\text{-}\alpha)$$

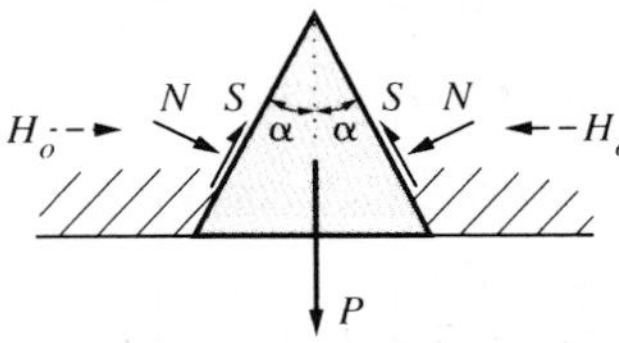

Figure 20.10 Analysis of a symmetric triangular roof prism.

Under the circumstances of a compressive rock stress, it is noted in the figure that there will only be restraint, i.e. $P > 0$, if $\alpha < \phi$. This accords with the ϕ_j theory which was discussed in Section 16.5, for very similar circumstances, i.e. slip around the boundary of an excavation. Also, the k_n and k_s referred to in Fig. 20.10 are the same as the k_{nn} and k_{ss} in Section 7.3.1, except that the second subscript has been omitted because there are no cross-stiffnesses in this case. In Fig. 20.11, we show a variation in $P/2H_o$, as a function of the semi-apical angle for different ratios of normal to shear discontinuity stiffness.

There are three interesting characterisitics indicated in Fig. 20.11:

(a) regardless of the ratio of discontinuity stiffness, P is always zero when $\alpha = \phi$;
(b) there is a tendency for the prism to be expelled from the surface when the semi-apical angle exceeds the angle of friction, which is exacerbated for high values of k_n/k_s; and,
(c) the relation between $P/2H_o$ and α becomes close to linear in the semi-apical angle range of 20–60°.

In the case of asymmetric triangular roof prisms, the analysis presented above has to be extended to include two different semi-apical angles, as shown in Fig. 20.12. A similar sensitivity analysis to the one illustrated in Fig. 20.11 can be generated to show how the components of the apical angle effect the stability of the roof prism.

Tetrahedral blocks. To consider the stability of a general tetrahedral block, through a three-dimensional analysis and taking into account Coulomb friction on the three faces of the block that are in contact with the rock mass, the earlier simple sliding analysis illustrated in Fig. 20.9 can be extended. This extended analysis is shown in Fig. 20.13.

The normal forces on each of the block faces shown in Fig. 20.13 can be obtained by transforming the stress state in the rock to obtain the normal

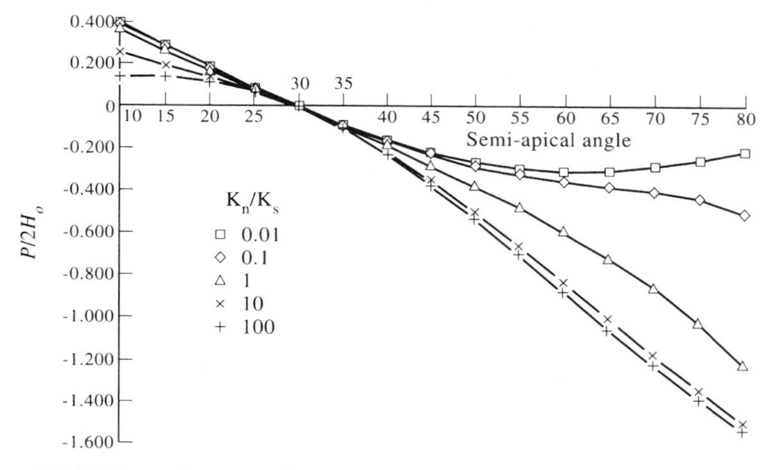

Figure 20.11 Effect of stress in the surrounding rock and discontinuity stiffness on the generation of constraining force for a symmetric triangular roof prism.

Taking into account both the two components α_1 and α_2 of the apical angle, the discontinuity stiffnesses and rock stress, we have for the limiting vertical load:

$$P_1 = \frac{H_0}{D_1}(ks_1\cos^2\alpha_1 + kn_1\sin^2\alpha_1)\sin(\phi_1 - \alpha_1) + \frac{H_0}{D_1}(ks_2\cos^2\alpha_2 + kn_2\sin^2\alpha_2)\sin(\phi_2 - \alpha_2)$$

where

$$D_1 = ks_1\cos\alpha_1\cos\phi_1 + kn_1\sin\alpha_1\sin\phi_1 \text{ and } D_2 = ks_2\cos\alpha_2\cos\phi_2 + kn_2\sin\alpha_2\sin\phi_2.$$

If $kn_1 \gg ks_1$ and $kn_2 \gg ks_2$, then we have

$$P_1 = \frac{H_0\sin\alpha_1\sin(\phi_1 - \alpha_1)}{\sin\phi_1} + \frac{H_0\sin\alpha_2\sin(\phi_2 - \alpha_2)}{\sin\phi_2}$$

which allows for different angles of friction on the two discontinuity surfaces.

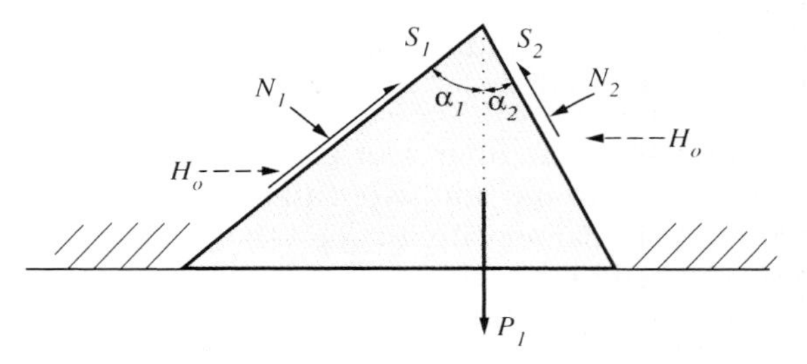

Figure 20.12 Analysis of an asymmetric triangular roof prism.

stresses, with the result that the normal forces can be obtained by multiplying this stress by the area of the face of the block. As shown in Fig. 20.13, the shear force is obtained from the normal force through the application of the Coulomb friction criterion, which defines the analysis as being one of limiting equilibrium.

In Figure 20.8, it has been seen how kinematic analysis may be used to determine the complete geometry of the block, and hence the orientations of the normals and the bisectors.

It is necessary to check the limiting condition on each face to ensure that all individual frictional components are assisting in the maintenance of stability, i.e. are negative (when reckoned in accordance with the axes shown in Fig. 20.13). If this is not the case, it is prudent to assume that progressive failure may take place through initial rotation of the block.

If the wedge is unstable, then the degree to which the weight is in excess of the constraining force can be used to indicate a factor of safety and the degree to which the block may require support.

20.1.4 Use of block theory

A major advance was made by Goodman and Shi (1985) in the application of mathematical topology to rock blocks and their removability from the rock surrounding an excavation. The advantages of a complete mathematical description of rock blocks are the ability to develop comprehensive

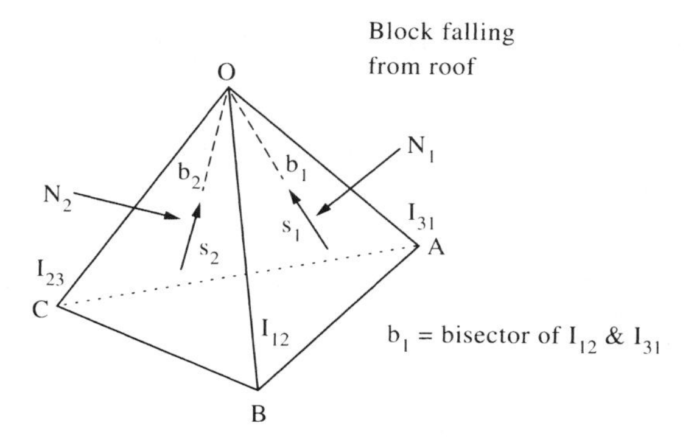

OAB = plane 1, OBC = plane 2, OCA = plane 3
ABC = excavation roof

Vertical force, $F_v = \Sigma \begin{pmatrix} \text{Vertical components of } S_1, S_2 \text{ and } S_3 \\ \text{Vertical components of } N_1, N_2 \text{ and } N_3 \end{pmatrix}$

The direction cosines of a line trend α and plunge β,
using right hand axes, are
$a_x = \cos\alpha\cos\beta$, $a_y = \sin\alpha\cos\beta$, $a_z = \sin\beta$
Thus, assuming a friction-only material, on any
face in contact with the rock there is a normal force N
and a shear force S, which can be expressed as $N\tan\phi$.
If we can examine the vertical components of these forces
for all three faces of the block, we have

$$F_z = \sum_{i=1}^{3} N_i(n_{z_i} + b_{z_i}\tan\phi_i)$$

where b_{z_i} are the vertical direction cosines of the
bisectors of the apical angle on each face in contact
with the rock, n_{z_i} is the vertical component of the
normal to the ith face. For wedge stability, $F_z + W < 0$,
otherwise the block will fall under the action of gravity.

Figure 20.13 Analysis of a tetrahedral block subject to *in situ* stresses and the action of gravity.

sophisticated analytical techniques based on computer methods and to incorporate all of the analyses we have discussed so far in one integrated approach.

The underlying principle of block theory is the recognition that blocks are formed from the intersection of a number of non-parallel and non-coincident planes. Any particular plane can be regarded as dividing the space occupied by the rock into two half-spaces: for the sake of simplicity these are called the 'upper half-space' and the 'lower half-space'. Thus, any great circle on a hemispherical projection, e.g. one of those in Fig. 20.8, also may be regarded as dividing space into these two half-spaces, and by convention they are coded with a numerical value of 0 for the upper half-space and 1 for the lower half-space. This idea stimulates the concept of extending the hemispherical projection beyond the customary boundary (which represents a horizontal plane) such that the upper and lower half-spaces may be studied concurrently. A 'spherical projection' is shown in Fig. 20.14, with the extensions of the great and small circles into the upper half-space being clearly identifiable.

The rock blocks in a rock mass are identified by numerical codes, according to how they are composed in terms of the upper and lower half-spaces produced by the various discontinuity planes in the rock mass. For example, consider the block 010—which is formed by the great circles associated with planes 1, 2 and 3 shown in Fig. 20.15. The first digit of zero means that the block is formed by the upper half-space defined by plane 1, i.e. is *outside* great circle 1 in the figure. Similarly, the second digit of 1 indicates that the block is formed by the lower half-space defined by plane 2, and hence is within great circle 2 in the figure. Finally, the third digit of zero represents the upper half-space defined by plane 3. In Fig. 20.15, all of the blocks defined by the three planes are shown, and it is clear from this diagram that block 111 resides within all three great circles, whereas block 000 resides outside all three great circles.

In the preceding discussion, the specific locations of the discontinuity planes are not considered, and so it is convenient to consider the geometry of the block as it would be defined if all of the planes intersected at a point. Under these conditions, blocks would exist as pyramidal shapes called 'joint

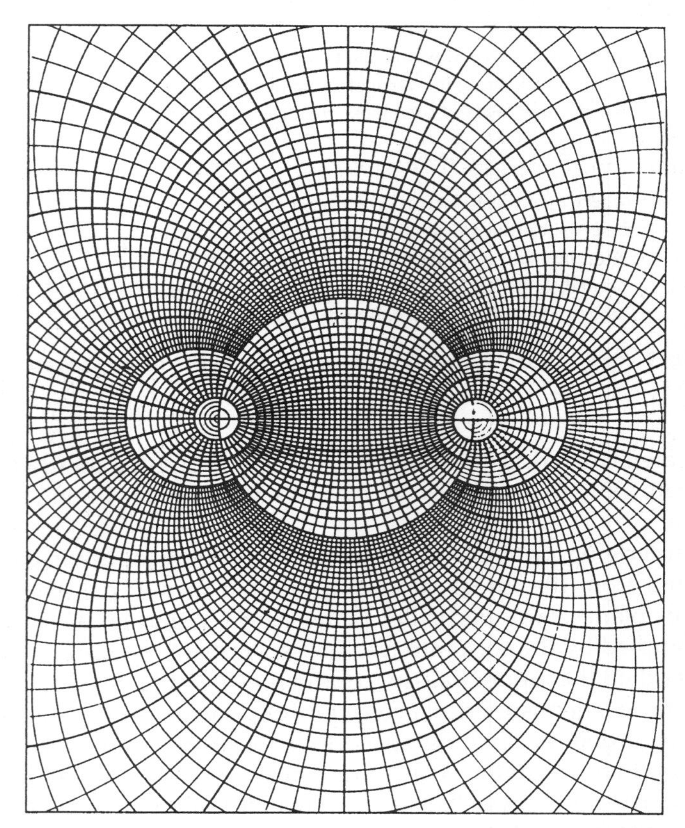

Figure 20.14 Composite upper- and lower-hemispherical projection, i.e. the spherical projection.

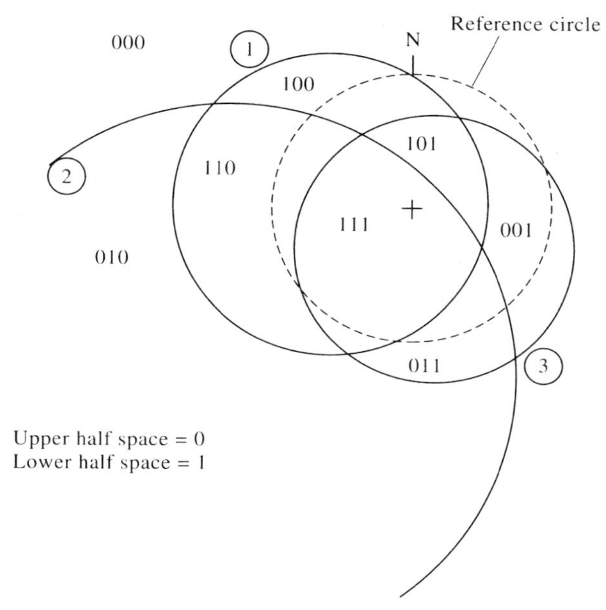

Figure 20.15 Illustration of rock blocks formed by the intersection of rock planes, using block theory notation.

pyramids', JP. Similarly, the planes that make up the boundary of an excavation can be considered in the same way, except that these planes divide space into rock and non-rock half-surfaces. By convention, when these planes are considered as intersecting at a point, the rock side is termed the excavation pyramid, EP. It follows that if the joint pyramid and the excavation pyramid do not intersect, i.e. JP $\cap$ EP = $\varnothing$, then the block is removable. Such a case is illustrated in Fig. 20.16.

At the left of Fig. 20.16, there are two discontinuity planes, 1 and 2, and two excavation planes, 3 and 4, which together delineate a rock block. From this diagram, and using the notation regarding upper and lower half-spaces presented earlier, the block is coded as 0100. If the diagram is transformed such that all of the planes intersect at a point, the diagram shown at the right of the figure is obtained. The joint and excavation pyramids are clearly

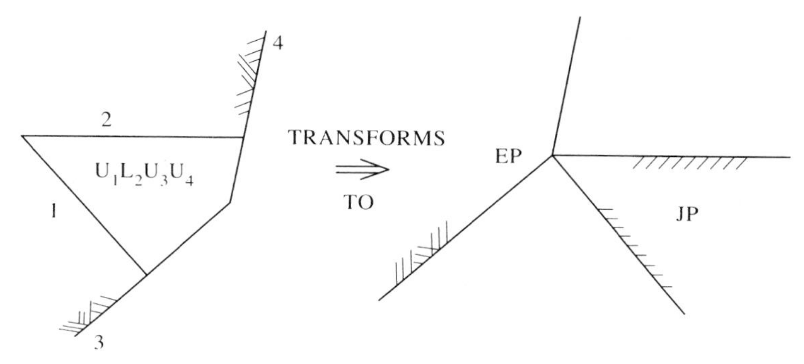

Figure 20.16 Example of block removability using the block theory concepts of joint and excavation pyramids.

shown in this diagram, and are mathematically defined as

$$U_1 \cap L_2 = \text{JP}$$
$$U_3 \cap U_4 = \text{EP}$$

and, from this diagram because JP and EP have no common sector, $\text{JP} \cap \text{EP} = 0$ and hence the block is removable. By an extension of this procedure, the removability of all blocks with respect to all potential excavation planes can be established. The power of the method lies in its ability to convert three-dimensional polyhedra (i.e. the blocks of rock) into mathematically defined sets, and to use mathematics to establish kinematic feasibility.

The mathematics of block theory is beyond the scope of this work but is well presented in the seminal book by Goodman and Shi (1985).

20.2 Design against stress-controlled instability

In the introduction to the chapter, we mentioned that rock instability around an excavation can occur due to block movement, stress effects or sometimes both mechanisms can occur concurrently. In this section, we describe design against stress-controlled instability through an understanding of the stress field around excavations, and how one can defend against the development of high stresses on the boundary of an excavation. Also described are the effect of rock bolting on the stress field and the use of the ground response curve to understand both the rock response to excavation and the potential need for installed support.

20.2.1 Zone of influence

When studying elastic stress distributions around underground openings, as described in Section 19.2.1, we note that the excavation affects the stresses and displacements for an infinite distance away from the opening. This is because, in the mathematical derivation of the various equations, the assumption has been made that material surrounding the opening extends to infinity. As engineers, we are only interested in *significant* changes to the stress field and displacements: below a certain level, it can be assumed that the changes have no significant engineering influence. This leads to the concept of the zone of influence, which is the zone around the excavation in which the stresses are perturbed from their *in situ* values by more than a defined amount.

For example, we could define the zone of influence around the excavation as the zone within which at least one component of the stress tensor is perturbed by greater than, say, 5% of its *in situ* value, expressed mathematically as

$$|\sigma_{\text{induced}} - \sigma_{\text{natural}}| \geq 0.05\sigma_{\text{natural}}$$

where σ_{induced} represents any component of induced stress, and the number 0.05 represents 5%—but may be any other percentage value relevant to the engineering objective.

Considering the stresses around a circular excavation (see Figs 19.10 and 19.11), the perturbation to the tangential stress component on the horizontal plane through the centre of the excavation can be calculated from the second of the equations shown in Fig. 19.10. For the example of $k = 1$, the equation reduces to

$$\sigma_\theta = p_z\,(1 + a^2/r^2)$$

and, by substituting this expression into the equation above, the 5% zone of influence is given by $r_{5\%} = a\sqrt{20}$. Thus, the 5% zone of influence is $4.47a$ (as measured from the centre of the excavation) or $3.47a$ measured from the wall of the excavation. In this case, the vertical and horizontal *in situ* stress components are equal, i.e. $k = 1$, and so this distance to the boundary of the zone of influence applies in all directions.

For other components of the stress field, and for other values of k, similar calculations can be made. For example, when $r = 5a$ and $k = 1$, $\sigma_r = 0.96p_z$ and $\sigma_\theta = 1.04p_z$, indicating that the 4% zone of influence (based on these components) then extends to $r = 5a$.

This principle of the zone of influence and the method of establishing its extent is directly applicable to any stress distribution, whether obtained by a closed form solution or numerically.

Elliptical approximation. In the case of a circular excavation when $k \neq 1$, the zone of influence is not circular in shape. The shapes of the zones of influence associated with each component of induced stress may be very different, as demonstrated by Fig. 20.17, but an approximation to the overall zone of influence may be found by drawing a circumscribed ellipse to the various perturbation contours. For the example shown in the figure, where $k = 0.5$, the major and minor axes of the ellipse are $11.76a$ and $7.98a$, respectively. The circumscribed ellipse in Fig. 20.17 does not indicate the magnitude of the stresses *per se*, but rather the magnitude of the perturbations to the *in situ* stress components. Thus, although the induced stresses would be expected to be greatest along the horizontal axis (for this value of k), this is not the case for the perturbations.

Similarly the value of the concept of the zone of influence is not in assessing the likelihood of inducing stresses which will lead to failure of the rock, but in determining—for the purposes of design—at which locations the induced stress field may be regarded as being unperturbed from the *in situ* stress field, and hence at what separations proximate excavations can be positioned. This is our next subject.

Multiple openings. In the case of adjacent circular openings, the stress distribution due to the two excavations can be approximated by summing the distributions due to the two single excavations. This provides two items of information for design: the stresses induced by multiple excavations; and the locations where the individual zones of influence overlap (or, are distinct).

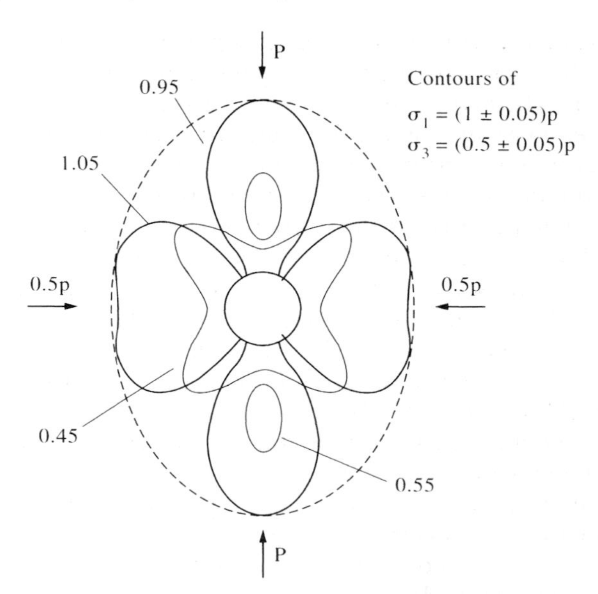

Figure 20.17 Elliptical approximation to the 5% zone of influence from the two-dimensional stress analysis of a circular excavation (from chapter by J. W. Bray in Brown, 1987).

Figure 20.18 shows the two main possibilities for the interaction of stresses between two proximate excavations. The first case shows how stresses may be amplified between excavations. The second case shows how stresses may be attenuated, with the production of a so-called 'stress shadow'. In the first of these cases, the stresses at a point between the excavations can be higher than the maximum stress induced by any single excavation, demonstrating an overlap of the zones of influence associated with the two excavations. This is also the case for the stress shadow, except then the overlap represents a reduction in the *in situ* stress.

For two proximate circular excavations with different diameters, the zones of influence associated with a given level of perturbation will have different extents for each excavation, and hence different effects on their neighbours. Consider the case of two circular excavations, one larger than

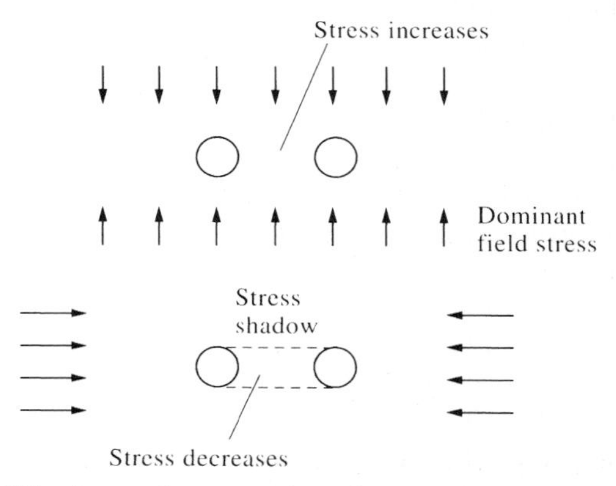

Figure 20.18 Amplification and attentuation of stresses between multiple excavations.

the other, as in Fig. 20.19. The zone of influence (in this case 5%) associated with the larger of the two excavations extends beyond the location of the smaller one, but this larger excavation is outside the zone of influence of the smaller excavation. Hence, Excavation I does influence the stresses around Excavation II, whereas Excavation II does not influence the stresses around Excavation I—at the 5% engineering level.

This concept suggests a means of obtaining a first approximation to the stress state that exists between the two excavations. The stresses induced by Excavation I can be calculated at both a point of interest and at the centre of Excavation II, and this latter stress state is used to calculate the stresses induced at the point of interest by Excavation II. It is essential to understand that when this procedure is undertaken, the *perturbation due to each excavation must be determined and added to the field stresses*, rather than adding the two absolute induced stresses. If the latter procedure is used, the field stresses are effectively duplicated.

Not only are these zones of influence helpful to designers in indicating zones of high and low stresses, they also indicate the optimal excavation sequencing of design layouts. For example, considering the two circular excavations in Fig. 20.19, the question to be answered is, 'Should we create Excavation I or Excavation II first?' The advantage of creating I first is that the final stress field acting on Excavation II will be in place before that excavation is made, and the process of creating Excavation II will not appreciably affect Excavation I. The advantage of creating Excavation II first is that the excavation is made in an unperturbed stress field, and the tunnel can be supported in anticipation of the stresses that will be induced following the creation of Excavation I.

This indicates two design alternatives, so through the use of the concept of the zone of influence the engineer has a method of considering the excavation sequencing alternatives. Of the two alternatives presented, the first is likely to be preferred, as both excavations will be created in stress fields that will not be subsequently disturbed. Very often, there can be a

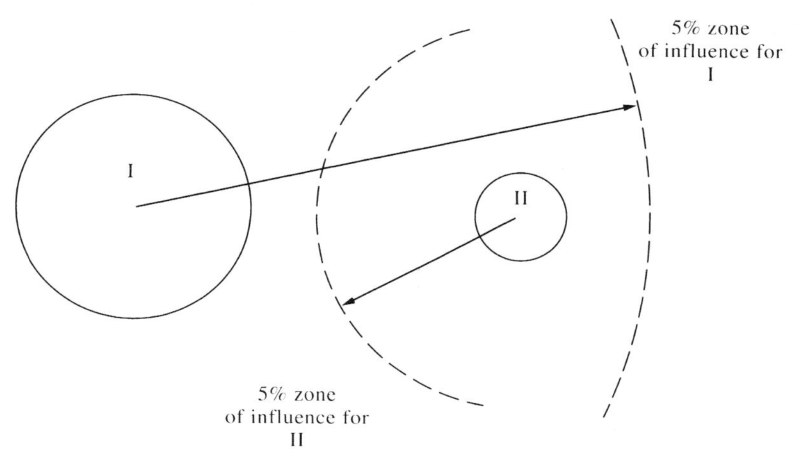

Figure 20.19 Mutual interaction between 5% zones of influence for two differently sized, circular excavations.

complex set of caverns and tunnels, so that these considerations become increasingly important. Operational requirements may dictate other, non-optimal (from the point of view of rock mechanics design) sequencing arrangements and the consequences of adopting these can be assessed using the concept of zone of influence.

In Figure 20.20, there are two cases of multiple excavation schemes, one referring to a common mining layout, and the other to a three-tunnel civil engineering railway scheme. In the mining layout, the footwall access drives have to be excavated before the process of stoping can begin. From the figure, it is directly evident that whilst the zones of influence of the footwall access drives are unlikely to significantly affect the stress field applied to the stope, the excavated stope will definitely significantly affect the stresses applied to the footwall access drives, which may have to be protected for this eventuality. In the case of the railway scheme, however, the primary criterion may be to accurately establish the ground conditions by excavating a small-diameter service tunnel before excavating the main large-diameter running tunnels. Such a procedure may be thought necessary if there were any doubt about the suitability of the rock for specific tunnel boring machines which may be used to excavate the running tunnels. Although the stresses applied to the service tunnel will change as the running tunnels are excavated, the known situation may be preferable to excavating the running tunnels in an unknown geomechanical environment.

Elliptical openings. Apart from a circle, the only other excavation shape for which a closed form stress solution is available is an ellipse. In a similar way to that illustrated in Fig. 20.17, contours of stress perturbation can be derived around excavations and hence the extent of the zone of influence can be determined. The calculation of the zone of influence on this basis is time consuming, and in a similar fashion to that illustrated in Fig. 20.17, we can adopt an elliptical approximation to the zone of influence for an elliptical opening.

In Fig. 20.21 are the equations for the circumscribing elliptical approximation for a zone of influence. The similarity between the diagram

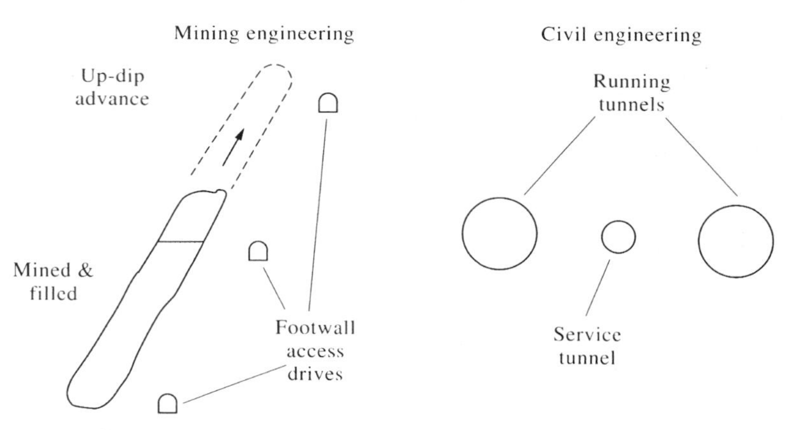

Figure 20.20 Illustration of zone of influence and excavation sequencing in different circumstances of multiple excavations.

in this figure and that shown in Fig. 20.17 is clear, except that now we are considering the excavation shape itself to be an ellipse, rather than a circle. The principle is that a percentage value, c, is chosen for the zone of influence, and then the width and height of the circumscribing zone of influence ellipse are determined from the equations in Fig. 20.21.

The value of c provides the value of A, and then the values of W_i and H_i can be directly evaluated from the equations in Fig. 20.21, using the values of k, q and α and the criteria given. Although in Fig. 20.21 the elliptical approximation to the zone of influence is indicated with its major axis in the vertical direction, this will not always be the case because the aspect ratio of this ellipse will depend on the parameters just described.

In Figure 20.22, two examples of this zone of influence are presented, both having a W/H value of 2, but with differing stress ratios k. The two cases have been chosen for comparison because they illustrate the use of the criteria presented in Fig. 20.21. In the left-hand diagram, the limits of the zone of influence are determined by the 5% contours—given by 1.05 and 0.95—associated with the vertical stress component. In the right-hand diagram, the limits of the zone of influence are determined by the 0.95 contours (associated with the vertical stress component) and, now, the 0.15 contour (associated with the horizontal stress component).

The 5% zone of influence produces the 0.95 and 1.05 contours for the perturbation to the vertical stress in both diagrams in Fig. 20.12. In the case of the horizontal stress component, we consider the criterion $|\sigma_3 - p_{\min}| > 0.05p_{\max}$ and so, because $p_{\min} = kp_{\max}$, the required contours are for

$$\sigma_3 > k + 0.05\, p_{\max}$$

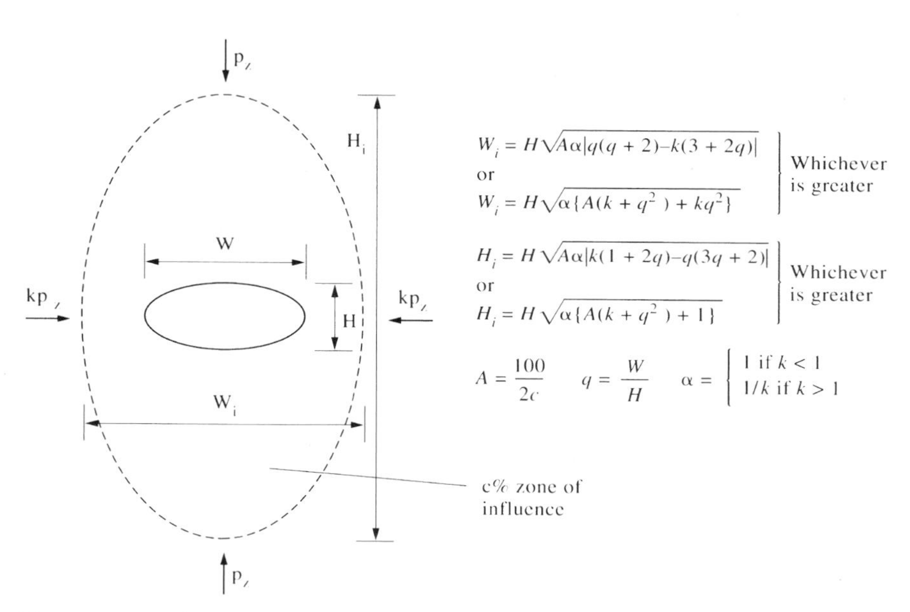

Figure 20.21 Elliptical approximation to the zone of influence around an elliptical excavation (from chapter by J. W. Bray in Brown, 1987).

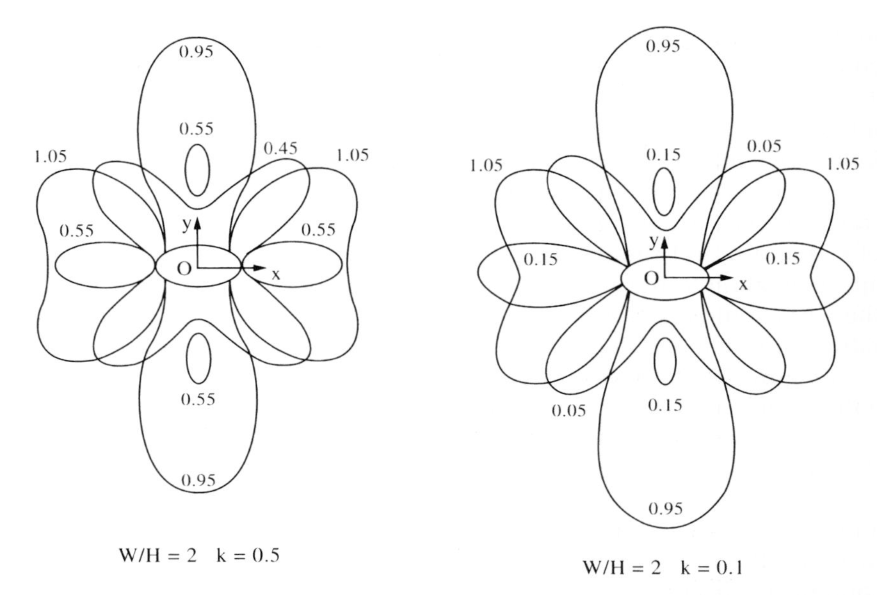

Figure 20.22 Illustration of the variation in the 5% stress perturbation contours for different ratios of vertical to horizontal stress for an elliptical opening (from chapter by J. W. Bray in Brown, 1987).

giving contour values of 0.45 and 0.55 in the left-hand diagram, and 0.05 and 0.15 in the right-hand diagram.

20.2.2 Approximations for other excavation shapes

The closed form solutions presented above, namely for circular and elliptical openings, can be used to give valuable engineering approximations for stress distributions in two important classes of problem: shapes other than truly circular or elliptical; and complicated boundary profiles.

Other excavation shapes. In Fig. 20.23, the upper diagram illustrates an ovaloid opening, in which the roof and floor are planar, and the ends are semi-cylindrical (but note that a vertical cross-section is being considered through a long excavation). Then, $W/H = 3$ and radii of curvature $\rho_A = H/2$ and $\rho_B = \infty$. As a method of approximately determining the circumferential stresses at A and B (and hence an indication of the maximum and minimum induced boundary stresses), the equations shown in Fig. 19.16 can be applied which give the stresses induced on the boundary of an elliptical excavation in terms of the radius of curvature of the boundary. For the stress at point A in terms of the radius of curvature at that point, the magnitude of the circumferential stress is $3.96p$.

By similar means, at point B the value is $-0.17p$ if we take a value for the radius of curvature appropriate for the ellipse inscribed to the ovaloid. As a means of determining a more exact answer to the boundary stresses for this geometry, the boundary element method was applied, with the result that the stresses at A and B were found to be $3.60p$ and $-0.15p$, respectively. Thus, the approximation is seen to be good for a preliminary estimation.

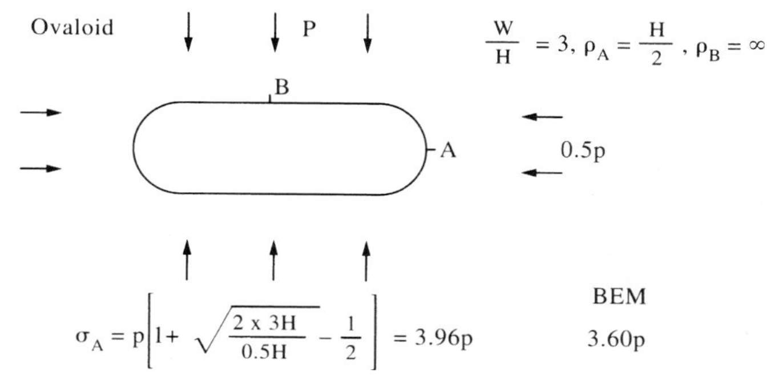

$\frac{W}{H} = 3, \rho_A = \frac{H}{2}, \rho_B = \infty$

BEM

$$\sigma_A = p\left[1 + \sqrt{\frac{2 \times 3H}{0.5H}} - \frac{1}{2}\right] = 3.96p \qquad 3.60p$$

For the inscribed ellipse,

$$\sigma_B = p[\,0.5\,(1 + (2 \times \frac{1}{3})) - 1] = -0.17p \qquad -0.15p$$

Square with rounded corners, k = 1

D

k = 1

A

$W = 1.25D$

$\rho_A = 0.2D$

BEM

$$\sigma_A = p\left[1 + \sqrt{\frac{2 \times 1.25D}{0.2D}} - 1\right] = 3.53p \qquad 3.14$$

Figure 20.23 Application of elliptical approximation to other excavation shapes.

Our second example concerns a square opening with rounded corners in a hydrostatic stress field, as shown in the lower diagram of Fig. 20.23. In this case of a hydrostatic stress field, we anticipate that the maximum stress will be associated with the smallest radius of curvature, i.e. at the rounded corners. Thus, using the geometry of the opening with $\rho_A = 0.2D$, we take $W = 1.25D$, and this gives $\sigma_A = 3.53p$. The more accurate value determined by the boundary element method was $3.14p$—again, the approximation gives a good preliminary estimate.

Approximation to complex boundary profiles. To show how the approach can be extended to complex boundary profiles, we show a typical underground hydroelectric scheme machine hall geometry in Fig. 20.24. From the equations shown in Fig. 19.16, one would expect:

(a) the radii of curvature at points A, B and C are very small, and hence the stress concentration will be very high at these points;
(b) the radius of curvature is negative at point D, and the induced stress might also be negative, i.e. tensile.

For the appropriately inscribed ellipse, the ellipse equations give the following values: sidewall stress = $1.83p_z$; and crown stress = $0.72p_z$.

When these values are compared to those in Fig. 20.24 determined from a boundary element analysis, the approximation is found to have provided a good early indication of the appropriate stress concentrations. In fact, from both this and the examples above, simple closed form solutions do provide a valuable insight into the stress distributions around complex excavation shapes.

20.2.3 Effect of rock bolting on the stress field

Rock bolts serve two purposes: they act to ensure that the rock around an opening behaves as a continuum; and they modify the stress field induced around the opening. Earlier in the book, we discussed the use of rock bolts to enhance the mechanical integrity of the rock mass. Here, we indicate the direct influence of installing a rock bolt on the stress field around a circular opening.

In Figure 20.25 the circumferential stress component on the boundary of the opening is shown, induced solely by a tensioned rock bolt anchored at points A and B. The geometry of the installation is shown in the upper left-hand diagram of Fig. 20.25, and the distribution of the induced tangential stress component is shown in the upper right-hand diagram. The maximum induced stress is at the bolt head, where a tensile stress with a magnitude of $0.99P/a$ is developed, and at an angular distance of 90° the induced stress has effectively diminished to zero. When a circular opening is subjected to an internal pressure of magnitude p, the corresponding induced tangential stress is $-p$. Thus, the effect of installing a rock bolt is similar, given that the applied load of P has been normalized by dividing by a, the radius of the opening.

In the lower part of Fig. 20.25, the tangential and radial stress distributions in the rock mass along the rock bolt length are shown . There is a

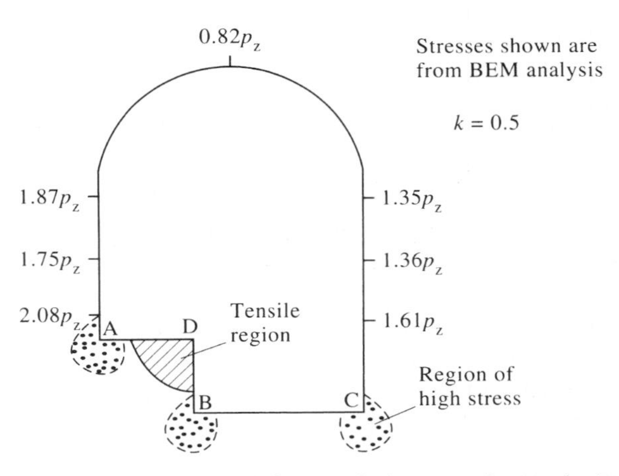

Figure 20.24 Application of closed form solutions to the analysis of a complex boundary shape.

high compressive radial stress induced beneath the bolt head (theoretically, the magnitude is infinite beneath a point load), but this dissipates rapidly as the distance along the bolt increases. At point A, the tangential stress is tensile, ameliorating the effects of a compressive tangential stress induced by the *in situ* stress field. At point B both the radial and tangential stresses are high, and the sign of these stresses changes passing from the left-hand side of the anchor point to the right-hand side. There are high deviatoric stresses in the rock at this point which may be sufficient to induce failure in the rock mass, a fact that is not often appreciated.

When rock bolts are used to counteract any anticipated structurally controlled instability, consideration should always be given to the stresses induced by the bolts, whether they be mechanically anchored or fully grouted bolts.

20.2.4 Ground response curve

In Section 16.4, we introduced the concepts of the ground response curve and available support lines, illustrated in Figs 16.6–16.8. The philosophy behind the ground response curve is that, under the action of the *in situ* stress field, stresses may be induced around an opening that cause failure of the rock material, either through the development of new discontinuities, by yield of the intact rock, or damage to existing discontinuities.

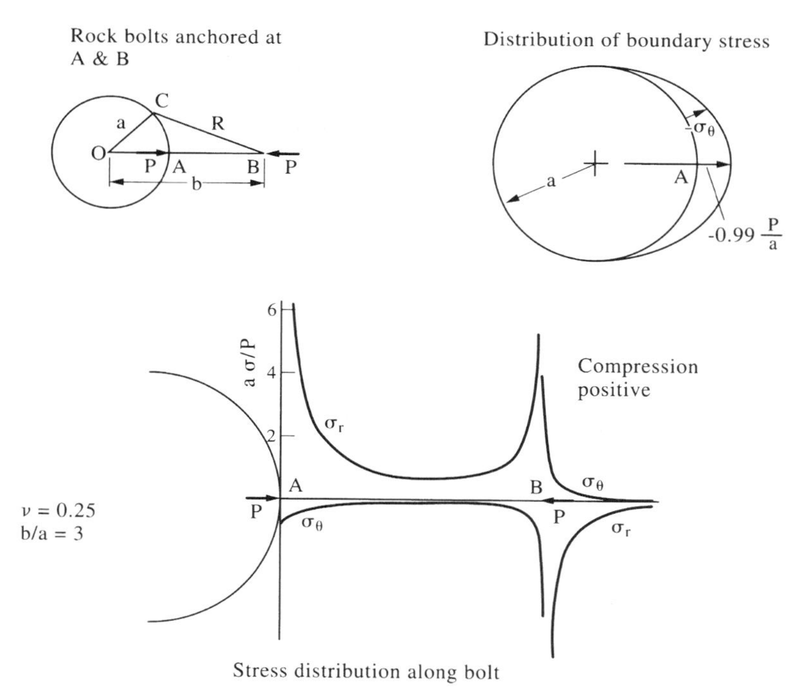

Figure 20.25 The influence of a tensioned rock bolt on the stress distribution around a circular opening (from chapter by J. W. Bray in Brown, 1987).

The design objective is to study the form of the ground response curve for any particular situation, and develop the support methods accordingly. We now discuss how to produce a ground response curve.

Development of a ground response curve. Consider the mechanical behaviour of the rock round a circular excavation in a hydrostatic stress field. We model the material first as an idealized elastic–brittle–plastic material, as shown by the complete stress–strain curve illustrated in Fig. 20.26. The associated relation between the major and minor principal strains and the related volumetric strain are also shown in the figure.

From the data in the figure, a ground response curve can be constructed by the following steps:

(a) substitute successive values of p_i into equation (2) to obtain a series of values for r_e;
(b) substitute r_e into equation (4) with $r = a$ to obtain values of r_i;
(c) plot values of p_i against corresponding values of $\delta_i = -u_i$ to obtain the ground response curve;
(d) the critical support pressure below which a fracture zone develops is given by equation (1);
(e) this procedure applies to the sidewalls. More support pressure is required to limit the measured displacments to these calculated values in the roof, and less in the floor. The roof and floor ground response curves are found from

$$\left.\begin{aligned} p_{\text{roof}} &= p_i + \gamma(r_e - a) \\ p_{\text{floor}} &= p_i - \gamma(r_e - a) \end{aligned}\right\} \text{ for each } p_i.$$

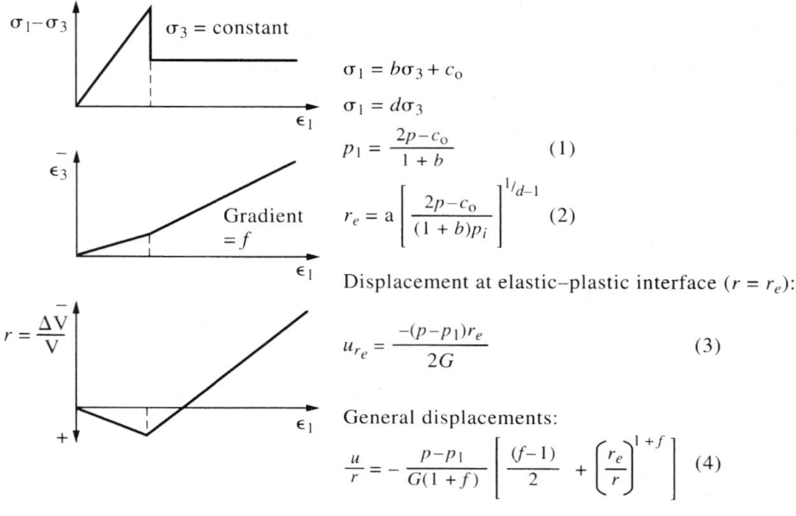

Figure 20.26 Material behaviour assumed in development of the ground response curve, and the related equations.

Following this procedure allows the production of three ground response curves (one each for the floor, sidewall and roof) in radial stress–radial displacement space. In itself, such a ground response curve is of limited use: to be of utility to an engineer, it is important to see how the ground response curve interacts with the curve representing the behaviour of a supporting element (see Brady and Brown, 1985, for more information).

Available support lines. For all elements that are used for reinforcement or support, it is possible to determine (either using closed form solutions, or by numerical calculation) the radial stress–radial displacement behaviour of the support system. To illustrate this, a concrete lining, as an example, provides a specific support line depending on its geometry and material properties. These support lines are commonly known as *available support lines*.

In Fig. 20.27, there is a cross-section through a shotcrete or plain concrete circular lining. By approximating this lining to a thick-walled elastic cylinder subject to an external pressure, a standard solution can be used to determine the radial stiffness of the lining, k_{conc}, and hence to determine the available support line through application of the formula $p_i = k_{conc}u_i$, where p_i is the support pressure and u_i is the support displacement. Such a lining has a maximum strength, and thus the maximum radial stress that the lining can withstand without crushing is also required. The terms in the formulae in Fig. 20.27 are: E_{conc} = Young's modulus of the shotcrete or concrete; ν_{conc} = Poisson's ratio of the shotcrete or concrete; t_{conc} = lining thickness; r_i = internal tunnel radius; and $\sigma_{c_{conc}}$ = uniaxial compressive strength of the shotcrete or concrete.

There are many different types of supporting elements—for example, blocked steel sets, rock bolts and other types of anchor—and support stiffness formulae can be established for all these. To present the full range of the associated formulae is beyond the scope of this book, but interested readers are referred to Hoek and Brown (1980) for a more comprehensive list. Using such formulae, the support pressure associated with a given ground response curve can be explicitly evaluated, and in Figure 20.28, the available support lines for five different types of support are shown, in conjunction with the ground response curves for the roof, sidewall and floor of a tunnel.

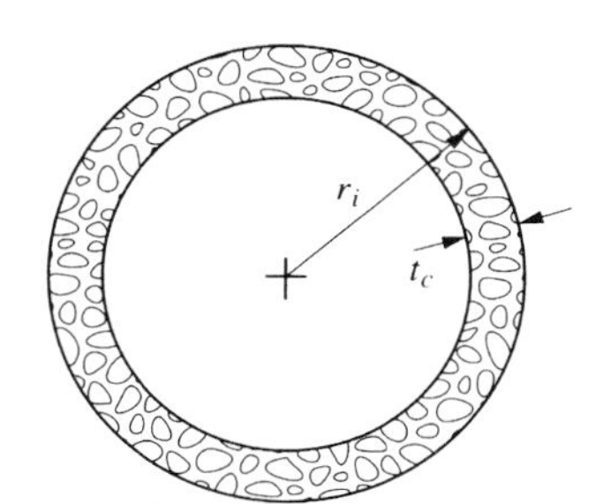

Support stiffness:

$$k_{conc} = \frac{1}{r_i}\frac{E_{conc}}{\left(1+\nu_{conc}\right)}\frac{\left[\left(r_i^2-\left(r_i-t_{conc}\right)^2\right]}{\left[\left(1-2\nu_{conc}\right)r_i^2+\left(r_i-t_{conc}\right)^2\right]}$$

Maximum support pressure:

$$p_{conc\,max} = \frac{\sigma_{c_{conc}}}{2}\left[1-\frac{\left(r_i-t_{conc}\right)^2}{r_i^2}\right]$$

Figure 20.27 Formulae for the available support line of a shotcrete or plain concrete circular excavation lining.

Figure 20.28 shows a wide range of principles associated with the ground response curve, as illustrated in the following points.

- The ground response curve for different locations on the periphery of the excavation is different.
- The ground response curves indicate that at, some locations, support is not necessary (because the displacements equilibriate at zero support pressure), and in other locations support is essential (because the ground response curve does not intersect the zero support pressure axis).
- Attempting to achieve zero radial displacement is impractial: to do so would require extremely high support pressures and support stiffnesses.
- The support cannot be installed at zero radial displacement, because the elastic response of the ground, on excavation, is instantaneous.
- Different types of support have different stiffnesses, and these result from the material of construction, the geometry of the support system, and the quality of construction: as a result, these different supports will offer different degrees of support to the excavation, and will support the rock with different support pressures.
- It is possible for supports to attain their peak strength (including some degree of yielding) and still be effective in supporting the excavation.

In conjunction with a given ground response curve, three variables determine the mechanical efficacy of a given support scheme: its time of emplacement, its stiffness, and its peak strength. The ground response curve itself can also be a function of construction techniques. Hence, the engineer has to optimize the overall interaction between the ground

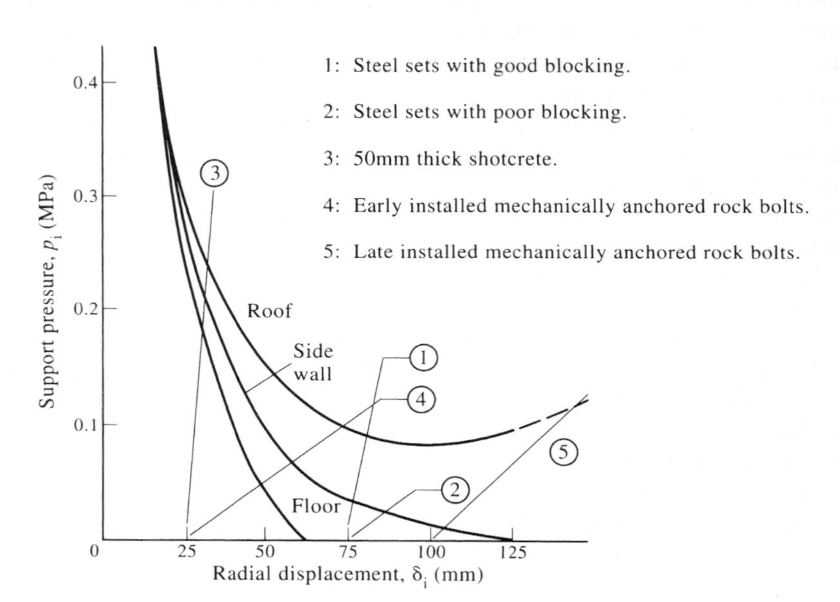

Figure 20.28 Available support lines and ground response curves (from Brady and Brown, 1985, and Hoek and Brown, 1980).

response curve and the available support line, such that practical support pressures are generated at tolerable radial displacements. With reference to Fig. 20.28:

(a) support type 3 (the shotcrete ring) may be both too stiff and installed too early, as it generates unnecessarily high support pressures;
(b) support type 4 (early installed rock bolts) is ideal for the roof;
(c) support type 1 (well installed steel sets) is similarly successful;
(d) support type 2 (poorly installed steel sets) is insufficent because the sets yield at a support pressure less than that required to support the roof;
(e) support type 5 (late installed rock bolts) is unsatisfactory because of the danger of the bolts being unable to maintain equilibrium of the excavation periphery at sufficiently low radial displacments, i.e. the support line may not intersect the ground response curve.

Note that in the preceding discussion, we have been discussing the effectiveness of the support with respect to roof stability, rather than the need to restrict sidewall and floor displacements. It is clear from Fig. 20.28 how one would use this technique to determine other such support criteria.

Pillar–country rock interaction. A natural extension to the analysis above is to consider other excavation shapes and natural support methods. Using the rock itself as the support element, rather than introducing artificial, and hence more expensive, materials is an elegant engineering solution to rock engineering projects. This is not always possible, but the concept of the ground response curve and available support lines can be successfully extended to the case of supporting the roof and floor during the excavation of a wide rectangular opening (as, for example, occurs during mining operations in a horizontal tabular ore body).

Consider the support of a slot-like excavation with a large width-to-height ratio, as illustrated in Fig. 20.29. We proceed according to the following steps:

(a) first, the displacement that would result should the entire excavation be opened are determined;
(b) second, the displacement induced by the application of a unit normal stress over the anticipated support area is determined;
(c) third, the two results are used to produce the ground response curve, assuming that the rock proximal to the excavation remains linearly elastic;
(d) fourth, the stress–strain behaviour of a natural pillar is considered, this being the supporting element;
(e) finally, the analysis described earlier is used to study the stability of the total structure, as shown in Fig. 20.30.

The ground response curve for the country rock is a straight line from a support pressure of 19.3 MPa at zero displacement, to a support pressure

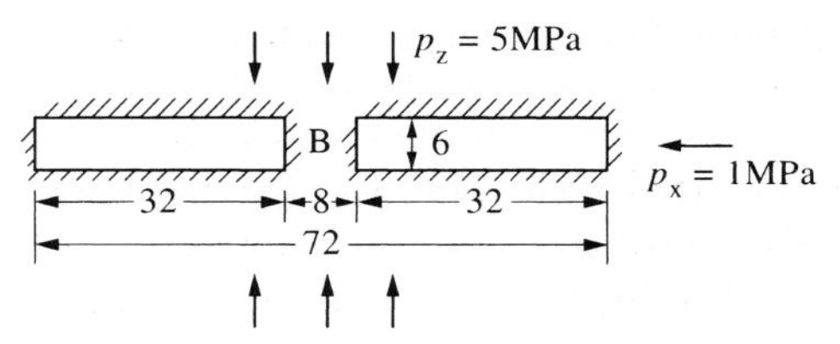

Pillar characteristic (determined by uniaxial loading under plane strain)

p(MPa)	5.0	8.0	10.0	11.1	11.4	10.9	10.0	8.7	3.0	0.5
$\epsilon_E \times 10^3$	0.5	1.0	1.5	2.0	2.5	3.0	3.5	4.0	7.0	12.0

Country rock characteristic (determined from BEM analysis)

(a) traction free slot $\frac{1}{2}\delta'_B$ δ'_B = 66.52 E-3 m

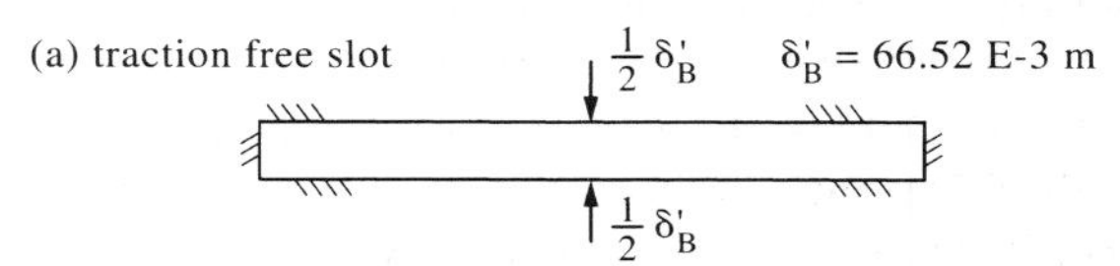

(a) 1MPa normal stress over pillar contact area

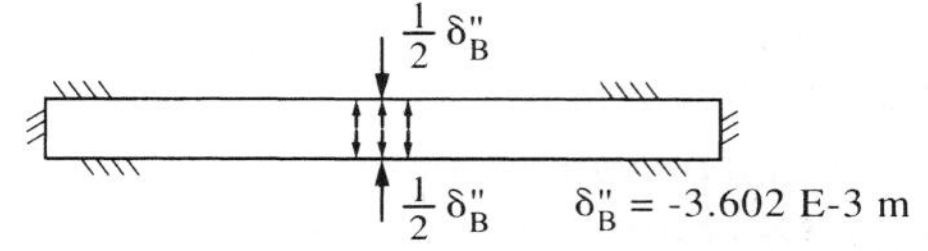

Solution : plot p-ϵ_z characteristics for pillar & country rock. Operating point is at the intersection of the curves.

Pillar

Initial state of strain is Plane strain : (5, 1)

$$\epsilon_{zo} = \frac{1-\nu^2}{E}\left[p_z - \frac{\nu}{1-\nu}p_x\right] = 0.5 \text{ E-3}$$

Final state of stress is and the resulting strain is ($5 + \Delta p_B$, 0)

$$\epsilon_z = \epsilon_{zo} + \frac{\delta_B}{6} \quad (1)$$

Country rock

The BEM results tell us that

$\delta_B = (66.52\text{-}3.602p_B)$ E-3

Substituting into (1)

$$\epsilon_z = (11.59\text{-}0.6003p_B) \text{ E-3}$$

This is the country rock characteristic.

Figure 20.29 The ground response curve concept illustrated through the analysis of a tabular excavation with and without a natural supporting pillar.

of zero at a strain of 0.0116, which is equivalent to a displacement of about 70 mm. The ground response curve is linear, because it has been developed on the basis of linear elasticity theory. The complete available support line is equivalent to the complete stress–strain for the pillar material, under conditions of plane strain (we discussed the complete stress–strain curve in Section 6.1, and noted the importance of the relative stiffnesses of the loading system and the descending, post-peak, portion of the curve).

The two curves in Fig. 20.30 now allow study of the stability of the entire structure. The operating point, indicated by the intersection of the two curves, represents a stage in the mechanical breakdown of the pillar that is almost complete. The displacement at the operating point is almost the displacement that would be reached without the pillar being present—when the excavation would be stable anyway. The conclusion is clear: the pillar is both ineffective and unnecessary.

There are many variations on this theme, and the way in which natural supporting elements in mining geometries can be used optimally to ensure stability whilst maximizing the amount of excavated material can be studied. Our purpose here is to demonstrate one specific case where the ground response curve analysis provides a clear conclusion, remembering that these analyses have been in two dimensions but rock engineering is always conducted in three dimensions.

20.2.5 Three-dimensional analysis

An additional level of complexity is introduced by the three-dimensional nature of the rock engineering geometry compared to studies for two-

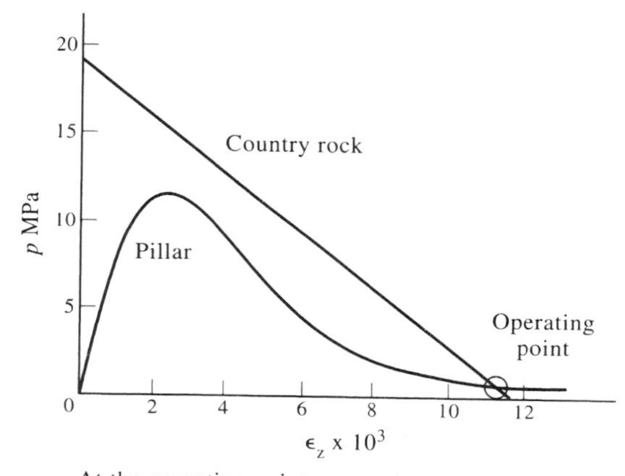

Figure 20.30 Ground response and available support lines for the tabular excavation illustrated in Fig. 20.29.

dimensional geometries. This is elegantly demonstrated by the stress distribution around a spherical opening in a uniaxial stress field, in which the magnitude of the induced stress on the boundary is given by the equation shown in Fig. 20.31. As an analogue to the maximum stress around a circular opening in a uniaxial stress field, the stress at

$$\theta = 0 \text{ is } \sigma_\theta = \frac{3}{2}\left(\frac{9-5\nu}{7-5\nu}\right)p$$

with numerical values of $2.00p$ when $\nu = 0.20$ and $2.02p$ when $\nu = 0.25$.

There are two key points to note. First, the stress concentration depends on one of the elastic constants, i.e. Poisson's ratio (note that in the two-dimensional case the maximum stress concentration was 3.00 for any isotropic elastic material, and independent of all elastic properties). Second, the stress concentration in the three-dimensional case is significantly different from that of the two-dimensional case. This means that one cannot validly approximate the three-dimensional geometry by a two-dimensional geometry—unless part of the three-dimensional geometry is well represented in two dimensions, which has tacitly been assumed in all of the two-dimensional solutions presented heretofore.

Even so, in cases where the geometry more accurately reflects engineering structures, and is therefore more complex, two-dimensional approximations can be successfully used in locations where these are likely to be valid. Two such cases are shown in Fig. 20.32.

The first of these, in the upper diagram, is a T-shaped intersection between two circular tunnels. At a distance of $3r$ from the centreline of the branch tunnel, the magnitude of the discrepancy between the maximum boundary stress computed using a three-dimensional analysis and a two-dimensional plane strain analysis is less than 10%. Moving further away from the intersection, to a distance of $5r$ from the centreline of the branch tunnel, the magnitude of the discrepancy has reduced to less than 5%. So, the two-dimensional approximation will be sufficient for engineering purposes at sufficiently large distances from the line of intersection.

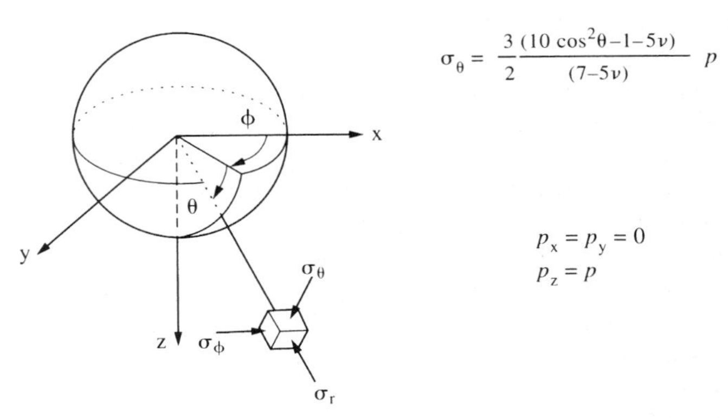

Figure 20.31 Boundary stress around a spherical opening in an isotropic material subjected to a uniaxial stress field.

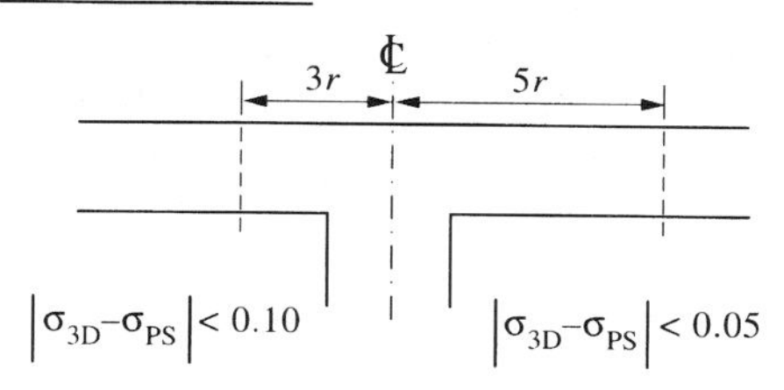

Effect of the end of a circular tunnel

x
z
y

$\nu = 0.25,\ 0 < k < 2$

At y = 0.75r

$|\sigma_{3D}-\sigma_{PS}| < 0.2$

At y = 4r

$|\sigma_{3D}-\sigma_{PS}| < 0.05$

Figure 20.32 Comparison between two-dimensional and three-dimensional stress analyses for two engineering geometries.

The second example, shown in the lower diagram, represents the conditions at the end of a borehole or circular tunnel. At a distance of 0.75r from the end of the tunnel, the discrepancy between the plane strain two-dimensional solution and the full three-dimensional solution is already less than 20%. At a distance of 4r from the end of the tunnel, this discrepancy is reduced to less than 5%. So, in the latter case, not only does the two-dimensional approximation provide an excellent estimate of the stresses over most of the tunnel length, it also indicates directly how rapidly the three-dimensional geometry effectively changes to a two-dimensional geometry during tunnel construction. This can be of use in determining design aspects such as the installation time of tunnel support elements and instrumentation.

In cases such as that shown in Fig. 20.32, where a two-dimensional approximation is adequate, there is benefit in restricting the analyses to two dimensions. However, there are circumstances where the intersections of such underground excavations—and the general engineering layout—cannot be adequately represented in two dimensions. For example, complex engineering structures such as hydroelectric schemes and most methods of mining cannot be adequately represented in two dimensions. We are fortunate today to have full three-dimensional analysis capabilities, both for discontinuous and continuous materials, readily available on desktop computers.

There are off-the-shelf codes now available for three-dimensional discrete element, finite element and boundary element methods of analysis. Moreover, there are also codes available for three-dimensional analysis of fluid flow through fracture networks. In addition to these off-the-shelf codes, there are manifold proprietary programs in which, for example, hybrid discrete element and finite element analyses are combined with the analysis of fluid flow.

In the early days of these programs, the computing was often difficult for the average user. We are now experiencing a major development in the ease of use of these methods through the use of improved graphical interfaces, so there is now every reason to apply such programs to all projects. However, it is of paramount importance to be certain that the rock mechanics and rock engineering principles are fully understood, and that the output from the computer programs is in accordance with these principles.

The nature of a complete rock engineering project is that it will contain components of many different kinds which need to be integrated. Our final remarks consider integrated design procedures and how they have evolved over the years.

20.3 Integrated design procedures

There is currently no overall standard procedure for the design and construction of a rock engineering project. In this book, we have presented a range of principles relating to engineering rock mechanics. The rock may well be inhomogeneous and anisotropic, in short, not always ideally suited for analysis. Indeed, we may well not have sufficient information about the site geology itself. A variety of contributing factors may influence our decisions: these can be in the areas of finance, environment, management and so on. Thus, we have a suite of well-understood supporting modules for the rock engineering (cf. Hudson, 1993), but not a universally utilized overall design methodology.

Techniques are presented in the books by Hoek and Brown (1980), Franklin and Dusseault (1989) and Bieniawski (1989). The Hoek and Brown methodology is concerned with identifying whether instability is likely to be the result of rock structure, rock stress, weathering or time. The Franklin methodology provides an extensive introduction to all the techniques available. The Bieniawski approach is more in line with classic management type charts. The rock engineering systems approach introduced in Chapter 14 enables the structure of the system to be generated and its operation to be studied in terms of critical mechanisms and hazards (Hudson and Jiao, 1996). It must be borne in mind that each of these methods has different advantages and values, depending on the engineering context and objectives.

Thus, the engineer should make an informed choice as to which, if any, of these methodologies is appropriate to their circumstances. In order to make an informed choice, the engineer must be fully conversant with the principles of engineering rock mechanics—which is what has been presented in this book.

References

Abramovitch M. and Stegun I. A. (eds) (1965) *Handbook of Mathematical Functions.* Dover, New York.

Almenara J. R. (1992) Investigation of the cutting process in sandstones with blunt PDC cutters, Ph.D. Thesis, Imperial College, London, 165pp.

Amadei B. (1983) *Rock Anisotropy and the Theory of Stress Measurements.* Springer, Berlin.

Antonio E. C. (1985) A study of discontinuity frequency in three dimensions, M.Sc. Thesis, Imperial College, London, 107pp.

Arnold P. N. (1993) The development of a rock engineering systems methodology, Ph.D. Thesis, Imperial College, London, 377pp.

Barton N. and Choubey V. (1977) The shear strength of rock joints in theory and practice. *Rock Mech.* **12**, 1–54.

Barros A., Nuno S. and Rottmann J. (1983) *Chile: La Tierra en queVivimos.* Editorial Antarctica S. A., San Francisco 116, Santiago, Chile, 506pp.

Barton N., Bandis S. and Bakhtar K. (1985) Strength, deformation and conductivity coupling of rock joints. *Int. J. Rock Mech. Min. Sci. & Geomech. Abtsr.* **22**, 3, 121–140.

Barton N. and Stephansson O. (eds) (1990) *Rock Joints.* A. A. Balkema, Rotterdam, 814pp.

Bear J. (1972) *Dynamics of Fluids in Porous Media.* Elsevier, New York.

Bieniawski Z. T. (1968) The effect of specimen size on compressive strength of coal. *Int. J. Rock Mech. Min. Sci.* **5**, 325–335.

Bieniawski Z. T. (1989) *Engineering Rock Mass Classifications.* John Wiley, New York, 251pp.

Bishop A. W. (1955) The use of the slip circle in the stability analysis of earth slopes. *Geotechnique,* **5**, 7–17.

Boussinesq J. (1883) *Applications des Potentiels a l'Etude de l'Equilibre et du Mouvement des Solides Elastiques.* Gauthier–Villars, Paris.

Brady B. H. G. and Brown E. T. (1985) *Rock Mechanics for Underground Mining.* George Allen & Unwin, London, 527pp.

Bray J. W. (1977) Unpublished note, Imperial College, London

Bray J. W. (1987) Some applications of elastic theory, in *Analytical and Computational Methods in Engineering Rock Mechanics* (E. T. Brown, ed.). Allen & Unwin, London, pp. 32–94.

Brown E. T. (ed.) (1981) *ISRM Suggested Methods.* Pergamon Press, Oxford, 211pp.

Brown E. T. (ed.) (1987) *Analytical and Computational Methods in Engineering Rock Mechanics.* Allen & Unwin, London, 259pp.

Brown E. T., Richards L. R. and Barr M. V. (1977) Shear strength characteristics of Delabole slates. *Proc. Conf. Rock Engng.,* Newcastle upon Tyne, pp. 33–51.

Burland J. B. and Lord J. A. (1969) The load deformation behaviour of middle chalk at Mundford, Norfolk. A comparison betwen full scale performances and *in situ* and laboratory measurements. *Proc. Conf. on* In-situ *Investigations in Soil and Rock.* Institute of Civil Engineers, London, pp. 3–15.

Cerruti V. (1882) *Acc. Lincei Mem fis mat, Roma,* **18**, 81.

Craig R. N. and Muir Wood A. M. (1978) *A Review of Tunnel Lining Methods in the United Kingdom.* Supplementary Report 335 of the Transport and Road Research Laboratory UK.

Creech T. (1683) *Titus Lucretius Carus—Six Books of Epicurean Philosophy*. Printed by Thomas Sawbridge at the Three Flower-de-Luces in Little Britain, and Anthony Stephens, Bookseller near the Theatre in Oxford, London, 270pp.

Cuisiat F. D. (1992) *In situ* rock stress measurement, Ph.D. Thesis, Imperial College, London, 165pp.

Cuisiat F. D. and Haimson B. C. (1992) Scale effects in rock mass stress measurements. *Int. J. Rock Mech. Min. Sci. & Geomech. Abstr.* **29**, 2, 99–118.

Daemen J. J. K. (1983) Slip zones for discontinuities parallel to circular tunnels or shafts. *Int. J. Rock. Mech. Min. Sci. & Geomech Abstr.* **20**, 3, 135–148.

Deere D. U. (1963) Technical description of rock cores for engineering purposes. *Rock Mech. Eng. Geol.*, **1**, 18–22.

Dowding C. H. (1985) *Blast Vibration Monitoring and Control*. Prentice-Hall, London, 297pp.

Dunnicliff J. (1988) *Geotechnical Instrumentation for Monitoring Field Performance*. John Wiley, Chichester, 577pp.

Dyke C. G. (1988) *In situ* stress indicators for rock at great depth, Ph.D. Thesis, Imperial College, London, 361pp.

Franklin J. A. (1979) Suggested method for determining water content, porosity, density, absorption and related properties. *Int. J. Rock Mech. Min. Sci. & Geomech. Abstr.* **16**, 2, 141–156.

Franklin J. A. (1985) Suggested method for determining point load strength. *Int. J. Rock Mech. Min. Sci. & Geomech. Abstr.* **22**, 2, 51–60.

Franklin J. A. and Dusseault M. B. (1989) *Rock Engineering*. McGraw-Hill, New York, 600pp.

Gaziev E. and Erlikhman S. (1971) Stresses and strains in anisotropic foundations. *Proceedings Symposium on Rock Fracture*. ISRM, Nancy, Paper II-1.

Gokay M. K. (1993) Developing computer methodologies for rock engineering decisions, Ph.D. Thesis, Imperial College, London, 396pp.

Goodman R. E. (1989) *Introduction to Rock Mechanics*, 2nd edn. John Wiley, Chichester, 562pp.

Goodman R. E. and Shi G.-H. (1985) *Block Theory and its Application to Rock Engineering*. Prentice-Hall, London, 338pp.

Griffith A. A. (1921) The phenomena of rupture and flow in solids. *Philos. Trans. R. Soc., London* **A221**, 163–198.

Harrison I. W. (1989) Development of a knowledge based system for open stope design, Ph.D. Thesis, Imperial College, London, 268pp.

Harrison J. P. (1993) Improved analysis of rock mass geometry using mathematical and photogrammetric methods, Ph.D. Thesis, Imperial College, London, 510pp.

Hendron A. J. (1977) Engineering of rock blasting on civil projects, in *Structural and Geotechnical Mechanics: A Volume Honoring Nathan M. Newmark* (W. J. Hall, ed.). Prentice-Hall, Englewood Cliffs, NJ, pp. 242–277

Heuze F. E. and Amadei B. (1985) The NX borehole jack: a lesson in trials and errors. *Int. J. Rock Mech. Min. Sci. & Geomech. Abstr.*, **22**, 2, 105–112.

Hoek E. (1990) Estimating Mohr–Coulomb friction and cohesion values from the Hoek–Brown failure criterion. *Int. J. Rock Mech. Min. Sci. & Geomach. Abstr.*, **27**, 227–229.

Hoek E. and Bray J. W. (1977) *Rock Slope Engineering*. Institution of Mining and Metallurgy, London, 402pp.

Hoek E. and Brown E. T. (1980) *Underground Excavations in Rock*. Institution of Mining and Metallurgy, London, 527pp.

Hoek E. and Brown E. T. (1988) The Hoek–Brown failure criterion—a 1988 update, *Proc. 15th Can. Rock Mech. Symp.* (J. H. Curran, ed.). Civil Engineering Department, University of Toronto, pp. 31–38

Hoek E., Kaiser P. K. and Bawden W. F. (1995) *Support of Underground Excavations in Hard Rock*. A. A. Balkema, Rotterdam, 215pp.

Hoek E., Wood D. and Shah S. (1992) Modified Hoek–Brown failure criterion for jointed rock masses. *Proc. ISRM Symposium: Eurock '92, Chester*. Thomas Telford, London, pp. 209–214.

Hood M., Nordland R. and Thimons E. (1990) A study of rock erosion using high pressure water jets. *Int. J. Rock Mech. Min. Sci. & Geomech. Abstr.*, **27**, 77–86.

Hudson J. A. (1989) *Rock Mechanics Principles in Engineering Practice*. CIRIA/ Butterworths, London, 72pp.

Hudson, J. A. (1992) *Rock Engineering Systems: Theory and Practice*. Ellis Horwood, Chichester, 185pp.

Hudson J. A. (ed.) (1993) *Comprehensive Rock Engineering*. Pergamon Press, Oxford, 4407pp.

Hudson J. A. and Cooling C. M. (1988) *In situ* rock stresses and their measurement in the U.K.—Part 1: The current state of knowledge. *Int. J. Rock Mech. Min. Sci. & Geomech. Abstr.* **25**, 363–370.

Hudson J. A., Crouch S. L. and Fairhurst C. (1972) Soft, stiff and servo-controlled testing machines: a review with reference to rock failure. *Engineering Geology* **6**, 155–189.

Hudson J. A. and Jiao Y. (1996) Analysis of Rock Engineering Projects. Imperial College Press, London, 320pp (in press).

Hudson J. A., McCaul C. and Priest S. D. (1977) A high load tunnel jacking test. *Ground Engineering* May, 22–26.

Hyett A. J. (1990) The potential state of stress in a naturally fractured rock mass, Ph.D. Thesis, Imperial College, University of London, 365pp.

Hyett A. J., Dyke C. G. and Hudson J. A. (1986) A critical examination of basic concepts associated with the existence and measurement if in situ stress, in *Proc. of the Int. Symp. on Rock Stress and Rock Stress Measurements*, Stockholm (O. Stephansson, ed.) pp. 387–396.

Ikegawa (1992) Three-dimensional geometrical analysis of rock mass structure, Ph.D. Thesis, Imperial College, London, 413pp.

Isherwood D. (1979) *Geoscience Data Base Handbook for Modelling a Nuclear Waste Repository*, Vol. 1. NUREG/CR-0912 V1. UCRL-52719. V1.

Jaeger J. C. and Cook N. G. W. (1979) *Fundamentals of Rock Mechanics*. Chapman & Hall, London, 593pp.

Janbu N. (1954) Application of composite slip circles for stability analysis. *Proceedings of the European Conference on Stability of Earth Slopes*, Stockholm, Vol. 3, pp. 43–49.

Jiao Yong (1995) Formalizing the systems approach to rock engineering, Ph.D. Thesis, Imperial College, London.

Kim K. and Franklin J. A. (1987) Suggested methods for rock stress determination. *Int. J. Rock Mech. Min. Sci. & Geomech. Abstr.* **24**, 1, 53–74.

Kimmance J. P. (1988) Computer aided risk analysis of open pit mine slopes in kaolin mineral deposits. Ph.D. Thesis, Imperial College, London, 393pp.

Kolsky H. (1963) *Stress Waves in Solids*. Dover, New York, 213pp.

La Pointe P. R. and Hudson J. A. (1985) Characterization and interpretation of rock mass jointing patterns. *Special Paper 199 of the Geological Society of America* (presented as a University of Wisconsin-Madison, Engineering Experiment Research Station Report, June 1981).

Langefors U. and Kihlstrom B. (1963) *Rock Blasting*. John Wiley, London, 403pp.

Long J. C. S. (1983) Investigation of equivalent porous medium permeability in networks of discontinuous fractures, Ph.D. Dissertation, University of California, Berkeley, CA, 277pp.

Lorig L. J. and Brady B. G. H. A hybrid computational scheme for excavation and support design in jointed rock, in (1984) *Design and Performance of Underground Excavations* (Brown E. T. and Hudson J. A., eds). British Geotechnical Society, London, pp. 105–112.

Loureiro-Pinto J. (1986) Suggested method for deformability determination using a large flat jack technique. *Int. J. Rock Mech. Min. Sci. & Geomech. Abstr.* **23**, 2, 131–140.

Matheson G. D. (1983a) *Rock Stability Assessment in Preliminary Investigations—Graphical Methods*. Department of the Environment, Department of Transport, Transport and Road Research Laboratory Report LR 1039.

Matheson G. D. (1983b) Presplit blashing for Highway Road Excavation, Department of the Environment, Department of Transport, Transport and Road Research Laboratory Report LR 1094.

Millar D. L. (1997) Parallel distributed processing in rock engineering systems. Ph.D. Thesis, Imperial College, London, (in press).

Milne D. M. (1988) Suggestion for standardization of rock mass classification, M.Sc. Thesis, Imperial College, London, 165pp.

Mirza U. A. (1978) Investigation into the design criteria for underground openings in rocks which exhibit rheological behaviour, Ph.D. Thesis, University of Newcastle-upon-Tyne.

New B. M. (1984) Explosively-induced ground vibration in civil engineering construction, Ph.D. Thesis, University of Durham, 371pp.

Nordqvist A. (1984) ROCKDISC—a microcomputer-based core logging system. *Int. J. Rock Mech. Min. Sci. & Geomech. Abstr.* **21**, 109–112.

Pan X. D. (1989) Numerical modelling of rock movements around mine openings, Ph.D. Thesis, Imperial College, London, 375pp.

Pande G. N., Beer G. and Williams J. R. (1990) *Numerical Methods in Rock Mechanics*. John Wiley, London.

Patton F. D. (1966) Multiple modes of shear failure in rock. *Proc. 1st Cong. ISRM*, Lisbon, Vol. 1, pp. 509–513.

Pierce F. T. (1926) Tensile tests for cotton yarns—V. The weakest link—theorems on the strength of long and composite specimens. *J. Textile Inst.* **17**, 355–368.

Pine R. J. and Batchelor A. S. (1984) Downward migration of shearing in jointed rock during hydraulic injections. *Int. J. Rock Mech. Min. Sci. & Geomech. Abstr.* **21**, 249–263.

Poulos H. G. and Davis E. H. (1974) *Elastic Solutions for Soil and Rock Mechanics*. 411pp.

Price N. J. and Cosgrove J. W. (1990) *Analysis of Geological Structures*. Cambridge University Press, Cambridge, 502pp.

Priest S. D. (1985) *Hemispherical Projection Methods in Rock Mechanics*. Allen & Unwin, London, 124pp.

Priest S. D. (1993) *Discontinuity Analysis for Rock Engineering*. Chapman & Hall, London, 473pp.

Priest S. D. and Brown E. T. (1983) Probability stability analysis of variable rock slopes. *Trans. Inst. Min. Metall. (Sect A: Min. Industry)* **92**, January.

Ramsay J. G. and Huber M. I. (1983) *The Techniques of Modern Structural Geology*, Vol. 1, *Strain Analysis*. Academic Press, 307pp.

Raudkivi A. J. and Callander R. A. (1976) *Analysis of Groundwater Flow*. Edward Arnold, London, 214pp.

Romana M. (1985) New adjustment ratings for application of Bieniawski classification to slopes. *Proc. Int. Symp. Rock. Mech. Excav. Min. Civ. Works*. ISRM, Mexico City, pp. 59–68.

Samaniego J. A. and Priest S. D. (1985) *Simulation of Fluid Flow in Fractured Rock: A Probabilistic Approach*. Imperial College, London, 37pp.

Sakurai S. and Shimizu N. (1987) Assessment of rock slope stability by fuzzy set theory. *Proceedings of the 6th International Conference on Rock Mechanics, Montreal*. A. A. Balkema, Rotterdam, pp. 503–506.

Serafim J. L. and Pereira J. P. (1983) Considerations of the geomechanical classifications of Bieniawski. *Proc. Int. Symp. Eng. Geol. Underground Constr.* A. A. Balkema, Boston, MA, pp. 33–43.

Siskind D. E., Stagg M. S., Kopp J. W. and Dowding C. H. (1980) *Structure Response and Damage Produced by Ground Vibrations from Surface Blasting*, U.S. Bureau of Mines, Report of Investigations 8507.

Tamai A. (1990) Internal Report, Imperial College, 85pp.
Terzaghi K. (1963) *Theoretical Soil Mechanics*. John Wiley, New York.
Turchaninov I. A., Iofis M. A. and Kasparyan E. V. (1979) *Principles of Rock Mechanics*. Terraspace, Rockville, 493pp.
Wawersik W. R. and Fairhurst C. (1970) A study of brittle rock fracture in laboratory compression experiments. *Int. J. Rock Mech. Min. Sci.* **7**, 561–575.
Wei Lingli (1992) Numerical studies of the hydro-mechanical behaviour of jointed rock, Ph.D. Thesis, Imperial College, London, 296pp.
Wei Z. Q. (1988) A fundamental study of the deformability of rock masses, Ph.D. Thesis, Imperial College, University of London, 268pp.
Windsor C. R. (1985) A study of reinforced discontinuity mechanics, M.Sc. Thesis, Imperial College, University of London, 243pp.
Worsey P. (1981) Geotechnical factors affecting the application of pre-split blasting to rock slopes, Ph.D. Thesis, University of Newcastle upon Tyne, 515pp.
Wu Bailin (1991) Investigation into the mechanical behaviour of soft rocks, Ph.D. Thesis, Imperial College, London, 485pp.

Appendix A: Stress and strain analysis

Stress analysis

We do not talk about 'internal forces' when we are dealing with solid bodies, we refer to **stresses** instead. The reason is simple. Consider a stack of concrete blocks of different sizes supporting a heavy weight.

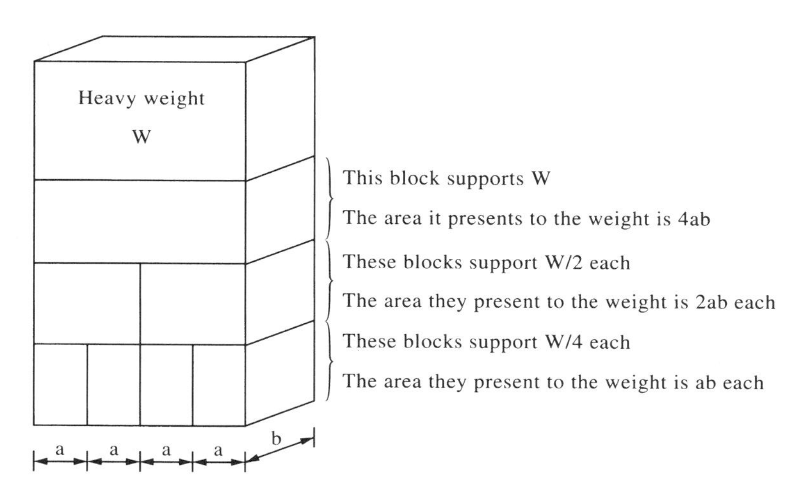

As we can see, when the size of the solid body being considered changes, so the force changes. **But** if we use stress, defined as

$$\text{stress} = \frac{\text{force}}{\text{area}}$$

then we can see that in the example above each block is subject to a stress of $W/4ab$—stress is independent of block size. Hence, if we were to divide a solid body into 'elements', providing we work in terms of stresses the size of the individual elements will not affect the stress values.

Notation

Stress is a property that needs three values to fully describe it in the two-dimensional case: the magnitude of the force, the direction of the force and the area it acts on.

It is known as a **tensor** property, cf. scalars with one value and vectors with two. Stress analysis is only possible if we work in components, like this:

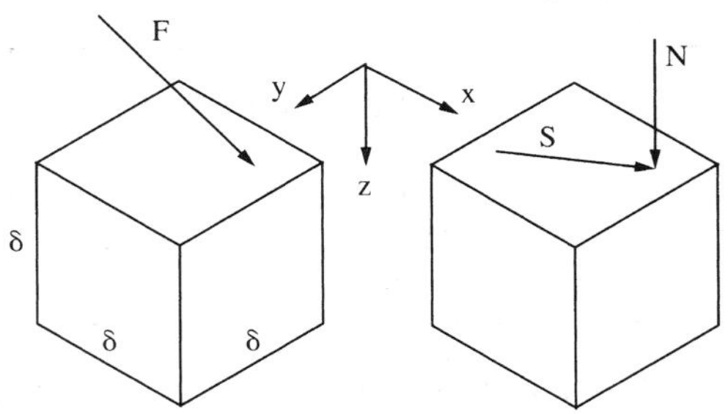

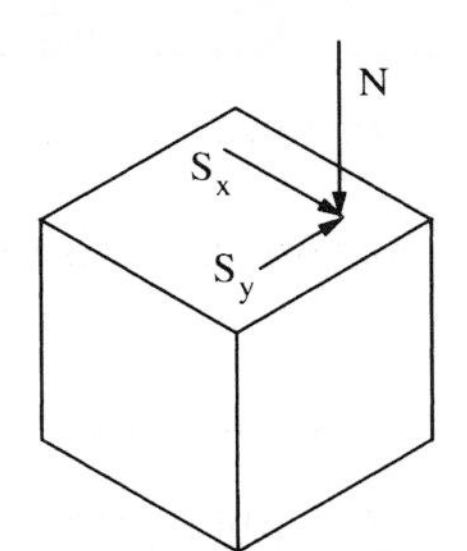

Force applied at an arbitrary angle to the face of an element.

Force resolved into normal and shear components.

Shear component resolved into two Cartesian components.

After we have resolved the force into three Cartesian components, we can define the associated stresses, **and not before.** In this example we have

$$\sigma_z = \frac{N}{\delta^2} \qquad \tau_{zx} = \frac{S_x}{\delta^2} \qquad \tau_{zy} = \frac{S_y}{\delta^2}.$$

and in general we have:

σ_x	σ_y	σ_z
normal stresses		

act normal to face

τ_{xy}	τ_{yz}	τ_{zx}
τ_{yx}	τ_{zy}	τ_{xz}
shear stresses		

act along a face, parallel to an edge of the element.

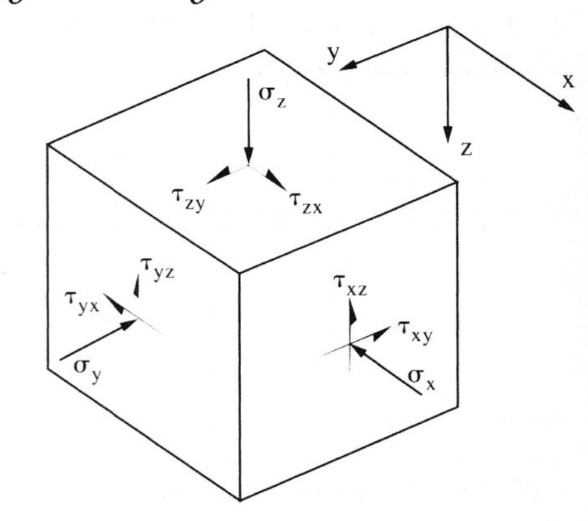

Only the stresses on the visible faces have been shown here

This is the geomechanics, or compression-positive, convention for right-handed axes.

Remember the notation:

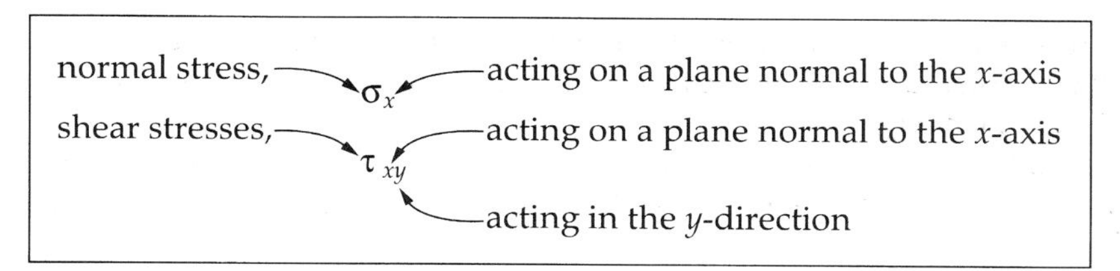

and also, remember the sign convention:

for normal stresses,	**compression is positive**;
for shear stresses,	**positive stresses** act in **positive directions** on **negative faces**.

The shear stress convention is difficult to remember. It is made easier by realizing there are always two pluses and a minus:

on a + face a + stress acts in a – direction
on a + face a – stress acts in a + direction
on a – face a + stress acts in a + direction.

If we consider moment equilibrium of the cube around the three axes, we find that

$$\tau_{xy} = \tau_{yx} \quad \tau_{yz} = \tau_{zy} \quad \tau_{zx} = \tau_{xz}$$

and so the **state of stress at a point** (as δ is reduced to zero) is defined by **six independent** qualities

$$\sigma_x, \sigma_y, \sigma_z, \tau_{xy}, \tau_{yz}, \tau_{zx}.$$

These stresses are usually written in matrix form:

$$\begin{bmatrix} \sigma_x & \tau_{yx} & \tau_{zx} \\ \tau_{xy} & \sigma_y & \tau_{zy} \\ \tau_{xz} & \tau_{yz} & \sigma_z \end{bmatrix} \textbf{ stress tensor.}$$

Note that because of the **complementary shear stresses** (i.e. $\tau_{xy} = \tau_{yx}$, etc.) the stress tensor is symmetric about the leading diagonal.

General stress field in three dimensions

It often happens that each component of the stress tensor varies in magnitude from point to point within a body—they are functions of x, y and z. If this is the case, then an element will be in equilibrium if

$$\frac{\partial \sigma_x}{\partial x} + \frac{\partial \tau_{yx}}{\partial y} + \frac{\partial \tau_{zx}}{\partial z} + X = 0$$

$$\frac{\partial \tau_{xy}}{\partial x} + \frac{\partial \sigma_y}{\partial y} + \frac{\partial \tau_{zy}}{\partial z} + Y = 0$$

$$\frac{\partial \tau_{xz}}{\partial x} + \frac{\partial \tau_{yz}}{\partial y} + \frac{\partial \sigma_z}{\partial z} + Z = 0$$

the equilibrium equations.

Each equation contains increments of the stress components in one direction. In these equations the vector (X, Y and Z) is the **body force vector,** that is the force (mass × acceleration) produced by the body itself. Normally we will be dealing with bodies at rest in the Earth's gravitional field, with the z-axis vertically downwards. In this case the body force vector is simply $(0, 0, \gamma_z)$.

Transformation of the stress tensor

It is often the case that we may know applied stresses relative to one set of axes (the **global** axes), but may wish to know the stress state relative to another set (the **local** axes). For example, suppose we are dealing with a discontinuity in a rock mass:

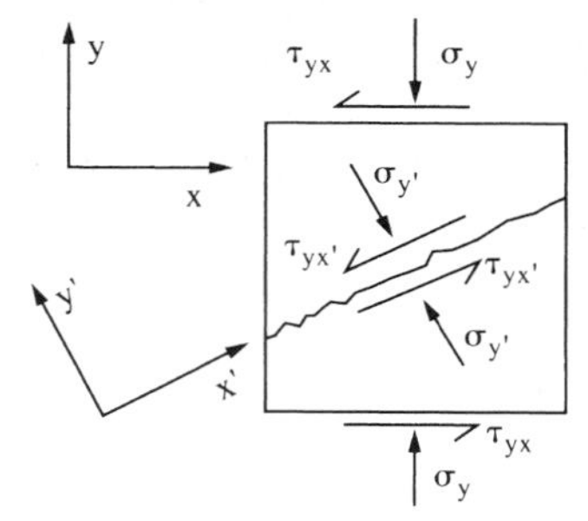

Given σ_y and τ_{yx}, what are $\sigma_{y'}$ and $\tau_{yx'}$?

Unfortunately, stresses are tensors, not vectors like forces, and so cannot be simply resolved: they must be **transformed.** We will limit ourselves to this case:

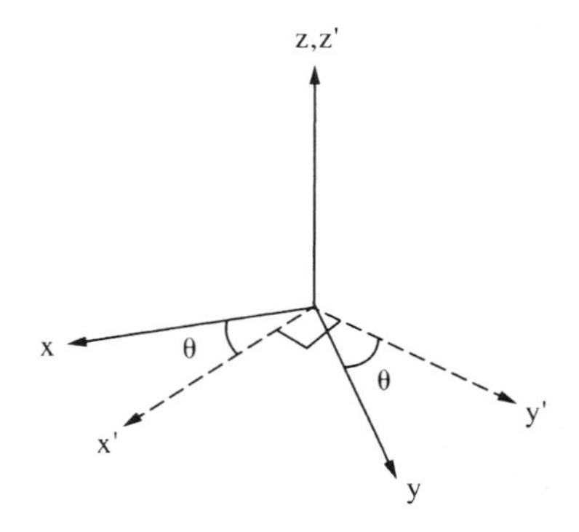

The global system is x, y, z.
The local system is x', y', z'.
In this case z and z' are coincident.

If the global stresses are

$$\begin{bmatrix} \sigma_x & \tau_{yx} & 0 \\ \tau_{xy} & \sigma_y & 0 \\ 0 & 0 & \sigma_z \end{bmatrix}$$

what are the local stresses?

$$\begin{bmatrix} \sigma_{x'} & \tau_{y'x'} & 0 \\ \tau_{x'y'} & \sigma_{y'} & 0 \\ 0 & 0 & \sigma_{z'} \end{bmatrix}.$$

If we think in terms of a small cubic element rotating about the z-axis, we can see that $\sigma_z = \sigma_{z'}$.

Calculating $\sigma_{x'}$, $\sigma_{y'}$ and $\tau_{x'y'}$ is more difficult and is done like this:

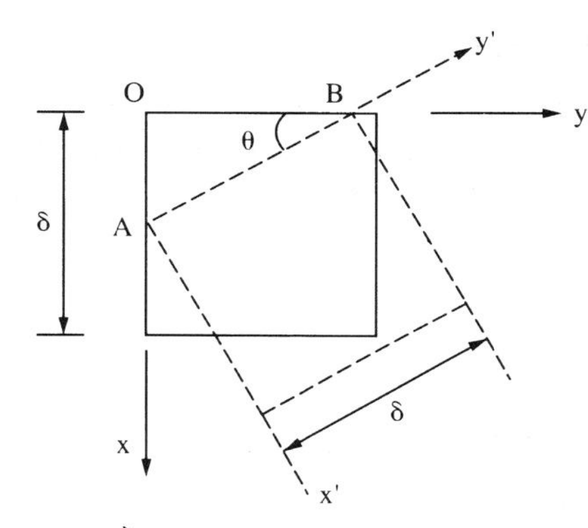

The rotated element is positioned over the original element such that the vertical **sides** (remember it is a cube) A and B touch two of the original faces. If we now cut off the prism OAB from the original element and examine the resulting free body diagram resolving stress components onto resolved areas:

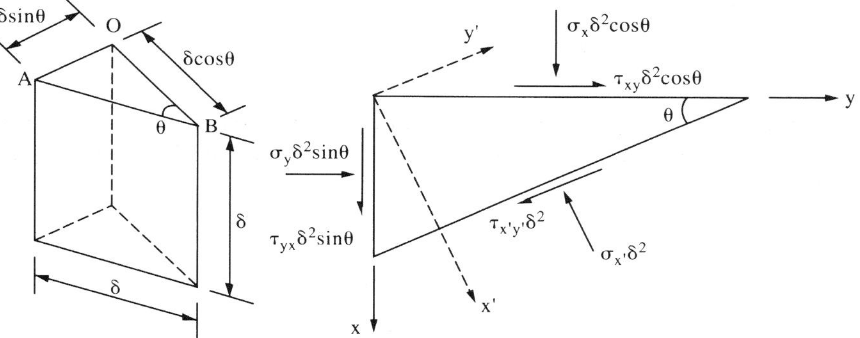

As usual, write down the equations of static equilibrium:

$$\Sigma F_{x'} = 0$$

$$(\sigma_x\ \delta^2\cos\ \theta)\cos\ \theta + (\sigma_y\ \delta^2\sin\ \theta)\sin\ \theta + (\tau_{xy}\ \delta^2\cos\ \theta)\sin\ \theta + (\tau_{yx}\ \delta^2\sin\ \theta)\cos\ \theta - \sigma_{x'}\delta^2 = 0.$$

Cancelling the δ^2 term, and remembering that $\tau_{xy} = \tau_{yx}$, then

$$\underline{\underline{\sigma_{x'} = \sigma_x \cos^2\theta + \sigma_y \sin^2\theta + 2\tau_{xy} \sin\theta\cos\theta.}}$$

Notice that each term has a trigonometric identity of order 2 associated with it: this is because in the transformation force is resolved once and area is resolved once.

$$\Sigma F_{y'} = 0$$

$$-(\sigma_x \delta^2 \cos\theta)\sin\theta + (\sigma_y \delta^2 \sin\theta)\cos\theta + (\tau_{xy} \delta^2 \cos\theta)\cos\theta - (\tau_{yx} \delta^2 \sin\theta)\sin\theta - \tau_{x'y'}\delta^2 = 0.$$

Again, cancelling the δ^2 term, and putting $\tau_{yx} = \tau_{xy}$, we find that

$$\tau_{x'y'} = -\sigma_x \cos\theta\sin\theta + \sigma_y \sin\theta\cos\theta + \tau_{xy}\cos^2\theta - \tau_{xy}\sin^2\theta$$

or

$$\underline{\underline{\tau_{x'y'} = \tau_{xy}(\cos^2\theta - \sin^2\theta) - (\sigma_x - \sigma_y)\sin\theta\cos\theta.}}$$

Determining $\sigma_{y'}$ can be done either by cutting a prism parallel to the x-axis, or simply by replacing θ with $(\theta + \pi/2)$ and $\sigma_{x'}$ with $\sigma_{y'}$ in the expression for $\sigma_{x'}$ on the previous page (this is valid because we know $\sigma_{y'}$ is perpendicular to $\sigma_{x'}$):

$$\sin[\theta + (\pi/2)] = \cos\theta \text{ and } \cos[\theta + (\pi/2)] = -\sin\theta$$

so

$$\sigma_{x'} = \sigma_x \cos^2\theta + \sigma_y \sin^2\theta + 2\tau_{xy}\sin\theta\cos\theta$$

becomes

$$\underline{\underline{\sigma_{y'} = \sigma_x \sin^2\theta + \sigma_y \cos^2\theta - 2\tau_{xy}\sin\theta\cos\theta}}$$

so the three equations are:

$$\sigma_{x'} = \sigma_x \cos^2\theta + \sigma_y \sin^2\theta + 2\,\tau_{xy}\sin\theta\cos\theta$$
$$\sigma_{y'} = \sigma_x \sin^2\theta + \sigma_y \cos^2\theta - 2\,\tau_{xy}\sin\theta\cos\theta$$
$$\tau_{x'y'} = \tau_{xy}(\cos^2\theta - \sin^2\theta) - (\sigma_x - \sigma_y)\sin\theta\cos\theta.$$

stress transformation equations

Example

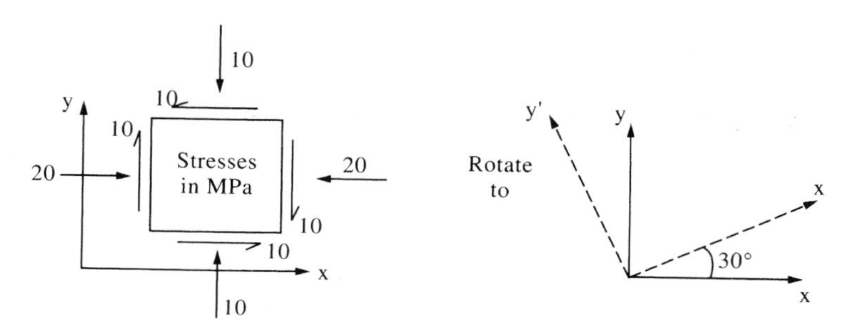

$$\theta = 30° \Rightarrow \sin\theta = 0.500, \cos\theta = 0.866$$

$$\therefore\ \sigma_{x'} = 20 \times (0.866)^2 + 10 \times (0.500)^2 + 2 \times 10 \times 0.500 \times 0.866 = \underline{26.16\text{ MPa}}$$

$$\sigma_{y'} = 20 \times (0.500)^2 + 10 \times (0.866)^2 - 2 \times 10 \times 0.500 \times 0.866 = \underline{3.84\text{ MPa}}$$

$$\tau_{x'y'} = 10\,(0.866^2 - 0.500^2) - (20 - 10) \times 0.866 \times 0.500 = \underline{0.67\text{ MPa}}.$$

An interesting check of the arithmetic is that

$$(\sigma_x + \sigma_y) = (\sigma_{x'} + \sigma_{y'}).$$

So in this case

$$\sigma_x + \sigma_y = 30\text{ MPa}$$

and

$$\sigma_{x'} + \sigma_{y'} = 26.16 + 3.84 = 30.00\text{ MPa}.$$

Notice that $\tau_{x'y'}$ is very low in comparison to τ_{xy}. We may be tempted to ask whether there exists for $x'y'$ an orientation such that $\tau_{x'y'} = 0$. Well there is.

Principal directions and principal stresses

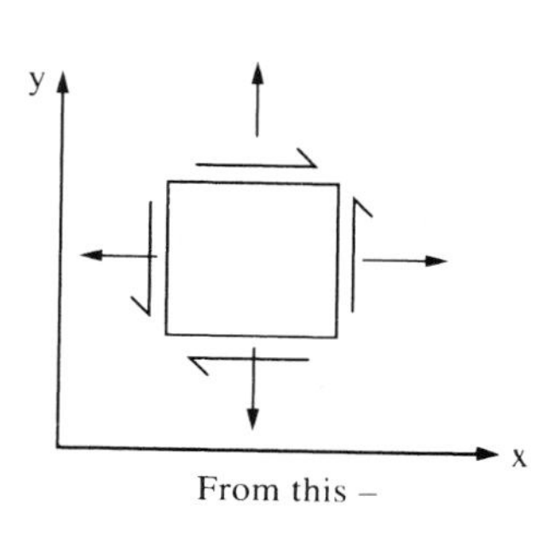

From this –

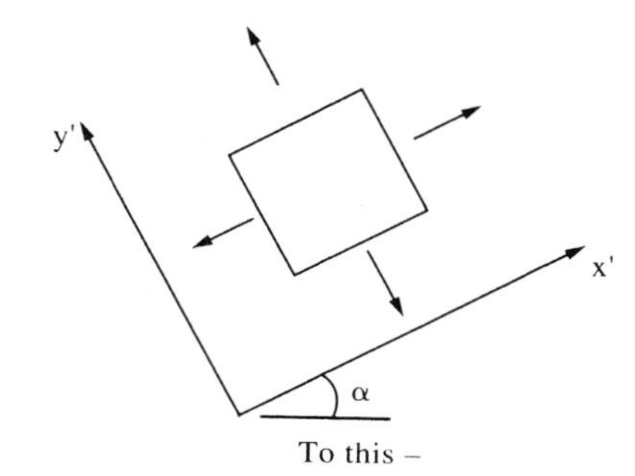

To this –

The **principal directions** are the directions of the co-ordinate axes such that for any given stress state the shear stresses are zero. The only stresses acting on the elemental cube in this new direction are the **principal stresses.**

In the rotated diagram above we want $\tau_{x'y'} = 0$, i.e.

$$\tau_{x'y'} = \tau_{xy}(\cos^2\alpha - \sin^2\alpha) - (\sigma_x - \sigma_y)\cos\alpha\sin\alpha = 0$$

$$\therefore\ \tau_{xy}(\cos^2\alpha - \sin^2\alpha) = (\sigma_x - \sigma_y)\cos\alpha\sin\alpha$$

or

$$\frac{\tau_{xy}}{(\sigma_x - \sigma_y)} = \frac{\cos\alpha\sin\alpha}{\cos^2\alpha - \sin^2\alpha} = \frac{\frac{1}{2}\sin 2\alpha}{\cos 2\alpha} = \frac{1}{2}\tan 2\alpha$$

inverting:

$$\boxed{\alpha = \tfrac{1}{2}\tan^{-1}\left(\frac{2\tau_{xy}}{\sigma_x - \sigma_y}\right).}$$

When $\tau_{x'y'} = 0$, x' and y' are the **principal directions** and $\sigma_{x'}$ and $\sigma_{y'}$ are the **principal stresses.**

Example. Continuing from before

$$\tau_{xy} = 10, \quad \sigma_x = 20, \quad \sigma_y = 10$$

so

$$\alpha = \tfrac{1}{2}\tan^{-1}\left(\frac{2\tau_{xy}}{\sigma_x - \sigma_y}\right) = \tfrac{1}{2}\tan^{-1}\left(\frac{2\times 10}{20-10}\right) = \tfrac{1}{2}\tan^{-1}(2)$$

i.e.

$$\underline{\underline{\alpha = 31.7^\circ}}.$$

This gives the principal directions. The principal stresses are found as before, with $\theta = 31.7°$.

$$\theta = 31.7^\circ \Rightarrow \sin\theta = 0.526, \cos\theta = 0.851$$

$$\sigma_{x'} = 20\times(0.851)^2 + 10\,(0.526)^2 + 2\times 10\times 0.526\times 0.851 = \underline{26.18\text{ MPa}}$$

$$\sigma_{y'} = 20\times(0.526)^2 + 10\times(0.851)^2 - 2\times 10\times 0.526\times 0.851 = \underline{3.82\text{ MPa}}.$$

$$\text{Check: } \sigma_{x'} + \sigma_{y'} = 26.18 + 3.82 = 30.00\text{ MPa}.$$

Note: the largest of these stresses is the **major principal stress,** σ_1. The smallest is the **minor principal stress,** σ_2.

Note that becauese $(\sigma_x + \sigma_y) = (\sigma_{x'} + \sigma_{y'}) = (\sigma_1 + \sigma_2)$ which is a constant, there can be no principal shear stresses, i.e. planes on which no normal stresses act

Mohr's circle of stress

This is a graphical method of transforming the stress tensor. It is easy to use and remember, and is the best way of remembering the transformation equations.

If we choose the global x- and y-axes to coincide with the principal directions (and because we can choose the axes arbitrarily there is nothing to prevent this), then the transformation equations become

$$\sigma_{x'} = \sigma_1 \cos^2\theta + \sigma_2 \sin^2\theta \qquad \text{(A)}$$

$$\sigma_{y'} = \sigma_1 \sin^2\theta + \sigma_2 \cos^2\theta \qquad \text{(B)}$$

$$\tau_{x'y'} = -(\sigma_1 - \sigma_2)\cos\theta\sin\theta \qquad \text{(C)}$$

where σ_1 and σ_2 are now the principal stresses, and θ is measured anticlockwise from the principal direction x to the local direction x'.

These new equations can be simplified still further, by making use of trigonometric identities.

Let

$$\phi = 2\theta = \theta + \theta$$

then

$$\sin\phi = \sin(\theta + \theta) = \sin\theta\cos\theta + \cos\theta\sin\theta = 2\sin\theta\cos\theta$$

$$\therefore \underline{\underline{\cos\theta\sin\theta = \tfrac{1}{2}\sin\phi}} \qquad \text{(D)}$$

and

$$\cos\phi = \cos(\theta + \theta) = \cos\theta\cos\theta - \sin\theta\sin\theta = \cos^2\theta - \sin^2\theta$$

but

$$\cos^2\theta + \sin^2\theta = 1$$

so

$$\cos\phi = \cos^2\theta - (1 - \cos^2\theta) = 2\cos^2\theta - 1$$

from which

$$\underline{\underline{\cos^2 \theta = \tfrac{1}{2}(1 + \cos \phi)}} \qquad \text{(E)}$$

or

$$\cos \phi = (1 - \sin^2 \theta) - \sin^2 \theta = 1 - 2 \sin^2 \theta$$

from which

$$\underline{\underline{\sin^2 \theta = \tfrac{1}{2}(1 + \cos \phi)}}. \qquad \text{(F)}$$

Substituting (D), (E) and (F) into equations (A), (B) and (C):

$$\sigma_{x'} = \sigma_1 (\tfrac{1}{2}(1 + \cos \phi)) + \sigma_2(\tfrac{1}{2}(1 + \cos \phi))$$

i.e.

$$\boxed{\begin{aligned} \sigma_{x'} &= \tfrac{1}{2}(\sigma_1 + \sigma_2) + \tfrac{1}{2}(\sigma_1 + \sigma_2) \cos \phi \\ \tau_{x'y'} &= -\tfrac{1}{2}(\sigma_1 + \sigma_2) \sin \phi. \end{aligned}}$$

These two equations are simply the equations of a circle centred at $\tfrac{1}{2}(\sigma_1 + \sigma_2)$ on the σ-axis in σ–τ space:

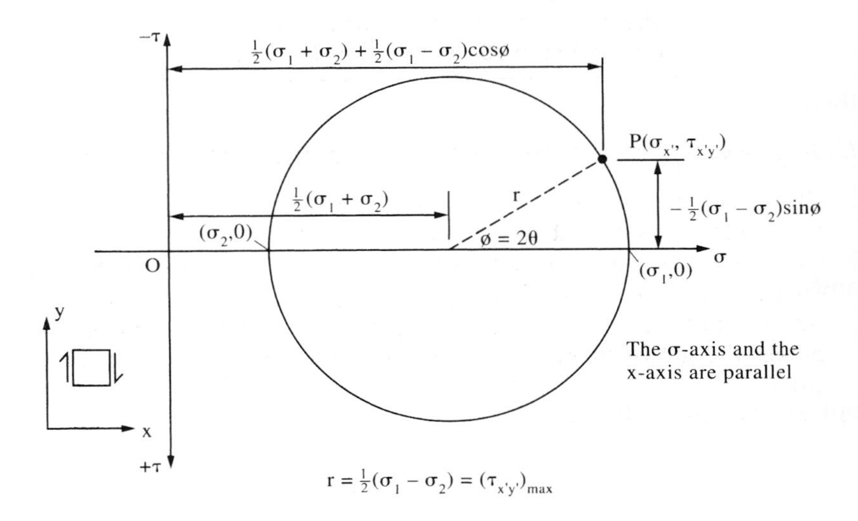

To use Mohr's circle you must **understand** and **remember**:

1. Positive shear stresses and positive rotations have been used in developing the equations for a point (σ, τ), but the τ co-ordinate is negative. This means the τ-axis is upside down:

positive shear stresses plot below the σ-axis.

2. The trigonometric relations used to simplify the equations resulted in $\phi = 2\theta$:

whatever rotation takes place in real life, twice the rotation takes place on Mohr's circle.

3. Each point on the circumference of the circle represents the (σ, τ) stress state on a plane of specific orientation. The points where the circle intersects the σ-axis represent planes on which $\tau = 0$: the principal planes. The associated σ-values are the principal stresses. Mohr's circle shows

the principal stresses are the maximum and minimum values of normal stress in the body.

4. The points representing the principal planes lie at opposite ends of a diameter: in real life planes are perpendicular.

Two perpendicular planes are represented on the circle by points at opposite ends of a diameter.

5. The maximum shear stress is given by ½ ($\sigma_1 - \sigma_2$) and occurs when $\phi = 90°$ (i.e. $\theta = 45°$). Thus

the planes of maximum shear stress are orientated at 45° to the principal planes.

Using Mohr's circle to determine principal stresses

1. Draw x–y-axes on the element, draw an element with positive normal and shear stresses on it, and so write down (σ_x, τ_{xy}) and (σ_y, τ_{yx}).
2. Draw σ–τ-axes (same scale on each) with the σ-axis parallel to, and in the same direction as, σ_x. Plot (σ_x, τ_{xy}) bearing in mind the positive shear stresses plot below the σ-axis. Then plot (σ_y, τ_{yx}) **on the other side of the** σ-axis. Draw the diameter between the two points, and then draw the circle.
3. Calculate the radius as

$$\frac{1}{2}\sqrt{\left(\sigma_x - \sigma_y\right)^2 + \left(2\tau_{xy}\right)^2}$$

and the σ-value of the centre as ½($\sigma_x + \sigma_y$).

4. Calculate the principal stresses and the maximum shear stress:

$$\sigma_1 = c + r, \quad \sigma_2 = c - r, \quad \tau_{max} = r.$$

5. Calculate the rotation angle and direction from σ_x to σ_1. Remember that rotations on the circle are twice real life rotations (ϕ = positive rotation).

$$\phi = \tan^{-1}\left(\frac{2\tau_{xy}}{\sigma_x - \sigma_y}\right)$$

but be aware that $0° < \phi < 180°$.

6. Finally, draw the element on which the principal stresses act, in the correct orientation.

Example

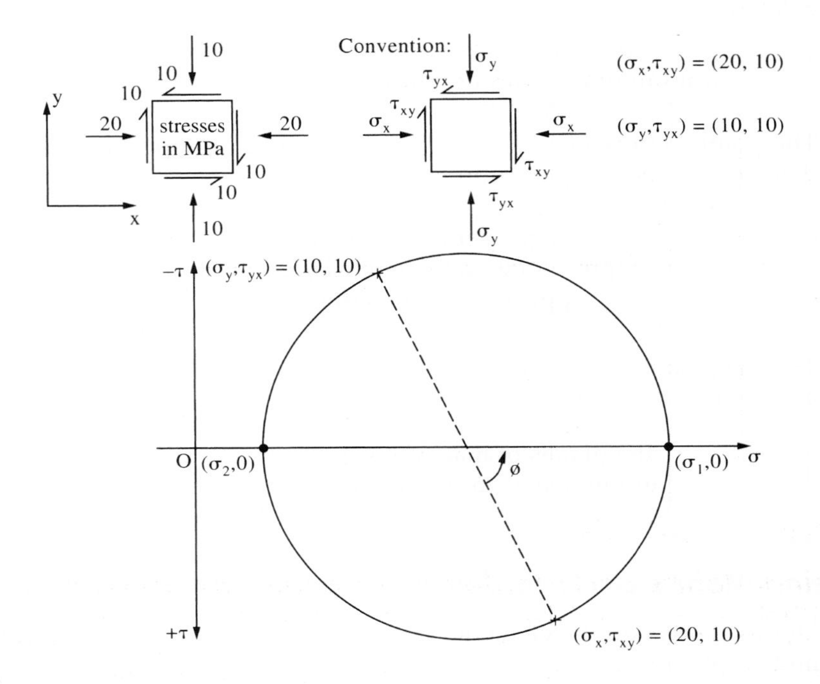

radius $= \frac{1}{2}\sqrt{(20-10)^2 + (2 \text{ x } 10)^2} = \underline{11.18\text{MPa}}$ centre $= \frac{1}{2}(20+10) = \underline{15\text{MPa}}$

$\sigma_1 = 15 + 11.18 = \underline{26.18\text{MPa}}$ $\sigma_2 = 15 - 11.18 = \underline{3.82\text{MPa}}$ $\tau_{max} = \underline{11.18\text{MPa}}$

$\text{ø} = \tan^{-1}\dfrac{2 \text{ x } 10}{20-10} = 63.43°$ $\therefore \underline{\theta = 31.72°}$

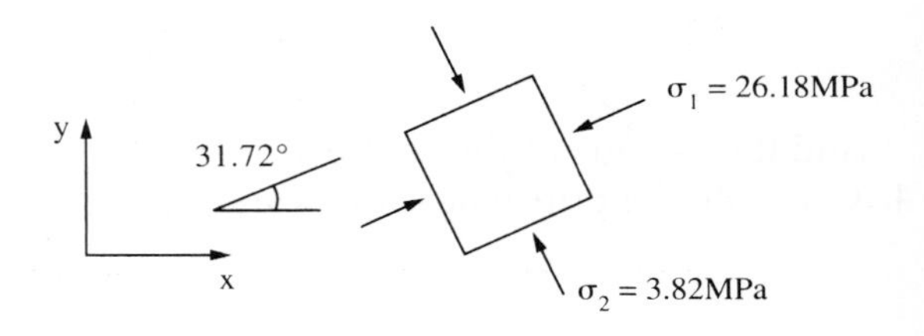

Using Mohr's circle to determine stresses on a plane

Follow points 1, 2, 3 and 5 of the method for determining principal stresses, then:

4. Draw an element of the correct orientation relative to the x–y-axes, and mark on positive $\sigma_{x'}$, $\sigma_{y'}$, $\tau_{x'y'}$ and $\tau_{y'x'}$. Write down the sense (positive anticlockwise) and the magnitude of the rotation x-axis to x'-axis.
5. Mark this rotation on the circle, measuring from the (σ_x, τ_{xy}) point, remembering that you do twice as much on the circle as you do in real life.
6. The new point is ($\sigma_{x'}$, $\tau_{x'y'}$). Draw the diameter to determine ($\sigma_{y'}$, $\tau_{y'x'}$).

Example. What are the stresses on an element rotated 30° anticlockwise relative to the element in the previous example?

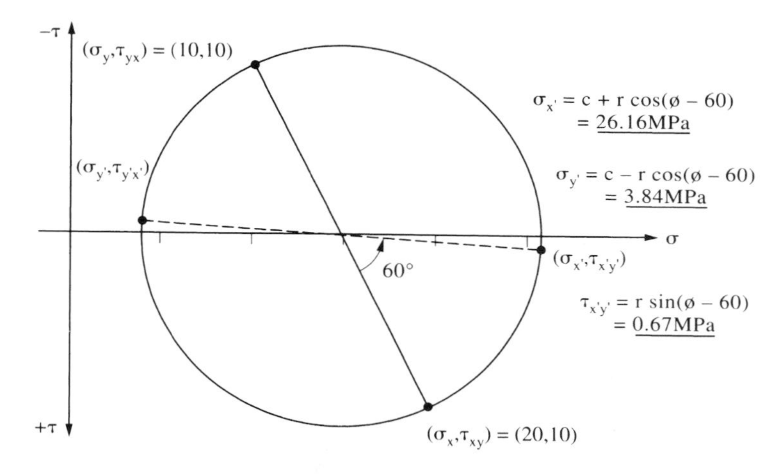

Strain analysis

If we apply stresses to a body, by how much does it deform? Obviously, it depends on the stress state, the material the body is composed of, and the size of the body. This last problem is resolved if, instead of absolute deformations, we talk about relative deformations, relative to the body. This is what strain is: deformation normalized to make it independent of the size of the body.

Analysis of displacement

Consider this situation:

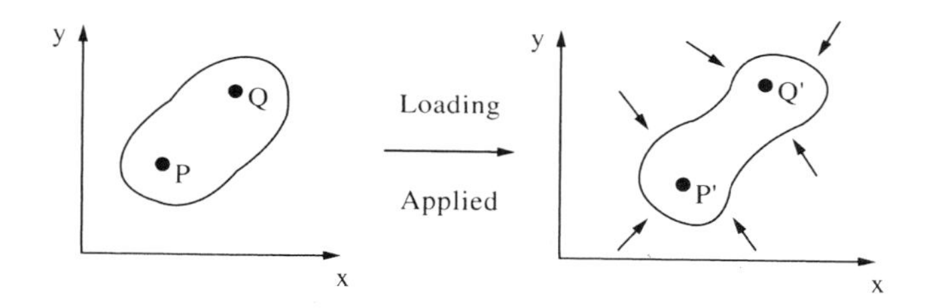

P moves to P′, Q moves to Q′. The vector P′Q′ may have a different magnitude and direction to the vector PQ. Is it possible for us to determine **P′Q′** knowing **PQ** and the general form deformation in this body takes? Providing we make some assumptions, yes.

Assume that displacement varies with position in the body—it is a function of x and y. Then say that:

$$\text{function describing displacements in the } x\text{-direction} = u(x, y)$$
$$\text{function describing displacements in the } y\text{-direction} = v(x, y).$$

To simplify the working, we will refer to these functions as u and v. Now, let us use these functions to analyse the displacements of P and Q:

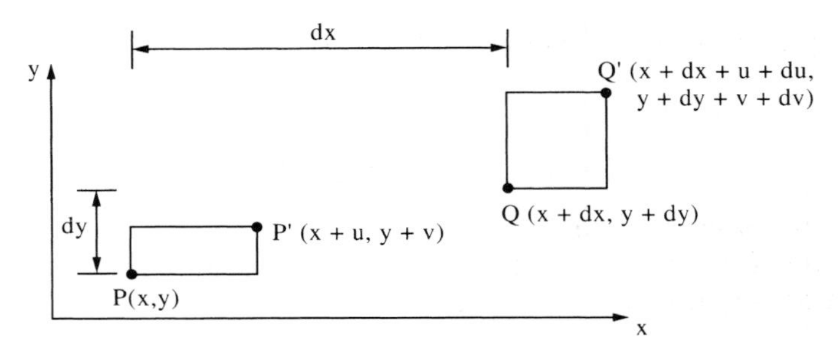

P and Q are separated by a small distance (dx, dy). The deformation causes P to move from (x, y) to $(x + u, y + v)$ [we calculate this by substituting the values of x and y into the functions u and v].

These deformations vary with position within the body, so they are different at Q: here they are (u + du) and (v + dv). The initial co-ordinates of Q are also a little different from those of P: they are (x + dx, y + dy). This means the co-ordinates of Q′ are:

([x + dx]	+	[u + du],	[y + dy]	+	[v + dv]).
initial X co-ordinate		X displacement	initial Y co-ordinate		Y displacement

Both u and v are functions of x and y, and so calculating the derivatives is awkward: the functions are surfaces, not curves, and each derivative contains components due to dx and dy. We can calculate du (and similarly dv) like this:

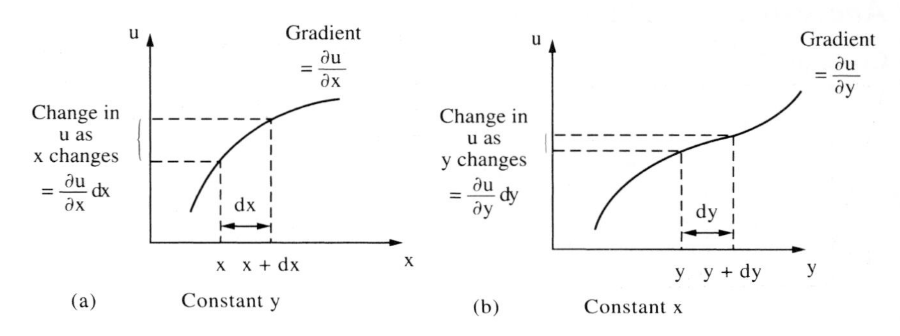

The total change in u (i.e. du) is given by the sum of these components:

$$du = (\text{change in } u \text{ as } x \text{ changes}) + (\text{change in } u \text{ as } y \text{ changes})$$

hence

$$du = \frac{\partial u}{\partial x}dx + \frac{\partial u}{\partial y}dy$$

similarly

$$dv = \frac{\partial v}{\partial x}dx + \frac{\partial v}{\partial y}dy.$$

We can put these equations into matrix form:

$$\begin{bmatrix} du \\ dv \end{bmatrix} = \begin{bmatrix} \partial u/\partial x & \partial u/\partial y \\ \partial v/\partial x & \partial v/\partial y \end{bmatrix} \begin{bmatrix} dx \\ dy \end{bmatrix}.$$

This shows how the displacements (du, dv) are functions of the original separation of P and Q (dx, dy). The problem is, we need the displacement **independent** of this separation: i.e. we need strain.

Strain in terms of displacement functions

In general, when a body deforms, the following components of deformation take place:

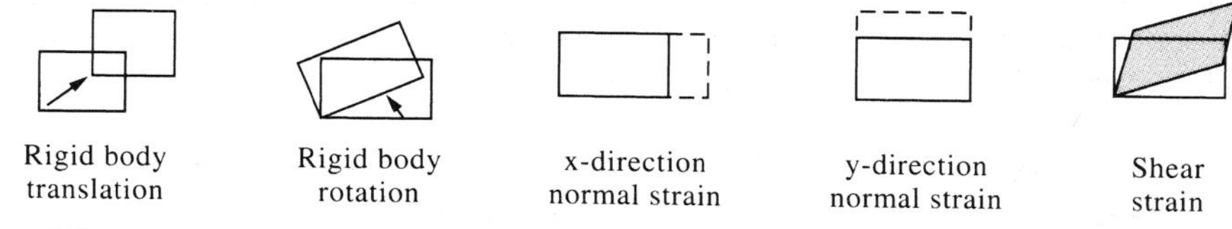

We can ignore rigid body rotation—we are only interested in the displacement of points relative to each other. Now all we have to do is analyse each component individually, and then combine them.

(a) Rigid body rotation.

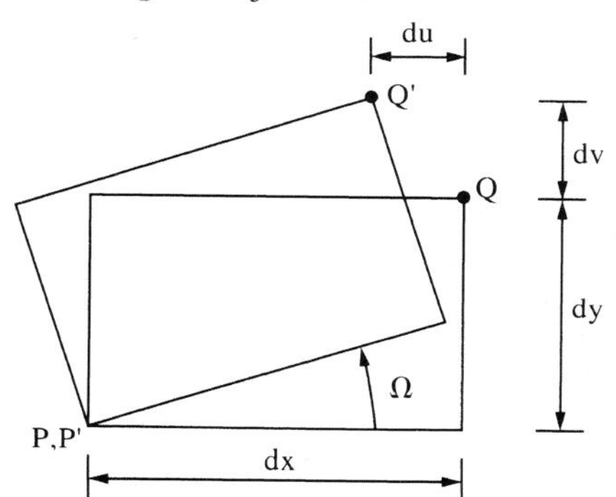

The element does not change shape: it rotates by Ω radians such that P and P′ remain coincident and Q moves to Q′. Hence

$$du = -\sin\Omega \cdot dy$$
$$dv = \sin\Omega \cdot dx.$$

For small rotations $\sin \Omega = \Omega$, so that

$$du = -\Omega\, dy$$
$$dv = \Omega\, dx$$

or

$$\begin{bmatrix} du \\ dv \end{bmatrix} = \begin{bmatrix} 0 & -\Omega \\ \Omega & 0 \end{bmatrix} \begin{bmatrix} dx \\ dy \end{bmatrix}.$$

(b) Normal strain

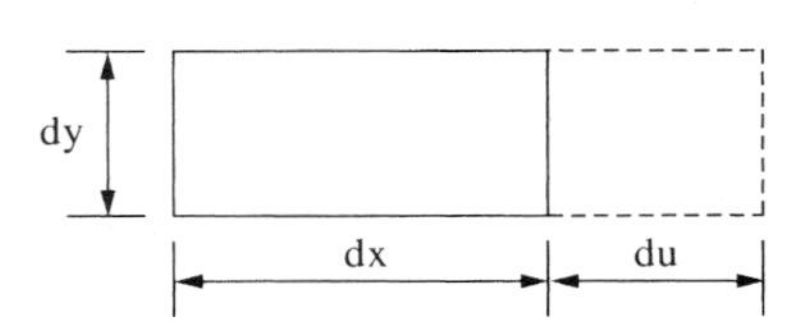

Because we use compression (and hence contraction) positive, we have

$$\varepsilon_x = -\frac{du}{dx}$$

from which we obtain

$$du = -\varepsilon_x\, dx$$
$$dv = -\varepsilon_y\, dy$$

or

$$\begin{bmatrix} du \\ dv \end{bmatrix} = \begin{bmatrix} -\varepsilon_x & 0 \\ 0 & -\varepsilon_y \end{bmatrix} \begin{bmatrix} dx \\ dy \end{bmatrix}.$$

(c) Shear strain

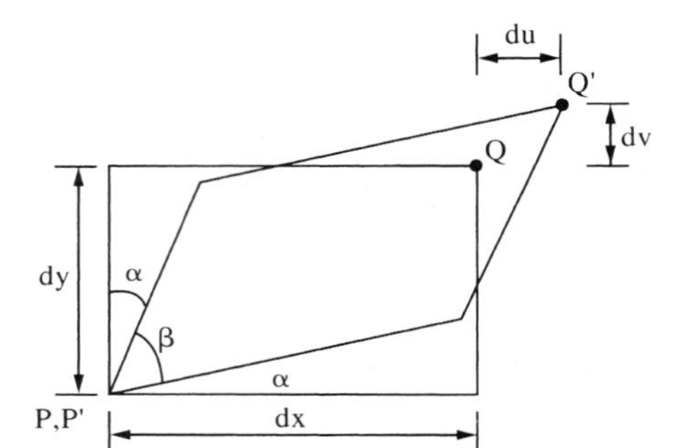

This is negative shear strain: P′Q′ is longer than PQ, and extension is negative. For small angles, to a good approximation we have

$$du = dy \sin \alpha + dx \cos \alpha - dx$$

but for small angles $\sin \alpha = \alpha$ and $\cos \alpha = 1$, so that $du = \alpha\, dy + dx - dx = \alpha\, dy$ and similarly $dv = \alpha dx$.

The definition of shear strain is the change in angle between two lines originally perpendicular to each other, i.e. $\gamma_{xy} = (\beta - \pi/2)$.

$$\pi/2 = \beta + 2\alpha \quad \Rightarrow \quad -2\alpha = \beta - \pi/2 = \gamma_{xy}.$$

Hence

$$du = -\tfrac{1}{2}\,\gamma_{xy}\,dy$$
$$dv = -\tfrac{1}{2}\,\gamma_{xy}\,dx$$

or

$$\begin{bmatrix} du \\ dv \end{bmatrix} = \begin{bmatrix} 0 & -\tfrac{1}{2}\gamma_{xy} \\ -\tfrac{1}{2}\gamma_{xy} & 0 \end{bmatrix} \begin{bmatrix} dx \\ dy \end{bmatrix}.$$

Note that the tensorial shear strain is half the engineering shear strain.

(d) Combined strain and rotation. We can now add together cases (a), (b) and (c) to form a single set of equations. It is useful to keep strain and rotation separate though:

This is because only the strain matrix represents **distortion.** The rotation

$$\begin{bmatrix} du \\ dv \end{bmatrix} = -\begin{bmatrix} \varepsilon_x & \tfrac{1}{2}\gamma_{xy} \\ \tfrac{1}{2}\gamma_{xy} & \varepsilon_y \end{bmatrix} \begin{bmatrix} dx \\ dy \end{bmatrix} + \begin{bmatrix} 0 & -\Omega \\ \Omega & 0 \end{bmatrix} \begin{bmatrix} dx \\ dy \end{bmatrix}.$$

strain matrix **rotation matrix**

matrix is just that: a rotation of the rigid body.

However, in the analysis of displacement we found that

$$\begin{bmatrix} du \\ dv \end{bmatrix} = \begin{bmatrix} \partial u/\partial x & \partial u/\partial y \\ \partial v/\partial x & \partial v/\partial y \end{bmatrix} \begin{bmatrix} dx \\ dy \end{bmatrix}$$

so

$$\begin{bmatrix} \partial u/\partial x & \partial u/\partial y \\ \partial v/\partial x & \partial v/\partial y \end{bmatrix} \begin{bmatrix} dx \\ dy \end{bmatrix} = -\begin{bmatrix} \varepsilon_x & \tfrac{1}{2}\gamma_{xy} \\ \tfrac{1}{2}\gamma_{xy} & \varepsilon_y \end{bmatrix} + \begin{bmatrix} 0 & -\Omega \\ \Omega & 0 \end{bmatrix}.$$

Writing these equations in full gives:

$$\partial u/\partial x = -\,\varepsilon_x \qquad \partial u/\partial y = \tfrac{1}{2}\,\gamma_{xy} - \Omega$$
$$\partial v/\partial x = \tfrac{1}{2}\,\gamma_{xy} + \Omega \qquad \partial v/\partial y = -\,\varepsilon_y.$$

From which we find

$$\partial v/\partial x + \partial u/\partial y = -\,\tfrac{1}{2}\,\gamma_{xy} + \Omega - \tfrac{1}{2}\,\gamma_{xy} - \Omega = -\,\gamma_{xy}$$

and

$$\partial v/\partial x - \partial u/\partial y = -\,\tfrac{1}{2}\,\gamma_{xy} + \Omega + \tfrac{1}{2}\,\gamma_{xy} + \Omega = 2\Omega.$$

Collecting these expressions together:

$$\varepsilon_x = -\partial u/\partial x, \qquad \varepsilon_y = -\partial v/\partial y$$

$$\gamma_{xy} = -\left(\frac{\partial v}{\partial x} + \frac{\partial u}{\partial y}\right), \quad \Omega = \frac{1}{2}\left(\frac{\partial v}{\partial x} - \frac{\partial u}{\partial y}\right).$$

These are the equations of strain in terms of displacement.

Example. Say that

$$u = \frac{1}{1000}\left(1 + 2xy^2 + 3x^2 y\right)$$

and

$$v = \frac{1}{1000}\left(x^2 + y^2\right).$$

What are the strains and rigid body rotation at (1, 1)? Taking the necessary derivatives and substituting $x = 1$ and $y = 1$:

$$\frac{\partial u}{\partial x} = \frac{1}{1000}\left(2y^2 + 6xy\right) = \frac{8}{1000}$$

$$\frac{\partial u}{\partial y} = \frac{1}{1000}\left(4xy + 3x^2\right) = \frac{7}{1000}$$

$$\frac{\partial v}{\partial x} = \frac{1}{1000}\left(2x\right) = \frac{2}{1000}$$

$$\frac{\partial v}{\partial y} = \frac{1}{1000}\left(2y\right) = \frac{2}{1000}.$$

Substituting into the analytical formulae:

$$\varepsilon_x = -\frac{\partial u}{\partial x} = -\frac{8}{1000} = \underline{-0.008}$$

$$\varepsilon_y = -\frac{\partial v}{\partial y} = -\frac{2}{1000} = \underline{-0.002}$$

$$\gamma_{xy} = -\left(\frac{\partial u}{\partial y} + \frac{\partial v}{\partial x}\right) = -\left(\frac{7}{1000} + \frac{2}{1000}\right) = -\frac{9}{1000} = \underline{-0.009}$$

$$\Omega = \frac{1}{2}\left(\frac{\partial v}{\partial x} - \frac{\partial u}{\partial y}\right) = \frac{1}{2}\left(\frac{2}{1000} - \frac{7}{1000}\right) = -\frac{5}{1000} = \underline{-0.0025 \text{ rad}}.$$

Note that strains are dimensionless.

Transformation of displacement (plane strain)

Say point P moves to P′ when deformation takes place. If we know u and v, can we calculate u' and v' on the new axes $x'y'$ in terms of θ?

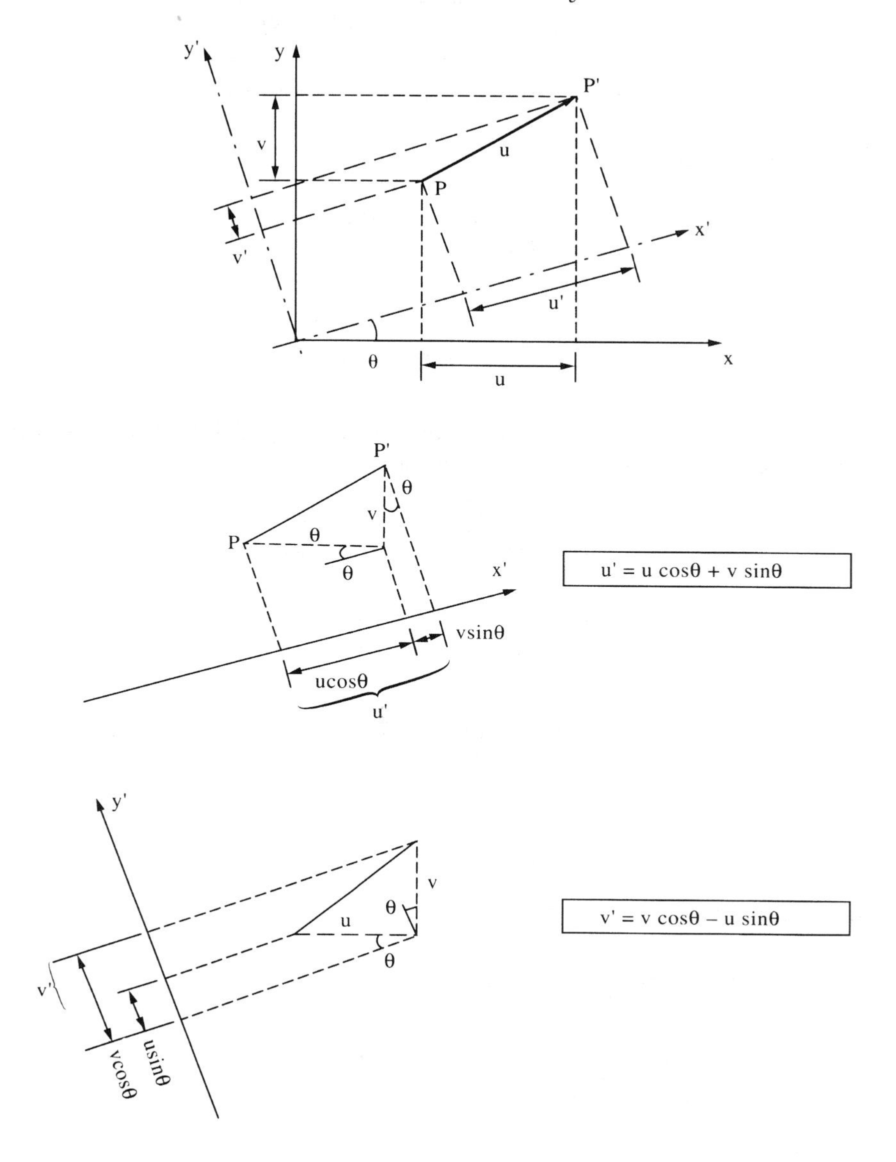

Transformation of strain (plane strain)

Given ε_x, ε_y and γ_{xy} (i.e. global components) how do we calculate ε_x', ε_y' and γ_{xy}' (the local components)?

Now

$$\varepsilon_x = -\frac{\partial u}{\partial x},\ \varepsilon_y = -\frac{\partial v}{\partial y} \text{ and } \gamma_{xy} = -\left(\frac{\partial u}{\partial x} + \frac{\partial v}{\partial y}\right),$$

so

$$\varepsilon_x' = -\frac{\partial u'}{\partial x'},\ \varepsilon_y' = -\frac{\partial v'}{\partial y'} \text{ and } \gamma_{xy}' = -\left(\frac{\partial u'}{\partial x'} + \frac{\partial v'}{\partial y'}\right).$$

By convention

ε_x' = decrease in length/unit length in x' direction

ε_y' = decrease in length/unit length in y' direction

γ'_{xy} = increase in angle between x' and y' axes.

Thus we need expressions for

$$\frac{\partial u'}{\partial x'},\ \frac{\partial v}{\partial y'},\ \frac{\partial u'}{\partial y'} \text{ and } \frac{\partial v'}{\partial x'}.$$

From the previous notes,

$$u' = u \cos\theta + v \sin\theta$$

and

$$v' = v \cos\theta + u \sin\theta.$$

Now, let f = any function of x and y. Then

$$\frac{\partial f}{\partial x'} = \frac{\partial f}{\partial x}\frac{\partial x}{\partial x'} + \frac{\partial f}{\partial y}\frac{\partial y}{\partial x'}$$

$$= \frac{\partial f}{\partial x}\cos\theta + \frac{\partial f}{\partial y}\sin\theta$$

and

$$\frac{\partial f}{\partial y'} = \frac{\partial f}{\partial x}\frac{\partial x}{\partial y'} + \frac{\partial f}{\partial y}\frac{\partial y}{\partial y'}$$

$$= \frac{-\partial f}{\partial x}\sin\theta + \frac{\partial f}{\partial y}\cos\theta.$$

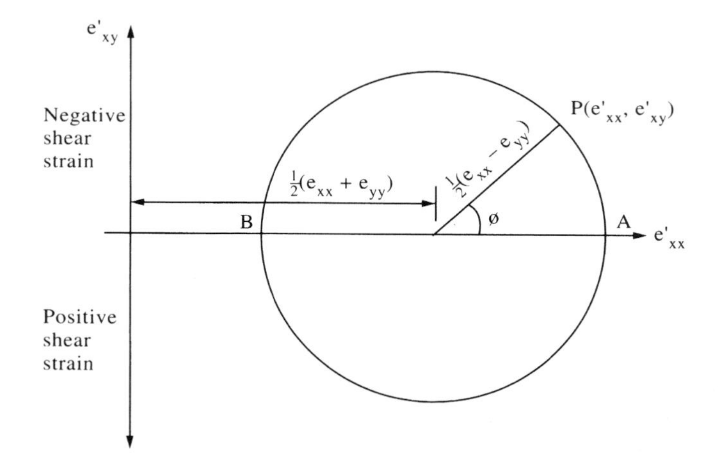

Now, if we replace f with u', v' and γ'_{xy} in turn:

$$\frac{\partial u'}{\partial x'} = \frac{\partial u'}{\partial x}\cos\theta + \frac{\partial u'}{\partial y}\sin\theta$$

$$= \cos\theta\frac{\partial}{\partial x}(u\cos\theta + v\sin\theta) + \sin\theta\frac{\partial}{\partial y}(u\cos\theta + v\sin\theta)$$

$$= \frac{\partial u}{\partial x}\cos^2\theta + \frac{\partial v}{\partial y}\sin^2\theta + \left(\frac{\partial v}{\partial x} + \frac{\partial u}{\partial y}\right)\cos\theta\sin\theta$$

$$\therefore\ \varepsilon'_x = \varepsilon_x\cos^2\theta + \varepsilon_y\sin^2\theta + \gamma_{xy}\cos\theta\sin\theta.$$

Note the similarity with σ'_x.
Similarly,

$$\varepsilon'_y = \varepsilon_x\sin^2\theta + \varepsilon_y\cos^2\theta - \gamma_{xy}\cos\theta\sin\theta$$

Note the similarity with σ'_y.

$$\gamma'_{xy} = \frac{\partial u'_x}{\partial y'} + \frac{\partial u'_y}{\partial x'} = \left(\frac{\partial u'_x}{\partial y}\cos\theta - \frac{\partial u'_x}{\partial x}\sin\theta\right) + \left(\frac{\partial u'_y}{\partial x}\cos\theta + \frac{\partial u'_y}{\partial y}\sin\theta\right)$$

but $u' = u\cos\theta + v\sin\theta$ and $v' = v\cos\theta - u\sin\theta$.
Substituting and rearranging:

$$\gamma'_{xy} = \cos\theta\frac{\partial}{\partial y}\left(u\cos\theta + v\sin\theta\right) - \sin\theta\frac{\partial}{\partial x}\left(u\cos\theta + v\sin\theta\right)$$

$$+\cos\theta\frac{\partial}{\partial x}\left(v\cos\theta - u\sin\theta\right) + \sin\theta\frac{\partial}{\partial y}\left(v\cos\theta - u\sin\theta\right)$$

$$= \frac{\partial u}{\partial y}\cos^2\theta + \frac{\partial v}{\partial y}\cos\theta\sin\theta - \frac{\partial v}{\partial x}\cos\theta\sin\theta - \frac{\partial v}{\partial x}\sin^2\theta + \frac{\partial v}{\partial x}\cos^2\theta$$

$$-\frac{\partial u}{\partial x}\cos\theta\sin\theta + \frac{\partial v}{\partial y}\cos\theta\sin\theta - \frac{\partial u}{\partial y}\sin^2\theta$$

$$= \left(\frac{\partial u}{\partial y} + \frac{\partial v}{\partial x}\right)\left(\cos^2\theta - \sin^2\theta\right) - 2\left(\frac{\partial u}{\partial x} - \frac{\partial v}{\partial y}\right)\cos\theta\sin\theta$$

$$\therefore\ \gamma'_{xy} = \gamma_{xy}(\cos^2\theta - \sin^2\theta) - 2(\varepsilon_x - \varepsilon_y)\cos\theta\sin\theta.$$

Note the similarity with τ'_{xy}.

The strain tensor

Let

$$e_{xx} = \varepsilon_x,\quad e_{yy} = \varepsilon_y,\quad \text{and } e_{xy} = \tfrac{1}{2}\gamma_{xy}$$

γ_{xy} is referred to as engineering shear strain.
e_{xy} is referred to as mathematical shear strain.

Then

$$e_{xx} = -\frac{\partial u}{\partial x},\quad e_{yy} = -\frac{\partial v}{\partial y},\quad e_{xy} = -\frac{1}{2}\left(\frac{\partial u}{\partial y} + \frac{\partial v}{\partial x}\right)$$

and

$$e'_{xx} = -\frac{\partial u'}{\partial x'},\quad e'_{yy} = -\frac{\partial v'}{\partial y'},\quad e'_{xy} = -\frac{1}{2}\left(\frac{\partial u'}{\partial y'} + \frac{\partial v'}{\partial x'}\right).$$

so

$$e'_{xx} = e_{xx}\cos^2\theta + e_{yy}\sin^2\theta + 2e_{xy}\cos\theta\sin\theta$$
$$e'_{yy} = e_{xx}\sin^2\theta + e_{yy}\cos^2\theta - 2e_{xy}\cos\theta\sin\theta$$
$$e'_{xy} = e_{xy}(\cos^2\theta - \sin^2\theta) - (e_{xx} - e_{yy})\cos\theta\sin\theta$$

Note that these are identical to the stress transformation equations.

Example. Given

$$\varepsilon_x = \frac{1}{125} = 8000\ \mu s \quad \text{or} \quad e_{xx} = 8000\ \mu s$$

$$\varepsilon_y = \frac{1}{500} = 2000\ \mu s \quad \text{or} \quad e_{yy} = 2000\ \mu s$$

$$\gamma_{xy} = \frac{9}{1000} = 9000\ \mu s \quad \text{or} \quad e_{xy} = 4500\ \mu s.$$

Compute the local components of strain if $\theta = 30°$ ($\cos\theta = 0.866$, $\sin\theta = 0.5$)

$$e'_{xx} = (8000 \times 0.750) + (2000 \times 0.250) + (2 \times 4500 \times 0.866 \times 0.500) = \underline{10400\ \mu s}$$

$$e'_{yy} = (8000 \times 0.250) + (2000 \times 0.750) - (2 \times 4500 \times 0.866 \times 0.500) = \underline{-400\ \mu s}$$

$$e'_{xy} = 4500(0.750 - 0.250) - (8000 - 2000) \times 0.866 \times 0.500 = \underline{-350\ \mu s}$$

or

$$\underline{\gamma'_{xy}} = \underline{-700\ \mu s}.$$

Principal directions and principal strains

Just as for stresses, does there exist a value of θ for which e'_{xy} (or γ'_{xy}) = 0? There is, and let us call us the angle β.

By analogy with the stress tensor,

$$\boxed{\beta = \frac{1}{2}\tan^{-1}\left(\frac{2e_{xy}}{e_{xx} - e_{yy}}\right).}$$

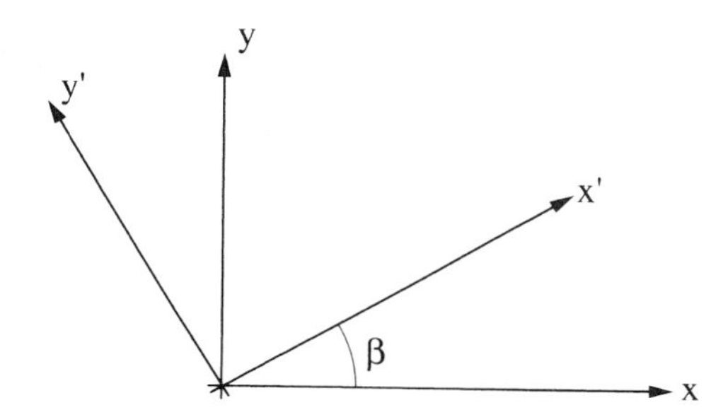

The directions x' and y' corresponding to this value of β are known as **principal directions of strain.** These directions are orthogonal. The longitudinal strains e'_{xx} and e'_{yy} in the principal directions are known as **principal strains.**
In the example,

$$\frac{2e_{xy}}{\left(e_{xx}-e_{yy}\right)}=\frac{2\times 4500}{\left(8000-2000\right)}=1.5$$

$\therefore\ \beta = \frac{1}{2}\tan^{-1} 1.5 = \underline{28.2^{\circ}}$.

Mohr's circle of strain

If the global axes x and y are chosen to coincide with the principal directions, the strain transformation equations become

$$\left.\begin{aligned} e'_{xx} &= e_{xx}\cos^2\theta + e_{yy}\sin^2\theta \\ e'_{yy} &= e_{xx}\sin^2\theta + e_{yy}\cos^2\theta \\ e'_{xy} &= -\left(e_{xx}-e_{yy}\right)\cos\theta\sin^2\theta \end{aligned}\right\}$$

compare to the stress transformation equations.

Now let $\phi = 2\theta$. By analogy with the stress tensor,

$$e'_{xx} = \tfrac{1}{2}(e_{xx}+e_{yy}) + \tfrac{1}{2}(e_{xx}-e_{yy})\cos\phi$$

$$e'_{xy} = -\tfrac{1}{2}(e_{xx}-e_{yy})\sin\phi.$$

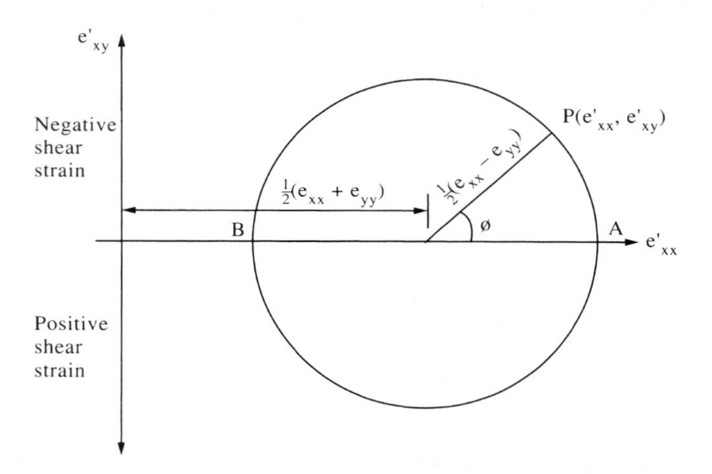

By analogy with Mohr's circle of stress, each point on the circle represents a direction in the material in which the longitudinal strain is e'_{xx}:

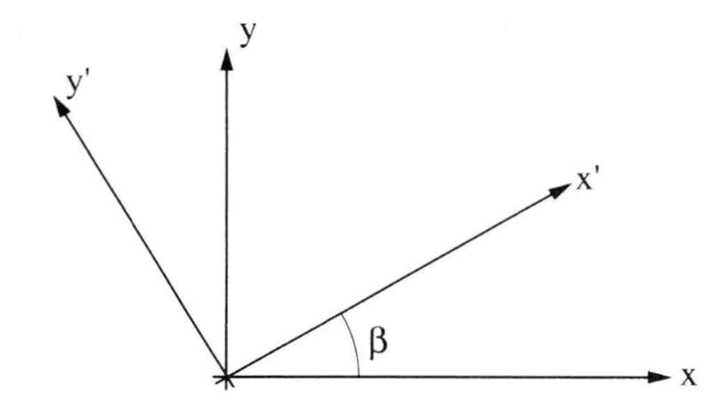

Example. Continuing from before, with $e_{xx} = 8000$ μs, $e_{yy} = 2000$ μs, $e_{xy} = 4500$ μs:

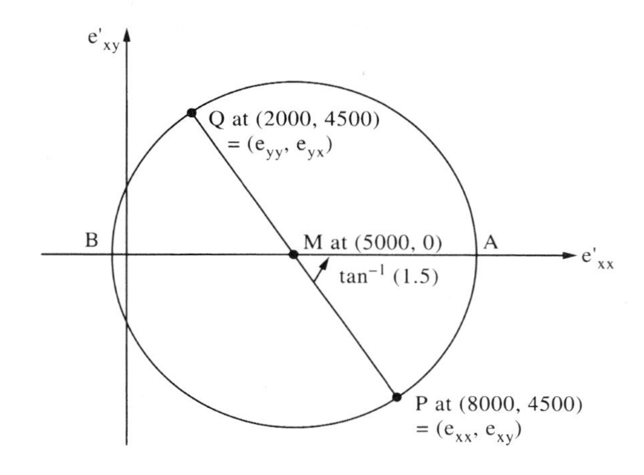

$$r = \sqrt{3000^2 + 4500^2} = \underline{5410\mu s} = \text{maximum mathematical shear strain}$$

Maximum engineering shear strain	=	2×5410	=	$\underline{10{,}820\ \mu s}$.
Principal strains are		$(5000 + 5410)$	=	$\underline{10{,}410\ \mu s}$
		$(5000 - 5410)$	=	$\underline{-410\ \mu s}$.

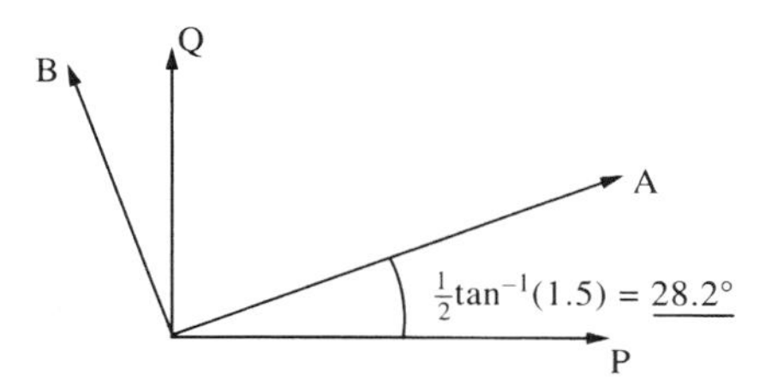

Determination of principal strains from measured strains

This is the practical use of the two-dimensional strain transformation equations.

It is not possible to measure shear strains in practice, so for two-dimensional strain three normal strains are measured at known relative angles. This permits solution of the strain transformation equation for three unknowns—ε_1, ε_2 and the angle between one gauge and ε_1.

Strain gauge rosettes are the devices used to measure strain. They consist of three strain gauges mounted on an epoxy backing, such that the mutual angles between the gauges is 45° or 60° (so-called **rectangular** and **delta** rosettes):

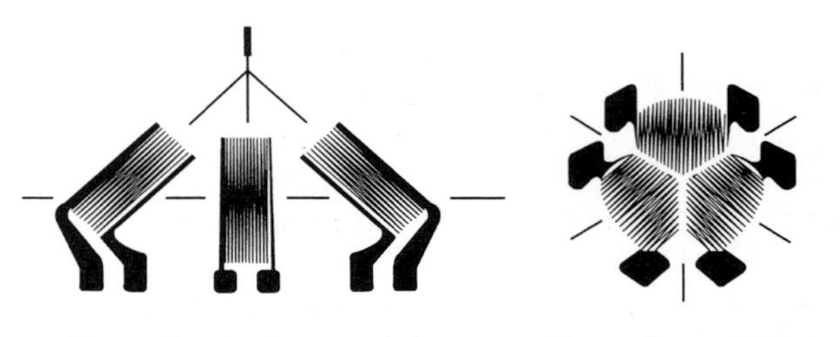

Three-element (rectangular) rosette, 45° planar, foil

Three-element (delta) rosette, 60° shear planar, foil

For the purposes of analysis we can assume this geometry, where

$\alpha = \beta = 45°$ for rectangular rosette
$\alpha = \beta = 60°$ for a delta rosette.

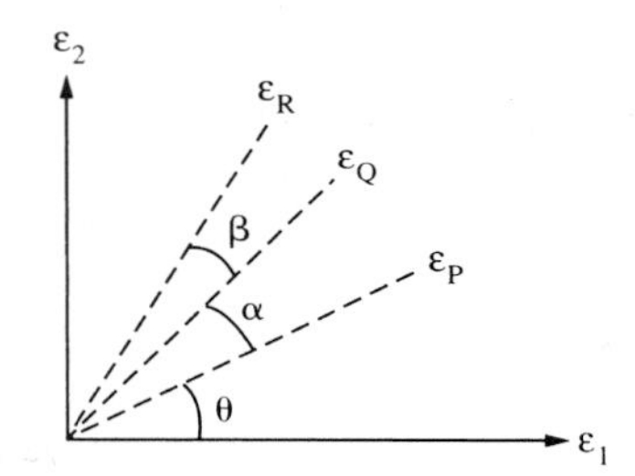

We measure ε_P, ε_Q and ε_R, we know α and β, and hence we can calculate ε_1, ε_2 and θ.

Substituting $\varepsilon_1 = \varepsilon_x$, $\varepsilon_2 = \varepsilon_y$ and $\gamma_{xy} = 0$ into the strain transformation equation for direct strain gives us three equations:

$$\varepsilon_P = \tfrac{1}{2}(\varepsilon_1 + \varepsilon_2) + \tfrac{1}{2}(\varepsilon_1 - \varepsilon_2)\cos 2\theta$$

$$\varepsilon_Q = \tfrac{1}{2}(\varepsilon_1 + \varepsilon_2) + \tfrac{1}{2}(\varepsilon_1 - \varepsilon_2)\cos 2(\theta + \alpha)$$

$$\varepsilon_R = \tfrac{1}{2}(\varepsilon_1 + \varepsilon_2) + \tfrac{1}{2}(\varepsilon_1 - \varepsilon_2)\cos 2(\theta + \alpha + \beta).$$

In general we have

$$\varepsilon_\phi = c + r\cos 2\phi$$

where c = centre distance of Mohr's circle = $\tfrac{1}{2}(\varepsilon_1 + \varepsilon_2)$
r = radius of Mohr's circle = $\tfrac{1}{2}(\varepsilon_1 - \varepsilon_2)$
ϕ = angle between gauge and ε_1.

(a) Delta rosette

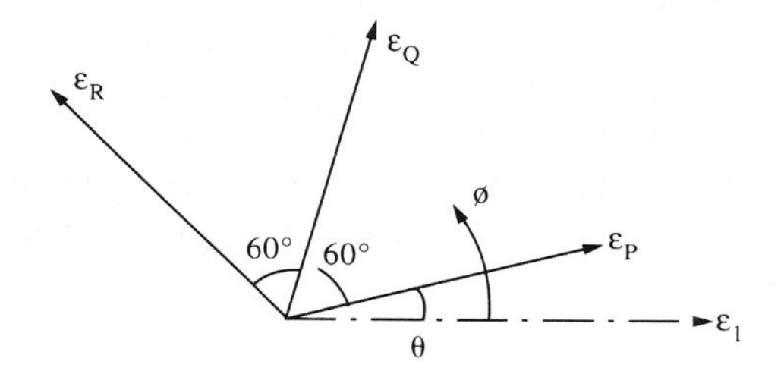

$$\begin{aligned}\varepsilon_P &= c + r\cos 2\theta \\ \varepsilon_Q &= c + r\cos(2\theta + 120) \\ \varepsilon_R &= c + r\cos(2\theta + 240). \qquad (1)\end{aligned}$$

Using the double angle formulae on the expressions for ε_Q and ε_R:

$$\begin{aligned}\varepsilon_Q &= c + r\{\cos 2\theta \cos 120 - \sin 2\theta \sin 120\} \\ &= c + r\{-\tfrac{1}{2}\cos 2\theta - \tfrac{\sqrt{3}}{2}\sin 2\theta\} \qquad (2)\end{aligned}$$

$$\begin{aligned}\varepsilon_R &= c + r\{\cos 2\theta \cos 240 - \sin 2\theta \sin 240\} \\ &= c + r\{-\tfrac{1}{2}\cos 2\theta + \tfrac{\sqrt{3}}{2}\sin 2\theta\}. \qquad (3)\end{aligned}$$

Adding (1), (2) and (3):

$$\underline{\underline{\varepsilon_P + \varepsilon_Q + \varepsilon_R = 3c}}$$

hence

$$c = \tfrac{1}{3}(\varepsilon_P + \varepsilon_Q + \varepsilon_R).$$

Subtracting (2) from (3):

$$\varepsilon_R - \varepsilon_Q = r\sqrt{3}\sin 2\theta \qquad (4)$$

$2 \times (1) - ((2) + (3))$:

$$2\varepsilon_P - (\varepsilon_Q + \varepsilon_R) = 3r\cos 2\theta. \qquad (5)$$

Dividing (4) by (5):

$$\frac{\varepsilon_R - \varepsilon_Q}{2\varepsilon_P - (\varepsilon_Q + \varepsilon_R)} = \frac{1}{\sqrt{3}}\tan 2\theta$$

hence

$$\tan 2\theta = \frac{\sqrt{3}(\varepsilon_R - \varepsilon_Q)}{2\varepsilon_P - (\varepsilon_Q + \varepsilon_R)}$$

rearranging (4) gives

$$r = \frac{\varepsilon_R - \varepsilon_Q}{\sqrt{3}\sin 2\theta}.$$

(b) Rectangular rosette

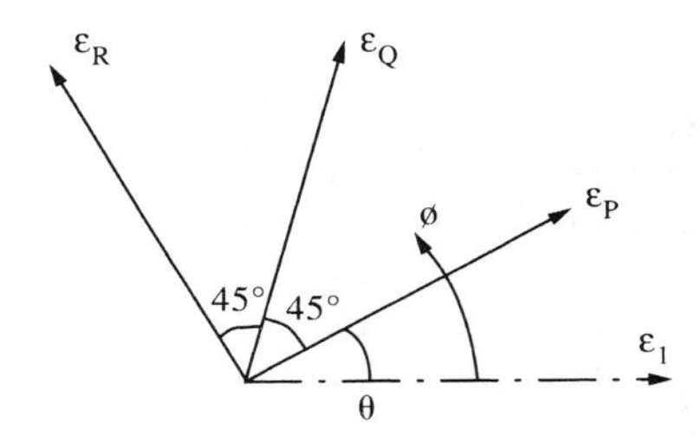

$$\begin{aligned}\varepsilon_P &= c + r\cos 2\theta \\ \varepsilon_Q &= c + r\cos(2\theta + 90) \\ \varepsilon_R &= c + r\cos(2\theta + 180).\end{aligned} \qquad (1)$$

Using the double angle formulae on the expressions for ε_Q and ε_R:

$$\varepsilon_Q = c + r\{\cos 2\theta \cos 90 - \sin 2\theta \sin 90\} = c + r\{-\sin 2\theta\} \qquad (2)$$

$$\varepsilon_R = c + r\{\cos 2\theta \cos 180 - \sin 2\theta \sin 180\} = c + r\{-\cos 2\theta\}. \qquad (3)$$

Adding (1) and (3):

$$\varepsilon_P + \varepsilon_R = 2c$$

hence

$$\underline{\underline{c = \tfrac{1}{2}(\varepsilon_P + \varepsilon_R)}}.$$

Subtracting (3) from (1):

$$\varepsilon_P - \varepsilon_R = 2r\cos 2\theta. \qquad (4)$$

((1) + (3)) – 2 × (2):

$$(\varepsilon_P + \varepsilon_R) - 2\varepsilon_Q = \cancel{2c} - \cancel{2c} + 2r\sin 2\theta \qquad (5)$$

Divide (5) by (4):

$$\frac{(\varepsilon_P + \varepsilon_R) - 2\varepsilon_Q}{\varepsilon_P - \varepsilon_R} = \tan 2\theta.$$

Rearranging (4):

$$r = \frac{\varepsilon_P - \varepsilon_R}{2\cos 2\theta}.$$

In summary then:

Delta rosette	**Rectangular rosette**
$c = \frac{1}{3}\left(\varepsilon_P + \varepsilon_Q + \varepsilon_R\right)$	$c = \frac{1}{2}\left(\varepsilon_P + \varepsilon_R\right)$
$\tan 2\theta = \frac{\sqrt{3}\left(\varepsilon_R - \varepsilon_Q\right)}{2\varepsilon_P - \left(\varepsilon_Q + \varepsilon_R\right)}$	$\tan 2\theta = \frac{\left(\varepsilon_P + \varepsilon_R\right) - 2\varepsilon_Q}{\varepsilon_P - \varepsilon_R}$
$r = \frac{\varepsilon_R - \varepsilon_Q}{\sqrt{3}\sin 2\theta}$	$r = \frac{\varepsilon_P - \varepsilon_R}{2\cos 2\theta}$

Example. In a delta rosette the three measured strains are ε_P = 8E–4, ε_Q = –8E–4 and ε_R = 2E–4. What are the principal strains and their orientation to ε_P?

$$\tan 2\theta = \frac{\sqrt{3}\left(2E-4--8E-4\right)}{2\times 8E-4-\left(-8E-4+2E-4\right)} = 0.7873$$

hence

$$2\theta = 38.2° \ \underline{\text{or}} \ -141.8°$$

remember
$-180 < \theta < 180$

$$c = \tfrac{1}{3}\,(8 - 8 + 2)E{-}4 = \underline{0.667E-4}$$

$$r = \frac{2E-4--8E-4}{\sqrt{3}\sin\left(38\cdot 2\right)} = 9.333E-4.$$

Use whichever value of 2θ gives positive r.
Hence

$$\varepsilon_1 = c + r = \underline{10.000E{-}4}$$

and

$$\varepsilon_2 = c - r = \underline{-8.667E{-}4}.$$

Now choose the value of θ which is compatible with these values of ε_1 and ε_2. In this case, θ = 19.1° puts ε_1 near to ε_P, which seems reasonable. θ = –70.9° puts ε_1 midway between ε_Q and ε_R, which cannot be correct (ε_1 = 10E–4, ε_Q = –8E–4, ε_R = 2E–4). Hence θ = 19.1° and the solution is:

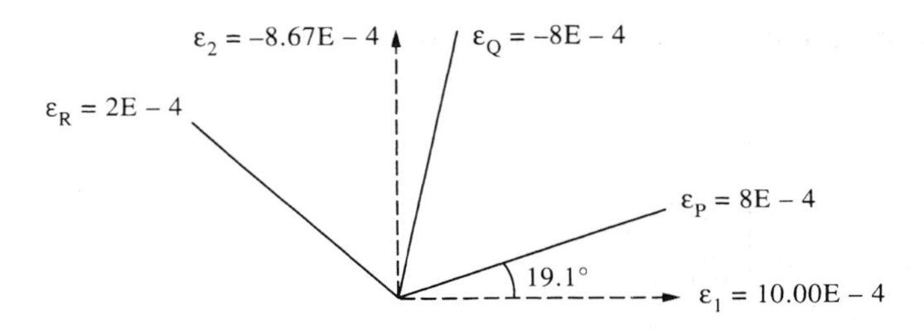

Mohr' circle of strain for strain gauge rosettes

Consider three arbitrarily orientated gauges, and their associated Mohr's circle:

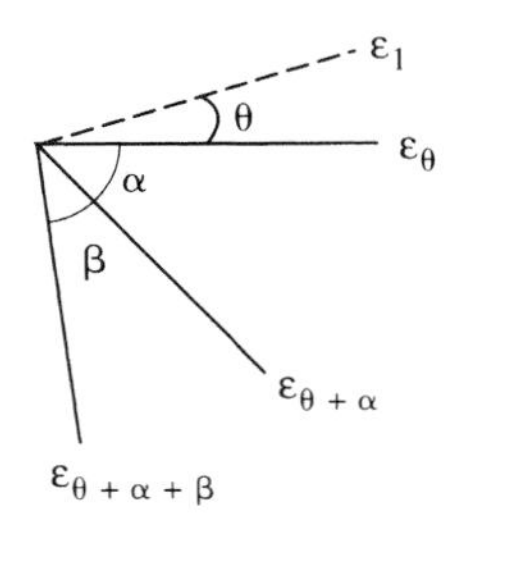

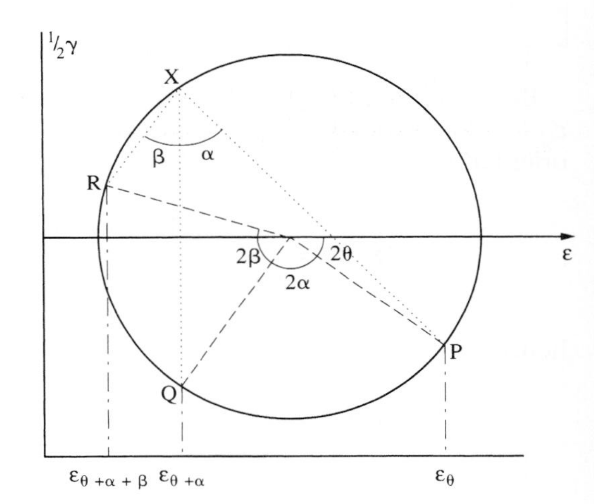

The points P, Q, R represent the state of strain in the directions of the gauges. Of particular interest are the dotted lines to the point X, and the angles PXQ = α and QXR = β. These angles arise from the geometry of a circle, and allow us to construct Mohr's circle when the angle θ is unknown. The procedure is:

(a) draw the shear strain axis and a temporary horizontal axis;
(b) mark off the strains ε_θ, $\varepsilon_{\theta+\alpha}$ and $\varepsilon_{\theta+\alpha+\beta}$ on this axis, and draw vertical construction lines through them;
(c) select any point X on the line through $\varepsilon_{\theta+\alpha}$, and draw lines inclined at α and β to intersect the other construction lines at P and R, respectively;
(d) draw the perpendicular bisectors to PX and RX, and mark their intersection. This is the centre of Mohr's circle;
(e) the true direct strain axis can be drawn through the centre of the circle, and ε_1, ε_2 and θ measured off.

Example. Solving the previous numerical example.

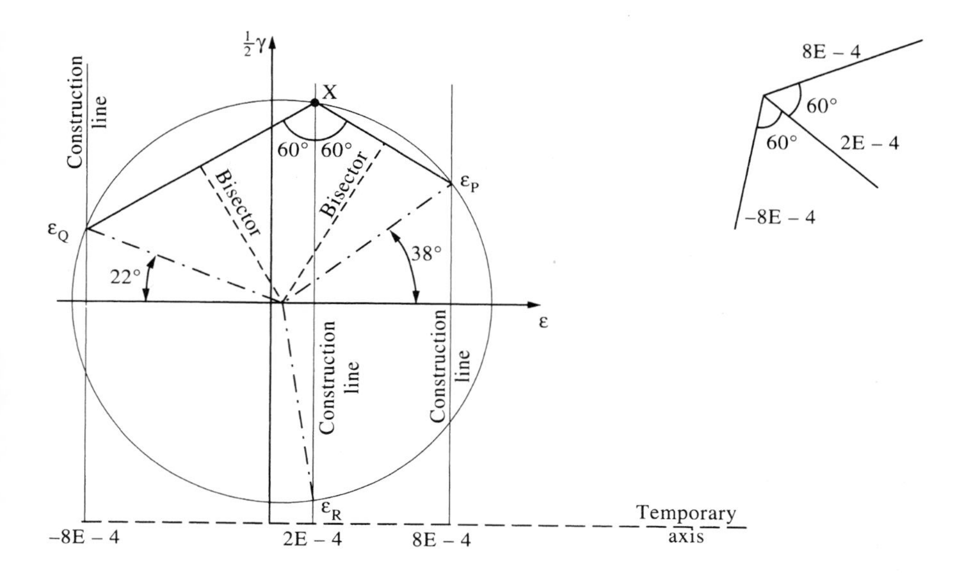

Appendix B: Hemispherical projection

Hemispherical projection methods

These methods enable three-dimensional orientation data to be displayed in two dimensions and manipulated graphically.

Fundamental geometry

Directions are **vectors** with **unit length**. We assume that these vectors emanate from the **origin** of a Cartesian co-ordinate system. It is convenient to use **east/north/down** for rock mechanics.

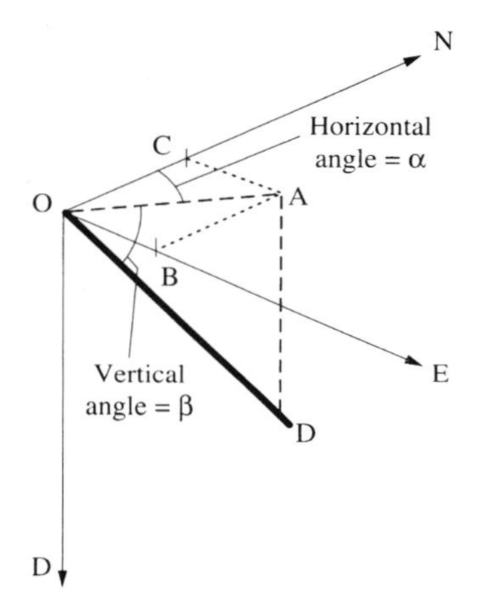

Directions are measured in terms of the angles

α = trend

β = plunge

α is measured with a compass,
β is measured with a clinometer.

Note that

$$OB = \sin\alpha\cos\beta$$
$$OC = \cos\alpha\cos\beta$$
$$AD = \sin\beta.$$

Because every vector has unit length, the tips lie on the surface of a sphere. We are usually only interested in downward-directed vectors, lying on the lower hemisphere.

Projection onto two dimensions

One way to form a two-dimensional plot associated with vectors on the lower hemisphere is to project the tips of the vectors onto the horizontal plane that passes through the origin (i.e. the centre of the sphere), assuming the point of projection is the North Pole of the sphere:

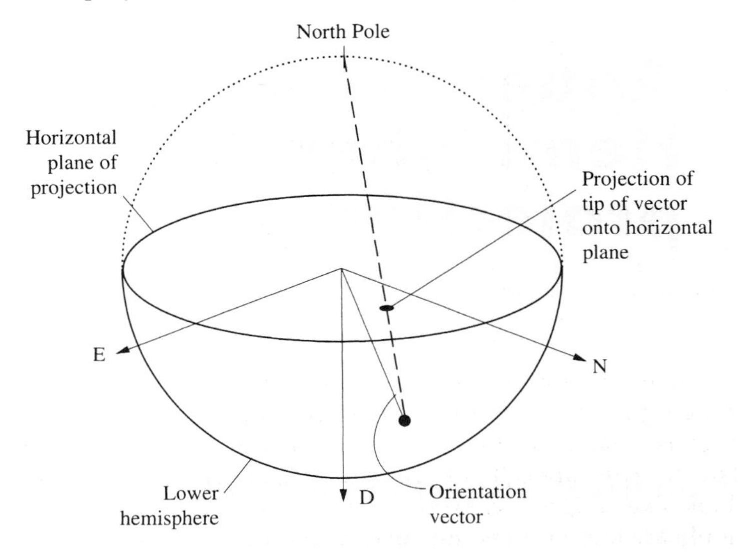

All points on the lower hemisphere can be projected in this way. This type of projection is called **equal angle projection** and is used exclusively in rock mechanics for engineering.

Equal angle projection of a plane: great circles

We regularly use planes in rock mechanics analyses and so it is important to determine the projection of such features.

The mathematics of the projection are tedious, but the result is simple: a circular arc is developed.

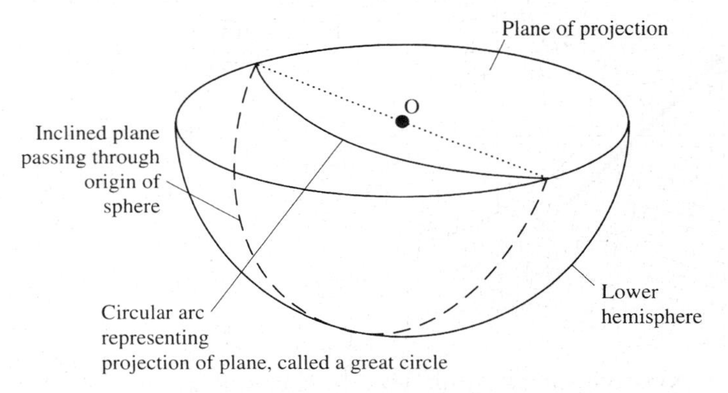

Generation of small circles

If, instead of the plane itself, we consider a single vector on the plane, we see that this vector traces a circular path on the surface of the hemisphere as the inclination of the plane changes. The projection of this trace is called a **small circle**.

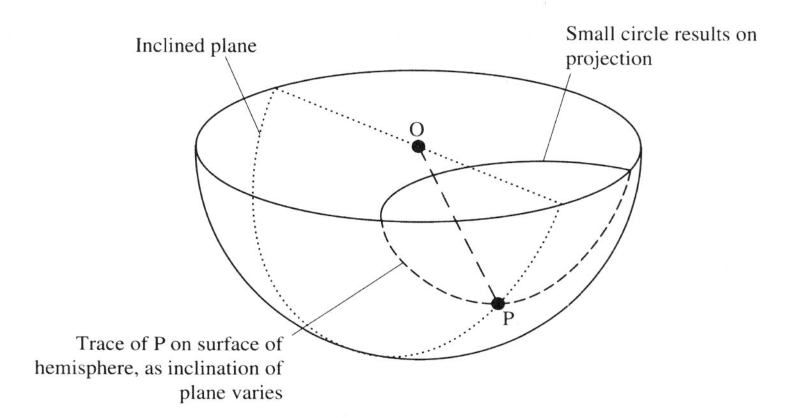

Rather than draw great and small circles for every application, we use pre-printed grids showing these circles at angular separations of, say, 2°. These grids enable us to plot and measure lines and planes of all orientations. They are called **equal angle equatorial projections** (because they look like a globe viewed from a point above the equator), but commonly are known as **hemispherical projection nets**.

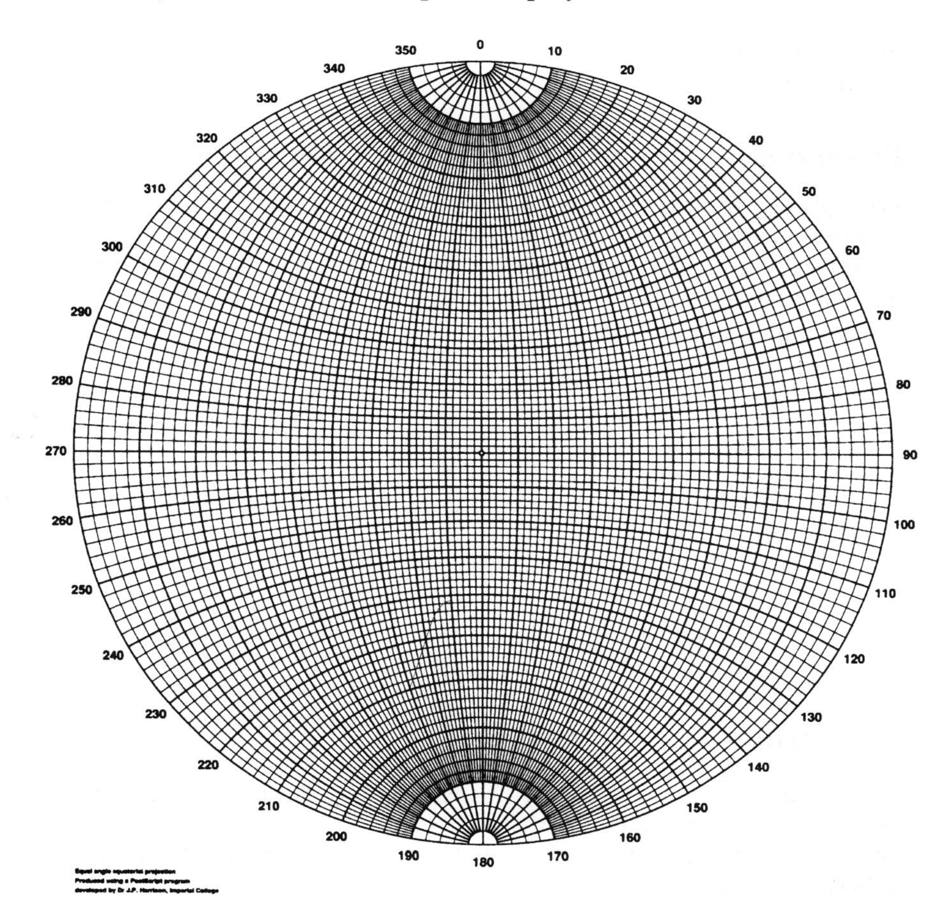

Using a hemispherical projection net: plotting vectors

Never write on the net: always use a piece of tracing paper over it. Carefully pierce the centre of the projection with a drawing pin, and then push the pin through from the reverse side of the net. Use the point of the pin to hold a sheet of tracing paper in place. Mark the north point as a datum.

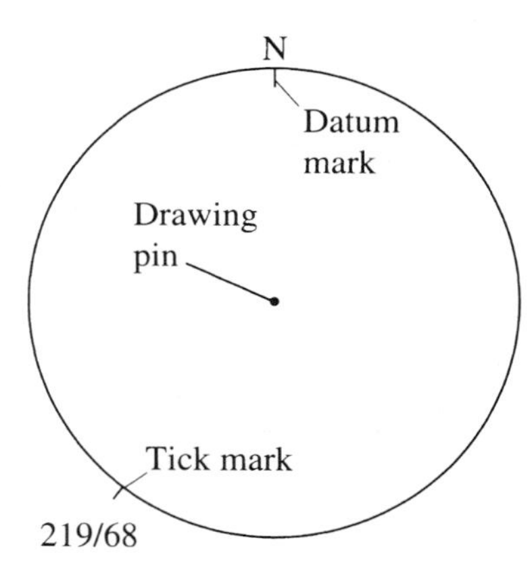

Mark, with a tick on the perimeter of the net at the correct azimuth, the vector to be plotted. Write on the projection the orientation. **Only write on the tracing paper, not the net.**

Rotate the tracing paper so that the tick is on the E–W line. Count in an amount equal to the dip of the vector. Mark the position of the vector. **Only write on the tracing paper, not the net.**

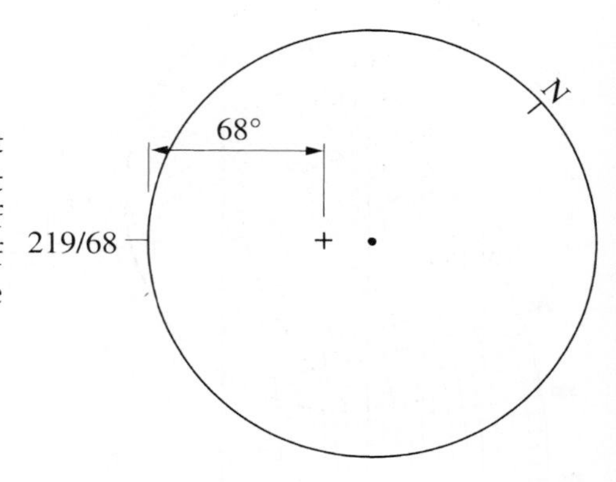

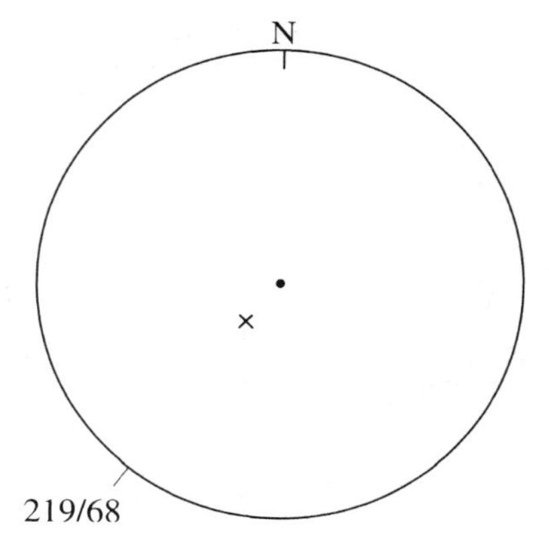

Rotate the tracing paper back to the datum: the position of the vector is now correct relative to north.

Using a hemispherical projection net: plotting planes

Start by marking a tick on the perimeter of the projection in the same way as for 'plotting a vector'. The azimuth should correspond to the **dip direction** of the planes.

Rotate the tracing paper so that the tick lies on the E–W line, and count in an amount equal to the **dip amount** of the plane. Count a further 90° along the E–W line, and mark this new position.

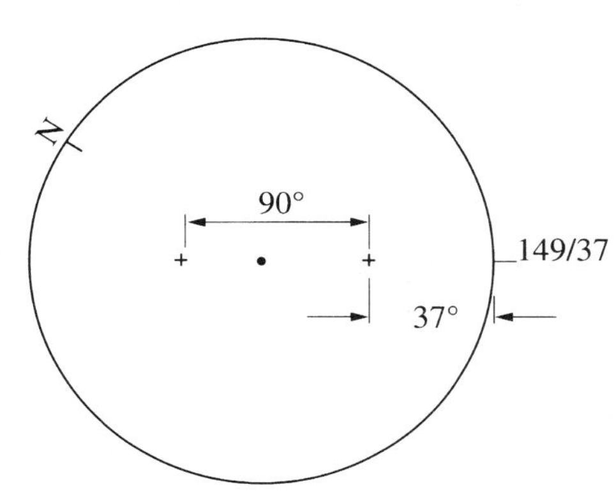

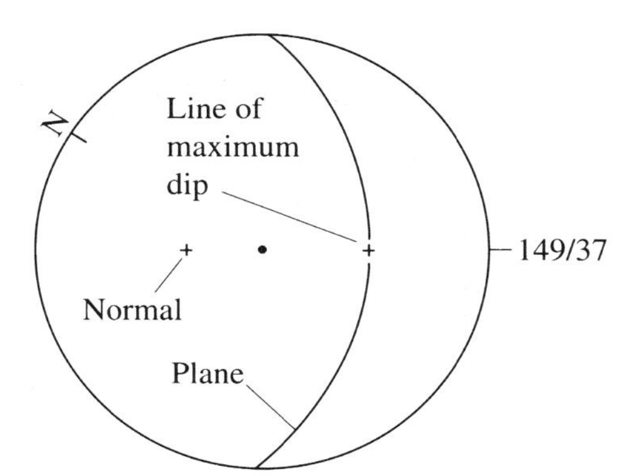

Trace the great circle that passes over the first point: this represents **the plane**. The second point represents **the normal to the plane** (i.e. the vector that is perpendicular to the plane). The first point represents the **line of maximum dip**; the second point is termed the **pole**.

Rotate the tracing paper back to the datum: the positions of the plane and the normal are now correct relative to north.

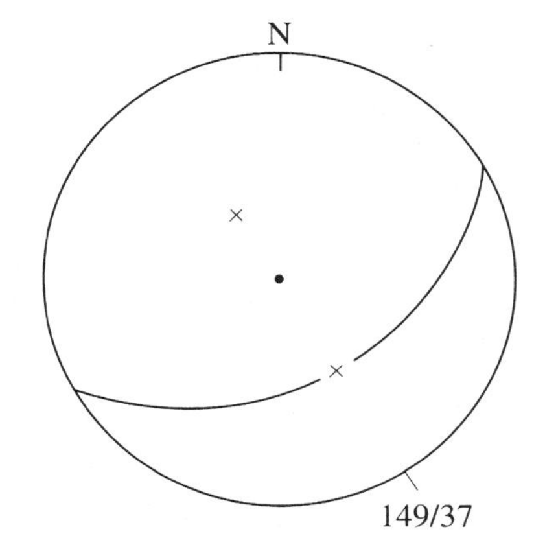

Determining the line of intersection of two planes

For any pair of planes, there is a line of intersection: it is where the two great circles cross. Finding its orientation is easy.

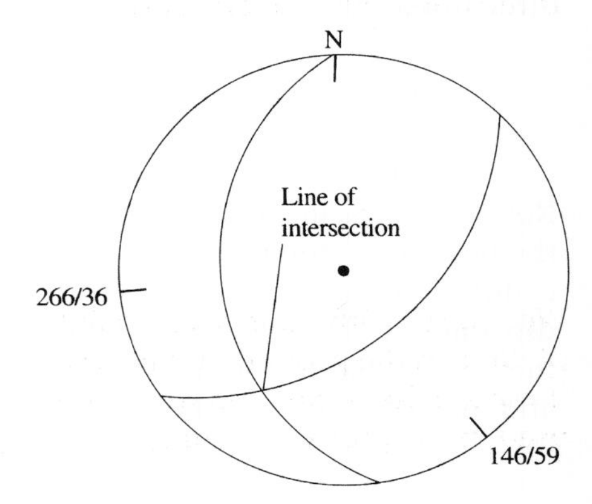

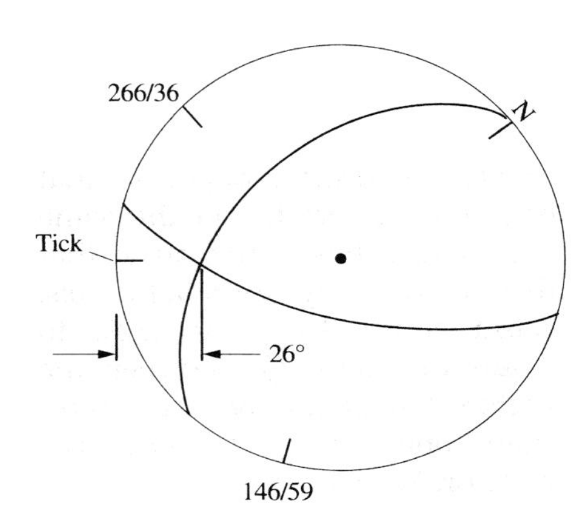

Rotate the tracing paper so that the intersection of the great circles lies on the E–W line. Mark the azimuth of this with a tick on the perimeter of the projection. Measure the plunge of the line by counting in from the perimeter, along the E–W line.

Rotate the tracing paper back to the datum, and measure the azimuth of the intersection. Thus we can see that the planes 266/36 and 146/59 have an intersection of 219/26. You should be able to measure graphically all such angles to the nearest degree.

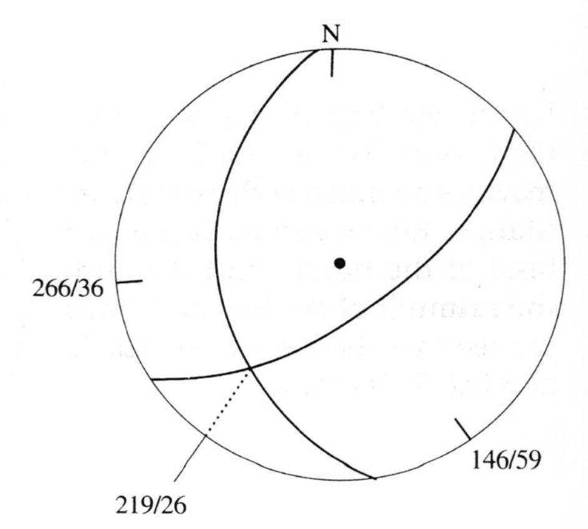

Determining the bisector of two vectors

Any two orientation vectors have a bisector (the line that is halfway between the two of them). Because the bisector must lie in the same plane as the other two vectors, finding its orientation is straightforward.

We start with the two points plotted on the projection (note that these are the normals to the planes used in the previous example).

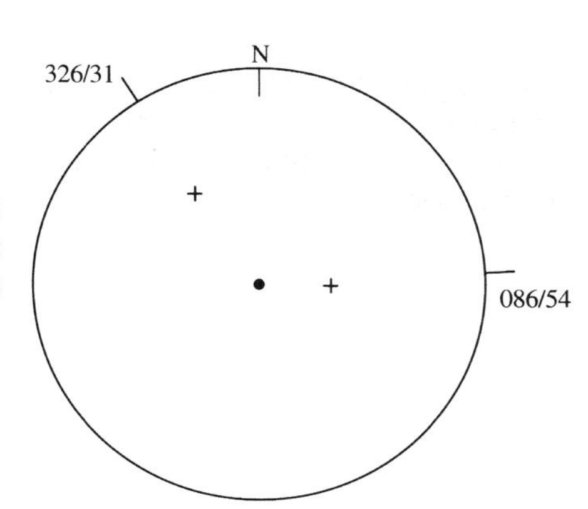

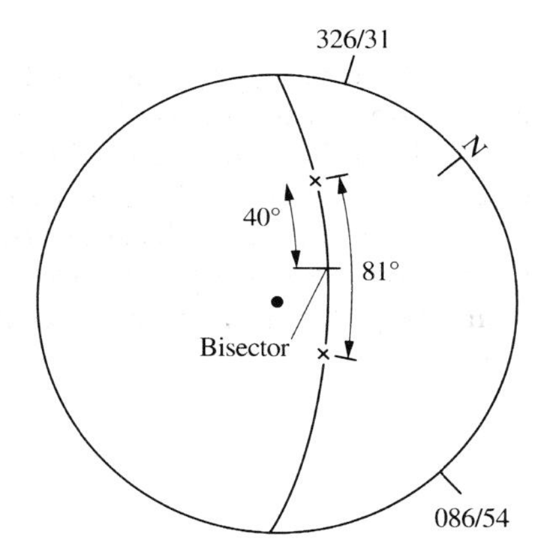

Rotate the tracing paper so that both the vectors lie on the same great circle: this is the plane that they lie in. Using the small circles, count along the great circle to determine the angle between the vectors. Divide this by 2, and count from one vector to find the bisector. Mark it.

Rotate the tracing paper so that the bisector lies on the E–W line, mark its azimuth and measure its plunge. Rotate the tracing paper back to the datum and measure the azimuth of the bisector. Thus we see that the bisector of 326/31 and 086/54 is 008/60.

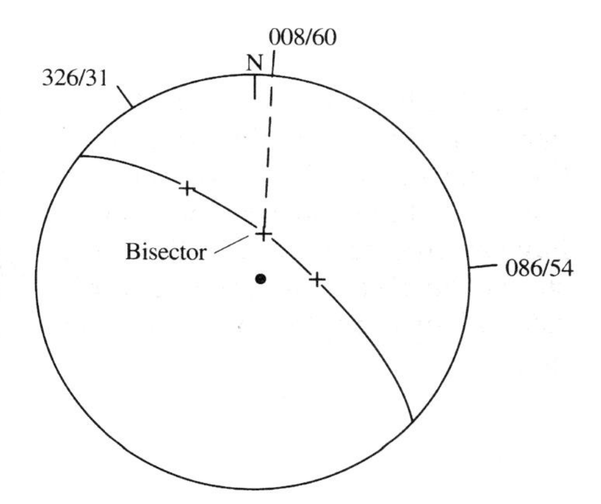

Rotation about an arbitrary axis

Consider this scenario: a borehole is drilled with a trend of 305° and a plunge of 65°, it intersects a plane with orientation 145/73, but the core has rotated through 55° clockwise (looking down the hole) during recovery. What is the apparent orientation of the discontinuity as it emerges from the hole? The simplest way of solving these sorts of problems is to use the net to perform the various rotations.

Start by plotting and labelling the data: here it is BH for the vector corresponding to the direction of the borehole, D corresponding to the line of maximum dip of the plane, and N for the normal to the plane.

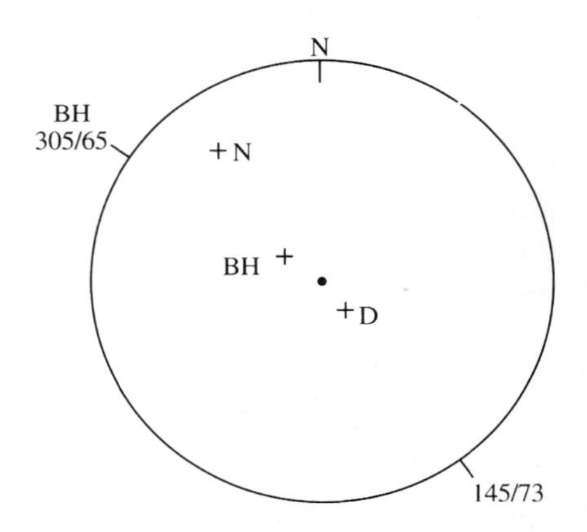

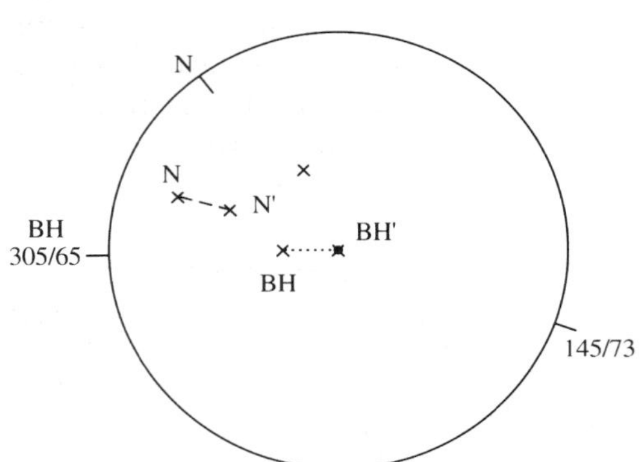

Rotate the tracing paper so that the borehole is on the E–W line. Then move BH to the centre of the net—in this case through an angle of 90 – 65 = 25°. Move the normal N through the same angle, **but along the small circle**. Label the new points BH′ and N′. In effect, the net has been inclined to be perpendicular to the borehole.

Rotate the tracing paper so that N′ is on the the E–W line. Mark its azimuth and measure its dip. Count round the perimeter the amount of the rotation (55° in this case), and put a new tick: call this N″.

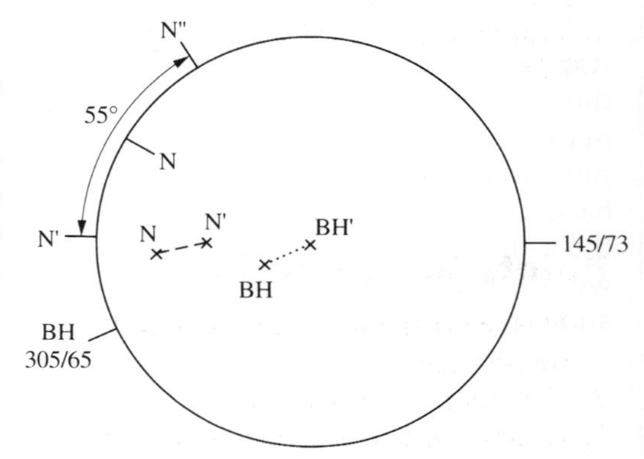

Rotate the tracing paper so that the tick for N″ is on the E–W line, and count in to get the rotated normal N″. In effect, we have modelled the rotation of the core.

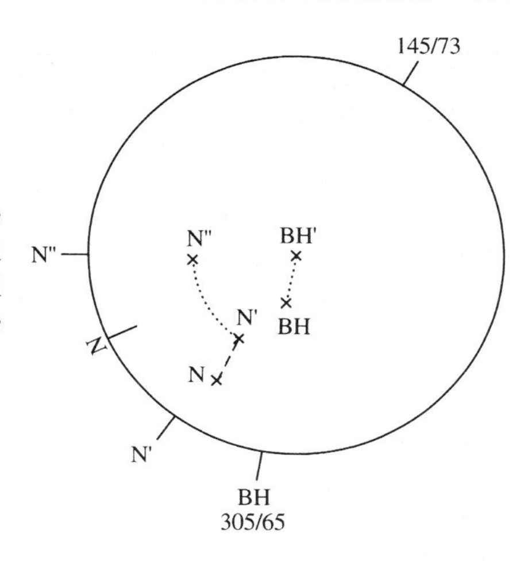

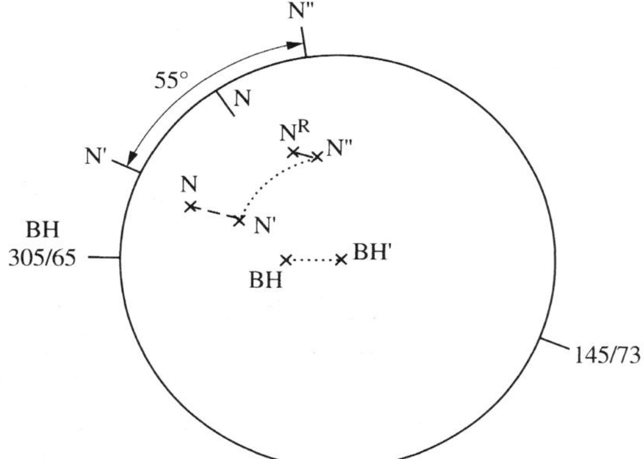

Now put the point BH on the E–W line, and move the point N″ along the small circle **by the same amount but in the opposite direction to that used in step 2**. This puts the projection back to its starting position, and N″ moves to N^R: the rotated normal.

Put N^R on the E–W line, count across 90°, and mark D_R. Measure the dip of D_R and mark its azimuth. Finally, put the tracing paper back to the datum and measure the azimuth of D_R. Thus, we see that the apparent orientation of the discontinuity is 187/58.

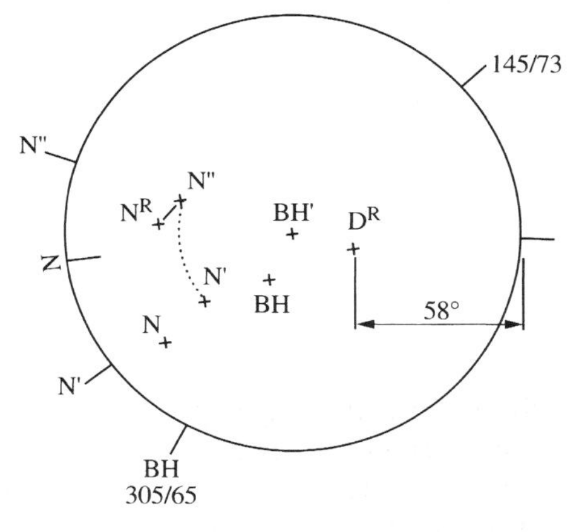

Points to remember

1. **Never write on the projection itself: you should always write on the tracing paper.**
2. Use pencil, not pen, as you will make mistakes that need to be erased.
3. Adopt—and use always—a simple but clear naming convention for

vectors. For example, N for normals, D for lines of maximum dip, superscript R for rotation (about the drawing pin), superscript prime or I for inclination (i.e. movement along the small circles), subscript numbers to identify particular vectors (or planes).

4. Make notes on the tracing paper as you go: this helps others understand what you've done, and gives you a valuable revision guide.
5. When rotating/inclining planes, always **use the normal to the plane**, never the line of maximum dip (this is because the normal is unique, whereas the line of maximum dip is arbitrary).
6. **Never write on the projection itself: you should always write on the tracing paper.**

Index